D1226762

FINITE ELEMENT ANALYSIS

Fundamentals

PRENTICE-HALL CIVIL ENGINEERING
AND ENGINEERING MECHANICS SERIES

N. M. Newmark and W. J. Hall,
editors

FINITE ELEMENT ANALYSIS

Fundamentals

RICHARD H. GALLAGHER

Department of Structural Engineering
Cornell University

PRENTICE-HALL, INC., Englewood Cliffs, New Jersey

Library of Congress Cataloging in Publication Data

GALLAGHER, RICHARD H.
Finite element analysis: fundamentals.

(Civil engineering and engineering mechanics series)
1. Finite element method. 2. Structures, Theory of.
I. Title
TA646.G33 624'.171 74-4339
ISBN 0-13-317248-1

10 9

Printed in the United States of America

PRENTICE-HALL INTERNATIONAL, INC., *London*
PRENTICE-HALL OF AUSTRALIA, PTY. LTD., *Sydney*
PRENTICE-HALL OF CANADA, LTD., *Toronto*
PRENTICE-HALL OF INDIA PRIVATE LIMITED, *New Delhi*
PRENTICE-HALL OF JAPAN, INC., *Tokyo*

To my wife
Terry

Contents

Preface

New fields generally progress through three periods of technological development. During the first period, the development progresses via the periodical literature and is coordinated only through the medium of occasional survey papers. Applications in practice are quite scarce. In the second phase, monograph-type texts appear and give a comprehensive view of the field for individuals actively engaged in further development of the field and applications are recorded by advanced technology groups in organizations with large resources. Finally, the applications spread to nearly all levels of practical activity and in the universities the subject matter of the field is offered as standard academic fare.

Finite element analysis has only recently emerged from the second of the phases above. A number of excellent monographic texts have appeared but a need exists for a text directed toward the conventional course offering and toward the individual with no prior acquaintance with the field. This book is intended to serve this purpose. It is oriented toward a graduate level course for students specializing in solid mechanics. This would include students enrolled in the fields of mechanical and aerospace engineering, naval architecture, engineering mechanics, and civil engineering. To the extent that a bias exists toward one of these fields it is directed toward civil engineers pointing toward structural engineering practice.

It is also hoped that this text will appeal to engineers already in practice, both to those seeking an introduction to a technology that was not found among courses of instruction during their period of formal education and to those who are routinely involved in finite element analysis and seek reference to basic proofs and formulative procedures. Much of the material contained herein in fact achieved realization in conjunction with numerous short courses offered by the author to engineers in practice.

The subject matter of this text requires some familiarity with the theory of elasticity and matrix structural analysis, and these imply an exposure to partial differential equations, the algebra of large-order equations, and the theory of structural analysis. Although it is the writer's belief that each of these topics is given adequate development from basic principles in the early chapters of the text, it is the experience of the author that the extent of coverage of finite element analysis in the usual course offering leaves very little room for a first-time assimilation of the prerequisities. We hasten to add, however, that sufficient background in theory of elasticity is normally found in the modern sophomore or junior year courses described by the title *continuum mechanics.*

The term *matrix structural analysis* requires clarification since it has been common to collect under this heading nearly all aspects of procedures related to digital computer applications in structural engineering. There has been a trend, however, toward the separation of procedures for constructing and solving the equations that describe the total structural problem, involving the connection of simple structural elements, from the formulative aspects of the elements. The former can be developed in large measure in terms of such elements as truss and framework members, whose theoretical basis can be established by only a very modest incursion into the latter area, and it is to this topic that the designation matrix structural analysis is applied.

Theoretical developments in finite element analysis have placed great dependence on the calculus of variations. We have chosen to exclude this topic from the group of prerequisite subjects since the writer has found it unrealistic to expect that students otherwise capable of initiating study in finite element analysis will have received a formal course in this topic prior to the study of finite element analysis.

The contents of this book are almost entirely devoted to the development of basic theoretical principles and, with the exception of Chapter 1, only scant attention is given to the practical application of finite element analysis. There is a wealth of such information available in the open literature, some of which can be found in the references given at the close of each chapter. In Chapter 1, in addition to portraying some representative applications of the method, we outline its developmental history, give a thumbnail sketch of the varieties of elements to be explored in later chapters of the text, and describe a motivating factor in the method, the concept of the *general-purpose* program.

Chapter 2 is devoted to the presentation of basic definitions, terminology, coordinate systems, and properties possessed by all finite element relationships independent of their mode of formulation. Chapter 3 details one method of constructing the equations of the complete structure from the equations of the individual elements—the direct stiffness method. Other methods for accomplishing this objective are outlined or mentioned in later chapters but, as we have noted, the objective of this text is to concentrate upon matters of element formulation.

Although we do not exclude one-dimensional members (e.g., axial members, beam segments) from our development, but in fact employ them liberally in the exemplification of basic theory, the prime motivation in finite element analysis is nevertheless the analysis of two- and three-dimensional continua. An understanding of the basic relationships of elasticity is therefore essential to the study of the method and we develop these from basic considerations in Chapter 4.

Two broad classifications of general procedures for formulation of element equations are covered in this text. *Direct methods*, described in Chapter 5, are appealing in their simplicity and rationale. The direct formulative process gives considerable insight into those conditions which are met by element formulation and those which are not. *Variational methods* (Chapter 6) are currently the more popular procedures for element formulation. Such methods, under well-defined circumstances, are valuable in furnishing assurances on convergence of the numerical solution and in ensuring that certain formulations yield upper or lower bounds at a given level of analysis refinement. In Chapter 6 we apply the variational methods to element formulation and in Chapter 7 we demonstrate the extension of the same ideas to the formulation of the complete structure. In this way we establish an alternative and broader view of system analysis than that given in Chapter 3.

At this point it is pertinent to take note of what the author believes to be a particular feature of this text. At the time of its preparation, all of practice and the majority of existing finite element theory dealt with finite element formulations in the class of *displacement-based* (i.e., stiffness or potential energy) procedures. Alternative formulations, based on assumed stress fields and even on both displacement and stress fields, hold considerable promise, however, and the author can foresee the possibility that all the alternatives will ultimately be on equal footing in practice. Thus, close attention is given to these alternatives in Chapters 5 through 7.

The portion of the text dealing with basic theoretical considerations concludes with Chapter 8, in which we examine procedures for the functional representation of element behavior and extend these same ideas to the representatation of element geometry. Concepts and formulations established in this chapter are probably the more generally useful than those dealt with in the prior chapters since they apply with equal force to nonstructural finite element analysis.

Specific forms of elements are given detailed examination in each of

Chapters 9 through 12. These encompass *plane stress elements* (Chapter 9), *solid elements* of general and special form (Chapters 10 and 11, respectively), and *plate flexure elements* (Chapter 12). By the same token greater emphasis is placed on references to the open literature than in the previous four chapters.

Chapter 13 treats a special form of behavior, *elastic instability*. The theory developed in this chapter applies equally well to all types of elements and for this reason it is expedient to employ once again the simplest types of elements—axial and frame members.

Three types of problems for assignment are presented. The first category comprises problems intended to exercise theoretical concepts and contains problems of the type assigned in traditional structural mechanics courses. A second category is devoted to intrinsically finite element problems but is intended for hand-calculation, e.g., the formulation of new element relationships or the solution of a structure describable by no more than three algebraic equations. Finally, we present data for problems with known classical or alternative solutions that are representable by the finite element method in terms of a relatively large number of equations. The assignment of such problems may be accomplished in many ways but one scheme the author has found effective has been to assign a different gridwork to each student in the class. Correlation of the results in class then gives the added benefit of describing the accuracy and rate of convergence of the finite element solution.

The finite element method is a technology that is wedded to digital computer analysis and it may appear surprising that no coded algorithms are presented herein. The author believes that few, if any, instructors or independent users of this text will find difficulty in obtaining access to widely distributed general-purpose finite element programs (e.g., STRUDL-II) suitable for the performance of the problem assignments discussed above. Alternatively, simpler finite element programs are found in numerous reports and texts.

It is conceivable that the subject matter of this text could be covered in the conventional three-hour per week, fifteen-week course. In the experience of the author this would require a higher proficiency in the prerequisite subjects (theory of elasticity, matrix structural analysis) and in the calculus of variations as well than would be brought to a course by the majority of students. The instructor may therefore choose to eliminate coverage of one or more of the last four chapters. In a trimester system, on the other hand, it would appear feasible to form a sequence of one ten-week course in matrix structural analysis and two succeeding ten-week finite element courses. The second finite element course would give latitude for coverage of such advanced topics as finite element theory and analysis for problems in soil mechanics, heat transfer, fluid flow, and other nonstructural applications; nonlinear problems; and the analysis of transients.

The author's thanks go out to very many former students and industrial colleagues who read various segments of the manuscript and offered both criticism and helpful suggestions. A debt of gratitude is owed to Professors J. T. Oden of the University of Texas and G. McNeice of Waterloo University for contributions to Chapter 9, and to Prof. K. Washizu of Tokyo University for his study and comments of Chapters 6 and 7. Special thanks are due to Professor Sidney Kelsey of the University of Notre Dame for his detailed review of nearly all chapters of the text and for his extensive and always perceptive comments, to Mr. James Bacci, Production Editor, Prentice-Hall and other members of the Prentice-Hall staff, including Barbara Cassel, for limitless patience in what proved to be a worldwide enterprise, and to Mrs. Helen Wheeler for her peerless typing of the manuscript and for being ever-watchful for incomplete sentences and grammatical miscues.

RICHARD H. GALLAGHER

List of Symbols

The following is a list of the principal symbols used in this text. Various other symbols are defined where they appear; this is often the case with symbols used to designate matrices (especially in Chapter 6), or which appear only in tables or figures. Symbols which have two distinctly different meanings are distinguished by subscripting (for example, L denotes length while L_i symbolizes an area or volume coordinate). Subscripts and superscripts applied to symbols with only one general meaning are not given below but are rather identified in the text where they appear.

Matrices are denoted by boldface letter within the symbols [ ] (for a rectangular matrix), { } (for a column vector) and ⌊ ⌋ (for a row vector). The definition of the boldface matrix symbol applies also to the lightface (that is, the non-boldface), subscripted, terms of the matrix. For example, the definition of the $n \times 1$ vector $\{\mathbf{a}\}$ applies to the individual components $a_1 \ldots a_i \ldots a_n$. The boldface symbol used to denote a matrix is often used in the non-boldface, unsubscripted form of a scalar with entirely different meaning, although in some cases the same meaning is preserved.

Overbars denote specified quantities. Primes denote differentiation.

A Area

$[\mathbf{A}]$ Matrix relating stresses to joint forces.

$[\mathcal{A}]$ Kinematics matrix. Coefficients of the relationships between element joint and global joint displacements.

a Dimension

{**a**} Vector of parameters in assumed displacement field.

[**B**] Matrix relating assumed displacement field parameters to joint displacements.

[$\mathfrak{B}$] Statics matrix. Coefficients of the relationships between element joint and global joint forces.

$b_{0_i}, b_{1_i}, b_{2_i}$ (i = 1, 2, 3). Coefficients in area coordinate equation.

C Constant in Poisson equation.

[**C**] Matrix relating assumed displacement field parameters to strain field.

$c_{0_i}, \ldots, c_{3_i}$ (i = 1, 2, 3, 4). Coefficients in volume coordinate equation.

D Plate flexural rigidity.

[**D**] Matrix relating joint displacements to strain field.

{**d**} Eigenvector.

d.o.f. Degree of freedom

E Elastic modulus

[**E**] Matrix of elastic constants

e Multiplier of variation.

{**F**} Vector of element nodal forces.

[$\mathfrak{F}$] Global flexibility matrix.

{**f**} Element flexibility matrix.

G Shear modulus.

[**G**] Constraint matrix.

I Moment of inertia.

[**I**] Identity matrix.

$\mathcal{I}$ Value of integral.

i, j, k Dummy subscripts or superscripts.

J St. Venant torsion constant.

[**J**] Jacobian matrix.

[**K**] Global stiffness matrix.

[**k**] Element stiffness matrix.

L Length.

L_i Area (i = 1, 2, 3) or volume (i = 1, 2, 3, 4) coordinate.

ℓ_x, ℓ_y, ℓ_z Direction cosines.

{**M**} Vector of joint bending moments.

$\mathfrak{M}, M_x, M_y, M_{xy}$ General internal moment vector in plate bending (line moments) and components.

m Order of polynomial series.

[**m**] Element mass matrix.

$\mathfrak{N}$ Number of sides of a polygon.

[**N**], $\lfloor$**N**$\rfloor$ Matrix of shape functions.

n Number of degrees of freedom.

[**O**], {**O**} Null matrix and vector.

$\{\mathbf{P}\}$ Vector of global nodal forces.

p Number of elements.

$\lfloor \mathbf{p} \rfloor$ Matrix of coefficients of a polynomial series.

Q_x, Q_y Shear line loads in plate flexure.

q Distributed load intensity.

R Residual.

$[\mathbf{R}]$ Static equilibrium matrix, relating element forces to each other.

$[\mathfrak{R}]$ Central matrix in generalization of one-dimensional interpolation function to two dimensions.

r Radial coordinate; number of constraint equations.

S, S_u, S_σ General surface and surfaces where displacements and stresses are prescribed, respectively.

$[\mathbf{S}]$ Stress matrix, connecting joint displacements to components of stress field.

$[\mathbb{S}]$ Stress matrix, connecting joint displacements to stresses at specified points.

s Coordinate.

$\{\mathbf{s}\}$ Vector of constants in constraint equations.

$\mathbf{T}, T_x, T_y, T_z$ Surface (boundary) traction vector and components.

t Thickness.

U, U^* Strain energy and complementary strain energy.

$\mathbf{u}$ Surface (boundary) displacement vector.

u, v, w Displacement components (in interior and of surface points).

V, V^* Potential and complementary potential of applied loads.

vol Volume.

W Work.

$\mathcal{W}$ Descriptor of displacement field variation.

$\mathbf{X}, X, Y, Z$ Body force vector and components.

x, y, z Cartesian coordinates.

Greek Symbols

α Coefficient of thermal expansion.

β Beta function (Sect. 8.3.1).

$\{\boldsymbol{\beta}\}$ Vector of parameters of assumed stress field.

$\boldsymbol{\Gamma}$ Gamma function (Sect. 8.3.1); warping constant (Sect. 13.3.2).

$\boldsymbol{\Gamma}_i$ Cross-derivative d.o.f. [Eq. (12.31)].

$[\boldsymbol{\Gamma}]$ Transformation matrix.

$\{\boldsymbol{\Delta}\}$ Vector of nodal point displacements.

δ Variational operator; infinitesmal change.

$\boldsymbol{\epsilon}$	General strain vector (includes both normal and shear strain).
$\epsilon_x, \epsilon_y, \epsilon_z$	Normal strains.
ξ, η, ζ	Nondimensional spatial coordinates.
θ	Angular displacement (angle of measure in Ch. 11).
$\boldsymbol{\kappa}, \kappa_x, \kappa_y, \kappa_{xy}$	Vector of curvatures in plate bending, and components.
$[\boldsymbol{\kappa}]$	Hessian matrix.
$\{\boldsymbol{\lambda}\}$	Vector of Lagrange multipliers.
μ	Poisson's ratio.
$\lfloor \mathbf{X}_i \rfloor$	Vector of stress field shape functions.
Π	General functional.
$\Pi_p, \Pi_p^{m_1}, \Pi_c,$	Energy functional (subscripts and superscripts denote specific type).
π	3.1416 . . .
$[\boldsymbol{\rho}]$	Material mass density matrix.
$\boldsymbol{\sigma}$	General stress field vector (includes both normal and shear stresses).
$\sigma_x, \sigma_y, \sigma_z$	Normal stresses.
$\{\boldsymbol{\sigma}\}$	Vector of node point stresses.
$\tau_{xy}, \tau_{yz}, \tau_{xz}$	Shear stresses.
Υ	Temperature change above ambient or stress-free state.
υ	Thermal conductivity.
Φ	Stress function
$\{\boldsymbol{\Phi}\}$	Vector of node point stress functions.
ϕ	Angle of measure or circumferential angular coordinate; Weighting factor in weighted residual integral.
Ω	Loading function for plate bending.
$[\boldsymbol{\Omega}]$	Mixed format force-displacement matrix.
$\{\boldsymbol{\omega}\}$	Vector of eigenvalues.

FINITE ELEMENT ANALYSIS

Fundamentals

1

Introduction

In problems of structural mechanics the design analyst seeks to determine the distribution, or *field*, of stresses throughout the structure to be designed. On occasion it is necessary to calculate the displacements at certain points of the structure to ensure that specified clearances are not violated. In some cases, especially where the loads and structural response are time-dependent, it is necessary to compute the entire distribution, or field, of displacements. The calculated stress field should represent a system of internal and external forces which are everywhere in *equilibrium* and simultaneously the displacements should be continuous (the condition of *compatibility*).

In initiating the determination of a system of stresses and displacements for a given design problem, the designer must first define the governing equations of the problem that, in one form or another, stipulate the satisfaction of the conditions of equilibrium and compatibility. A basic difficulty in this regard, quite apart from the solvability of the chosen equations, is the ability of the equations to represent the design conditions. Complications in geometry, loading, and material properties enter into this consideration.

Taking note of the discrepancies likely to accrue from the sources above, the analyst moves ahead with the process of solution of the chosen equations. When the behavior being examined is of a two- or three-dimensional nature, these are of the form of partial differential equations. Rarely do exact solutions exist for such equations and only slightly more often does it prove feasible to construct an adequate solution with an approximation consisting of a few terms. Many such terms are needed for an accurate solution.

The appearance of the electronic digital computer altered radically the capability for solution of partial differential equations. Thus, numerical solutions are today within the grasp of most practicing analysts. The number of terms that can be employed in a series representation of the stress or displacement field is very great. The technique of finite differences, in which the differential equation is approximated by discrete values of the variables at selected points, is likewise applicable. These methods have the advantage of a long history of development resulting in theoretical proofs of convergence. In addition they often result in algebraic equations to be solved that are of especially simple form.

The finite element method is an analytical procedure whose active development has been pursued for a relatively short period of time. The basic concept of the method, when applied to problems of structural analysis, is that a continuum (the total structure) can be modeled analytically by its subdivision into regions (the finite elements) in each of which the behavior is described by a separate set of assumed functions representing the stresses or displacements in that region. These sets of functions are often chosen in a form that ensures continuity of the described behavior throughout the complete continuum. In other cases, the chosen fields do not ensure continuity and nevertheless enable satisfactory solutions but do not feature the rigorous assurances of convergence possessed by the fully continuous analytical models. If the behavior of the structure is characterized by a single differential equation, then the finite element method, in common with the series and finite difference schemes, represents an approach to the approximate solution of that equation. If the total structure is heterogeneous, being composed of many separate structural forms in each of which the behavior is described by a distinct differential equation, the finite element approach continues to be directly applicable.

In common with the alternative procedures for the accomplishment of numerical solutions for practical problems in structural mechanics, the finite element method requires the formation and solution of systems of algebraic equations. The special advantages of the method reside in its suitability for automation of the equation formation process and in the ability to represent highly irregular and complex structures and loading situations.

The rapid advancement of the finite element method has already been noted. The method has grown from a virtually insignificant area of activity as recently as 1955 to one of the presently most active fields of endeavor in the numerical analysis of problems of mathematical physics. We use the term *mathematical physics* to denote a broad range of analytical problems—structural analysis, heat transfer, fluid flow, electromagnetic wave phenomena—and do not intend it as a reference to activities that are remote from reality and design situations. The interest in and the popularity of the method, in fact, are due principally to the aforementioned ability to cope with the complexities of the practical design analysis problem.

The development of practical finite element analysis is an outgrowth of the technological advances of the mid-1950's. A basic premise of the method, implied by our previous comments, is that large-order systems of algebraic equations can be formed automatically and solved efficiently. The introduction of electronic digital computers during the 1950's enabled the satisfaction of these requirements. The theoretical concepts of finite element analysis crystallized during the same period. It is of interest to review next the process of development of these concepts.

1.1 A Brief History of the Finite Element Method†

Although it was immediately preceded by the great accomplishments of the school of French elasticians, such as Navier and St. Venant, the period from 1850 to 1875 is a logical starting point for our review. The concepts of framework analysis emerged during this period, due to the efforts of Maxwell,[1.1] Castigliano,[1.2] and Mohr,[1.3] among others. The concepts represent the cornerstone of the methodology of matrix structural analysis, which did not take form until nearly 80 years later and which itself underlies finite element analysis.

Progress in the development of theory and the analytical techniques subsidiary to finite element analysis was particularly slow from about 1875 to 1920. This was due, in large measure, to practical limitations on the solvability of algebraic equations with more than a few unknowns. We might note parenthetically that for the structures of primary interest in that period—trusses and frames—an analysis approach based upon assumed stress distributions, with force parameters as unknowns, was almost universally employed.

In approximately 1920, due to the efforts of Maney[1.4] in the United States and Ostenfeld[1.5] in Denmark, the basic ideas of a truss and framework analysis approach based on displacement parameters as unknowns took form. These ideas represent the forerunners of matrix structural analysis concepts in vogue today. Severe limitations on the size of problems which could be handled by either force or displacement unknowns nevertheless continued to prevail until 1932, when Hardy Cross introduced the method of moment distribution.[1.6] This method made feasible the solution of structural analysis problems that were an order of magnitude more complex

†For section, reference, figure, and equation cross-references, the first number refers to the chapter and the following numbers to the sequence within the chapter.

than the most sophisticated problems treatable by the prior approaches. Moment distribution became the staple of structural analysis practice for the next 25 years.

Digital computers appeared first in the early 1950's but their real significance to both theory and practice did not become widely apparent immediately. Certain individuals did foresee this impact, however, and undertook the codification of the well-established framework analysis procedures in a format suited to the computer, the matrix format. Such contributions, which are too numerous to review in depth here, have been surveyed by Argyris and Patton.[1.7] Two noteworthy developments were the publications authored by Argyris and Kelsey[1.8] and Turner, Clough, Martin, and Topp.[1.9] These publications wedded the concepts of framework analysis and continuum analysis and cast the resulting procedures in a matrix format. They represent the predominant influence on the development of the finite element method in the subsequent years. It would be inaccurate to ascribe the introduction of all basic aspects of the finite element method to these works, however, because key features of the method had surfaced even before 1950 in papers by Courant,[1.10] McHenry,[1.11] and Hrenikoff.[1.12] The work of Courant is particularly significant because of its concern with problems governed by equations applicable to other than structural mechanics situations. We again refer to this aspect of finite element analysis only in passing, however, because our attention is fixed upon problems in structural analysis.

The technology of finite element analysis has advanced through a number of indistinct phases in the period since the mid-1950's. A detailed review of this progress is given by Zienkiewicz.[1.13] Motivated by the specific formulation of elements for plane stress, researchers established element relationships for solids, plates in bending, thin shells, and other structural forms. Once these had been established for the purpose of linear, static, elastic analysis, attention turned to special phenomena such as dynamic response, buckling, and material and geometric nonlinearities. It was necessary to extend not only the element formulations but also the general framework of analysis. These developments were followed by a period of rather intensive development of "*general-purpose*" computer programs, intended to place the capabilities of the method in the hands of the practitioner.

General-purpose programs for finite element analysis are now extensively dispersed in practice. The ready availability of such programs at a modest cost of acquisition accounts for the abundance of practical applications of the finite element method. On the developmental side, many researchers continue to be occupied with the formulation of new elements and the further development of improved formulations and algorithms for special phenomena and in the construction of new programs. The establishment of finite element representations of nonstructural and interdisciplinary phenomena is of major interest. Familiar examples of the latter include thermostructural analyses, where the calculation of thermal stress is best integrated with the calculation of transient temperature distribution, and the interaction of fluid and structural response in cases of hydroelasticity.

Although we have given emphasis to certain distinctions and special advantages of the finite element method in structural mechanics, the method in its present form is unlikely to be the last word in analysis procedures. It should be viewed as but one phase in the continuing development of structural design analysis tools. Such books as Timoshenko's readable *History of Strength of Materials*[1.14] are an invaluable component of the education of the structural engineering practitioner and new books in this vein or technical books with an emphasis on history (e.g., Ref. 1.15) continue to gain publication and deserve his attention.

1.2 Types of Elements

Elements which are commonly employed in practice and which are given explicit consideration in this text are illustrated in Fig. 1.1.

The *simple* framework element, Fig. 1.1(a), is a member of the total family of finite elements. When used in combination with elements of exclusively the same type, it describes truss and space frame structures. When in combination with elements of a different type, especially plate elements, it is usually intended to describe stiffening members. We do not devote a specific section of the text to the formulation of this element as we do with other types of elements since the theoretical relationships associated with it are well known. Rather, we employ it in the early chapters of the text as a vehicle for illustration of many of the basic features of element formulative procedures.

The *basic* elements in finite element analysis are the thin plates loaded by forces in their own plane (the condition of *plane stress*). Triangular and quadrilateral plane stress elements are illustrated in Fig. 1.1(b). Many other geometric shapes are feasible in this class of element but such other forms generally serve very specialized purposes. These elements are termed *basic* not only because of their usefulness for a wide range of practical design analysis situations but also on account of their priority in the history of development of finite element analysis. Theoretical work during the early years of development of the finite element method was almost exclusively concerned with this type of element.

The *solid elements*, Fig. 1.1(c), are three-dimensional generalizations of the plane stress elements. The tetrahedron and hexahedron are the most common shapes of three-dimensional elements and are essential to analytical models of soil and rock mechanics problems and of structures for nuclear power. It is pertinent to observe that virtually no other approach to struc-

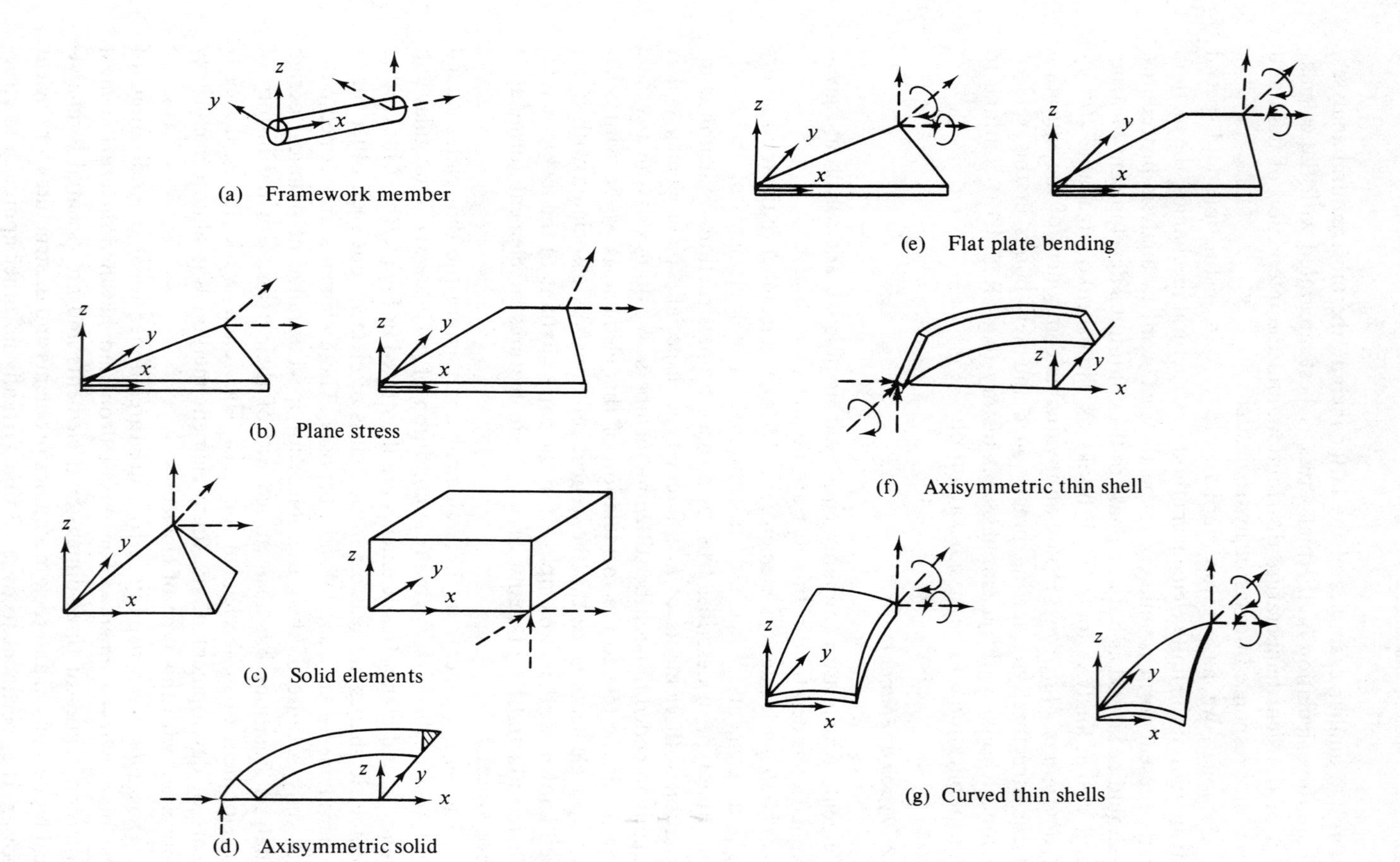

Fig. 1.1 Types of finite elements.

tural analysis has yet been able to deal with practical problems of three-dimensional stress analysis.

One of the most important fields of application of the finite element method is in the analysis of *axisymmetric solids*, Fig. 1.1(d). A great variety of engineering problems fall in this category, including concrete and steel tanks, nuclear containment vessels, rotors, pistons, shafts, and rocket nozzles. The loading, as well as the geometry, is usually axisymmetric. We show here only the triangular element although the general quadrilateral shape, similar to Fig. 1.1(b), is also useful.

Thin *flat plates* in *bending* are employed not only in connection with the behavior of flat plates, per se, but also in "faceted" representations of shells and for thin-walled members as well. The scope of element geometric forms parallels that of the plane stress elements, with the greatest emphasis likewise placed on the triangular and quadrilateral shapes, Fig. 1.1(e).

Axisymmetric thin shell structures, Fig. 1.1(f), have the same range of significance in practical application as the axisymmetric solids although here the governing relationships derive from the simplifying assumptions of thin shell theory. Axisymmetric thin shell formulations bridge the gap between flat plate bending and stretching and general thin shell elements and they serve to identify key problems arising in the latter.

When a thin shell structure is in fact curved, it would appear preferable to employ curved *thin shell elements* for the analytical model. The advantages include the ability to describe more accurately the geometry of the actual shell surface and a proper representation of the coupling of stretching and bending within each element. Typical doubly curved thin shell elements are sketched in Fig 1.1(g). A large number of alternative formulations for this type of element are in existence.

1.3 Some Applications of Finite Elements

We now examine a few representative structural design applications of the finite element method in order to illustrate the way in which the elements described in the previous section are employed and the scale and complexity of the design analysis problems to which they are addressed.

The development of the finite element method owes much to the early work of individuals involved in aerospace structural design, and it is not surprising that this field continues to lead in the practical application of the method. Figure 1.2 shows many features of the finite element analysis of a portion of the Boeing 747 aircraft.[1.16] Aircraft fuselage structures consist of thin sheet metal (the *skin*) wrapped around a framework consisting of members termed *frames* and *stiffeners*. The wing framework members are termed *spars* and *ribs*.

Experience has shown that when determining overall stress fields the

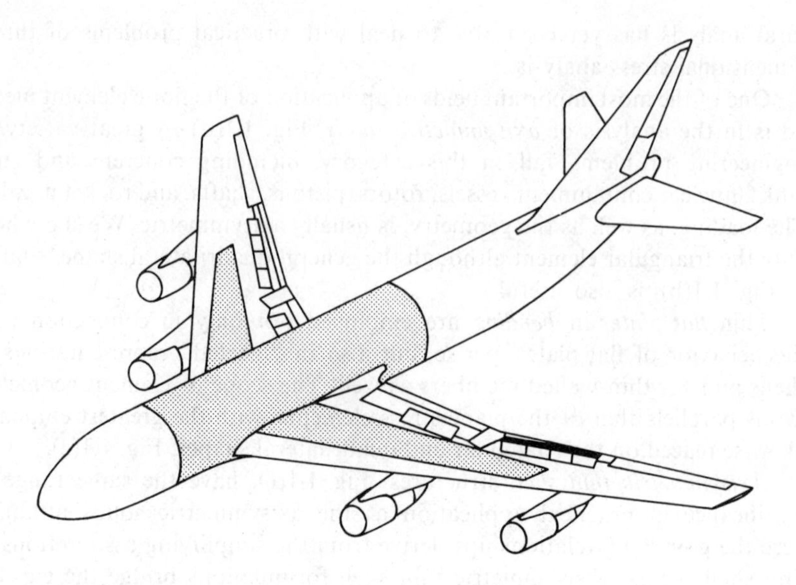

a. Boeing 747 Aircraft. (Cross-hatched area indicates portion of the airframe analyzed by finite element method.)

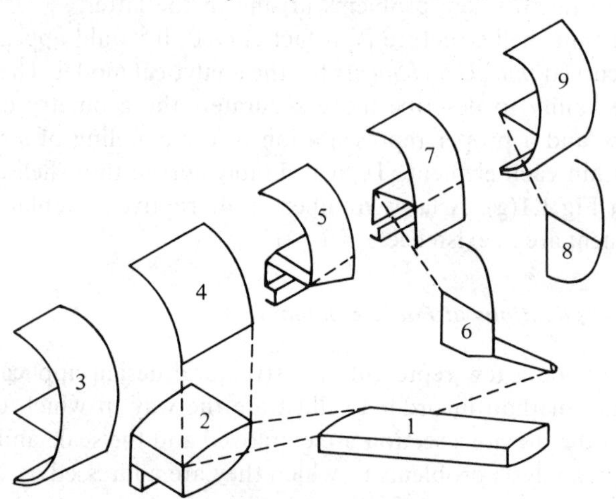

b. Substructures for finite element analysis of cross-hatched region.

Fig. 1.2 Finite element analysis of 747 aircraft (from Ref. 1.16).

local bending effects in the analytical description of the skin of an aircraft can be neglected. Thus, the skin can be assumed to consist of plane stress elements, such as the triangles and quadrilaterals of Fig. 1.1(b) and (c), while the framework is idealized with use of members of the type shown in Fig. 1.1(a). The finite element analysis of the 747 wing-body region shown crosshatched in Fig. 1.2(b) required a total of over 7,000 unknowns. This many unknowns is unwieldy from the standpoint of data handling and with respect to the identification of sources of error in the analysis. It is therefore common practice to divide the structure into regions, or *substructures*, and each of these is analyzed by the finite element method in such a way as to produce a *super element*. The super elements are tied together via a conventional finite element procedure in the final phase of analysis. The substructuring scheme for the 747 analysis is shown in Fig. 1.1(b) and details are listed in Table 1.1.

Table 1.1 Summary of Finite Element Idealization of Center Section of 747 Aircraft (from Ref. 1.16).

Substructure	*Description*	*Nodes*	*Unit* Loadcases*	*Beams*	*Plates*	*Interact† Freedoms*	*Total Freedoms*
1	Wing	262	14	355	363	104	796
2	Wing Center	267	8	414	295	198	880
3	Body	291	7	502	223	91	1,026
4	Body	213	5	377	185	145	820
5	Body	292	7	415	241	200	936
6	Bulkhead	170	10	221	103	126	686
7	Body	285	6	392	249	233	909
8	Bulkhead	129	10	201	93	148	503
9	Body	286	7	497	227	92	1,038
Total		2,195	63	3,374	1,979	555	7,594

*Some unit loadcases involve more than one substructure.
†Several freedoms of substructure interface contribute to a single interaction freedom.

As is customary in the design phase of major aircraft, tests were conducted on the 747. Figure 1.3 shows some results of these tests and compares them with the finite element solution. It is fair to say that no solution based on simplified concepts in structural engineering could have predicted these results accurately, as was done by the finite element method.

It should be added that the dynamic response of an aircraft is important from the standpoint of both structural integrity and riding qualities and that elastic instability is an important mode of failure. Neither of these phenomena can be dealt with adequately using simplified analysis concepts but their analysis by use of the finite element method is quite feasible.

A practical design problem of similar form is encountered in naval architecture. Figure 1.4 shows a portion of the midship of a modern ship structure.[1.17] The thrust toward increased size of tanker ships has produced much concern regarding their structural integrity and design efficiency. Supertankers have, in fact, sustained considerable structural damage while in service.

The finite element idealization of a ship bears many similarities to the representation of aerospace structures. The plating is represented by elements

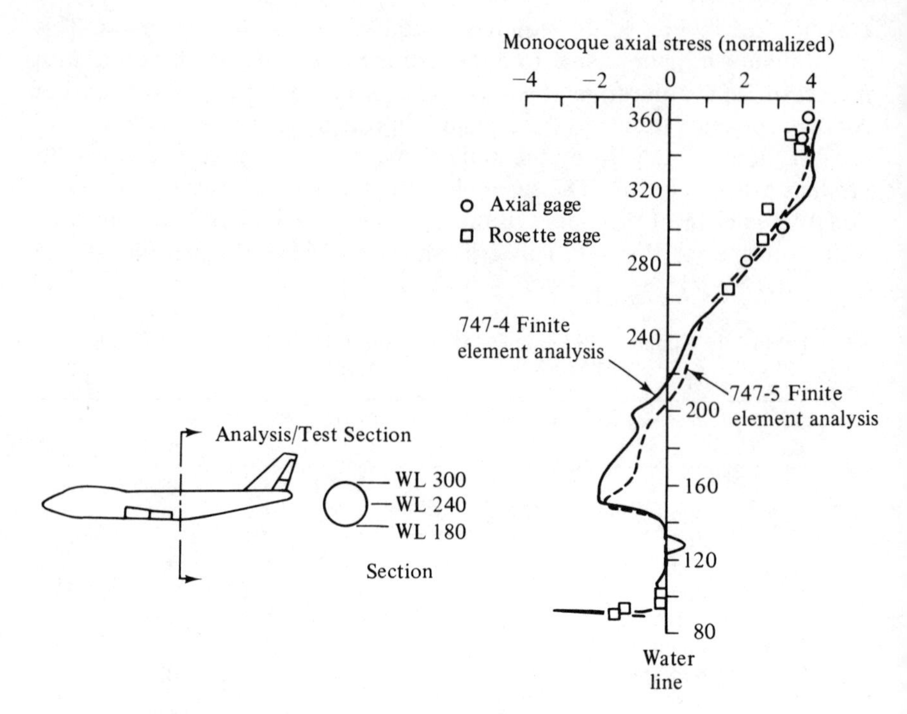

Fig. 1.3 Correlation with test data of 747 aircraft finite element analysis (from Ref. 1.16).

in plane stress. Framework elements are employed in representation of the internal structural gridwork. The total number of unknowns to be determined for structurally important portions of a ship may range up to 50,000 and again it is customary to subdivide the problem into many subregions with fewer unknowns, i.e., to substructure.

Requirements of safety in nuclear reactor structural design have caused that industry to rely heavily on finite element analysis. Figure 1.5(a) shows a prestressed concrete reactor vessel.[1.18] Due to symmetry it is possible to analyze only one-twelfth of the total structure, Fig. 1.5(b). This volume is conceived as an assemblage of tetrahedronal and hexahedronal elements, Fig. 1.5(c). In problems of this type the number of unknowns ranges upward of 20,000 and it is common for the analysis to extend to the treatment of inelastic phenomena.

Not all problems to which the finite element method is applied are of monumental proportions. Figures 1.6 and 1.7 show applications to some basic problems in civil engineering structures. One way of increasing the design efficiency of rolled structural steel sections is by cutting the web in the sawtooth manner of Fig. 1.6(a), placing one portion atop the other and welding the portions together as shown in Fig. 1.6(b). One then obtains a

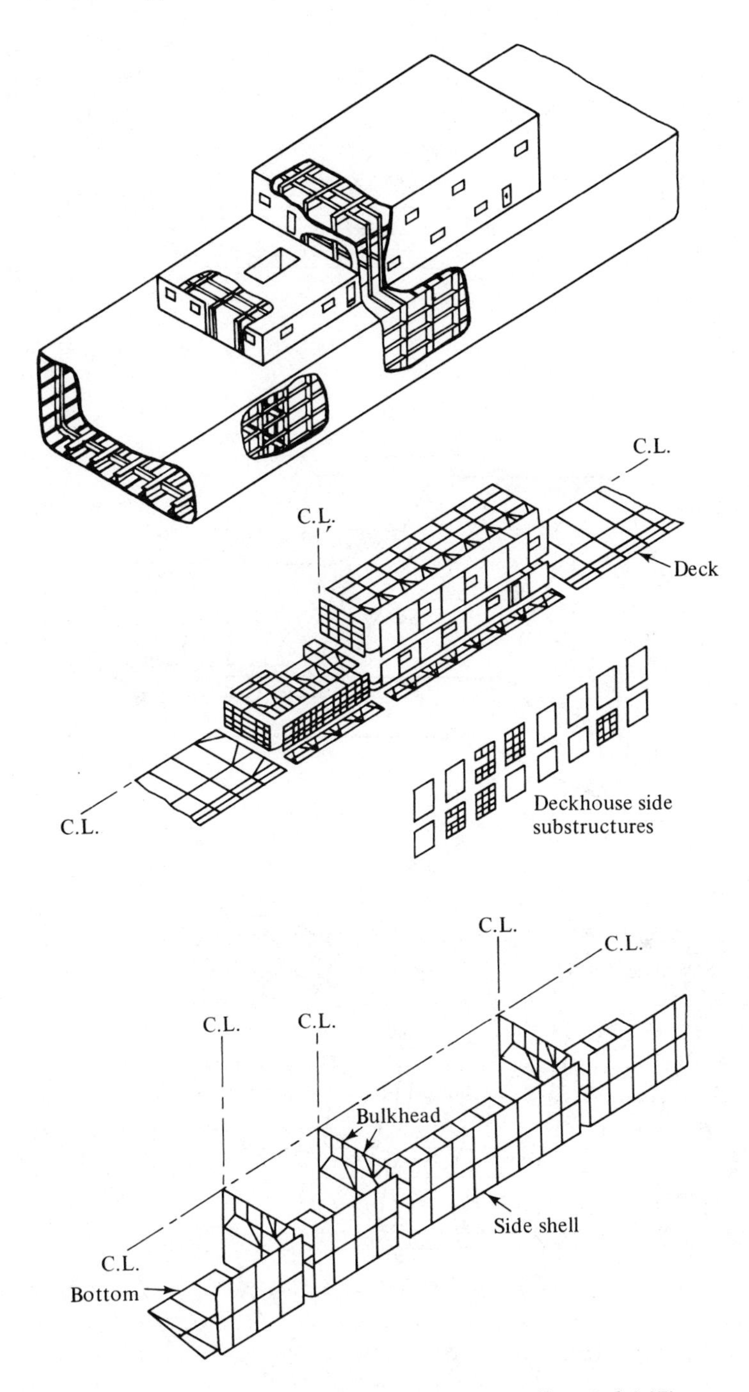

Fig. 1.4 Finite element analysis of ship structure (from Ref. 1.17).

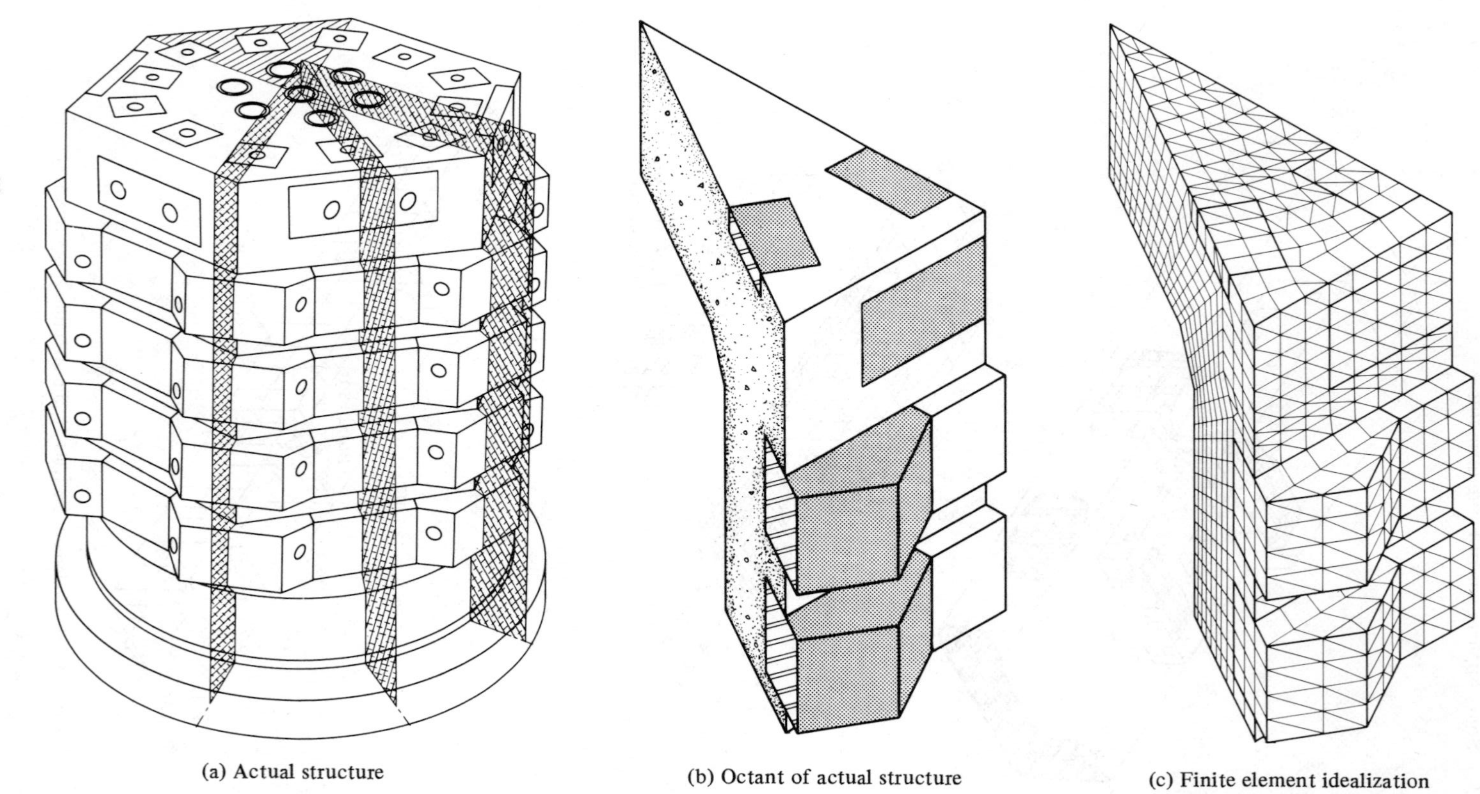

Fig. 1.5 Finite element analysis of prestressed concrete reactor vessel (from Ref. 1.18).

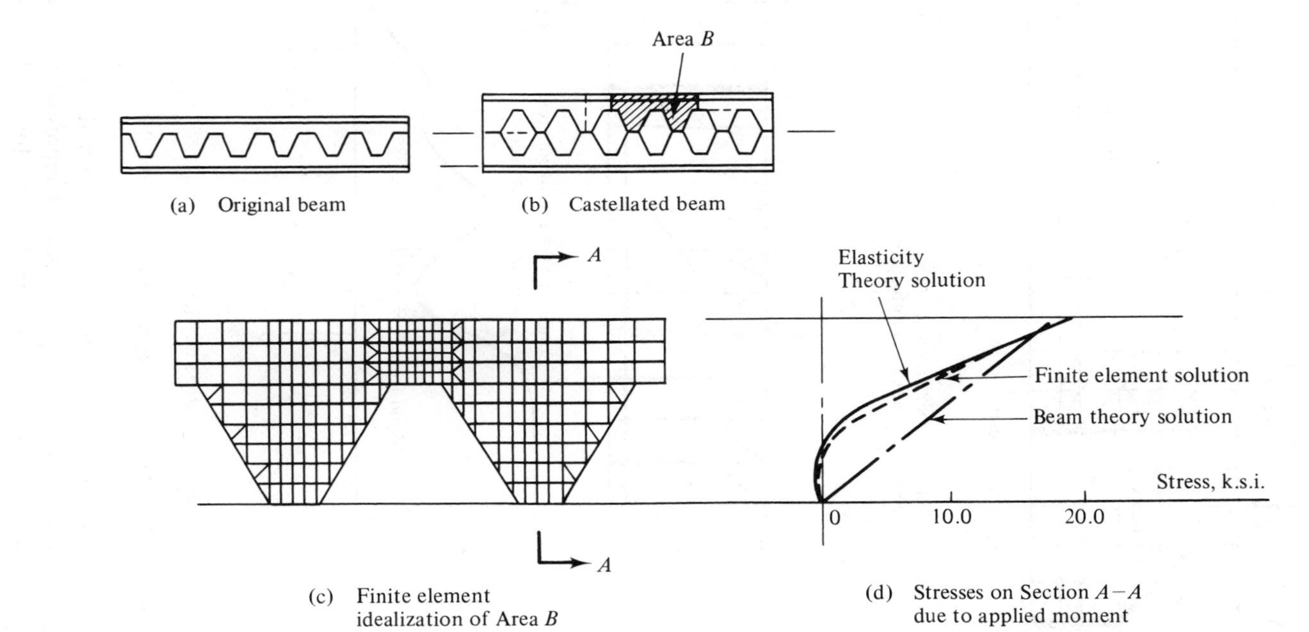

(a) Original beam

(b) Castellated beam

(c) Finite element idealization of Area B

(d) Stresses on Section $A-A$ due to applied moment

Fig. 1.6 Finite element analysis of castellated beam (from Ref. 1.19).

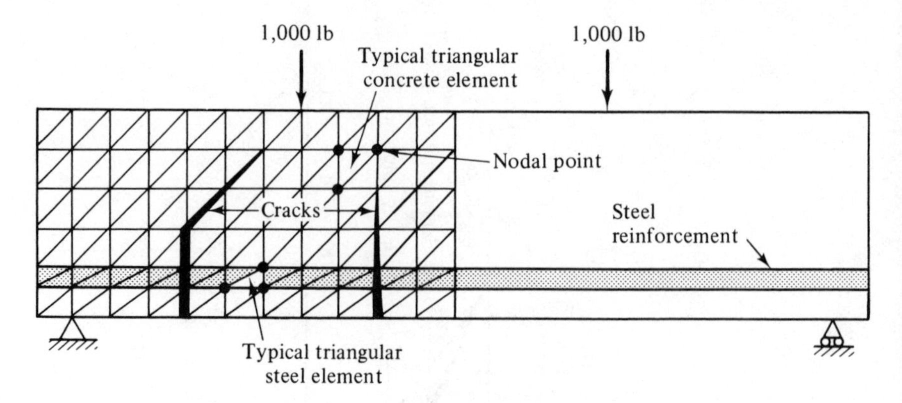

(a) Finite element representation of concrete beam and reinforcement

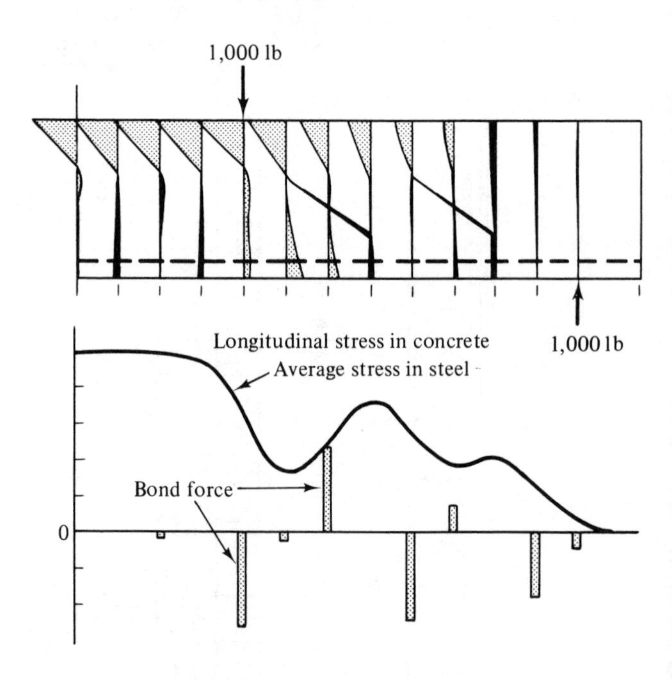

(b) Calculated stress distributions

Fig. 1.7 Finte element anaysis of concrete beam (from Ref. 1.20).

"castellated" form of beam for which approximate methods of analysis are available. Figure 1.6(c) illustrates a finite element representation of one such design, using triangular and rectangular elements, formed for the purpose of verifying the approximate theory.[1.19] The results, Fig. 1.6(d), show that the approximate theory is indeed a valid approach for calculation of extreme fiber stress and that the more expensive finite element method or other more sophisticated procedures are not needed in routine design exercises for this structural form.

An even more common problem is that of the reinforced concrete beam, Fig. 1.7, for which little is known regarding the detailed nature of the bond between the concrete and reinforcing rods and of the growth and pattern of cracks in the concrete as loading progresses. This problem is one of fundamental research in structural engineering. Figure 1.7(a) portrays the finite element representation and the analytically described crack trajectories employed in a study by Ngo and Scordelis.[1.20] Stress profiles are shown in Fig. 1.7(b).

As a group, the relatively few examples described above demonstrate that the finite element method can be used to advantage in any situation requiring the prediction of internal strains and stresses, displacements, vibration and buckling mode shapes, and a host of other solution parameters. These situations arise in a number of fields that are traditionally regarded as separate engineering disciplines, e.g., civil, mechanical, and aerospace engineering and naval architecture. The finite element method furnishes a unified analysis technology for these and other fields.

Our intent in this text is a development of fundamental theoretical considerations and we shall not return again to discussion of specific problems of practical character. A compendium of such problems would fill many volumes. The reader is urged to consult finite element conference proceedings for descriptions of a significant number and variety of applications.

1.4 General-Purpose Programs

We have indicated that the equations of the finite element method are of a form so generally applicable that it is possible in theory to write a single computer program that will solve all the aforementioned problems and almost a limitless variety of other problems in structural mechanics. Computer programs that attempt this objective, even on a restricted scale, are termed *general-purpose* programs. The advantage of general-purpose programs is not merely this capability but the unity afforded in the instruction of prospective users, input and output data interpretation procedures, and documentation.

The cost of development of a general-purpose program is usually very high so that amortization of investment in every way possible is essential. Certain general-purpose programs are coded in a computer language that is general enough to allow the program to be operated by many different organizations at widely separated geographic locations. Other programs are intended for use only by the component activities of a single governmental or industrial organization and achieve economic feasibility by limiting the scope of the program capabilities. We note, therefore, that not every such program has the same degree of generality. Surveys of general-purpose programs have been published,[1.21, 1.22] but the reader should note that this is one of the fastest changing aspects of a field that is growing rapidly in all its component facets.

The four components shown in the flow chart of Fig. 1.8 are common to virtually all general-purpose finite element analysis programs. As a minimum, the *input* phase should require of the user no information beyond that relating to the material of construction, geometric description of the finite element representation (including support conditions), and the conditions of loading. The more sophisticated general-purpose programs facilitate this input process through such features as prestored material property schedules, schemes for the computer generation of finite element gridworks, and perspective and stereographic plotting of the finite element idealization so that errors in input may be detected prior to performance of calculations.

The phase comprised of the *library of finite elements* is of principal interest to this text. Herein resides the coded formulative process for the individual elements. Many general-purpose programs contain all the elements portrayed in Fig. 1.1 as well as certain others and alternative formulations for a given type of element, for instance the triangle in bending, may be present. Ideally, the element library is open-ended and is capable of accommodating new elements to any degree of complexity.

The element library phase receives the stored input data and establishes

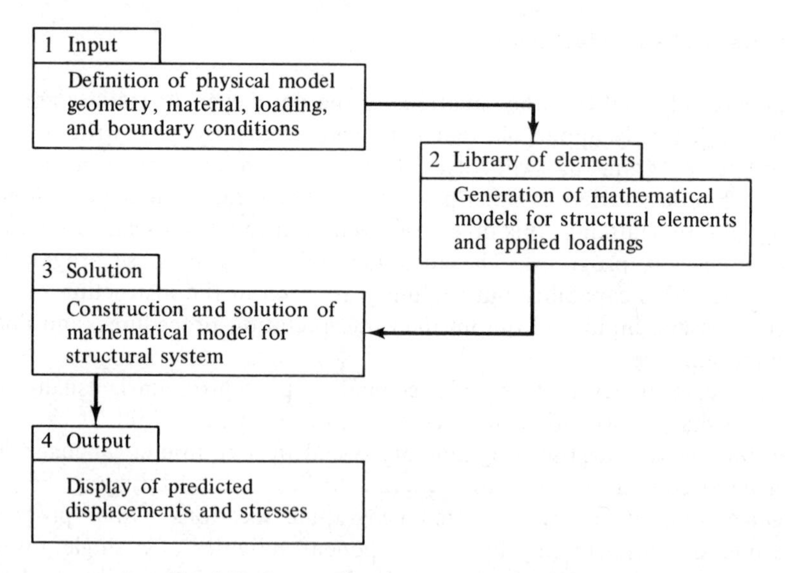

Fig. 1.8 Structural analysis computational flow.

the element algebraic relationships by application of the relevant coded formulative process. This phase of the general-purpose program also includes all operations necessary to position the element algebraic relationships for connection to the neighboring elements and the connection process itself. The latter operations thereby produce the full set of algebraic equations of the finite element representation of the complete structure.

The *solution* phase of a general-purpose program operates on the equations of the problem formed in the prior phase. In the case of a static structural analysis problem this may mean no more than the solution of a set of linear algebraic equations for a known right-hand side. Solutions for dynamic response will require very extensive computations over a time-history of applied loads. In still other cases it is necessary to operate on subdivided portions of the total structure, as indicated by the 747 analysis referred to in Section 1.3, or to perform special operations on the originally constructed equations. Included in this phase are the back-substitution operations needed to obtain all desired aspects of the solution.

The *output* phase presents the analyst with a record of the solution upon which he can base decisions regarding the proportioning of the structure and other design questions. The record is commonly a printed list of stresses at points or within the respective finite elements, of displacements at specified points, and of other desired information. As in the input phase there is a strong trend toward graphical representation of data, such as plots of trajectories of the principal stresses or modes of buckling and vibration.

At this juncture it is pertinent to emphasize that the contents of this text deal exclusively with analysis. That is, it is assumed that all the data of the problem, by way of material properties, geometric data, and member sizes, are specified in the input phase. Modern developments in the field of mathematical programming and similar disciplines[1.23] have brought into being the capability for selection of such design parameters, within the computer, of a form that meets the goal of minimum weight or minimum cost. These developments are beyond the scope of this text.

The phases of the general-purpose program above have one feature in common—the *modularity* of their component aspects. A well-constructed program should permit the user to insert new component operations as the need arises or as more efficient ways of performing these operations are identified. These operations include, for example, new elements, improved methods for the solution of algebraic equations, and various forms of computer plots of output. Some programs permit the user to do this by following written instructions without personal reference to the individuals who originally coded the program.

Insofar as it is possible, the procedures of this text are formulated in a manner and form consistent with procedures found in widely distributed general-purpose programs.

REFERENCES

1.1 MAXWELL, J. C., "On the Calculations of the Equilibrium and Stiffness of Frames," *Phil. Mag.* (4), **27**, 294, 1864.

1.2 CASTIGLIANO, A., *Theorie de l'equilibre des Systemes Elastiques*, Turin, 1879 (English Translation by Dover Publications).

1.3 MOHR, O., "Beitrag zur Theorie der Holz—und Eisen Konstruktionen," *Zeit. des Architekten und Ingenieur Verienes zu Hannover*, 1868.

1.4 MANEY, G. B., *Studies in Engineering*—No. 1, U. of Minnesota, Minneapolis, Minn., 1915.

1.5 OSTENFELD, A., *Die Deformationsmethode*, Springer-Verlag OHG, Berlin, 1926.

1.6 CROSS, H., "Analysis of Continuous Frames by Distributing Fixed-End Moments," *Trans. ASCE*, **96**, 1932, pp. 1–10.

1.7 ARGYRIS, J. H. and P. C. PATTON, "Computer Oriented Research in a University Milieu," *Appl. Mech. Rev.*, **19**, No. 12, Dec., 1966, pp. 1029–1039.

1.8 ARGYRIS, J. and S. KELSEY, *Energy Theorems and Structural Analysis*, Butterworth Scientific Publications, London, 1960.

1.9 TURNER, M., R. CLOUGH, H. MARTIN, and L. TOPP, "Stiffness and Deflection Analysis of Complex Structures," *J. Aero. Sci.*, **23**, No. 9, Sept., 1956, pp. 805–823.

1.10 COURANT, R., "Variational Methods for the Solution of Problems of Equilibrium and Vibration," *Bull. Am. Math. Soc.*, **49**, pp. 1–43, 1943.

1.11 MCHENRY, D., "A Lattice Analogy for the Solution of Plane Stress Problems," *J. Inst. Civil Eng.*, **21**, pp. 59–82, 1943.

1.12 HRENIKOFF, A., "Solution of Problems in Elasticity by the Framework Method," *J. Appl. Mech.*, **8**, pp. 169–175, 1941.

1.13 ZIENKIEWICZ, O. C., "The Finite Element Method: From Intuition to Generality," *Appl. Mech. Rev.*, **23**, No. 23, Mar., 1970, pp. 249–256.

1.14 TIMOSHENKO, S., *History of Strength of Materials*, McGraw-Hill Book Co., New York, N.Y., 1953.

1.15 VOLTERRA, E. and J. E. GAINES, *Advanced Strength of Materials*, Prentice-Hall, Inc., Englewood Cliffs, N.J., 1971.

1.16 MILLER, R. E. and S. D. HANSEN, "Large Scale Analysis of Current Aircraft," *On General Purpose Finite Element Computer Programs*, P. V. Marcal (ed.), *ASME Special Publication*, New York, N.Y., 1970.

1.17 SMITH, C. S. and G. MITCHELL, "Practical Considerations in the Application of Finite Element Techniques to Ship Structures," Proc. of Symposium on Finite Element Techniques, U. of Stuttgart, Stuttgart, Germany, June, 1969.

1.18 CORUM, J. M., and N. KRISHNAMURTHY, "A Three-Dimensional Finite

Element Analysis of a Prestressed Concrete Reactor Model," Proc. of Symposium on Application of Finite Element Methods in Civil Engineering, Vanderbilt University, Nashville, Tennessee, Nov. 1969.

1.19 CHENG, W. K., M. U. HOSAIN, and V. V. NEIS, "Analysis of Castellated Beams by the Finite Element Method," Proc. of Conf. on Finite Element Method in Civil Eng., McGill U., Montreal, Canada, 1972, pp. 1105–1140.

1.20 NGO, D. and A. C. SCORDELIS, "Finite Element Analysis of Reinforced Concrete Beams," *J. Am. Concrete Inst.*, **64**, No. 3, Mar., 1967, pp. 152–163.

1.21 GALLAGHER, R. H., "Large-Scale Computer Programs for Structural Analysis" in *On General Purpose Finite Element Computer Programs*, P. V. Marcal (ed.), *ASME Special Publication*, 1970, pp. 3–34.

1.22 MARCAL, P. V., "Survey of General Purpose Programs for Finite Element Analysis," in *Advances in Computational Methods in Structural Mechanics and Design*, J. T. Oden, et al. (ed.), U. of Alabama Press, University, Ala., 1972.

1.23 GALLAGHER, R. H. and O. C. ZIENKIEWICZ, *Optimum Structural Design*, John Wiley & Sons, Inc., New York, N.Y., 1973.

2

Definitions and Basic Element Operations

The present chapter, as well as Chapters 3 and 4, are in most respects preliminary to the subject matter of finite element analysis. Here and in Chapter 3 we treat definitions, notions, and procedures that are dealt with in much greater detail in a course devoted to the matrix analysis of framework structures, a subject with which the reader is assumed to be familiar. (By the same token it is assumed that the reader is familiar with the symbolism and basic operations of matrix algebra.) All fundamental aspects of matrix structural analysis relevant to our development of finite element analysis are nevertheless presented here and in Chapter 3, albeit without the extensive numerical examples one would expect to find in such texts as Refs. 2.1–2.4. The various symbols and operations in matrix algebra are also defined where they first appear.

We begin this chapter with the description of the principal coordinate system and sign convention to be used throughout. The relationship between the finite element analytical representation and the behavior of the region of the actual structure covered by the finite element is examined next. We then proceed to the definition of influence coefficients of structural members whose load-versus-displacement response is determined at discrete points on the member. This leads naturally to the definition of the notions of work and energy, expressed in terms of influence coefficients, and to proof of the property of symmetry possessed by these coefficients when they refer to linear elastic behavior.

A rather extensive section of this chapter is devoted to the transformation of a set of influence coefficients of one format, say the format of *displacement* influence coefficients, into the opposite format, which for the case cited would be *force* influence coefficients. Such operations, which are of small importance in framework analysis, are of major significance in finite element analysis. The question of transformation matrices is treated in the two subsequent sections, where it is broadened beyond the usual application to coordinate axis changes to permit the reduction in the number of solution parameters. Finally, we describe a procedure for operating on the element stiffness matrix to define the number of rigid body modes contained therein.

2.1 Coordinate System

In this text we shall most often work with a set of orthogonal axes identified by the symbols x, y, and z, as shown in Fig. 2.1.

In an elastic structure the imposition of loads causes displacement of points of the body, including displacement of the points relative to one another. The present chapter is concerned only with the overall behavior of the structural elements as defined by the displacement of points on a body under the action of systems of forces applied to these points. The detailed study of the relative displacement of points within the body (the strains) and the distribution of forces per unit area with the body (the stresses) is the subject of later chapters, particularly Chapters 4 through 6, as are such conditions as distributed applied loads and initial strains due to thermal expansion.

A set of axes is defined for a structure in the undeformed state, in the manner of Fig. 2.1. These axes remain fixed throughout the deformation of the structure and the displacements of points of the body are referenced to them. Consider a particle located at the point g when the structure is in the unloaded, undeformed state [Fig. 2.1(a)]. A force vector, with components F_{x_g}, F_{y_g}, F_{z_g} is applied to this particle. Under the action of this force the particle displaces to a point in space defined by the symbol h. The *translational displacements* of the particle are then given by $u_g = x_h - x_g$, $v_g = y_h - y_g$, $w_g = z_h - z_g$. In correspondence with the designation of the force vector components, we show these as vector components located at the relevant point of the undeformed body. Positive values of the force and displacement components correspond to the positive sense of the coordinate axes.

The subject matter of this text, except for topics discussed in the final chapter, is limited to linear behavior. As far as the above description of translational force and displacement are concerned, this means that the components of the force vector remain unchanged as the particle moves from g to h. Also, the mechanical aspects of this behavior, such as the work done by the forces F_{x_g}, F_{y_g}, F_{z_g} acting through the displacements u_g, v_g, w_g are not dependent upon the path taken to point h.

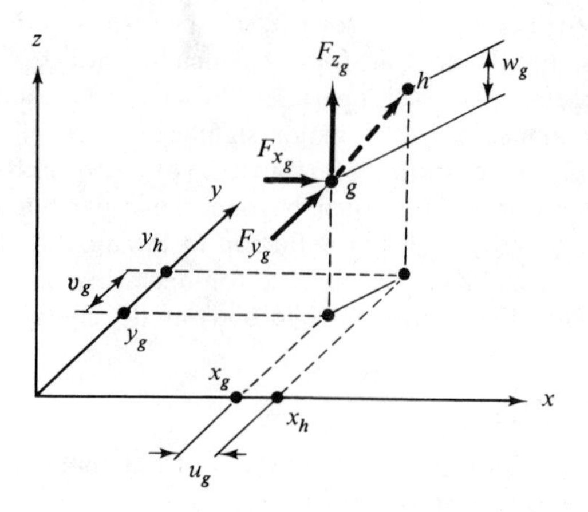

(a) Displacement from point g to point h

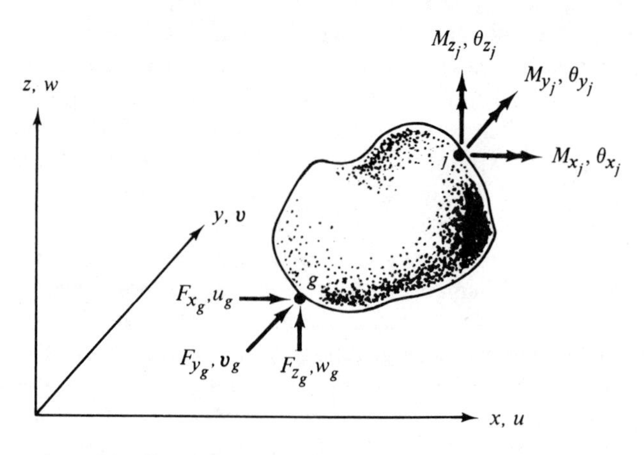

(b) Forces and moments and corresponding displacements

Fig. 2.1 Force and displacement conventions.

Translational displacements *per se* might not be sufficient for the complete description of the displacements of a structure. In problems of beam and frame structures, thin plates, and thin shells, analysts generally make the simplifying assumption that a line drawn normal to the middle line (for the beam and frame) or middle surface (for plates and shells) of the undeformed structure remains normal to the middle line or surface as the structure deforms under loads. A measure of displacement at the points of such structures is the rotation (θ) of the normal with respect to the undeformed state and this rotation is often assumed equal to the slope of the middle line or middle surface. These *angular displacements*, when described in the coordinate system of Fig. 2.1(b) for the point k of a prismatic member that lies along the x-axis are represented by

$$\theta_{x_k} = \left.\frac{\partial w}{\partial y}\right|_k - \left.\frac{\partial v}{\partial z}\right|_k; \qquad \theta_{y_k} = \left.\frac{\partial u}{\partial z}\right|_k - \left.\frac{\partial w}{\partial x}\right|_k; \qquad \theta_{z_k} = \left.\frac{\partial v}{\partial x}\right|_k - \left.\frac{\partial u}{\partial y}\right|_k$$

(For simplicity of presentation we have excluded representation of translational forces at point j). Note that we define a positive rotation in terms of the right-hand rule; i.e., when the right hand grasps an axis with the thumb pointing in the positive direction of the axis, the fingers wrap around the axis in the sense of a positive rotation. This accounts for the negative signs in the definitions of θ's where a positive rotation gives a negative displacement. The force quantities corresponding to the angular displacements are the moment vectors M_{x_j}, M_{y_j}, and M_{z_j}.

It should be emphasized that the derivatives of the displacements can be defined as measures of structural behavior at specified points without the assignment of physical meaning to such quantities. Indeed, higher derivatives of displacement (e.g., $\partial^2 w/\partial x^2$, $\partial^3 w/\partial x^3$) can be and are employed in this manner in finite element analysis. The physical meaning of these measures is often obscure and there are similar difficulties in the definition of the counterpart force quantities, but these factors may be counterbalanced by improved efficiency in analysis.

As already indicated, the overall description of the behavior of a structural body—in our case the individual finite element—is accomplished through the medium of force and displacement components at designated points on the body. Such points are customarily called *node points.* They are also called *joints* because in most cases of practical application of the finite element method they represent points of connection of the elements forming the total, or *global*, analytical model of the complete structure. There are many cases where the points do not have this physical significance but we shall nevertheless use the terms *joint* and *node* interchangeably throughout the text.

The developments of this chapter require a concise form of designation of the forces and displacements at the node points of a given structural element. The forces and displacements are listed in column matrices $\{\mathbf{F}\}$ and $\{\mathbf{\Delta}\}$, respectively. (Braces, { }, denote column vectors). For the element shown in Fig. 2.1(b), for example, with translational forces at point g and moments at point j.

$$\{\mathbf{F}\} = \begin{Bmatrix} F_{x_g} \\ F_{y_g} \\ F_{z_g} \\ M_{x_j} \\ M_{y_j} \\ M_{z_j} \end{Bmatrix} \qquad \{\mathbf{\Delta}\} = \begin{Bmatrix} u_g \\ v_g \\ w_g \\ \theta_{x_j} \\ \theta_{y_j} \\ \theta_{z_j} \end{Bmatrix}$$

Clearly the listing of the contents of a vector in columnar fashion is inefficient from a typographical standpoint and for this reason we shall customarily list these contents in a row vector, $\lfloor \quad \rfloor$. The transpose of a matrix is defined as the matrix obtained by interchanging the rows and columns of the matrix being transposed; in accordance with this definition, when a column matrix is transposed, it produces a row vector and vice versa. Thus, denoting the transpose by the Roman superscript T, we can alternatively write the vectors above in the form

$$\{\mathbf{F}\} = \lfloor F_{x_g}\ F_{y_g}\ F_{z_g}\ M_{x_j}\ M_{y_j}\ M_{z_j}\rfloor^{\mathrm{T}}$$

$$\{\boldsymbol{\Delta}\} = \lfloor u_g\ v_g\ w_g\ \theta_{x_j}\ \theta_{y_j}\ \theta_{z_j}\rfloor^{\mathrm{T}}$$

An individual entry of a typical vector of n joint displacements $\{\boldsymbol{\Delta}\} = \lfloor \Delta_1 \ldots \Delta_i \ldots \Delta_n \rfloor^{\mathrm{T}}$, say Δ_i, is termed the *ith degree-of-freedom*. Degrees-of-freedom are referred to throughout this text by the abbreviated designation d.o.f.

The first step in the formation of the force and displacement vectors is the definition of the node points and their location with respect to coordinate axes. In finite element analysis we must distinguish among global, local, and joint axes. The *global* axes are those established for a complete structure described by many finite elements. The *local* (or *element*) axes are fixed to the respective elements and since the elements are in general differently oriented within a structure (a situation that can be seen in the illustrations of aircraft, ship, and nuclear reactor structural analysis in Chapter 1), these axes will in general be differently oriented from one element to the next. The element axes are identified by primes in Fig. 2.2(a). Finally, *joint* axes are defined at the element connection points and in general have an orientation different from some or all of the elements meeting at the joints. These axes are denoted with double primes. The primed and double-primed designations are used only when the different types of axes are being compared or appear in the same portion of the text. No primes are employed when any of these axes is used alone.

We use element axes in most of our formulations of element equations, and the convention for referencing such axes to the elements and numbering the joints of the element is described below. Global axes figure prominently in Chapter 3 in our development of the equations of the complete structure (the global equations), in Sections 3.1 and 3.2. The significance and manner of treatment of joint axes are dealt with in Section 3.5.3.

Our custom in defining element axes and in numbering joints for planar finite elements is shown in Fig. 2.2(b). The joint at or closest to the origin of coordinates is designated as joint 1. Then proceeding in the positive x direction in the $x - y$ plane, the next point is designated as point 2. The designations proceed in a counterclockwise direction. We define planar elements (plates in plane stress or bending or plane strain situations) in the $x - y$

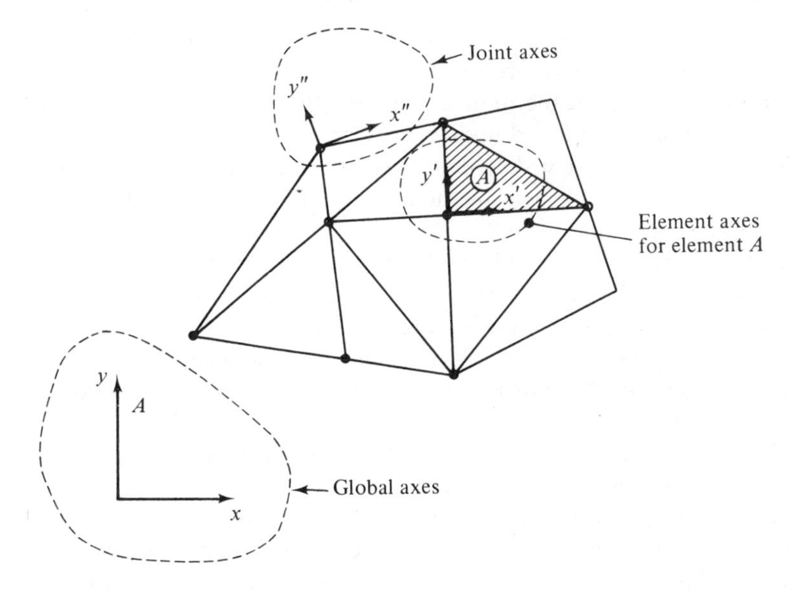

(a) Types of coordinate axes

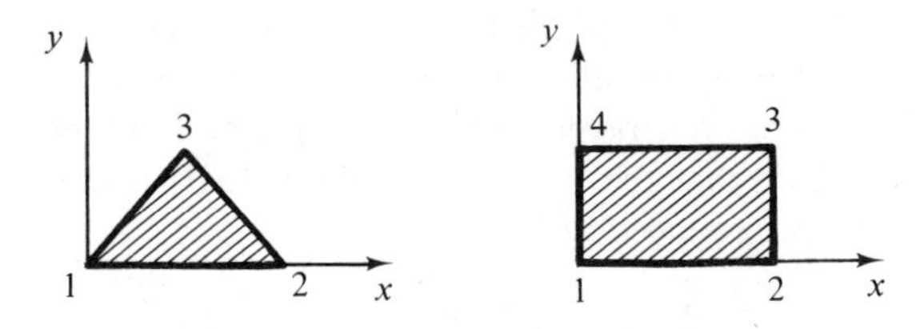

(b) Joint numbering schemes in element axes

Fig. 2.2 Coordinate axes and joint numbering schemes.

plane. Exceptions to this are the cross sections of axisymmetric solid elements. Rules for the numbering of joints for three-dimensional (solid) elements parallel this convention.

Given the basic elastic properties of a finite element or structure in terms of local coordinate directions, the transformations of forces and displacements from these directions to global directions can be constructed easily. The procedure is outlined following the definition of various basic considerations pertinent to finite element elastic properties.

2.2 The Basic Element Idealization

In order to appreciate the significance of the many alternative approaches to the formulation of finite element relationships, it is useful to examine the relationship between a simple model of element behavior and the behavior of the real structure. This model is a valid characterization of element behavior, although there are other equally valid ways of looking at the underlying concepts of the finite element method. Indeed, the approach to finite element analysis based upon energy or variational principles (Chapter 6) is perhaps the most widely employed scheme but views the underlying concepts of the finite element method from a different perspective than presented in this section.

In the present simple model each finite element in the real structure sustains a distributed state of stress, but for the purpose of forming a *mathematical model* the state of stress is represented by forces—*generalized forces*—at the *element joints* or *nodes*. Correspondingly, the displacements of these points—*the degrees-of-freedom*—are employed in the characterization of the displaced state of the element.

The real and idealized behavior of a typical element is compared in Fig. 2.3. The actual variation of element edge stress is sketched in Fig. 2.3(a); the actual displacements vary in a similarly irregular manner. Figure 2.3(b) shows the fundamental idealization, a direct representation of the assumed element behavior. The stress, strain, and displacement states are assumed to exist in simplified form. Finally, Fig. 2.3(c) shows the idealization required by the total framework of analysis, one in which the distributed edge stresses have been replaced by the node point generalized forces. The formulative process in finite element analysis thus begins with the definition of the state shown in Fig. 2.3(b) and after algebraic manipulation it advances to the mathematical model portrayed in Fig. 2.3(c). The fundamental idealization must be chosen so that the behavior of the actual structure is approached as the elements are reduced in size.

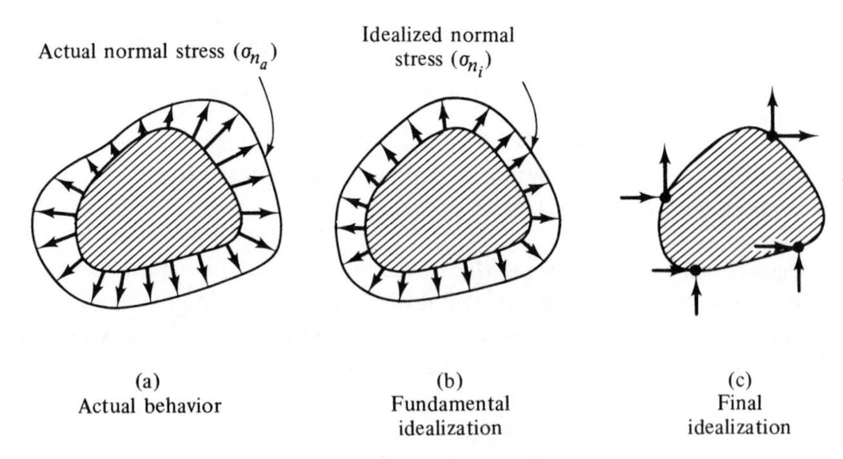

Fig. 2.3 Finite element behavior representations.

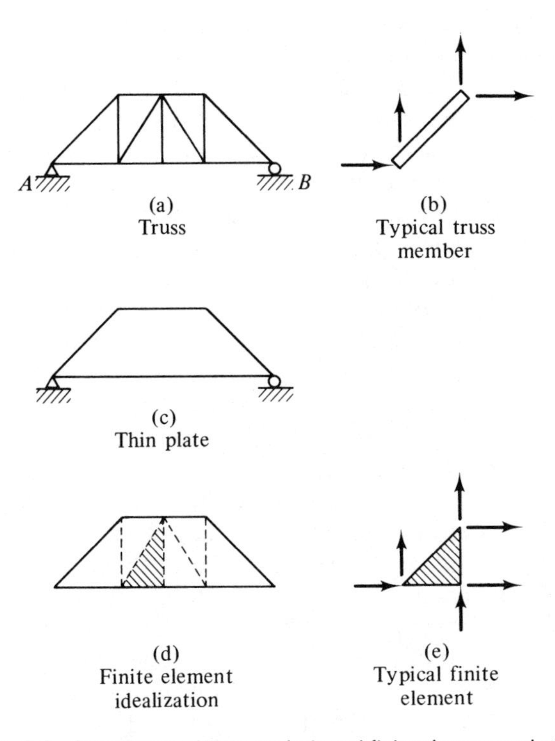

Fig. 2.4 Comparison of truss analysis and finite element analysis.

We can use the design analysis problem shown in Fig. 2.4 to describe the motivation for casting the element relationships in the form portrayed in Fig. 2.3(c). If the distance from points *A* to *B* is to be spanned by the truss structure of Fig. 2.4(a), then analysis is easily accomplished in terms of matrix structural analysis concepts that, as we have already noted, are assumed to be familiar to the reader. The individual truss elements are isolated, as for the typical axial member, Fig. 2.4(b), relationships are formed between the joint forces and displacements, and the truss is reconstructed analytically by examining the equilibrium of member forces at each joint.

Now, suppose the span is to be made with a thin plate structure, Fig. 2.4(c). The procedure described above is applicable to this problem as well if, as shown in Fig. 2.4(d), the structure is idealized as a system of triangular elements of the type shown in Fig. 2.4(e), where again the definition of relationships between element joint forces and displacements is indicated. The analytical reconstruction of the plate structure then follows the procedure for truss analysis identically. Chapter 3 is devoted to a review of this procedure.

There are noteworthy distinctions between the representation of the truss and that of the plate. By summing up the truss element forces at each joint in each coordinate direction and equating the resultants to the corresponding applied loads, the conditions of equilibrium within the truss are fully accounted for. The joining of the truss members at the joints also ensures that the truss displaces as a structural entity without any discontinuities in the pattern of displacement. The truss solution is *exact* within the confines of our assumption that the joints are pinned and no flexural behavior takes place. If the individual truss members were subdivided into smaller elements and an analysis performed of this more refined representation, the solution would not change.

The finite element solution for a continuum structure, such as the thin plate of Fig. 2.4(e), or solids, plates in bending, and shells, is not exact however. To illustrate this point we shall assume that the triangular elements of Fig. 2.4(d) have been formulated on the basis of displacement fields that vary quadratically over the surface of the elements. Figure 2.5(a) portrays the displaced states of two representative elements. When the elements are joined as described above, there is in general a lack of continuity of the edge displacements of the elements along the juncture line, Fig. 2.5(b). The joining of the elements at the vertices enforces the continuity of displacements only at those points. A quadratic function is defined uniquely by three points, and since only the two end points of the element interface participate in defining the shape of displacement along an edge, the displacements of the edges of the respective elements will differ except for special cases. As more elements are used, Fig. 2.5(c), the disparity of the displacements of the adjacent element edges is reduced and the error in the solution due to this condition is reduced. This error is finite for any finite grid of elements and the solution is approximate.

A similar argument holds for the conditions of equilibrium. The joint forces are the statical equivalents of the *edge forces* or *tractions*. In the planar case, Fig. 2.5(d), there are two edge traction components: T_n, which is normal to the edge, and T_s, which is tangential to the edge. Suppose, as shown in Fig. 2.5(d), the normal tractions T_n^A and T_n^B of the adjacent elements A and B are constant on their interfaces and each such edge traction is defined by parameters at the vertices of the element with which it is associated. Consequently, the edge tractions of the adjacent elements will in general differ and equilibrium conditions will not be satisfied along the element juncture line. The same situation may exist with respect to the tangential tractions T_s. Thus, we see that the joint forces represent the *approximate* satisfaction of the equilibrium conditions at discrete points and again we are dealing with an analytical error that one seeks to reduce through the device of refining the gridwork of elements.

We should note that a successive refinement of a grid of elements, each of which is formulated on the basis of the same assumptions on stress or displacement, is not the only way that one can approach the convergent

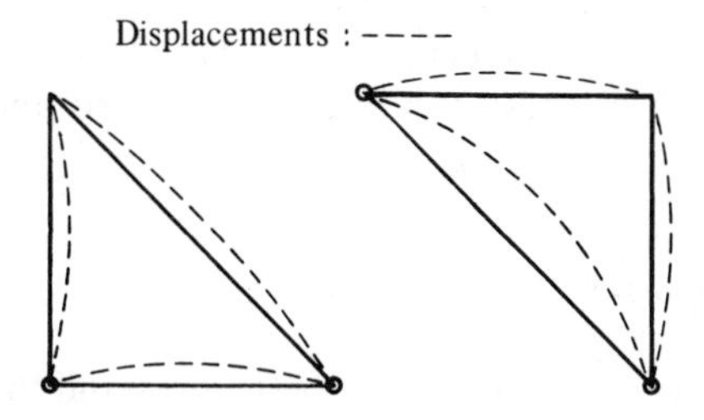

(a) Deflected shapes of individual elements

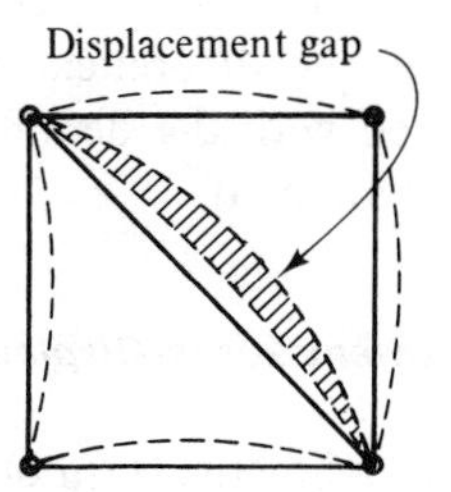

(b) Disparity of displacements along juncture line of connected elements

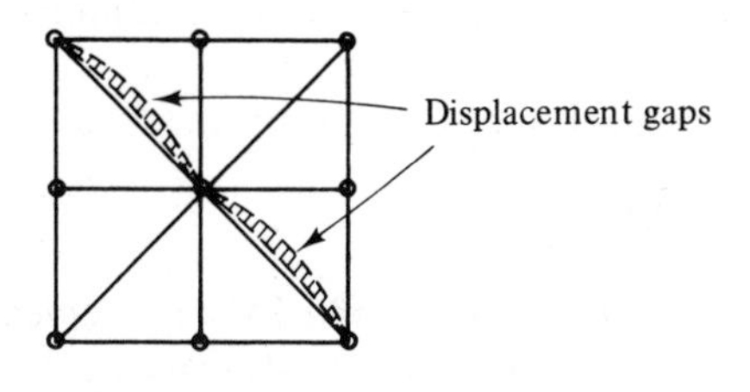

(c) Reduction of displacement gap by refinement of gridwork

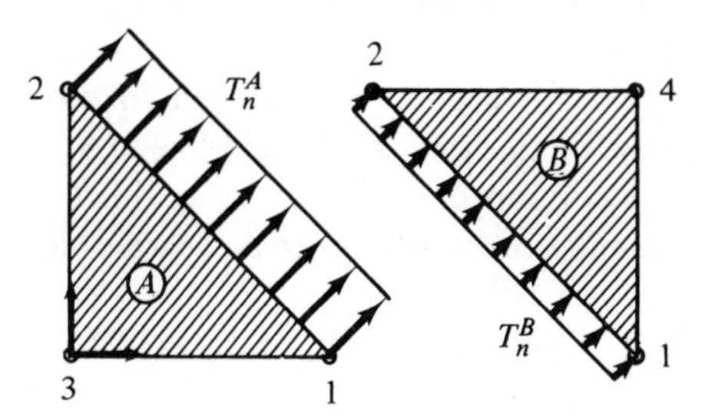

(d) Normal edge tractions on individual elements

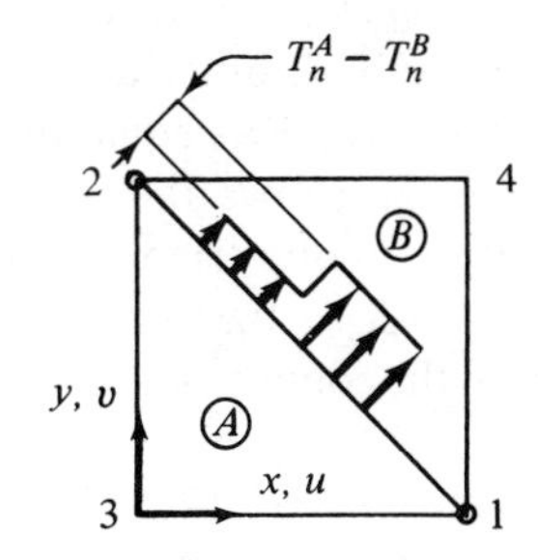

(e) Discontinuity of element edge tractions along juncture line of connected elements

Fig. 2.5 Sources of analysis error in the finite element method.

solution. It is also possible to hold the size of the elements fixed and successively refine the assumptions employed in the element. The formulations that are more refined than the simplest possible formulation for the element are known as *higher-order* elements.

The two conditions, equilibrium and continuity of displacements, characterize the sources of error in the finite element analysis. Most of the existing formulations attempt to meet the conditions on displacement continuity and for these the analysis error can be ascribed to deficiencies in meeting the conditions of equilibrium. The overall technology of finite element analysis, however, allows for all combinations of these two sources of error. As we shall see, the theoretical study of the finite element method is very much concerned with the definition of which of these conditions is satisfied and which is violated.

2.3 Element Force-Displacement Properties

We now define the form of the relationships between the joint forces and the joint displacements of finite elements, the so-called *force-displacement equations*. Element force-displacement equations are written in one of three general forms: (1) stiffness equations, (2) flexibility equations, or (3) mixed force-displacement equations.

Element stiffness equations are linear algebraic equations in the form

$$\{\mathbf{F}\} = [\mathbf{k}]\{\mathbf{\Delta}\} \tag{2.1}$$

The matrix [**k**] is the *element stiffness matrix* and {**F**} and {**Δ**} are element force and displacement vectors, respectively. Note that we designate a rectangular matrix with the symbol [ ]. An individual term of the [**k**] matrix, k_{ij}, is an *element stiffness coefficient*. When the displacement Δ_j is imposed at unit value and all other d.o.f. are held fixed against displacement ($\Delta_k = 0$, $k \neq j$), the force F_i is equal in value to k_{ij}.

Figure 2.6 shows the imposition of such a unit displacement in d.o.f. 1 (i.e., $\Delta_1 = 1$) of a triangular element with all other d.o.f. held fixed ($\Delta_2 = \Delta_3 = \ldots = \Delta_6 = 0$). Consequently, the column of joint forces for this case is equal to the column of stiffness coefficients corresponding to Δ_1 and we have

$$\{\mathbf{F}\} = \{\mathbf{k}_{i1}\} \qquad (i = 1, \ldots, 6)$$

where

$$\{\mathbf{F}\} = \lfloor F_1 \ldots F_6 \rfloor^{\mathrm{T}}$$

$$\{\mathbf{k}_{i1}\} = \lfloor k_{11} \ldots k_{61} \rfloor^{\mathrm{T}}$$

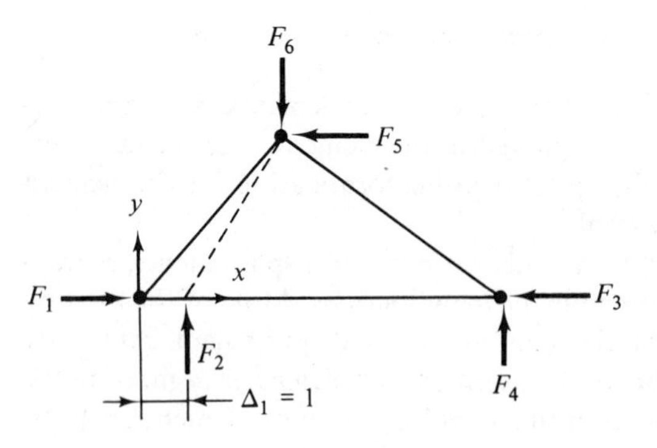

Fig. 2.6 Triangular plate element.

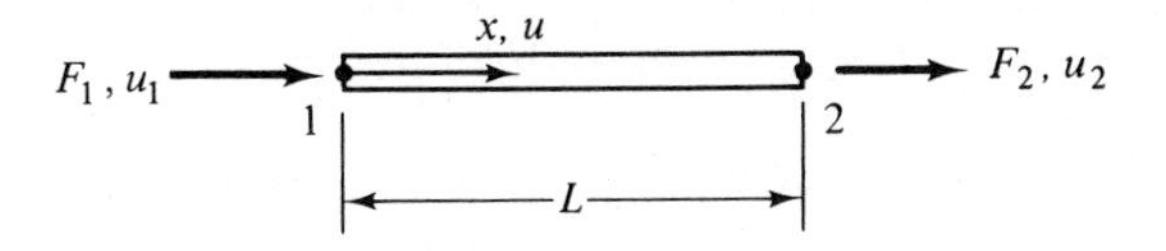

Fig. 2.7 Axial element.

Clearly, $F_1 = k_{11}$ is the force required to impose the unit value of Δ_1 and $F_2 = k_{21}$, etc., are the reactive forces. Thus, the column of stiffness coefficients $\{\mathbf{k}_{i1}\}$ represents *a system of equilibrated forces acting on the element.* A similar interpretation can be placed on each of the other columns of the element stiffness matrix.

As an example of an element stiffness matrix, we have the familiar relationships for an axial member, Fig. 2.7.

$$\begin{Bmatrix} F_1 \\ F_2 \end{Bmatrix} = \frac{AE}{L}\begin{bmatrix} 1 & -1 \\ -1 & 1 \end{bmatrix}\begin{Bmatrix} u_1 \\ u_2 \end{Bmatrix}$$

Here, the pertinent equilibrium equation is $\Sigma F_x = 0$ and it is seen that this involves a zero sum of terms in each column.

Another example is the simple flexural element, Fig. 2.8(a), for which the stiffness matrix can be written (for detailed formulation of this matrix, see Section 5.2):

$$\begin{Bmatrix} F_1 \\ M_1 \\ F_2 \\ M_2 \end{Bmatrix} = \frac{2EI}{L^3}\begin{bmatrix} 6 & -3L & -6 & -3L \\ -3L & 2L^2 & 3L & L^2 \\ -6 & 3L & 6 & 3L \\ -3L & L^2 & 3L & 2L^2 \end{bmatrix}\begin{Bmatrix} w_1 \\ \theta_1 \\ w_2 \\ \theta_2 \end{Bmatrix}$$

in which the angular displacements

$$\theta_1 = -\left.\frac{dw}{dx}\right|_1, \qquad \theta_2 = -\left.\frac{dw}{dx}\right|_2$$

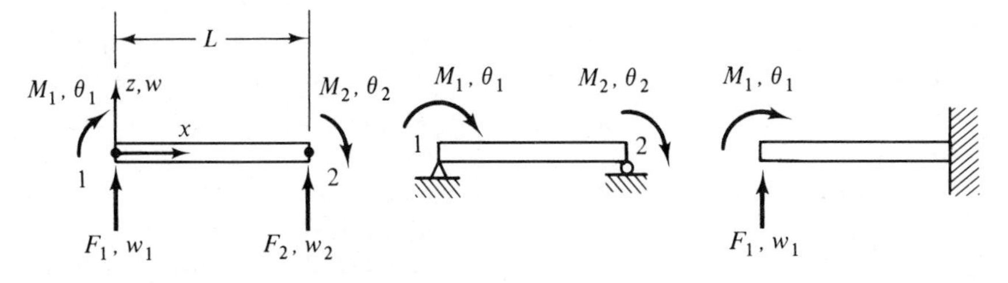

Fig. 2.8 Beam element.

As we noted in Section 2.1, the minus sign is required by the fact that positive (clockwise) rotations (θ_1, θ_2) of the end sections cause a negative displacement w.

There are a number of points to observe in this algebraic representation. For one, the forces acting on this element are both direct forces (F_1, F_2) and moments (M_1, M_2) and the displacements are correspondingly translatory (w_1, w_2) and angular (θ_1, θ_2) d.o.f. Thus, when we employ the general terms *force* and *displacement*, they may refer not only to direct forces and displacements but also to moments and rotations, higher derivatives of the displacement (e.g., d^2w/dx^2) and associated force quantities and even to *generalized* displacement and force quantities lacking physical meaning.

A second noteworthy factor is that the condition of equilibrium of forces of each column of the stiffness matrix is not simply ascertained by a zero sum of such columns. The stiffness coefficients relating to the z-direction forces, F_1, F_2, do indeed sum to zero in accordance with the equilibrium condition $\Sigma F_z = 0$, but for the remaining coefficients moment equilibrium must be taken into account. For column 1, for example, taking moments about point 2,

$$\Sigma M_2 = (6L - 3L - 3L) = 0$$

Third, the order of presentation of the components of the force and displacement vectors gives all quantities at joint 1 followed by those at joint 2. Alternatively, one might array these vectors so that all z-direction forces appear first, followed by all moment quantities, i.e., in the form $\lfloor F_1 \; F_2 \; M_1 \; M_2 \rfloor$ and correspondingly for the displacements. One or the other of these formats might be most advantageous from the view of computer operations and simplicity of presentation. Both formats are employed in this text.

Also, if we compare this stiffness matrix with that of the axial member, it is observed that all terms of the latter are constants whereas the former has both constants and terms which are dependent on the length, e.g., 6, 3L, 2L^2. The ratio of these terms may be quite large, a factor which can have serious effects on numerical accuracy in the solution of the algebraic equations formed of the stiffness coefficients. Apart from questions of numerical accuracy, it is clear that great convenience and computational efficiency can be realized if the coefficients of the algebraic form of the element stiffness matrix are independent of the specific dimensions of the element, i.e., in *nondimensional* form.

The nondimensional form of the flexural element stiffness matrix is readily established by redefinition of the joint force and displacement parameters. Thus, the angular displacements are supplanted with linear displacements $\theta_1 L$ and $\theta_2 L$ and the moments are supplanted with direct forces M_1/L and

M_2/L. In this way the quantity L is factored out of the second and fourth columns and rows, leaving a nondimensional set of stiffness coefficients. The scalar multiplier of the matrix is dependent, nevertheless, on the specific dimensions and mechanical properties of the element. Most of the stiffness matrices of this text are presented with dimensioned coefficients, but the nondimensional form can generally be constructed through either redefinition of displacement parameters or factorization. The latter procedure will be described for the triangular element in Section 5.2.

Finally, a *complete* set of stiffness equations for an element connects all the element node point forces and d.o.f. When this is the case, the d.o.f. include *rigid-body-motion* terms. In the case of the beam element, for example, the suppressed displacements associated with any of the support conditions portrayed in Fig. 2.8(b) and (c) are sets of displacements associated with rigid body motion. Such d.o.f. and forces, when extracted, enable a more concise description of the element stiffness properties but also require specially defined procedures in the construction of the complete analytical model, as described in Chapter 7.

Element flexibility equations express, for elements supported in a stable manner, the joint displacements ($\{\mathbf{\Delta}_f\}$) as a function of the joint forces ($\{\mathbf{F}_f\}$); i.e.,

$$\{\mathbf{\Delta}_f\} = [\mathbf{f}]\{\mathbf{F}_f\} \tag{2.2}$$

where $[\mathbf{f}]$ is the element flexibility matrix. An individual flexibility coefficient, f_{ij}, is the value of the displacement Δ_i caused by a unit value of the force F_j. The subscripts f on the force and displacement vectors denote that the force and displacement vectors exclude components related to the support conditions. For simplicity, we do not use the subscripts $_{ff}$ on the matrix $[\mathbf{f}]$.

Flexibility relationships can be written only for elements supported in a stable manner because rigid body motion of undefined (infinite) amplitude would otherwise result from application of applied loads. Thus, certain of the element node point d.o.f. are absent from Eq. (2.2). Flexibility relationships could of course be derived for elements supported in a statically indeterminate manner but would not be of general use in finite element analysis since, with certain exceptions, they could not be combined with other elements in the representation of a complex structure.

Flexibility equations for an element can be defined in as many ways as there are stable and statically determinate support conditions. Two of the alternative forms for the beam element are as follows. *Simple support*, Fig. 2.8(b):

$$\begin{Bmatrix} \theta_1 \\ \theta_2 \end{Bmatrix} = \frac{L}{6EI}\begin{bmatrix} 2 & -1 \\ -1 & 2 \end{bmatrix}\begin{Bmatrix} M_1 \\ M_2 \end{Bmatrix}$$

Cantilever, Fig. 2.8(c):

$$\begin{Bmatrix} w_1 \\ \theta_1 \end{Bmatrix} = \frac{L}{6EI}\begin{bmatrix} 2L^2 & 3L \\ 3L & 6 \end{bmatrix}\begin{Bmatrix} F_1 \\ M_1 \end{Bmatrix}$$

Although the coefficients of flexibility matrices differ, we shall see in the next section that a fundamental property of each form—the complementary strain energy—is identical.

The *mixed force-displacement* format defines a relationship between vectors containing both force and displacement. If the forces and corresponding d.o.f. of an element are divided into two groups, designated by subscripts $_s$ and $_f$, then the general form of a mixed representation can be written as

$$\begin{Bmatrix} \mathbf{F}_f \\ \mathbf{\Delta}_f \end{Bmatrix} = [\mathbf{\Omega}]\begin{Bmatrix} \mathbf{F}_s \\ \mathbf{\Delta}_s \end{Bmatrix} \tag{2.3}$$

One form of mixed force-displacement relationship is the *transfer matrix* format, in which the forces and displacements at one end of a member ($\lfloor \mathbf{F}_f \; \mathbf{\Delta}_f \rfloor$) are *transferred* to the opposite end ($\lfloor \mathbf{F}_s \; \mathbf{\Delta}_s \rfloor$) via the matrix $[\mathbf{\Omega}]$. In the cantilever beam of Fig. 2.8(c), for example, $\lfloor \mathbf{F}_f \; \mathbf{\Delta}_f \rfloor = \lfloor F_1 \; M_1 \; w_1 \; \theta_1 \rfloor$ and $\lfloor \mathbf{F}_s \; \mathbf{\Delta}_s \rfloor = \lfloor F_2 \; M_2 \; w_2 \; \theta_2 \rfloor$. In this case the coefficients of the matrix $[\mathbf{\Omega}]$ can be constructed using equations representing static equilibrium of the element and the element flexibility matrix ($[\mathbf{f}]$). We shall demonstrate this development in Section 2.6. Other forms of element mixed force-displacement equations can be derived directly, through application of basic concepts of element formulation. These are described in Chapter 6.

2.4 Work and Energy

The work (W) of a force is the product of that force and the displacement of its point of application in the direction of the force. Thus, for a vector of forces $\{\mathbf{F}\}$ and the corresponding vector of displacements $\{\mathbf{\Delta}\}$ we have the product

$$W = \tfrac{1}{2}\lfloor \mathbf{\Delta} \rfloor \{\mathbf{F}\} = \tfrac{1}{2}\lfloor \mathbf{F} \rfloor \{\mathbf{\Delta}\} \tag{2.4}$$

where the $\frac{1}{2}$ multiplier designates a loading situation in which the load increases *gradually* from zero to its full intensity (i.e., inertia forces due to dynamic behavior are negligible). We see that in the relationship between a single force F_i and its corresponding displacement Δ_i (Fig. 2.9), the work is described by the cross-hatched area.

Equation (2.4) can be transformed into expressions exclusively in terms

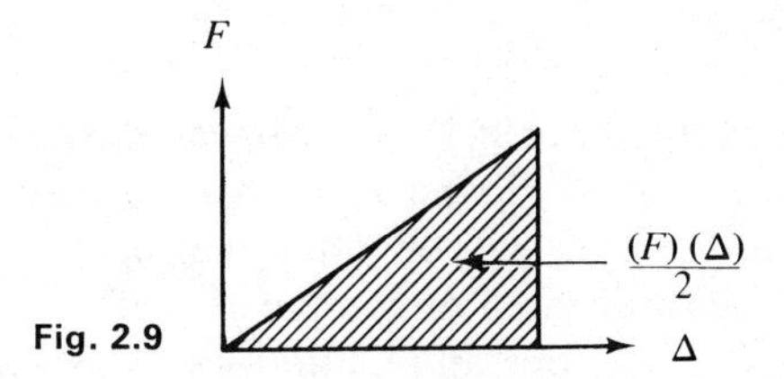

Fig. 2.9

of either the forces or the displacements by substitution of the stiffnesses [Eq. (2.1)] or flexibilities [Eq. (2.2)], respectively. Thus

$$W = \tfrac{1}{2}\lfloor \mathbf{\Delta} \rfloor [\mathbf{k}]\{\mathbf{\Delta}\} = U \tag{2.4a}$$

or

$$W = \tfrac{1}{2}\lfloor \mathbf{F}_f \rfloor [\mathbf{f}]\{\mathbf{F}_f\} = U^* \tag{2.4b}$$

As shown in subsequent chapters, such quantities describe the element *strain energy* of deformation (U) and *complementary strain energy* of deformation (U^*). It can be observed that both U and U^* are quadratic functions of the parameters $\lfloor \mathbf{\Delta} \rfloor$ and $\lfloor \mathbf{F}_f \rfloor$, respectively.

As we have indicated in Section 2.3, the alternative forms of flexibility matrix for a given element each represent the same complementary strain energy content. For illustration, consider again the beam element. When the beam is simply supported [see Fig. 2.8(b)], we have

$$\begin{aligned} U^* &= \frac{\lfloor M_1 \; M_2 \rfloor}{2}\left(\frac{L}{6EI}\right)\begin{bmatrix} 2 & -1 \\ -1 & 2 \end{bmatrix}\begin{Bmatrix} M_1 \\ M_2 \end{Bmatrix} \\ &= \frac{L}{6EI}(M_1^2 + M_2^2 - M_1 M_2) \end{aligned}$$

To compare with the cantilever beam we must first compute the shear force F_1 in terms of the moments M_1 and M_2. From equilibrium of moments about the right support, point 2,

$$F_1 = -\frac{(M_1 + M_2)}{L}$$

Now, the complementary strain energy for the cantilever is

$$U^* = \frac{\lfloor F_1 \; M_1 \rfloor}{2}\left(\frac{L}{6EI}\right)\begin{bmatrix} 2L^2 & 3L \\ 3L & 6 \end{bmatrix}\begin{Bmatrix} F_1 \\ M_1 \end{Bmatrix}$$

and, by substitution of the expression for F_1 above and expansion of the expression for U^*, we obtain an expression for U^* in terms of M_1 and M_2 that is identical to the one given above.

2.5 Reciprocity

Flexibility and stiffness coefficients for linear elastic behavior have the property of *reciprocity* ($f_{ij} = f_{ji}$ and $k_{ij} = k_{ji}$). Cognizance of this property is important from the view of computational efficiency and may also be of value in checking formulated or computed coefficients. To prove reciprocity and thereby define its scope and limitations, consider the work done on the supported structure in Fig. 2.10 as load F_1 is first applied, followed by F_2.

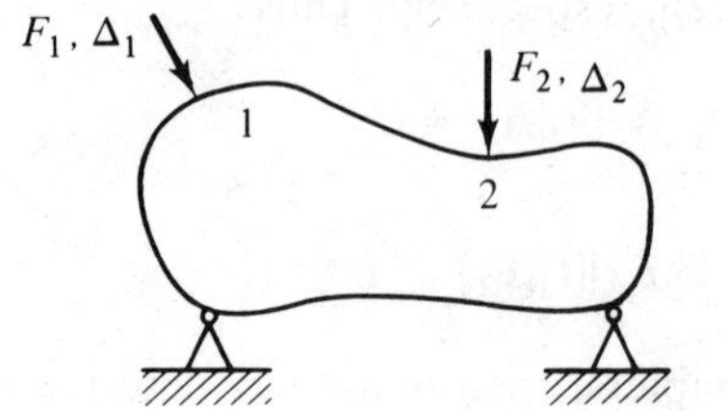

Fig. 2.10

We designate this work as W_I. For (gradual) application of F_1,

$$W_{I_1} = \tfrac{1}{2}(\Delta_1)_1 F_1 = \tfrac{1}{2}(f_{11}F_1)F_1 \tag{2.5}$$

where the subscript on W_I and on the right parenthesis on Δ_1 means *due to force 1*. Applying now F_2, with F_1 held constant, and using similar designations

$$W_{I_2} = \tfrac{1}{2}(\Delta_2)_2 F_2 + (\Delta_1)_2 F_1 = \tfrac{1}{2}(f_{22}F_2)F_2 + (f_{12}F_2)F_1 \tag{2.6}$$

so that the total work W_I is

$$W_I = W_{I_1} + W_{I_2} = \tfrac{1}{2}f_{11}(F_1)^2 + \tfrac{1}{2}f_{22}(F_2)^2 + f_{12}F_2F_1 \tag{2.7}$$

Reversing now the sequence of application and recomputing the components of work, we have for initial application of F_2 (the work in this second sequence is identified as W_{II})

$$W_{II_2} = \tfrac{1}{2}(\Delta_2)_2 F_2 = \tfrac{1}{2}f_{22}(F_2)^2 \tag{2.5a}$$

followed by application of F_1

$$W_{II_1} = \tfrac{1}{2}(\Delta_1)_1 F_1 + (\Delta_2)_1 F_2 = \tfrac{1}{2}f_{11}(F_1)^2 + f_{21}F_1F_2 \tag{2.6a}$$

so that

$$W_{II} = W_{II_1} + W_{II_2} = \tfrac{1}{2}f_{22}(F_2)^2 + \tfrac{1}{2}f_{11}(F_1)^2 + f_{21}F_1F_2 \tag{2.7a}$$

Since, for a linear system, the sequence of application of loads is immaterial from the standpoint of the work performed, we can equate the two expressions for W and after canceling terms

$$f_{21} = f_{12} \tag{2.8}$$

and, in general,

$$f_{ij} = f_{ji} \tag{2.9}$$

This algebraic statement is known as *Maxwell's reciprocal theorem.*

Since the inverse of a symmetric matrix is symmetric and the stiffness matrix is the inverse of the flexibility matrix,

$$k_{ij} = k_{ji} \tag{2.10}$$

Maxwell's reciprocal theorem is usually defined as a special case of *Betti's law*, which states that the work done by a system of forces $\{\mathbf{P}_1\}$, when acting through the displacements $\{\mathbf{\Delta}_2\}$ caused by a system of forces $\{\mathbf{P}_2\}$, is equal to the work done by the forces $\{\mathbf{P}_2\}$ acting through displacements $\{\mathbf{\Delta}_1\}$ corresponding to $\{\mathbf{P}_1\}$.

2.6 Flexibility-Stiffness Transformation

Given one form of element force-displacement relationships, it is possible to obtain the alternate forms through simple operations. Consider first the transformation of stiffness into flexibility. We illustrate this case by examination of the planar element shown in Fig. 2.11(a). As noted in Section 2.3, an element must be supported in a stable, statically determinate manner if flexibility relationships are to be constructed. Such supports for the element under study are shown in Fig. 2.11(b).

Quantities pertinent to the supports are assigned the subscript s; those related to the remaining d.o.f. have the subscript f.

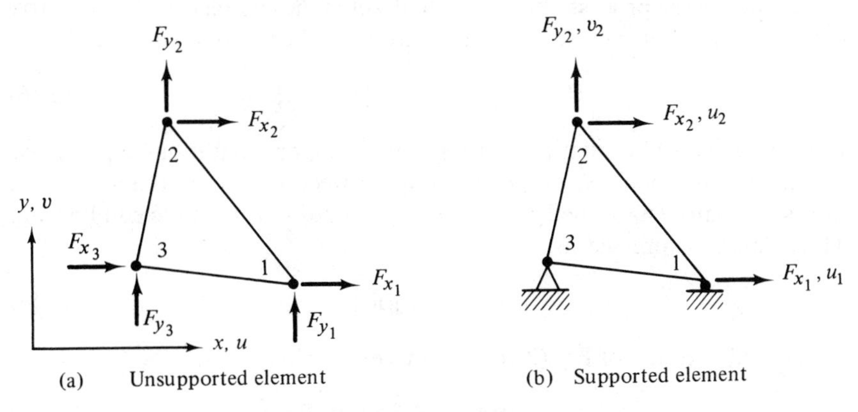

Fig. 2.11 Planar element.

Thus, we partition the stiffness matrix as follows:

$$\begin{Bmatrix} \mathbf{F}_f \\ \hline \mathbf{F}_s \end{Bmatrix} = \left[\begin{array}{c|c} \mathbf{k}_{ff} & \mathbf{k}_{fs} \\ \hline \mathbf{k}_{sf} & \mathbf{k}_{ss} \end{array}\right] \begin{Bmatrix} \mathbf{\Delta}_f \\ \hline \mathbf{\Delta}_s \end{Bmatrix} \tag{2.11}$$

where, for the condition shown in Fig. 2.11(b), each of the submatrices ($[\mathbf{k}_{ff}]$, etc.) is a 3×3 matrix and

$$\begin{aligned} \{\mathbf{F}_f\} &= \lfloor F_{x_1}\ F_{x_2}\ F_{y_2} \rfloor^{\mathrm{T}} & \{\mathbf{F}_s\} &= \lfloor F_{x_3}\ F_{y_1}\ F_{y_3} \rfloor^{\mathrm{T}} \\ \{\mathbf{\Delta}_f\} &= \lfloor u_1\ u_2\ v_2 \rfloor^{\mathrm{T}} & \{\mathbf{\Delta}_s\} &= \lfloor u_3\ v_1\ v_3 \rfloor^{\mathrm{T}} \end{aligned} \tag{2.12}$$

Now, since $\{\mathbf{\Delta}_s\} = 0$ due to the conditions of support,

$$\begin{Bmatrix} \mathbf{F}_f \\ \hline \mathbf{F}_s \end{Bmatrix} = \begin{bmatrix} \mathbf{k}_{ff} \\ \hline \mathbf{k}_{sf} \end{bmatrix} \{\mathbf{\Delta}_f\} \tag{2.13}$$

The equations above the matrix partitioning line are an independent set of equations relating the external forces $\{\mathbf{F}_f\}$ to the corresponding permissible node point displacements. Solution of these by inversion of $[\mathbf{k}_{ff}]$ yields

$$\{\mathbf{\Delta}_f\} = [\mathbf{f}]\{\mathbf{F}_f\} \tag{2.14}$$

where

$$[\mathbf{f}] = [\mathbf{k}_{ff}]^{-1} \tag{2.15}$$

(Note that we designate the operation of matrix inversion with the -1 superscript.) The matrix $[\mathbf{f}]$ is, by definition, the desired set of element flexibility coefficients. Thus, the flexibilities are derived from the stiffnesses by merely defining a stable, statically determinate support system, removing from the stiffness matrix the rows and columns corresponding to the support components, and inverting the remainder.

To reverse the process above, i.e., to develop the complete stiffness matrix given the flexibility matrix, we begin by inverting the flexibility matrix. Hence,

$$\{\mathbf{F}_f\} = [\mathbf{f}]^{-1}\{\mathbf{\Delta}_f\} = [\mathbf{k}_{ff}]\{\mathbf{\Delta}_f\} \tag{2.16}$$

In view of the stable, statically determinate support conditions represented by the flexibility matrix, the relationship between the external and support forces is readily established by applying the conditions of static equilibrium. These can be expressed as

$$\{\mathbf{F}_s\} = [\mathbf{R}]\{\mathbf{F}_f\} \tag{2.17}$$

and, by substitution of Eq. (2.16) into this expression,

$$\{\mathbf{F}_s\} = [\mathbf{R}][\mathbf{f}]^{-1}\{\mathbf{\Delta}_f\} = [\mathbf{k}_{sf}]\{\mathbf{\Delta}_f\} \tag{2.18}$$

so that

$$[\mathbf{k}_{sf}] = [\mathbf{R}][\mathbf{f}]^{-1} \tag{2.19}$$

For development of the other terms in the complete stiffness equations we examine the anticipated form of the final result, Eq. (2.11). The work done by the external loads $\{\mathbf{F}_f\}$ acting through their corresponding displacements $\{\boldsymbol{\Delta}_f\}$ must be equal to the work done by the remaining forces $\{\mathbf{F}_s\}$ acting through their associated displacements $\{\boldsymbol{\Delta}_s\}$ under the circumstances of a reversal of the support conditions (i.e., the forces $\{\mathbf{F}_f\}$ become support reactions). This can be stated in matrix form as

$$\tfrac{1}{2}\lfloor \mathbf{F}_s \rfloor \{\boldsymbol{\Delta}_s\} = \tfrac{1}{2}\lfloor \boldsymbol{\Delta}_f \rfloor \{F_f\} \tag{2.20}$$

Now, since by transpose of Eq. (2.18) $\lfloor \mathbf{F}_s \rfloor = \lfloor \boldsymbol{\Delta}_f \rfloor [\mathbf{k}_{sf}]^{\mathrm{T}}$, we can write Eq. (2.20) as

$$\tfrac{1}{2}\lfloor \boldsymbol{\Delta}_f \rfloor [\mathbf{k}_{sf}]^{\mathrm{T}} \{\boldsymbol{\Delta}_s\} = \tfrac{1}{2}\lfloor \boldsymbol{\Delta}_f \rfloor \{\mathbf{F}_f\} \tag{2.20a}$$

Thus,

$$\{\mathbf{F}_f\} = [\mathbf{k}_{sf}]^{\mathrm{T}} \{\boldsymbol{\Delta}_s\} = [\mathbf{k}_{fs}]\{\boldsymbol{\Delta}_s\} \tag{2.21}$$

Hence, in view of Eq. (2.19)

$$[\mathbf{k}_{fs}] = [\mathbf{f}]^{-1}[\mathbf{R}]^{\mathrm{T}} \tag{2.22}$$

Again considering the format of the resulting stiffness matrix [Eq. (2.11)], we can write, from equilibrium, the same relationships as are given by Eq. (2.17). Substitution of Eq. (2.21) into Eq. (2.17), using the result of Eq. (2.22), yields

$$\{\mathbf{F}_s\} = [\mathbf{R}][\mathbf{f}]^{-1}[\mathbf{R}]^{\mathrm{T}}\{\boldsymbol{\Delta}_s\} = [\mathbf{k}_{ss}]\{\boldsymbol{\Delta}_s\} \tag{2.23}$$

Thus, the constructed stiffness matrix assumes the form

$$[\mathbf{k}] = \left[\begin{array}{c|c} [\mathbf{f}]^{-1} & [\mathbf{f}]^{-1}[\mathbf{R}]^{\mathrm{T}} \\ \hline [\mathbf{R}][\mathbf{f}]^{-1} & [\mathbf{R}][\mathbf{f}]^{-1}[\mathbf{R}]^{\mathrm{T}} \end{array}\right] \tag{2.24}$$

In summary, it is seen that the stiffness matrix is constructed from the inverse of the flexibility matrix and a matrix [**R**] that derives from the element static equilibrium relationships. The matrix [**f**] is at the outset a symmetric matrix. Since $[\mathbf{k}_{fs}]$ is the transpose of $[\mathbf{k}_{sf}]$ the symmetry property of these portions of the resulting stiffness matrix is ensured. We see that the $[\mathbf{k}_{ss}]$ portion of the result is given by a matrix triple product in which the pre-multiplier of the central matrix is equal to the transpose of the post-multiplying matrix. This triple product, known as a *congruent transformation*, produces a symmetric matrix when the central matrix is symmetric. In this case [**f**] is a symmetric matrix so that [$\mathbf{k}_{ss}$] is assured to be symmetric. Equation (2.24) represents a general formula for transformation from flexibility to a stiffness that includes rigid-body-motion d.o.f. The number (s) of support forces is dictated by the requirements of stable statically determinate support, but there is no limit on the number (f) of external forces (i.e., on the order of the flexibility matrix).

To illustrate this procedure, consider the cantilever beam of Fig. 2.8(c). The flexibility matrix [**f**] has been previously given and in this case the reaction equations are

$$\begin{Bmatrix} F_2 \\ M_2 \end{Bmatrix} = \begin{bmatrix} -1 & 0 \\ -L & -1 \end{bmatrix} \begin{Bmatrix} F_1 \\ M_1 \end{Bmatrix}$$

so that the 2×2 matrix on the right side is the matrix [**R**]. The reader can verify that the use of [**f**] and [**R**] in Eq. (2.24) yields the beam element stiffness matrix given earlier.

All component relationships are also now available to construct a mixed format of element relationships as defined in Section 2.3. That representation sought to express $\lfloor F_f \; \Delta_f \rfloor$ in terms of $\lfloor F_s \; \Delta_s \rfloor$. First, by solution of Eq. (2.17), we have

$$\{\mathbf{F}_f\} = [\mathbf{R}]^{-1}\{\mathbf{F}_s\} \tag{2.17a}$$

Next, we write the upper part of Eq. (2.11) in equation form, using the results of Eq. (2.24) (top partition)

$$\{\mathbf{F}_f\} = [\mathbf{f}]^{-1}\{\boldsymbol{\Delta}_f\} + [\mathbf{f}]^{-1}[\mathbf{R}]^{\mathrm{T}}\{\boldsymbol{\Delta}_s\} \tag{2.24a}$$

Solving for $\{\boldsymbol{\Delta}_f\}$,

$$\{\boldsymbol{\Delta}_f\} = [\mathbf{f}]\{\mathbf{F}_f\} - [\mathbf{R}]^{\mathrm{T}}\{\boldsymbol{\Delta}_s\} \tag{2.24b}$$

and by substitution of Eq. (2.17a) for $\{\mathbf{F}_f\}$

$$\{\boldsymbol{\Delta}_f\} = [\mathbf{f}][\mathbf{R}]^{-1}\{\mathbf{F}_s\} - [\mathbf{R}]^{\mathrm{T}}\{\boldsymbol{\Delta}_s\} \tag{2.24c}$$

Thus, by forming an array with both Eqs. (2.17a) and (2.24c), we have

$$\begin{Bmatrix} \mathbf{F}_f \\ \hline \boldsymbol{\Delta}_f \end{Bmatrix} = \left[\begin{array}{c|c} \mathbf{R}^{-1} & \mathbf{0} \\ \hline [\mathbf{f}][\mathbf{R}]^{-1} & -\mathbf{R}^{\mathrm{T}} \end{array}\right] \begin{Bmatrix} \mathbf{F}_s \\ \hline \boldsymbol{\Delta}_s \end{Bmatrix} \tag{2.3a}$$

The square matrix on the right side is of the form of the matrix [$\boldsymbol{\Omega}$] of Eq. (2.3). The zero symbol in the upper right corner of Eq. (2.3a) designates a *null* submatrix, i.e., a matrix completely populated with zeroes.

2.7 Transformation of d.o.f.

A system of equations is often formulated in terms of certain d.o.f. $\{\mathbf{\Delta}'\}$ and a requirement arises to express the system in terms of other d.o.f. $\{\mathbf{\Delta}\}$. The most common example occurs when the original d.o.f. are referred to one set of axes and it is desired that the equations of the problem be referred to d.o.f corresponding to a different set of axes; i.e., one seeks a *coordinate axis* transformation. In the general case the transformed d.o.f. might not have physical meaning and are not necessarily equal in number to the original d.o.f. The relationships connecting the two sets of d.o.f. are written as

$$\{\mathbf{\Delta}'\} = [\mathbf{\Gamma}]\{\mathbf{\Delta}\} \tag{2.25}$$

Suppose the equations to be transformed are

$$[\mathbf{k}']\{\mathbf{\Delta}'\} = \{\mathbf{F}'\} \tag{2.26}$$

Also, we assume that each force component F'_i of the vector $\{\mathbf{F}'\}$ produces the work $\frac{1}{2}F'_i\,\Delta'_i$ during the displacement Δ'_i and no work under any other displacement component in $\{\mathbf{\Delta}'\}$. Such force and displacement vectors are termed *conjugate vectors*, a condition which holds when the components act along orthogonal axes. Both sets, $\{\mathbf{\Delta}'\}$, $\{\mathbf{F}'\}$ and $\{\mathbf{\Delta}\}$, $\{\mathbf{F}\}$ are taken to be conjugate sets of vectors. In order that work remain invariant under the imposed transformation, we have

$$\lfloor \mathbf{F}' \rfloor\{\mathbf{\Delta}'\} = \lfloor \mathbf{F} \rfloor\{\mathbf{\Delta}\}$$

and, from Eq. (2.25)

$$\lfloor \mathbf{F}' \rfloor[\mathbf{\Gamma}]\{\mathbf{\Delta}\} = \lfloor \mathbf{F} \rfloor\{\mathbf{\Delta}\}$$

thus,

$$\lfloor \mathbf{F}' \rfloor[\mathbf{\Gamma}] = \lfloor \mathbf{F} \rfloor$$

or, by transposition

$$[\mathbf{\Gamma}]^{\mathrm{T}}\{\mathbf{F}'\} = \{\hat{\mathbf{F}}\} \tag{2.27}$$

where the caret (^) designates the set of forces obtained by the transformation of $\{\mathbf{F}'\}$.

Hence, we see that transformation of displacements, Eq. (2.25), implies the transformation of forces given by Eq. (2.27). The force and displacement transformations are termed *contragradient* under the stipulated condition of conjugacy. It follows that if the force transformation is first defined, then the displacement transformation matrix is given by the transpose of the force transformation matrix. The principle of contragradience is of considerable importance when the displacement (or force) transformation is readily constructed from physical meaning but the formation of the transformation of the conjugate vector is not readily perceived. This occurs, for example, when the condensation of d.o.f. is accomplished via the transformation process, as discussed in Section 2.8.

To perceive the effect of the above considerations on the element stiffness relationships it is convenient to deal with strain energy and external work, quantities that were introduced in Section 2.4. Again, we require that the work remain invariant under the imposed transformation, permitting direct substitution of Eq. (2.25) into Eqs. (2.4a) and (2.4). Thus

$$U = \frac{\lfloor \mathbf{\Delta}' \rfloor}{2}[\mathbf{k}']\{\mathbf{\Delta}'\} = \frac{\lfloor \mathbf{\Delta} \rfloor}{2}[\mathbf{\Gamma}]^{\mathrm{T}}[\mathbf{k}'][\mathbf{\Gamma}]\{\mathbf{\Delta}\} = \frac{\lfloor \mathbf{\Delta} \rfloor}{2}[\hat{\mathbf{k}}]\{\mathbf{\Delta}\} \tag{2.4c}$$

$$W = \frac{\lfloor \mathbf{\Delta}' \rfloor}{2}\{\mathbf{F}'\} = \frac{\lfloor \mathbf{\Delta} \rfloor}{2}[\mathbf{\Gamma}]^{\mathrm{T}}\{\mathbf{F}'\} = \frac{\lfloor \mathbf{\Delta} \rfloor}{2}\{\mathbf{F}\} \tag{2.4d}$$

Hence, the transformed stiffness matrix, designated by the caret (^), is given by

$$[\hat{\mathbf{k}}] = [\mathbf{\Gamma}]^{\mathrm{T}}[\mathbf{k}'][\mathbf{\Gamma}] \tag{2.28}$$

The force vector is of course transformed in the manner described by Eq. (2.27). The transformation of $[\mathbf{k}']$ into $[\hat{\mathbf{k}}]$, represented by Eq. (2.28) is of the form of a congruent transformation so that if $[\mathbf{k}']$ is symmetric, the transformed matrix $[\hat{\mathbf{k}}]$ will be symmetric.

In the case of an orthogonal coordinate axis transformation these formulas can be established by a more direct, if lengthier, procedure. We assume that the transformation of displacement components is obtained by direct examination of the relationship between vectors of displacements $\{\mathbf{\Delta}'\}$ and $\{\mathbf{\Delta}\}$. Instead of accepting Eq. (2.27) as a transformation of the force vectors, however, we assume that this transformation is independently formed, also by direct examination, of the relationship between the vectors of forces $\{\mathbf{F}'\}$ and $\{\mathbf{F}\}$. We write this as

$$\{\mathbf{F}'\} = [\mathbf{\Gamma}]\{\mathbf{F}\} \tag{2.29}$$

Thus, by substitution of Eq. (2.25) and (2.29) into Eq. (2.26), there is obtained

$$[\mathbf{k}'][\mathbf{\Gamma}]\{\mathbf{\Delta}\} = [\mathbf{\Gamma}]\{\mathbf{F}\}$$

or

$$[\mathbf{\Gamma}]^{-1}[\mathbf{k}'][\mathbf{\Gamma}]\{\mathbf{\Delta}\} = \{\mathbf{F}\} \tag{2.30}$$

The coordinate transformation for orthogonal coordinate axes possesses the property

$$[\mathbf{\Gamma}][\mathbf{\Gamma}]^{\mathrm{T}} = [\mathbf{I}]$$

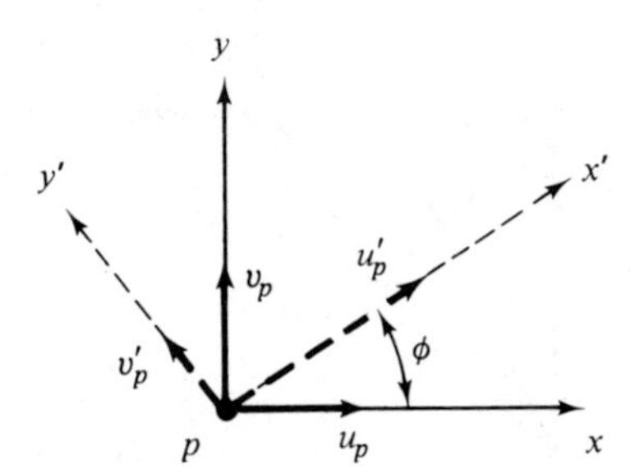

Fig. 2.12

where [I] is the *identity matrix*, a diagonal matrix all terms of which are unity. Since, by definition of the inverse matrix $[\mathbf{\Gamma}][\mathbf{\Gamma}]^{-1} = [\mathbf{I}]$, it follows that

$$[\mathbf{\Gamma}]^T = [\mathbf{\Gamma}]^{-1} \tag{2.31}$$

when a matrix possesses the property defined by Eq. (2.31), i.e., if its transpose is equal to its inverse, it is termed an *orthogonal matrix*. Substitution of Eq. (2.31) into Eq. (2.30) results in the definition of $[\hat{\mathbf{k}}]$ given by Eq. (2.28).

Suppose, for example, that a planar-element stiffness matrix, formulated with respect to the x' and y' axes shown Fig. 2.12, is to be transformed to expressions related to the x and y axes. We have, for the vectors at any point p of the element, the transformation

$$\begin{Bmatrix} u'_p \\ v'_p \end{Bmatrix} = \begin{bmatrix} \cos\phi & \sin\phi \\ -\sin\phi & \cos\phi \end{bmatrix} \begin{Bmatrix} u_p \\ v_p \end{Bmatrix} = [\mathbf{\Gamma}_p] \begin{Bmatrix} u_p \\ v_p \end{Bmatrix}$$

and

$$[\mathbf{\Gamma}_p]^{-1} = \begin{bmatrix} \cos\phi & -\sin\phi \\ \sin\phi & \cos\phi \end{bmatrix} = [\mathbf{\Gamma}_p]^T$$

Thus, if the total element consists of $n/2$ joints (i.e., n d.o.f. for this planar case), the resulting transformation matrix of the total element is of the form

$$[\mathbf{\Gamma}] = \left\lceil \begin{matrix} [\mathbf{\Gamma}_1] & & & & & & \\ & \cdot & & & & & \\ & & \cdot & & & & \\ & & & [\mathbf{\Gamma}_2] & & & \\ & & & & \cdot & & \\ & & & & & \cdot & \\ & & & & & & [\mathbf{\Gamma}_{n/2}] \end{matrix} \right\rfloor$$

($\lceil \quad \rfloor$ symbolizes a diagonal matrix in this text).

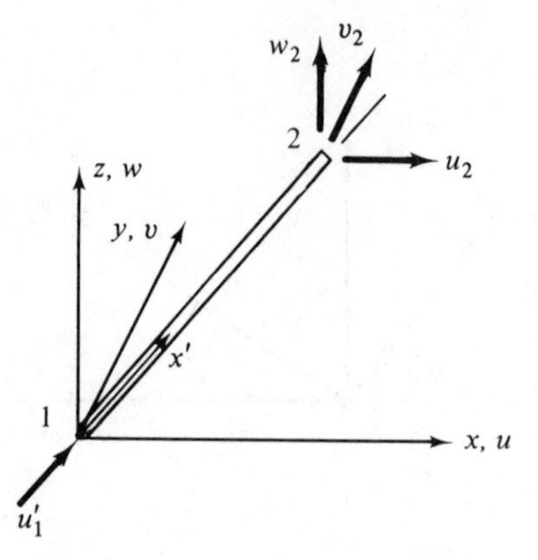

Fig. 2.13

Since the inversion of the transformation matrix is not required and only transposition is needed, it is permissible to define nonsquare coordinate axis transformation matrices. The stiffness matrix of the axial member (Section 2.3) features the two axial displacements. This element in global coordinates (see Fig. 2.13) is described by six displacement components. Denoting the direction cosines of the axis of the member with respect to the x, y, and z axes as $l_{x'x}$, $l_{x'y}$, etc., we have the coordinate axis transformation

$$\begin{Bmatrix} u'_1 \\ u'_2 \end{Bmatrix} = \begin{bmatrix} l_{x'x} & l_{x'y} & l_{x'z} & 0 & 0 & 0 \\ 0 & 0 & 0 & l_{x'x} & l_{x'y} & l_{x'z} \end{bmatrix} \begin{Bmatrix} u_1 \\ v_1 \\ w_1 \\ u_2 \\ v_2 \\ w_2 \end{Bmatrix}$$

[For simplicity, only global (unprimed) components are shown at point 2 and only the element axis component u'_1 at point 1.]

2.8 Condensation

The term *condensation* refers here to the contraction in size of a system of equations by elimination of certain d.o.f. The condensed equations are to be expressed in terms of preselected d.o.f. $\{\mathbf{\Delta}_c\}$ that are to be retained, together with the *surplus* d.o.f. $\{\mathbf{\Delta}_b\}$, to comprise the total original set of d.o.f., i.e., $\lfloor \mathbf{\Delta} \rfloor = \lfloor \lfloor \mathbf{\Delta}_b \rfloor \lfloor \mathbf{\Delta}_c \rfloor \rfloor$. The original equations are of the form

$$\begin{bmatrix} \mathbf{k}_{bb} & \mathbf{k}_{bc} \\ \mathbf{k}_{cb} & \mathbf{k}_{cc} \end{bmatrix} \begin{Bmatrix} \mathbf{\Delta}_b \\ \mathbf{\Delta}_c \end{Bmatrix} = \begin{Bmatrix} \mathbf{F}_b \\ \mathbf{F}_c \end{Bmatrix} \tag{2.32}$$

and are to be condensed to the form

$$[\hat{\mathbf{k}}_{cc}]\{\mathbf{\Delta}_c\} = \{\hat{\mathbf{F}}_c\} \tag{2.33}$$

We adopt an approach to condensation that employs the concept of a coordinate transformation. Thus, our objective is to construct the relationships

$$\begin{Bmatrix} \mathbf{\Delta}_b \\ \mathbf{\Delta}_c \end{Bmatrix} = [\mathbf{\Gamma}_0]\{\mathbf{\Delta}_c\} \tag{2.34}$$

where $[\mathbf{\Gamma}_0]$ is the desired transformation matrix. To do so we first solve the upper partition of Eq. (2.32):

$$\{\mathbf{\Delta}_b\} = -[\mathbf{k}_{bb}]^{-1}[\mathbf{k}_{bc}]\{\mathbf{\Delta}_c\} + [\mathbf{k}_{bb}]^{-1}\{\mathbf{F}_b\} \tag{2.35}$$

Since the second term on the right-hand side is constant for given loads, the stiffness relationship between the d.o.f. ($\{\mathbf{\Delta}_c\}$ and $\{\mathbf{\Delta}_b\}$) is given by $-[\mathbf{k}_{bb}]^{-1}[\mathbf{k}_{bc}]$. Noting also that $\{\mathbf{\Delta}_c\} = [\mathbf{I}]\{\mathbf{\Delta}_c\}$, we can write the following transformation of coordinates†

$$\begin{Bmatrix} \mathbf{\Delta}_b \\ \mathbf{\Delta}_c \end{Bmatrix} = \begin{bmatrix} -[\mathbf{k}_{bb}]^{-1}[\mathbf{k}_{bc}] \\ \hdashline \mathbf{I} \end{bmatrix}\{\mathbf{\Delta}_c\} = [\mathbf{\Gamma}_0]\{\mathbf{\Delta}_c\} \tag{2.36}$$

Applying this to Eq. (2.32) in the manner of a conventional transformation of coordinates, Eq. (2.33) is obtained where

$$[\hat{\mathbf{k}}_{cc}] = [[\mathbf{k}_{cc}] - [\mathbf{k}_{cb}][\mathbf{k}_{bb}]^{-1}[\mathbf{k}_{bc}]] \tag{2.37}$$

$$\{\hat{\mathbf{F}}_c\} = [\mathbf{\Gamma}_0]^{\mathrm{T}}\begin{Bmatrix} F_b \\ F_c \end{Bmatrix} = \{\mathbf{F}_c\} - [\mathbf{k}_{cb}][\mathbf{k}_{bb}]^{-1}\{\mathbf{F}_b\} \tag{2.38}$$

Note that this transformation, constructed on the basis of the relationship

†It might first appear that the transformation given by Eq. (2.36) should include the constant vector in Eq. (2.35), $\{\mathbf{\Delta}'_b\} = [\mathbf{k}_{bb}]^{-1}\{\mathbf{F}_b\}$, that it should read

$$\begin{Bmatrix} \mathbf{\Delta}_b \\ \mathbf{\Delta}_c \end{Bmatrix} = \begin{bmatrix} -[\mathbf{k}_{bb}]^{-1}[\mathbf{k}_{bc}] \\ \hdashline \mathbf{I} \end{bmatrix}\{\mathbf{\Delta}_c\} + \begin{Bmatrix} \mathbf{\Delta}'_b \\ \hdashline \mathbf{0} \end{Bmatrix} \tag{2.36a}$$

It can be proved, however, that the vector $\lfloor \mathbf{\Delta}'_b 0 \rfloor$ has no influence on the transformation. The vector $\lfloor \mathbf{\Delta}'_b 0 \rfloor$ represents a rigid body motion. Although a transformation that includes a rigid body motion will change the total energy of a system, the algebraic equations that govern structural behavior [e.g., the stiffness equations, Eq. (2.1)] derive from a stationary condition imposed on the energy and this condition is not affected by rigid body motion. One may verify this consideration by substitution of Eq. (2.36a) into the potential energy expression

$$\Pi_p = \frac{\lfloor \mathbf{\Delta}_b\ \mathbf{\Delta}_c \rfloor}{2}\begin{bmatrix} \mathbf{k}_{bb} & \mathbf{k}_{bc} \\ \mathbf{k}_{cb} & \mathbf{k}_{cc} \end{bmatrix}\begin{Bmatrix} \mathbf{\Delta}_b \\ \mathbf{\Delta}_c \end{Bmatrix} - \lfloor \mathbf{\Delta}_b\ \mathbf{\Delta}_c \rfloor\begin{Bmatrix} \mathbf{F}_b \\ \mathbf{F}_c \end{Bmatrix}$$

Substitution of Eq. (2.36a), followed by differentiation with respect to $\{\mathbf{\Delta}_b\}$, gives the same result as use of the transformation $[\mathbf{\Gamma}_0]$. Concepts that explain this sequence of operations are taken up in Chapters 6 and 7.

between the d.o.f. alone, serves also to transform the r.h.s. (right-hand-side) vector.

These results would of course have followed directly by substitution of Eq. (2.35) into the lower partition of Eq. (2.32), but the notion of condensation by means of a transformation of d.o.f. ($[\mathbf{\Gamma}_0]$) proves extremely valuable in dynamic and elastic stability analysis and may prove convenient from a programming standpoint even in linear static analysis.

In illustration we again refer to the cantilever beam, Fig. 2.8(c), and remove the d.o.f. θ_1 by condensation. The basic stiffness matrix is obtained from that presented in Section 2.3 by removing the third and fourth rows and columns. Thus, we have

$$\frac{2EI}{L^3}\begin{bmatrix} 2L^2 & -3L \\ -3L & 6 \end{bmatrix}\begin{Bmatrix} \theta_1 \\ w_1 \end{Bmatrix} = \begin{Bmatrix} M_1 \\ F_1 \end{Bmatrix}$$

Since we are removing the top row, $k_{bb} = 4EI/L$, $k_{bc} = -6EI/L^2$. Thus, the transformation matrix for condensation is

$$[\mathbf{\Gamma}_0] = \begin{bmatrix} -\dfrac{L}{4EI}\cdot\dfrac{-6EI}{L^2} \\ \hdashline 1 \end{bmatrix} = \begin{bmatrix} \dfrac{3}{2L} \\ \hdashline 1 \end{bmatrix}$$

and, applying this to the stiffness matrix in the product $[\mathbf{\Gamma}_0]^T[\mathbf{k}][\mathbf{\Gamma}_0]$ and to the right-hand-side as $[\mathbf{\Gamma}_0]^T\{\mathbf{F}\}$, we obtain

$$\frac{3EI}{L^3}w_1 = F_1 + \frac{3}{2L}M_1$$

or, by solution for w_1,

$$w_1 = \frac{F_1L^3}{3EL} + \frac{M_1L^2}{2EI}$$

i.e., the correct flexibility equation for this structure.

It is of interest to observe that condensation of a stiffness matrix means satisfaction of the equilibrium conditions corresponding to the elements eliminated.

We shall have occasion to use the approach to condensation above in Section 3.5 and for a number of other purposes throughout the text.

2.9 Detection of Rigid Body Modes

The conventional approach to the formulation of certain types of elements, particularly those that are curved, makes it difficult to ascertain the number and form of rigid body modes contained in the resulting stiffness matrix. The present section defines the algebraic operations to be performed on a finite element stiffness matrix in order to determine this information.

The inclusion of rigid body motion d.o.f. in a set of element stiffness equations represents the *linear dependence* of certain of the equations on the others. A single linear dependence exists in a set of equations when one of the equations can be written as a linear combination of other equations in the set. A geometric interpretation can also be placed upon linear dependency: An nth-order system of equations can be viewed as a set of n vectors whose components are described by n axes. When two of the vectors—in this case the sets of coefficients of two of the equations—are collinear, linear dependency exists.

The coefficients of a set of element stiffness equations are in general coupled; i.e., the off-diagonal terms are nonzero. Thus, each row is a vector with components in more than one of the n principal directions. These rows can be transformed into vectors corresponding to the n principal directions, which then have only a nonzero term on the main diagonal and as a group form a diagonal matrix. If a pair of the original vectors is collinear, then one of the diagonal terms will be zero (there is one less principal direction than was indicated by the size of the original set). If there are s sets of collinear vectors, there will be s zero values on the main diagonal of the matrix of principal directions.

In view of the above, the number of rigid body motions contained in an element stiffness can be determined by transforming the stiffness matrix into diagonal form (the principal directions); the number of resulting zero diagonal terms will equal the number of such rigid body motions. To effect this transformation we begin by establishing the *eigenvectors* and associated *eigenvalues* of the stiffness matrix. To define these quantities we must first define the *characteristic equation* of a matrix $[\mathbf{k}]$. The characteristic equation of a matrix $[\mathbf{k}]$ is the polynomial equation obtained from expansion of the determinant $|\mathbf{k} - \omega\mathbf{I}| = 0$. When $[\mathbf{k}]$ is of order $n \times n$, the polynomial equation is of degree n in ω and the roots of the equation $\omega_1, \ldots \omega_i, \ldots, \omega_n$ are the *eigenvalues* of $[\mathbf{k}]$. The *eigenvector* corresponding to the eigenvalue ω_i is the nonzero vector $\{\mathbf{d}_i\}$ satisfying the equation $[\mathbf{k}]\{\mathbf{d}_i\} = \{\mathbf{d}_i\}\omega_i$.

Under the conditions prevailing in linear structural analysis the eigenvectors possess the property of *orthogonality* with respect to the matrix $[\mathbf{k}]$. In accordance with this property, for two arbitrary eigenvectors $\{\mathbf{d}_i\}$, $\{\mathbf{d}_j\}$ $(i \neq j)$

$$\lfloor \mathbf{d}_i \rfloor [\mathbf{k}] \{\mathbf{d}_j\} = 0 \tag{2.39a}$$

Also, when $\{d_i\}$ is normalized so that $\lfloor \mathbf{d}_i \rfloor \{\mathbf{d}_i\} = 1$

$$\lfloor \mathbf{d}_i \rfloor [\mathbf{k}] \{\mathbf{d}_i\} = \omega_i \tag{2.39b}$$

Consider now the formation of an $n \times n$ matrix $[\mathbf{\Gamma}_d]$ whose columns are the eigenvectors of $[\mathbf{k}]$; i.e.

$$[\mathbf{\Gamma}_d] = [\{\mathbf{d}_1\} \ldots \{\mathbf{d}_i\} \ldots \{\mathbf{d}_n\}] \tag{2.40}$$

If we form a congruent transformation of $[\mathbf{k}]$ using $[\mathbf{\Gamma}_d]$ as the transformation matrix, then the properties given by Eq. (2.39) ensure that a diagonal stiffness matrix, which we shall designate as the *modal stiffness matrix* $\lceil \mathbf{k}_m \rfloor$, will be produced. Thus

$$\lceil \mathbf{k}_m \rfloor = [\mathbf{\Gamma}_d]^T[\mathbf{k}][\mathbf{\Gamma}_d] \tag{2.41}$$

The number of independent equations in $[\mathbf{k}]$ *is evidenced by the number of non-zero terms in the modal stiffness matrix* $\lceil \mathbf{k}_m \rfloor$. The number of zero main diagonal terms gives the number of rigid body modes. The eigenvectors corresponding to these rows describe geometrically the form of the associated rigid body displacement. Since the main diagonal terms are also the eigenvalues, the search for rigid body modes can also be described as a search for zero eigenvalues.

To test this assertion, consider the stiffness matrix for the axial member, Fig. 2.7. In this case the relevant eigenvalue problem may be written as

$$\begin{vmatrix} \left(\dfrac{AE}{L} - \omega\right) & -\dfrac{AE}{L} \\ -\dfrac{AE}{L} & \left(\dfrac{AE}{L} - \omega\right) \end{vmatrix} = 0$$

and, by expansion of this determinant,

$$\omega^2 - \frac{2AE}{L}\omega = 0$$

so that $\omega = 0, 2AE/L$ and the normalized eigenvectors are

$$[\hat{\mathbf{\Gamma}}_d] = \frac{1}{\sqrt{2}}\begin{bmatrix} 1 & 1 \\ 1 & -1 \end{bmatrix}$$

Finally, applying the transformation of Eq. (2.41),

$$\lceil \mathbf{k}_m \rfloor = \frac{AE}{L}\begin{bmatrix} 0 & 0 \\ 0 & 2 \end{bmatrix}$$

In this problem the axial member possesses one rigid body motion d.o.f. There is, correspondingly, one zero eigenvalue and the associated eigenvector represents rigid body motion of the element (axial displacement of the element as a whole).

An alternative interpretation can be placed upon the procedure above. A rigid body motion should produce zero strain energy since there is no deformation. In terms of the principal directions the contributions to the strain energy are given by $\frac{1}{2}\lfloor \mathbf{d}_i \rfloor \lceil \mathbf{k}_m \rfloor \{\mathbf{d}_i\}$. The contributions due to the eigenvectors associated with rigid body motion must be zero and in order for this to be so the corresponding eigenvalues must be zero.

REFERENCES

2.1 BEAUFAIT, F., W. H. ROWAN, P. G. HOADLEY, and R. M. HACKETT, *Computer Methods of Structural Analysis*, Prentice-Hall, Inc., Englewood Cliffs, N.J., 1970.

2.2 MEEK, J. L., *Matrix Structural Analysis*, McGraw-Hill Book Co., New York, N.Y., 1971.

2.3 WANG, C. K., *Matrix Methods of Structural Analysis*, 2nd ed., International Textbook Co., Scranton, Pa., 1970.

2.4 WILLEMS, N. and W. LUCAS, *Matrix Analysis for Structural Engineers*, Prentice-Hall, Inc., Englewood Cliffs, N.J., 1970.

PROBLEMS

2.1 Construct the mixed form of the beam element force-displacement relationships [Eq. (2.3)].

2.2 Given the following beam element flexibility matrix, verify that the complementary strain energy content equals that of the simply supported element

$$\begin{Bmatrix} F_2 \\ M_1 \end{Bmatrix} = \frac{L}{6EI}\begin{bmatrix} 2L^2 & -3L \\ -3L & 6 \end{bmatrix}\begin{Bmatrix} w_2 \\ \theta_1 \end{Bmatrix}$$

Fig. P2.2

2.3 The flexibility matrix for triangular plate element in plane stress (see Fig. P2.3) is given below. Calculate the element stiffness matrix and verify through comparison with the stiffness matrix shown in Fig. 5.4.

$$\begin{Bmatrix} u_2 \\ u_3 \\ v_3 \end{Bmatrix} = \frac{2}{Etx_2y_3}\left[\begin{array}{c|c|c} x_2^2 & x_2x_3 & -\mu x_2y_3 \\ \hline x_2x_3 & 2(1+\mu)y_3^2 + x_3^2 & -\mu x_2y_3 \\ \hline -\mu x_2y_3 & -\mu x_2y_3 & y_3^2 \end{array}\right]\begin{Bmatrix} F_{x_2} \\ F_{x_3} \\ F_{y_3} \end{Bmatrix}$$

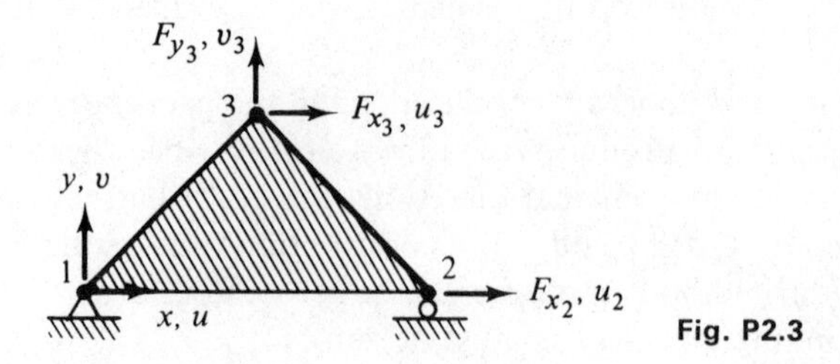

Fig. P2.3

2.4 The following triangular element flexibility matrix is defined for the condition $u_2 = v_2 = v_3 = 0$. Prove that the complementary strain energy content is the same as for the flexibility matrix of Prob. 2.3.

$$\begin{Bmatrix} u_1 \\ v_1 \\ u_3 \end{Bmatrix} = \frac{2}{Etx_2y_3}\begin{bmatrix} (x_2)^2 & & (SYM) \\ \dfrac{\mu x_2^2 y_3}{x_{3-2}} & \dfrac{y_3^2 x_2^2}{(x_{3-2})^2} & \\ \dfrac{\mu x_2 y_3^2}{x_{3-2}} - x_2 x_{3-2} & -\mu x_2 y_3 + \dfrac{y_3^3 x_2}{(x_{3-2})^2} & 2y_3^2 + \dfrac{x_{3-2}^4 + y_3^4}{(x_{3-2})^2} \end{bmatrix}\begin{Bmatrix} F_{x_1} \\ F_{y_1} \\ F_{x_3} \end{Bmatrix}$$

$(x_{3-2}) = (x_3 - x_2)$

2.5 The flexibility matrix of the cantilever beam, Fig. 2.8(c), can be modified to approximate the effect of transverse shear deformation by adding L/A_sG to the flexibility coefficient that relates w_1 to F_{z_1} [i.e., $f_{11} = (L^3/3EI + L/A_sG)$] where A_s is the shear area of the cross section (an equivalent area in constant shear stress that gives the same total shear force as the beam-theory shear stress distribution on the actual area) and G is the shear modulus. Compute the corresponding element stiffness matrix.

2.6 The flexibility matrix for a curved beam loaded in its own plane is given in Fig. P2.6. Construct the element stiffness matrix.

$$\begin{Bmatrix} u_1 \\ v_1 \\ \theta_1 R \end{Bmatrix} = \frac{R^2}{EI}\begin{bmatrix} \dfrac{3\beta}{2} - 2\sin\beta + \dfrac{\sin 2\beta}{4} & (SYM) & \\ \cos\beta + \dfrac{\sin^2\beta}{2} - 1 & \dfrac{\beta}{2} - \dfrac{\sin 2\beta}{4} & \\ \beta - \sin\beta & \cos\beta - 1 & \beta \end{bmatrix}\begin{Bmatrix} F_1 R \\ Q_1 R \\ M_1 \end{Bmatrix}$$

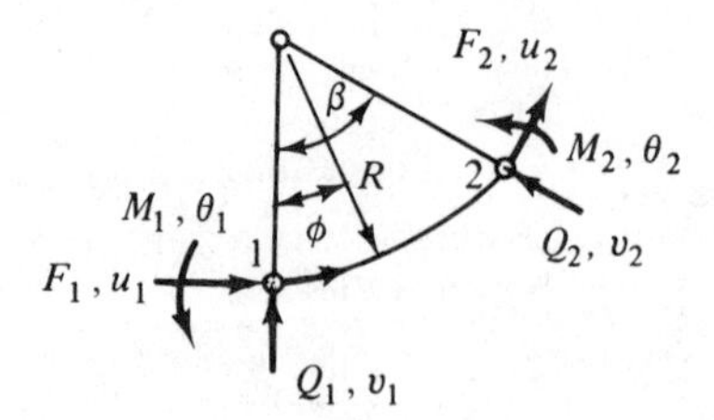

Fig. P2.6

2.7 Establish the equilibrium matrix **[R]** for the curved beam element lying in the x-y plane, as shown in Fig. P2.7.

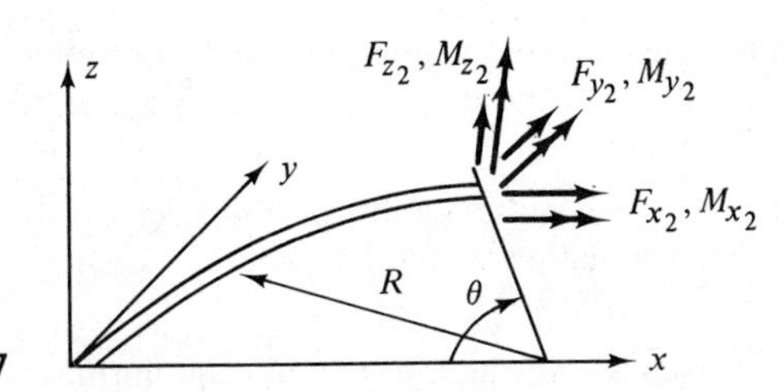

Fig. P2.7

2.8 Check the satisfaction of the equilibrium conditions for the third and fourth columns of the triangular plane stress element stiffness matrix given in Fig. 5.4.

2.9 Check the satisfaction of the equilibrium conditions for the first and sixth columns of the rectangular plane stress element stiffness matrix given in Fig. 9.13.

2.10 Check the satisfaction of the equilibrium conditions for the first two columns of the 12-term rectangular plate-bending element stiffness matrix given in Table 12.1.

2.11 The stiffness matrix for a three-node axial member is as shown in Fig. P2.11. Condense this representation to a set of stiffness equations in terms of u_1 and u_3.

$$\begin{Bmatrix} F_{x_1} \\ F_{x_3} \\ F_{x_2} \end{Bmatrix} = \frac{AE}{6L}\left[\begin{array}{c:c:c} 7 & 1 & -8 \\ \hdashline 1 & 7 & -8 \\ \hdashline -8 & -8 & 16 \end{array}\right]\begin{Bmatrix} u_1 \\ u_3 \\ u_2 \end{Bmatrix}$$

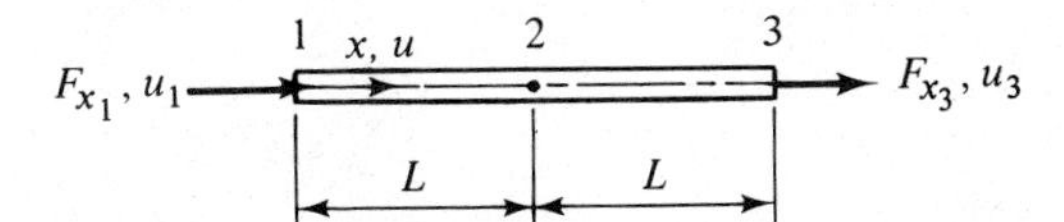

Fig. P2.11

2.12 The stiffness matrix for the triangular plate element in plane stress is given in element (x', y') coordinates by $\{\mathbf{F}\} = [\mathbf{k}]\{\boldsymbol{\Delta}\}$, where

$$\lfloor \boldsymbol{\Delta} \rfloor = \lfloor u_1' \; u_2' \; u_3' \; v_1' \; v_2' \; v_3' \rfloor.$$

For the element shown in Fig. P2.12, compute the transformation matrix to the indicated global (x', y', z') axes.

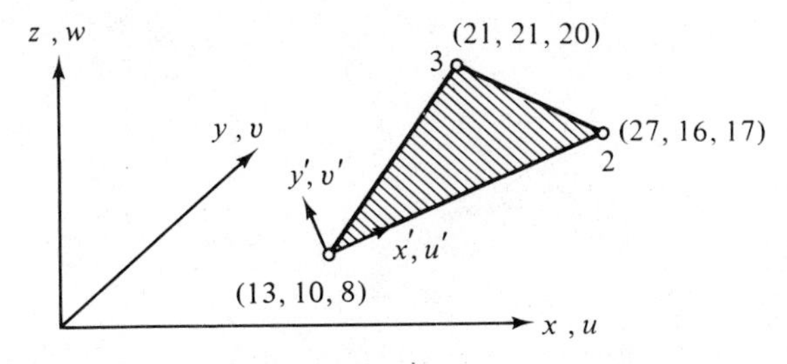

Fig. P2.12

2.13 Calculate the eigenvalues and eigenvectors of the stiffness matrix for the simple flexural member and interpret the results in terms of rigid body modes.

2.14 Prove Betti's law by partitioning of the structural flexibility matrix and introduction of the reciprocal theorem.

2.15 It was pointed out in Section 2.8 that condensation of a stiffness matrix means satisfaction of the equilibrium conditions corresponding to the displacements eliminated. Discuss the meaning of the condensation of a flexibility matrix.

2.16 An axial member stiffness matrix **[k]** is established with reference to orthogonal axes x and y and is to be transformed to relationships referenced to skew axes x' and y'. Develop the transformed stiffness matrix.

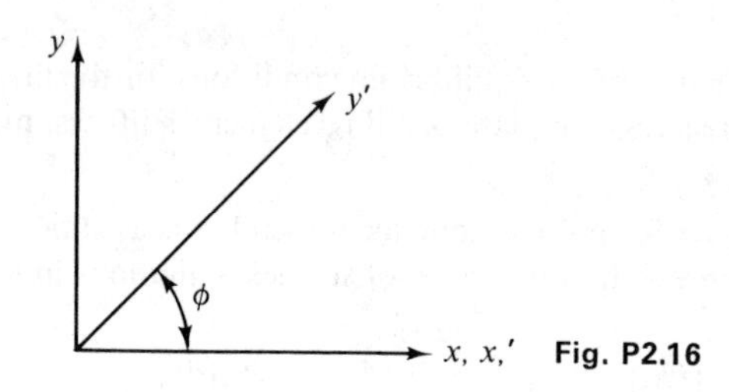

Fig. P2.16

3

Global Analysis Procedures

There are three principal classes of methods for constructing the algebraic equations of the complete (global) finite element representation: the displacement (stiffness), force (flexibility), and mixed methods. The form of the associated equations parallels the form of the element equations defined in Section 2.3. These classes of methods also correspond to alternative forms of energy principles, and we shall find it advantageous in later chapters to develop the methods from an energy standpoint. In this chapter we examine two different approaches to the construction of just one final form of global equations, the *stiffness* form, where joint displacements play the role of unknowns. To establish these approaches we require only the notions and algebraic form of the finite element stiffness matrix given in Section 2.3. The approaches themselves merely involve the application of the conditions of equilibrium and of continuity of displacement at the joints of the assembled analytical model.

Our objective in this development is to furnish the reader with sufficient means of forming global equations using the element relationships we shall establish in detail in subsequent chapters, rather than to conduct a thorough review of feasible means of forming equations in finite element analysis. The stiffness format is chosen for description because, in the view of the writer, it is the simplest and most powerful of the alternative representations. We should hasten to add that the use of a stiffness format places few, if any, restrictions on the mode of formulation of the individual finite element equations since it was shown in Section 2.6 that if these equations are constructed in one format (e.g., flexibility), they can be transformed into another format (in this example, stiffness).

There exist many variants in the details of construction of the global stiffness equations. The approaches examined in this chapter are the *direct stiffness* and *congruent transformation* methods. After completion of these developments we pause, in Section 3.4, to survey the advantages (and some limitations) of the finite element method as a general procedure for structural analysis. In Section 3.5 we move to the treatment of some special operations on global stiffness equations that are either necessary or convenient. These include substructuring, the imposition of constraints, and the use of joint coordinates.

We shall return to questions of global analysis in Chapter 7, where the development of stiffness equations is examined from other perspectives, some properties of the solution that cannot be perceived in the present development are explained, and alternative forms of the global equations (e.g., the global flexibility equations) are studied. Since the present text is devoted primarily to questions of element formulation, no emphasis is given to detailed examples of global formulation. The reader interested in such details should consult one of the many available texts in matrix structural analysis (e.g., Refs. 3.1–3.4).

3.1 Direct Stiffness Method—Its Basic Concept

A complete set of force-displacement relationships for an element with n d.o.f. is, from Eq. (2.1),

$$\begin{matrix} F_1 = k_{11}\Delta_1 + k_{12}\Delta_2 + \cdots + k_{1j}\Delta_j + \cdots + k_{1n}\Delta_n \\ \vdots \\ F_i = k_{i1}\Delta_1 + k_{i2}\Delta_2 + \cdots + k_{ij}\Delta_j + \cdots + k_{in}\Delta_n \\ \vdots \\ F_n = k_{n1}\Delta_1 + k_{n2}\Delta_2 + \cdots + k_{nj}\Delta_j + \cdots + k_{nn}\Delta_n \end{matrix} \tag{3.1}$$

It is assumed that coordinate transformations have already been applied so that the d.o.f. refer to the global axes of the structure.

The numbers $1 \ldots i \ldots n$ identify the d.o.f. at the joints of the element and in this example they correspond to the global numbering system of these same joints. All d.o.f. appear in each of the rows of Eq. (3.1); the element does not have a defined support condition.

Once the element force-displacement relationships have been numerically evaluated for each element of the structure, application of the direct stiffness method consists of their combination in algebraic form in a manner dictated by the requirements of juncture point equilibrium and compatibility. These operations produce a set of force-displacement equations for the juncture points of the elements of the assembled analytical model.

To illustrate this procedure, consider the formulation of the force-displacement equation of the point q in the x direction of the assembled analytical model shown in Fig. 3.1. We designate quantities in the x direction at point q by the subscript i. The elements shown, three triangles and a quadrilateral, all lie in the $x - y$ plane. For convenience we have identified the d.o.f. at each joint appearing in Fig. 3.1 but the only force which has been identified is the external applied load (P_i) which acts in the x direction at point q.

Under the condition of juncture point equilibrium, the applied load P_i is equal to the sum of the internal forces acting upon the respective elements common to the point.† In order to clarify this operation we show, in Fig. 3.2, the elements separated from the joint. We have, from the condition of x-direction equilibrium,

$$P_i = F_i^A + F_i^B + F_i^C + F_i^D \tag{3.2}$$

where F_i^A is the x-direction (internal) force on element A, etc. The force-displacement equations for the elements, each of the form of Eq. (3.1), yield

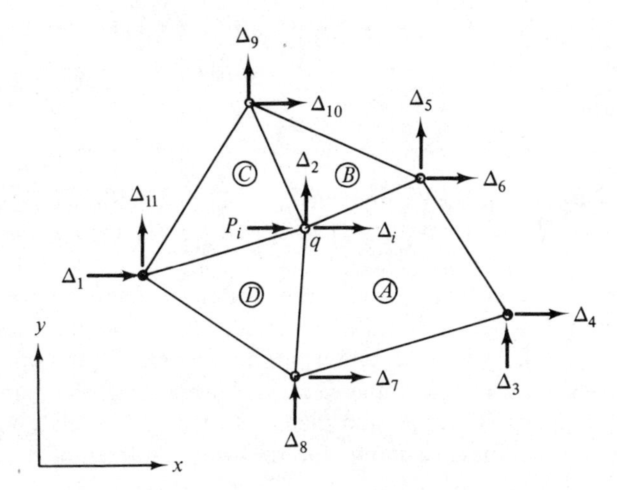

Fig. 3.1 Representative joint in interior of a planar structure.

†At this juncture we should take note of an important feature of nomenclature. Internal (or element) joint forces are identified by the symbol F and external joint forces are symbolized by P, appropriate subscripts or superscripts being assigned in each case. We do not introduce distinct symbols for the element and applied joint moments since no occasion arises herein where both quantities are dealt with in the same problem.

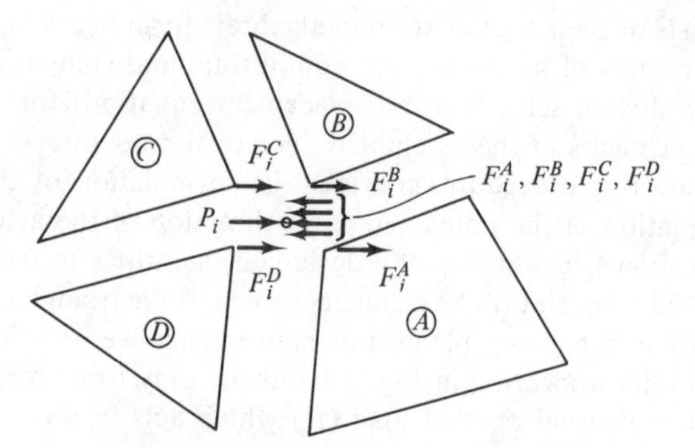

Fig. 3.2 Study of equilibrium in $P_i - \Delta_i$ direction.

expressions for $F_i^A \ldots F_i^D$ in terms of the corresponding element d.o.f. $\Delta_i^A \ldots \Delta_{11}^D$ and the substitution of such expressions into Eq. (3.2) results in

$$P_i = (k_{ii}^A \Delta_i^A + k_{i2}^A \Delta_2^A + \cdots + k_{i8}^A \Delta_8^A) + (k_{ii}^B \Delta_i^B + k_{i2}^B \Delta_2^B + \cdots + k_{i10}^B \Delta_{10}^B) + (k_{ii}^C \Delta_i^C + k_{i1}^C \Delta_1^C + \cdots + k_{i11}^C \Delta_{11}^C) + (k_{ii}^D \Delta_i^D + k_{i1}^D \Delta_1^D + \cdots + k_{i11}^D \Delta_{11}^D) \tag{3.3}$$

and since by the condition of compatibility the displacement Δ_i is the same for A, B, C, and D in each d.o.f. ($\Delta_i^A = \Delta_i^B = \Delta_i^C = \Delta_i^D = \Delta_i$)

$$P_i = (k_{ii}^A + k_{ii}^B + k_{ii}^C + k_{ii}^D)\,\Delta_i + (k_{i1}^C + k_{i1}^D)\,\Delta_1 + (k_{i2}^A + k_{i2}^B + k_{i2}^C + k_{i2}^D)\,\Delta_2 + \cdots + (k_{i11}^C + k_{i11}^D)\,\Delta_{11} \tag{3.4}$$

or

$$P_i = K_{ii}\,\Delta_i + K_{i1}\,\Delta_1 + K_{i2}\,\Delta_2 + \cdots + K_{i11}\,\Delta_{11}$$

This is the final form of the desired equation. The capitalized terms K_{ii}, K_{i1}, $K_{i2} \ldots K_{i11}$ are *global stiffness coefficients* and Eq. (3.4) is a *global stiffness equation.*

It is important to note that each of the four elements meeting at the indicated juncture point possesses stiffness coefficients with common subscripts (e.g., k_{ii}^A, k_{ii}^B, k_{ii}^C, k_{ii}^D). When the subscripts of coefficients of two or more different elements are identical, the elements have a d.o.f. in common, designated by the second subscript, and such coefficients are added to form one coefficient of the stiffness equation for the force represented by the first subscript.

3.2 The Direct Stiffness Method— The General Procedure

In view of the foregoing reasoning, the following *automatic* approach to calculating the applied load-versus-displacement equations for the complete structure suggests itself.

1. Each element stiffness coefficient, upon numerical evaluation, is assigned a double subscript (k_{ij}). The first subscript (i) designates the force for which the equation is written, while the second (j) designates the associated d.o.f.

2. Provision is made for an array (square matrix), whose size is equal to the number of d.o.f. in the complete system, with the possibility that each force will be related to every displacement in the system. Each term in the array is identified by two subscripts. The first subscript (row) pertains to the force equation; the second (column), to the d.o.f. in question. The array is illustrated in Fig. 3.3 for a two-dimensional structure with a total of n

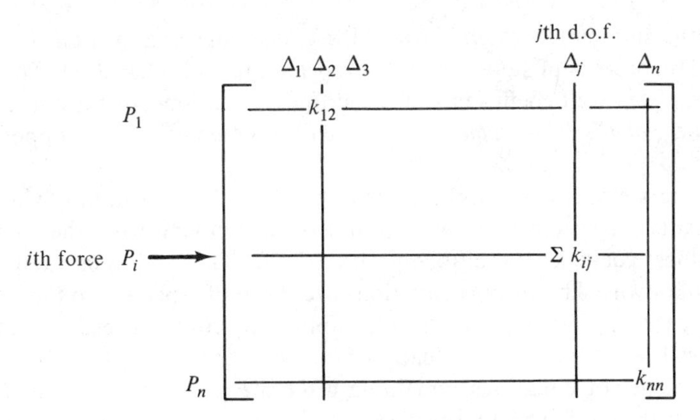

(a) Mode of formation of a global stiffness coefficient k_{ij} and insertion of k_{12}

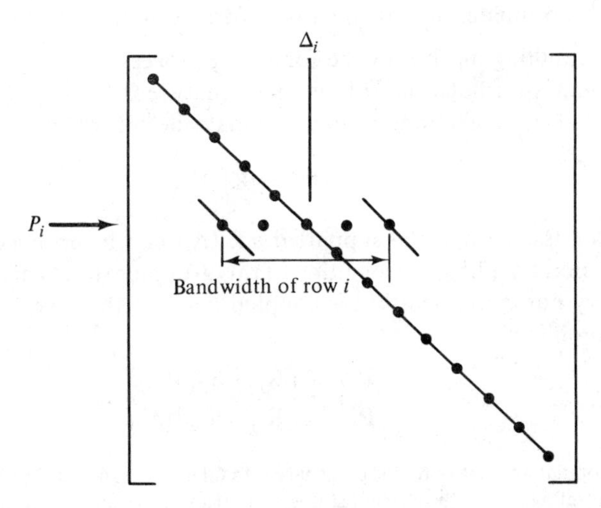

(b) Typical final form of row of global matrix

Fig. 3.3 Key aspects of global stiffness matrix.

d.o.f. First, a search is made through the respective element coefficients. When an element is reached whose designation contains a 1, it is placed in a location in row 1, in the column indicated by its second subscript. For example, k_{12} is positioned as illustrated in Fig. 3.3(a).

3. The procedure of step 2 is continued for d.o.f. 1 until all elements have been searched. Each time a coefficient is placed in a location where a value has already been placed, it is added to the latter. At the completion of this step all terms in the first row have achieved their final value. Hence, for the ith d.o.f.,

$$K_{1i} = \Sigma\, k_{1i}$$

where the summation extends over all elements meeting at d.o.f. i.

4. The process of steps 2 and 3 is repeated for all other d.o.f. The result is a complete set of coefficients of the stiffness equations for the entire structure (the *global stiffness equations*) but with no recognition of support conditions.

5. The support conditions are accounted for by first noting which displacements are specified to be zero and then removing from the equations the stiffness coefficients multiplying these d.o.f.‡ The result is more equations than unknowns. The surplus equations are those that pertain to the external loads at the support points, i.e., the support reactions. These are removed and saved for subsequent evaluation.

6. The set of equations remaining after the performance of step 5 are solved for the d.o.f. The internal forces acting at the element joints are determined by substituting the solved d.o.f. back into the element force-displacement equations. These quantities may require a retransformation from global to local coordinates and finally a transformation into stress.

The resulting algebra of the foregoing process will now be summarized into a matrix formulation. It is assumed that steps 1–4 have been performed and that the global stiffness equations have been formed.

$$\{\mathbf{P}\} = [\mathbf{K}]\{\mathbf{\Delta}\} \tag{3.5}$$

Next, we assume that the support d.o.f. $\{\mathbf{\Delta}_s\}$ can be grouped together and we partition Eq. (3.5) in recognition of this grouping (in practice this is neither necessary nor convenient but is adopted here for the sake of clarity in presentation). Thus

$$\begin{Bmatrix} \mathbf{P}_f \\ \mathbf{P}_s \end{Bmatrix} = \left[\begin{array}{c|c} \mathbf{K}_{ff} & \mathbf{K}_{fs} \\ \hline \mathbf{K}_{sf} & \mathbf{K}_{ss} \end{array}\right] \begin{Bmatrix} \mathbf{\Delta}_f \\ \mathbf{\Delta}_s \end{Bmatrix} \tag{3.6}$$

‡A popular alternative to this procedure is to take cognizance of the support conditions at the outset, i.e., form the element stiffness matrices only with respect to the unsupported d.o.f. Steps 2–4 then lead directly to the *reduced* stiffness matrix and step 5 is eliminated.

Note that the subscripting corresponds to that defined in Section 2.6. Since $\{\mathbf{\Delta}_s\} = 0$, we have

$$\{\mathbf{P}_f\} = [\mathbf{K}_{ff}]\{\mathbf{\Delta}_f\} \tag{3.7a}$$

$$\{\mathbf{P}_s\} = [\mathbf{K}_{sf}]\{\mathbf{\Delta}_f\} \tag{3.7b}$$

The general solution to Eq. (3.7a) is obtained, *symbolically*, by

$$\{\mathbf{\Delta}_f\} = [\mathbf{K}_{ff}]^{-1}\{\mathbf{P}_f\} = [\mathfrak{F}]\{\mathbf{P}_f\} \tag{3.8}$$

where the matrix $[\mathfrak{F}]$ is the set of *global displacement influence coefficients.* We emphasize that the operation of matrix inversion is symbolic. In practice, when relatively few different load conditions $\{\mathbf{P}_f\}$ are to be examined, it will be most efficient to view this process as one of equation-solving for a known right-hand side.

The support reactions $\{\mathbf{P}_s\}$ are found by substitution of Eq. (3.8) into Eq. (3.7b)

$$\{\mathbf{P}_s\} = [\mathbf{K}_{sf}][\mathfrak{F}]\{\mathbf{P}_f\} \tag{3.7c}$$

To obtain the internal force distribution in the ith element, one may substitute the calculated d.o.f. for that element, designated herewith by $\{\mathbf{\Delta}^i\}$, into the element stiffness matrix $[\mathbf{k}^i]$, resulting in the numerical evaluation of the joint forces $\{\mathbf{F}^i\}$ of that element. To obtain the stresses rather than the joint forces from the displacements, the usual procedure is to construct at the outset of an analysis the direct relationships between the element stresses and the associated node point d.o.f.

$$\{\boldsymbol{\sigma}^i\} = [\mathfrak{S}^i]\{\mathbf{\Delta}^i\} \tag{3.9}$$

where $\{\boldsymbol{\sigma}^i\}$ are the stress values that characterize the state of stress within the ith element and $[\mathfrak{S}^i]$ is the corresponding *element stress matrix.* $\{\boldsymbol{\sigma}^i\}$ is a vector that lists values of stress at specified points in the element. Thus, in this approach, when the displacement vector for the element is evaluated, it is pre-multiplied by the corresponding stress matrix to obtain the solution for stress.

Certain key features of the assembled stiffness equations have escaped attention in the development above. For one, the symmetry properties of the element stiffness coefficients are transferred to the global stiffness equations so that only the main diagonal and terms to one side of the main diagonal need be stored.

Second, we note that the stiffness (equilibrium) equation for a given d.o.f. is influenced by the d.o.f. associated with the elements connecting to that d.o.f. The elements shown in Fig. 3.1 may comprise a small region of what is actually a very large finite element idealization; elements that might exist beyond elements A, B, C, and D have no effect on Eq. (3.4). In other words, the nonzero terms on a given row of a stiffness matrix consist only of the main diagonal term and the terms corresponding to d.o.f. at that joint and at joints on the elements meeting at that joint. All other terms in the row are zero. When there are many d.o.f. in the complete finite element idealization the stiffness matrix may contain relatively few nonzero terms, in which case it is characterized as *sparse* or *weakly populated.*

Clearly, it would be advantageous in the solution phase of analysis to cluster all nonzero terms as close to the main diagonal as possible [see Fig. 3.3(b)], thereby isolating the zero terms and facilitating their removal in the solution process. This can be done by numbering the d.o.f. in such a way that the columnar distance of the term most remote from the main diagonal term in each row is minimized, i.e., by minimizing the *bandwidth.*

Bandwidth minimization is but one scheme for achieving efficiency in the equation-solving phase. Whatever the approach taken, it is essential to the economy of the solution process that account be taken of the symmetry and sparseness of the stiffness matrix in large-scale applications. A discussion of equation-solving algorithms is beyond the scope of this text; the reader is advised to consult Ref. 3.5 for a comprehensive view of this question.

All the details of the direct stiffness approach above are illustrated in Fig. 3.4, which studies the case of a stiffened triangular plate element.

As a final note we call attention to the fact that not all the element d.o.f. appearing in Eq. (3.1) will necessarily be joined analytically to counterpart d.o.f. of adjacent elements. A simple example is given in Fig. 3.5 by the case of a beam with interior hinge at joint i. The angular displacements of the members A and B (θ_i^A and θ_i^B) are independent and must not be joined. These will occupy separate column headings in the global stiffness matrix. It is shown in Chapter 6 that the basic theoretical requirements in the formulation of finite elements bring about *admissibility* conditions on the joining of d.o.f. of adjacent elements. Some elements possess d.o.f. in excess of those required by admissibility. In some cases it is very desirable, nevertheless, to join these d.o.f. but for other cases (notably for certain plate and shell elements) it is not permissible to effect this joining. This question is taken up again in Chapter 12.

3.3 Congruent Transformation Approach to Stiffness Analysis

One is not restricted to the procedure described in Section 3.2 in the formation of the global stiffness matrix. One alternative is to form an unconnected array of all element stiffness matrices and then effect their connection via construction and application of a transformation of coordinates in which

Fig. 3.4 Illustrative example—direct stiffness method in application to stiffened triangular plate.

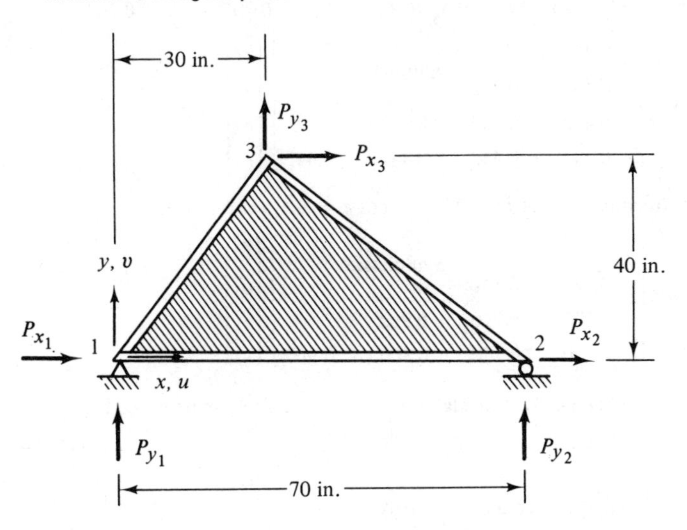

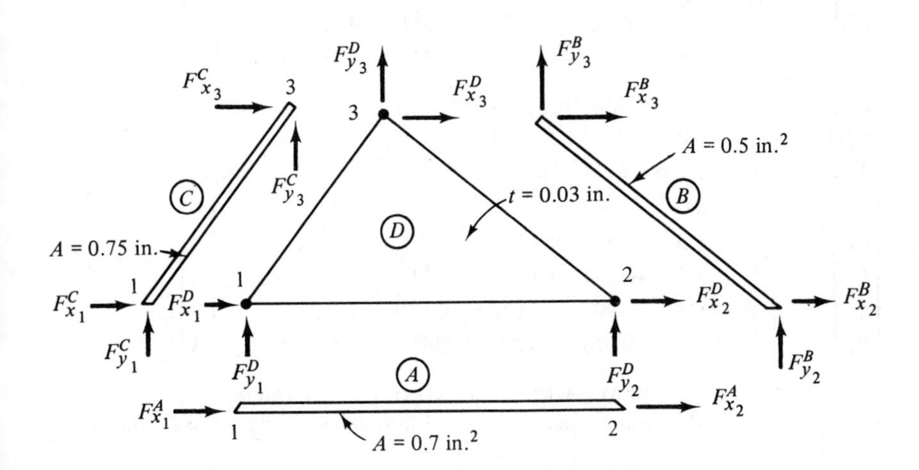

ELEMENT STIFFNESS EQUATIONS

$E = 10^7$ p.s.i., $\mu = 0.3$. All values are hand calculated.

Element A (member 1-2) $A/L = 0.7/70$. For stiffness matrix see Section 2.3. Transform in accordance with procedures of Section 2.7. $\cos\phi = 1$, $\sin\phi = 0$

$$\begin{Bmatrix} F^A_{x_1} \\ F^A_{x_2} \end{Bmatrix} = 10^5 \begin{bmatrix} 1.000 & -1.000 \\ -1.000 & 1.000 \end{bmatrix} \begin{Bmatrix} u_1 \\ u_2 \end{Bmatrix}$$

Fig. 3.4 *(Cont.)*

Element B (member 2-3) $A/L = 0.5/56.56$, $\cos\phi = -0.707$, $\sin\phi = 0.707$

$$\begin{Bmatrix} F^B_{x_2} \\ F^B_{x_3} \\ F^B_{y_2} \\ F^B_{y_3} \end{Bmatrix} = 10^5 \begin{bmatrix} 0.442 & & \text{(Symmetric)} & \\ -0.442 & 0.442 & & \\ -0.442 & 0.442 & 0.442 & \\ 0.442 & -0.442 & -0.442 & 0.442 \end{bmatrix} \begin{Bmatrix} u_2 \\ u_3 \\ v_2 \\ v_3 \end{Bmatrix}$$

Element C (member 1-3) $A/L = 0.75/50$, $\cos\phi = 0.6$, $\sin\phi = 0.8$

$$\begin{Bmatrix} F^C_{x_1} \\ F^C_{x_3} \\ F^C_{y_1} \\ F^C_{y_3} \end{Bmatrix} = 10^5 \begin{bmatrix} 0.540 & & \text{(Symmetric)} & \\ -0.540 & 0.540 & & \\ 0.720 & -0.720 & 0.960 & \\ -0.720 & 0.720 & -0.960 & 0.960 \end{bmatrix} \begin{Bmatrix} u_1 \\ u_3 \\ v_1 \\ v_3 \end{Bmatrix}$$

Element D (member 1-2-3) (For algebraic form of stiffness matrix, see Fig. 5.4)

$$\begin{Bmatrix} F_{x_1} \\ F_{x_2} \\ F_{x_3} \\ F_{y_1} \\ F_{y_2} \\ F_{y_3} \end{Bmatrix} = 10^5 \begin{bmatrix} 1.272 & & & & \text{(Symmetric)} & \\ -0.695 & 1.127 & & & & \\ -0.577 & -0.433 & 1.010 & & & \\ 0.613 & -0.035 & -0.577 & 1.272 & & \\ -0.118 & -0.459 & 0.577 & 0.377 & 0.860 & \\ -0.495 & 0.495 & 0 & -1.649 & -1.237 & 2.886 \end{bmatrix} \begin{Bmatrix} u_1 \\ u_2 \\ u_3 \\ v_1 \\ v_2 \\ v_3 \end{Bmatrix}$$

FORMATION OF GLOBAL STIFFNESS MATRIX

By summation of equations above, we obtain

$$\begin{Bmatrix} P_{x_1} \\ P_{x_2} \\ P_{x_3} \\ P_{y_1} \\ P_{y_2} \\ P_{y_3} \end{Bmatrix} = 10^5 \begin{bmatrix} 2.812 & & & & \text{(Symmetric)} & \\ -1.695 & 2.569 & & & & \\ -1.117 & -0.875 & 1.992 & & & \\ 1.333 & -0.035 & -1.297 & 2.232 & & \\ -0.118 & -0.901 & 1.019 & 0.377 & 1.302 & \\ -1.215 & 0.936 & 0.278 & -2.609 & -1.678 & 4.288 \end{bmatrix} \begin{Bmatrix} u_1 \\ u_2 \\ u_3 \\ v_1 \\ v_2 \\ v_3 \end{Bmatrix}$$

APPLICATION OF DISPLACEMENT BOUNDARY CONDITIONS

$u_1 = v_1 = v_2 = 0$. Remove the first, fourth, and fifth columns and set aside the corresponding rows:

$$\begin{Bmatrix} P_{x_2} \\ P_{x_3} \\ P_{y_3} \end{Bmatrix} = 10^5 \begin{bmatrix} 2.569 & -0.875 & 0.936 \\ -0.875 & 1.992 & 0.278 \\ 0.936 & 0.278 & 4.288 \end{bmatrix} \begin{Bmatrix} u_2 \\ u_3 \\ v_3 \end{Bmatrix}$$

STIFFNESS MATRIX INVERSION AND CALCULATION OF DISPLACEMENTS

and, by inversion, and with $P_{x_2} = 4{,}000$ lb, $P_{x_3} = 10{,}000$ lb, $P_{y_3} = 2{,}000$ lb,

$$\begin{Bmatrix} u_2 \\ u_3 \\ v_3 \end{Bmatrix} = 10^{-6} \begin{bmatrix} 5.203 & 2.466 & -1.296 \\ 2.466 & 6.235 & -0.943 \\ -1.296 & -0.943 & 2.677 \end{bmatrix} \begin{Bmatrix} P_{x_2} \\ P_{x_3} \\ P_{y_3} \end{Bmatrix} = \begin{Bmatrix} 0.04288 \text{ in.} \\ 0.07033 \text{ in.} \\ -0.00926 \text{ in.} \end{Bmatrix}$$

Fig. 3.4 *(Cont.)*

CALCULATION OF SUPPORT REACTIONS
From first, fourth, and fifth rows of global stiffness equations (with corresponding columns removed)

$$\begin{Bmatrix} P_{x_1} \\ P_{y_1} \\ P_{y_2} \end{Bmatrix} = 10^5 \begin{bmatrix} -1.695 & -1.117 & -1.215 \\ -0.035 & -1.297 & -2.609 \\ -0.901 & 1.019 & -1.678 \end{bmatrix} \begin{Bmatrix} 0.04288 \\ 0.07033 \\ -0.00926 \end{Bmatrix} = \begin{Bmatrix} -14{,}000 \\ -6{,}857 \\ 4{,}857 \end{Bmatrix}$$

(Forces shown are in pounds.) These values are in agreement with the values calculated from statics of the total structure.

CALCULATION OF AXIAL MEMBER FORCES
Member A—from the element stiffness matrix, with column corresponding to $u_1 = 0$
$F^A_{x_1}(= -F^A_{x_2}) = -10^5 \times u_2 = -10^5 \times 0.04288 = -4{,}288$ lb
Member B (v_2 column removed)

$$\begin{Bmatrix} F^B_{x_2} \\ \\ F^B_{y_2} \end{Bmatrix} = 0.442 \times 10^5 \begin{bmatrix} 1 & -1 & 1 \\ & & \\ -1 & 1 & -1 \end{bmatrix} \begin{Bmatrix} 0.04288 \\ 0.07033 \\ -0.00926 \end{Bmatrix} = \begin{Bmatrix} -1{,}623 \text{ lb} \\ \\ 1{,}623 \text{ lb} \end{Bmatrix}$$

Resultant axial load $= \sqrt{(-1{,}623)^2 + (1{,}623)^2} = 2{,}294$ lb
Member C (u_1, v_1 columns removed)

$$\begin{Bmatrix} F^C_{x_1} \\ F^C_{y_1} \end{Bmatrix} = -10^5 \begin{bmatrix} 0.540 & 0.720 \\ 0.720 & 0.960 \end{bmatrix} \begin{Bmatrix} 0.07033 \\ 0.00926 \end{Bmatrix} = -\begin{Bmatrix} 3{,}131 \\ 4{,}175 \end{Bmatrix}$$

Resultant axial load $= \sqrt{(3{,}131)^2 + (4{,}175)^2} = 5{,}218$ lb
CALCULATION OF STRESSES IN MEMBER D
Using stress matrix for element given in Fig. 5.5, with columns for u_1, v_1, and v_2 removed

$$\begin{Bmatrix} \sigma_x \\ \sigma_y \\ \tau_{xy} \end{Bmatrix} = 3.925 \times 10^3 \begin{bmatrix} 40 & 0 & 21 \\ 12 & 0 & 70 \\ -10.5 & 24.5 & 0 \end{bmatrix} \begin{Bmatrix} 0.04288 \text{ in.} \\ 0.07033 \text{ in.} \\ -0.00926 \text{ in.} \end{Bmatrix} = \begin{Bmatrix} 5{,}969 \text{ p.s.i.} \\ -525 \text{ p.s.i.} \\ 4{,}996 \text{ p.s.i.} \end{Bmatrix}$$

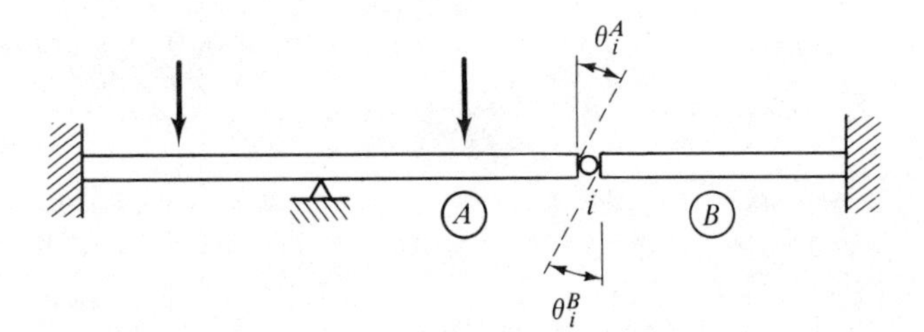

Fig. 3.5

the element and joint d.o.f., respectively, comprise the vectors being transformed. We term this approach the *congruent transformation method.*

Consider first a structure described by a total of p finite elements whose individual element stiffness equations are written in the form of Eq. (3.1).

We now array these element stiffness equations as follows:

$$\{\mathbf{F}^e\} = \lceil \mathbf{k}^e \rfloor \{\boldsymbol{\Delta}^e\} \tag{3.10}$$

where $\{\mathbf{F}^e\}$ and $\{\boldsymbol{\Delta}^e\}$ are vectors that list the d.o.f. of the respective elements; i.e.,

$$\{\mathbf{F}^e\} = \lfloor \lfloor \mathbf{F}^1 \rfloor \lfloor \mathbf{F}^2 \rfloor \dots \lfloor \mathbf{F}^i \rfloor \dots \lfloor \mathbf{F}^p \rfloor \rfloor^{\mathrm{T}} \tag{3.11}$$

$$\{\boldsymbol{\Delta}^e\} = \lfloor \lfloor \boldsymbol{\Delta}^1 \rfloor \lfloor \boldsymbol{\Delta}^2 \rfloor \dots \lfloor \boldsymbol{\Delta}^i \rfloor \dots \lfloor \boldsymbol{\Delta}^p \rfloor \rfloor^{\mathrm{T}} \tag{3.12}$$

and $\lceil \mathbf{k}^e \rfloor$ is, consequently, a diagonal array of submatrices in which each entry on the diagonal is one of the element stiffness matrices

$$\lceil \mathbf{k}^e \rfloor = \begin{bmatrix} [\mathbf{k}^1] & & & & & & \\ & \cdot & & & & & \\ & & \cdot & & & & \\ & & & [\mathbf{k}^i] & & & \\ & & & & \cdot & & \\ & & & & & \cdot & \\ & & & & & & [\mathbf{k}^p] \end{bmatrix} \tag{3.13}$$

$\lceil \mathbf{k}^e \rfloor$ is termed the *unassembled global stiffness*.

We must now effect the connection of the elements. To do so, one merely considers the continuity of displacements at the joints of the structure, represented algebraically by the equations

$$\{\boldsymbol{\Delta}^e\} = [\mathcal{A}]\{\boldsymbol{\Delta}\} \tag{3.14}$$

where $\{\boldsymbol{\Delta}\}$ lists the *global joint displacements* and $[\mathcal{A}]$ is termed the *global kinematics* or *connectivity matrix*. (We shall subsequently illustrate the form of $[\mathcal{A}]$ through the medium of a simple example.) On the basis of work considerations (see Section 2.4) we can establish the corresponding force transformations. First, we denote this transformation symbolically:

$$\{\mathbf{P}\} = [\mathcal{B}]\{\mathbf{F}^e\} \tag{3.15}$$

where $[\mathcal{B}]$ is the *global statics matrix* since it is apparent that it represents the equations defining the equilibrium of external ($\{\mathbf{P}\}$) and internal forces ($\{\mathbf{F}^e\}$). Thus, each row is of the form of Eq. (3.2).

Using the definition of work given in Section 2.4 we can express the external work as

$$W_{\text{ext}} = \tfrac{1}{2}\lfloor \mathbf{P} \rfloor \{\boldsymbol{\Delta}\} \tag{3.16}$$

and, using Eq. (3.15),

$$W_{\text{ext}} = \tfrac{1}{2}\lfloor \mathbf{F}^e \rfloor [\mathcal{B}]^{\mathrm{T}} \{\boldsymbol{\Delta}\} \tag{3.16a}$$

Also, the internal work is given by

$$W_{\text{int}} = \tfrac{1}{2}\lfloor \mathbf{F}^e \rfloor \{\boldsymbol{\Delta}^e\} \tag{3.17}$$

and, from Eq. (3.14),

$$W_{\text{int}} = \tfrac{1}{2}\lfloor \mathbf{F}^e \rfloor [\mathcal{A}]\{\boldsymbol{\Delta}\} \tag{3.17a}$$

Noting the condition of equality of internal and external work and comparing Eqs. (3.16a) and (3.17a), we have

$$[\mathcal{B}]^{\mathrm{T}} = [\mathcal{A}] \tag{3.18}$$

We now apply this consideration directly to the transformation of Eq. (3.10). On the basis of transformation procedures established in Section 2.7 we construct the global stiffness equations in the conventional form, i.e., Eq. (3.5), in which

$$[\mathbf{K}] = [\mathcal{A}]^{\mathrm{T}} \lceil \mathbf{k}^e \rfloor [\mathcal{A}] \tag{3.19}$$

It would appear that the congruent transformation approach is less efficient than the direct stiffness method. The congruent transformation approach involves the formation of the $\lceil \mathbf{k}^e \rfloor$ and $[\mathcal{A}]$ matrices, each of which is larger than the stiffness matrix $[\mathbf{K}]$, as well as the performance of the matrix product defined by Eq. (3.19). On the other hand, the effort to construct the unassembled stiffness matrix is minimal. For one, the element matrices need not include the rigid body displacement modes; i.e., d.o.f. that correspond to a statically determinate stable support condition may be excluded. The portion of the stiffness matrix to be retained for insertion in $\lceil \mathbf{k}^e \rfloor$ is given by $[\mathbf{k}_{ff}]$ in Eq. (2.11). A more rigorous justification for this is given in Section 7.1 but for the present it is sufficient to note that the transformation process described by Eq. (3.19) has the effect of releasing the individual elements from their respective supports.

The operations represented by Eq. (3.19) are also very simple because of the nature of the matrix $[\mathcal{A}]$. Let us examine the construction of this matrix. In general, if the elements A, B, C, and D are joined at a d.o.f. Δ_i, the requirement for compatible displacements dictates that

$$\Delta_i = \Delta_i^A = \Delta_i^B = \Delta_i^C = \Delta_i^D$$

This equation produces a column of the $[\mathcal{A}]$ matrix in which a unit value appears at each location corresponding to $\Delta_i^A, \ldots, \Delta_i^D$ and zero appears in all other locations in this column.

We should further note that the use of element stiffness matrices expressed in terms of global coordinates results in unit values of the terms of $[\mathcal{A}]$. A matrix thus populated is termed a *Boolean matrix* and it is clear that it lends itself to highly efficient algorithms in the performance of the matrix product, Eq. (3.19). If the element stiffness matrix is written only in terms of element coordinate axes, Eq. (3.14) must be revised to include the transformation from element to global coordinate axes. In this case the coefficients of the $[\mathcal{Q}]$ matrix are no longer exclusively unitary and $[\mathcal{Q}]$ is not of Boolean form. At worst, however, $[\mathcal{Q}]$ will be a sparse matrix with unit values, direction cosines, and lengths. Furthermore, as demonstrated in Section 7.1, Eq. (3.19) need not be a formal matrix product algorithm.

To illustrate this approach we again examine the stiffened triangular plate example of Fig. 3.4. The matrices $\lceil \mathbf{k}^e \rfloor$ and $[\mathcal{Q}]$ are presented in Fig. 3.6. The element stiffness submatrices exclude rigid-body-motion terms. Because of the groundwork laid in the calculations given in Fig. 3.4, the element joint vectors are written in terms of global directions and consequently the matrix $[\mathcal{Q}]$ contains only unit values.

The congruent transformation, developed here through a process of

Fig. 3.6 Congruent transformation approach to the construction of the stiffness equations for the illustrative example.

Unassembled stiffness equations $\{\mathbf{F}^e\} = \lceil \mathbf{k}^e \rfloor \{\mathbf{\Delta}^e\}$
(Statically determinate support conditions are imposed on each element.)

$$\begin{Bmatrix} F^A_{x_2} \\ F^B_{x_2} \\ F^B_{x_3} \\ F^B_{y_3} \\ F^C_{x_3} \\ F^C_{y_3} \\ F^D_{x_2} \\ F^D_{x_3} \\ F^D_{y_3} \end{Bmatrix} = 10^5 \begin{bmatrix} 1.000 & & & & & & & & \\ 0 & 0.442 & & & & \text{Symmetric} & & & \\ 0 & -0.442 & 0.442 & & & & & & \\ 0 & 0.442 & -0.442 & 0.442 & & & & & \\ 0 & 0 & 0 & 0 & 0.540 & & & & \\ 0 & 0 & 0 & 0 & 0.720 & 0.960 & & & \\ 0 & 0 & 0 & 0 & 0 & 0 & 1.127 & & \\ 0 & 0 & 0 & 0 & 0 & 0 & -0.433 & 1.010 & \\ 0 & 0 & 0 & 0 & 0 & 0 & 0.495 & 0 & 2.886 \end{bmatrix} \begin{Bmatrix} u^A_2 \\ u^B_2 \\ u^B_3 \\ v^B_3 \\ u^C_3 \\ v^C_3 \\ u^D_2 \\ u^D_3 \\ v^D_3 \end{Bmatrix}$$

Connectivity relationships $\{\mathbf{\Delta}^e\} = \{\mathcal{Q}\}\{\mathbf{\Delta}\}$

$$\begin{Bmatrix} u^A_2 \\ u^B_2 \\ u^B_3 \\ v^B_3 \\ u^C_3 \\ v^C_3 \\ u^D_2 \\ u^D_3 \\ v^D_3 \end{Bmatrix} = \begin{bmatrix} 1 & 0 & 0 \\ 1 & 0 & 0 \\ 0 & 1 & 0 \\ 0 & 0 & 1 \\ 0 & 1 & 0 \\ 0 & 0 & 1 \\ 1 & 0 & 0 \\ 0 & 1 & 0 \\ 0 & 0 & 1 \end{bmatrix} \begin{Bmatrix} u_2 \\ u_3 \\ v_3 \end{Bmatrix}$$

Formation of the product $[\mathcal{Q}]^T \lceil \mathbf{k}^e \rfloor [\mathcal{Q}]$ produces the stiffness matrix (with support conditions taken into account) that was constructed in Fig. 3.4.

direct reasoning, can also be established via application of an energy principle. This alternative is developed in Section 7.2. It will be shown that the alternative view of the problem opens up the possibility of forming the properties of the global analysis without actually forming the global matrices. The scheme is known as *direct energy minimization.*[3.6]

Before concluding this section we examine certain important properties of the system statics matrix $[\mathcal{B}]$ (and, of course, of its transpose, the system kinematics matrix $[\mathcal{A}]$). These permit the identification of any possible *kinematic instability* of the structural representation and the definition of *redundant forces.* A structure is kinematically unstable when it evidences rigid body motion under applied load. Redundant forces are those in excess of the number that would comprise a statically determinate force system.

In order to establish the procedure for operating on the statics equations so as to identify the factors above, we study a planar truss representation consisting of n d.o.f. (two d.o.f. per joint and therefore $n/2$ joints), p members, and t support reaction forces. Extension to the more general case presents no difficulty.

First, we construct the force vector $\{\mathbf{F}^e\}$ in such a way that it contains the independent element internal forces (i.e., rigid-body-motion terms are excluded) and the support reaction forces of the total structure. These forces are sufficient to describe the conditions of equilibrium with a single set of external loads $\{\mathbf{P}\}$. Eq. (3.15) is then again applicable as a description of these conditions. We rewrite Eq. (3.15) as follows:

$$[\mathcal{B} \mid -\mathbf{I}]\begin{Bmatrix} \mathbf{F}^e \\ \mathbf{P} \end{Bmatrix} = 0 \tag{3.15a}$$

The matrix $[\mathcal{B} \mid -\mathbf{I}]$ is termed the *augmented matrix*. Since this matrix is produced by construction of the two equilibrium equations at each of the $n/2$ joints of the structure, it possesses n rows. With p member forces and t support reactions, the vector $\{\mathbf{F}^e\}$ contains $(p + t)$ terms and for indeterminate structures this exceeds n. The excess, $r = (p + t) - n$, comprises the number of redundants. The basic objective in identifying redundant forces and/or kinematic instabilities is to isolate r terms in the vector $\{\mathbf{F}^e\}$. These terms, $\{\mathbf{F}^r\}$ are the redundant forces. Then, we solve for the nonredundant forces $\{\mathbf{F}^0\}$ in terms of the redundant forces $\{\mathbf{F}^r\}$ and the applied loads $\{\mathbf{P}\}$. This can be accomplished by application of the Gauss-Jordan elimination procedure.

Application of Gauss-Jordan elimination to $[\mathcal{B} \mid -\mathbf{I}]$ is accomplished as follows:

1. All terms in the first row of the augmented matrix are divided by the value of the coefficient in the first column. (Precede this step with an interchange of columns if the first column contains a zero.)

2. The modified first row is multiplied by the value of the coefficient in the first column of the second row and is subtracted from the second row. This produces a modified second row with a zero in the first column. The same procedure is applied to all other rows to produce zeros in all other locations in the first column.

3. Steps 1 and 2 are repeated for the second column in such a way as to produce a unit value in the main diagonal location and zero values in all other locations in this column. This procedure is applied to n columns in all, resulting in a unit matrix $[\mathbf{I}]$ of order $n \times n$.

In consequence of the foregoing Eq. (3.15a) is transformed to

$$[\mathbf{I} \mid \mathfrak{C}_2 \mid \mathfrak{C}_1]\begin{Bmatrix} \mathbf{F}^0 \\ \hline \mathbf{F}^r \\ \hline \mathbf{P} \end{Bmatrix} = 0 \tag{3.15b}$$

or

$$\{\mathbf{F}^0\} = -[\mathfrak{C}_1]\{\mathbf{P}\} - [\mathfrak{C}_2]\{\mathbf{F}^r\} \tag{3.15c}$$

Finally, expanding this so that we have $\{\mathbf{F}^e\}$ on the left side ($\{\mathbf{F}^e\} = \lfloor \mathbf{F}^0 \ \mathbf{F}^r \rfloor^{\mathrm{T}}$)

$$\{\mathbf{F}^e\} = [\mathfrak{D}_1]\{\mathbf{P}\} + [\mathfrak{D}_2]\{\mathbf{F}^r\} \tag{3.15d}$$

where

$$[\mathfrak{D}_1] = \begin{bmatrix} -\mathfrak{C}_1 \\ \hline \mathbf{0} \end{bmatrix}, \qquad [\mathfrak{D}_2] = \begin{bmatrix} -\mathfrak{C}_2 \\ \hline \mathbf{I} \end{bmatrix}$$

We first observe, in regard to the procedure above, that the columns associated with $\{\mathbf{F}^0\}$ need not be the first n columns of the initially defined $[\mathfrak{B} \mid -\mathbf{I}]$ matrix. This would cause the redundants to be identified by the rather arbitrary initial selection of the columns of this matrix. Preferably, prior to beginning the normalization of the main diagonal term of a given column, a search should be undertaken to determine the available column with the "best" coefficient in the corresponding row. Then, such a column would be interchanged with the vector presently in the given column and the normalization, etc. (step 2), performed. There are a number of views regarding the criterion for the "best" coefficient in a given row. The simplest of these chooses the column with the largest numerical coefficient.

The second observation regarding the procedure above is that *kinematic instability of the analytical model will be identified through the generation of as many zero rows as there are* d.o.f. *associated with the instability*. The Gauss-Jordan elimination process produces a diagonalized matrix. Recall from

Section 2.9 that rigid body motion d.o.f. in the element stiffness matrix are identified when the stiffness matrix is transformed to a diagonal form and zero diagonal terms are produced. The nonzero terms of the diagonal matrix in the present development likewise comprise the coefficients of all independent equations.

Figure 3.7 illustrates the identification of the kinematic instability of a simple truss structure through application of the procedure above. In order to simplify the algebra of the presentation we have excluded the support points 1 and 4 from the computation.

Fig. 3.7 Identification of kinematic instability by application of Gauss-Jordan elimination to joint equilibrium equations.

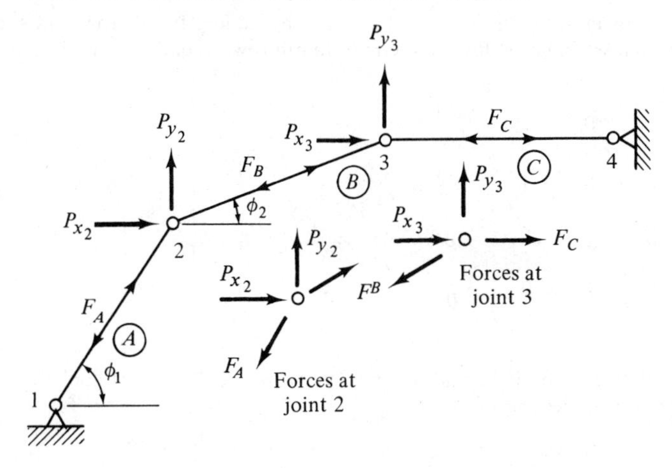

The equations of equilibrium at joints 2 and 3 are as follows ($c_2 = \cos\phi_2$, $s_2 = \sin\phi_2$, etc.):

$$\begin{array}{c} \\ \sum F_{x_2} \\ \sum F_{y_2} \\ \sum F_{x_3} \\ \sum F_{y_3} \end{array} \begin{array}{c} \begin{array}{cccc|cccc} F^A & F^B & F^C & & P_{x_2} & P_{y_2} & P_{x_3} & P_{y_3} \end{array} \\ \left[\begin{array}{ccc|cccc} -c_1 & c_2 & 0 & 1 & 0 & 0 & 0 \\ -s_1 & s_2 & 0 & 0 & 1 & 0 & 0 \\ 0 & -c_2 & 1 & 0 & 0 & 1 & 0 \\ 0 & -s_2 & 0 & 0 & 0 & 0 & 1 \end{array}\right] \end{array}$$

Pivot on term in first row, first column (i.e., divide first row by $-c_1$).

$$\left[\begin{array}{ccc|cccc} 1 & -\dfrac{c_2}{c_1} & 0 & -\dfrac{1}{c_1} & 0 & 0 & 0 \\ -s_1 & s_2 & 0 & 0 & 1 & 0 & 0 \\ 0 & -c_2 & 1 & 0 & 0 & 1 & 0 \\ 0 & -s_2 & 0 & 0 & 0 & 0 & 1 \end{array}\right]$$

Fig. 3.7 *(Cont.)*

Eliminate terms in first column beneath row 1 by multiplying first row by s_1 and adding to second row. Also, divide resulting second and fourth rows by s_1 and third row by c_1.

$$\left[\begin{array}{ccc|cccc} 1 & -\frac{c_2}{c_1} & 0 & -\frac{1}{c_1} & 0 & 0 & 0 \\ 0 & \left(\frac{s_2}{s_1}-\frac{c_2}{c_1}\right) & 0 & -\frac{1}{c_1} & \frac{1}{s_1} & 0 & 0 \\ 0 & -\frac{c_2}{c_1} & 1 & 0 & 0 & \frac{1}{c_1} & 0 \\ 0 & -\frac{s_2}{s_1} & 0 & 0 & 0 & 0 & \frac{1}{s_1} \end{array}\right]$$

Eliminate term in second row, second column, by adding fourth row and subtracting third row from second row. Pivot on term in fourth row, second column (i.e., multiply by $-s_1/s_2$).

$$\left[\begin{array}{ccc|cccc} 1 & 0 & -1 & -\frac{1}{c_1} & 0 & -\frac{1}{c_1} & 0 \\ 0 & 0 & -1 & -\frac{1}{c_1} & \frac{1}{s_2} & -\frac{1}{c_1} & \frac{1}{s_1} \\ 0 & 1 & -\frac{c_1}{c_2} & 0 & 0 & -\frac{1}{c_2} & 0 \\ 0 & 1 & 0 & 0 & 0 & 0 & -\frac{1}{s_2} \end{array}\right]$$

Eliminate all other terms in second column. Only row 3 is affected. Multiply row 4 by -1 and add to row 3, yielding

$$\left[\begin{array}{ccc|cccc} 1 & 0 & -1 & -\frac{1}{c_1} & 0 & -\frac{1}{c_1} & 0 \\ 0 & 0 & -1 & -\frac{1}{c_1} & \frac{1}{s_2} & -\frac{1}{c_1} & \frac{1}{s_1} \\ 0 & 0 & -\frac{c_1}{c_2} & 0 & 0 & -\frac{1}{c_2} & \frac{1}{s_2} \\ 0 & 1 & 0 & 0 & 0 & 0 & -\frac{1}{s_2} \end{array}\right]$$

Pivot on term in third row, third column (i.e., multiply by c_2/c_1).

$$\left[\begin{array}{ccc|cccc} 1 & 0 & -1 & -\frac{1}{c_1} & 0 & -\frac{1}{c_1} & 0 \\ 0 & 0 & -1 & -\frac{1}{c_1} & \frac{1}{s_2} & -\frac{1}{c_1} & \frac{1}{s_1} \\ 0 & 0 & 1 & 0 & 0 & -\frac{1}{c_1} & -\frac{c_2}{c_1 s_2} \\ 0 & 1 & 0 & 0 & 0 & 0 & -\frac{1}{s_2} \end{array}\right]$$

Fig. 3.7 *(Cont.)*

Eliminate all other terms in third column by adding row 3 to them.

$$\left[\begin{array}{ccc|cccc} 1 & 0 & 0 & -\frac{1}{c_1} & 0 & -\frac{2}{c_1} & -\frac{c_2}{c_1 s_2} \\ 0 & 0 & 0 & -\frac{1}{c_1} & \frac{1}{s_2} & -\frac{2}{c_1} & \left(\frac{1}{s_1} - \frac{c_2}{c_1 s_2}\right) \\ 0 & 0 & 1 & 0 & 0 & -\frac{1}{c_1} & -\frac{c_2}{c_1 s_2} \\ 0 & 1 & 0 & 0 & 0 & 0 & -\frac{1}{s_2} \end{array}\right]$$

A unit matrix has been established in correspondence with the internal forces. The indicated zero row identifies a kinematic instability in one d.o.f.

3.4 Resumé of Finite Element Advantages

Figure 3.8 represents a fairly general structural analysis situation. All aspects of this hypothetical situation appear in one or more of the real design situations pictured in Chapter 1. The overall geometry cannot be described by a single mathematical expression and the presence of cutouts and the orientation of stiffening members precludes the use of a regular gridwork. Both the support conditions, which relate to both forces and displacements, and the loading conditions would be difficult to treat by classical methods even for the case of a highly simplified structural form. Such factors, relating to the *analytical*, *geometric*, *boundary condition* representations, mandate the application of the finite element method to the solution of this type of prob-

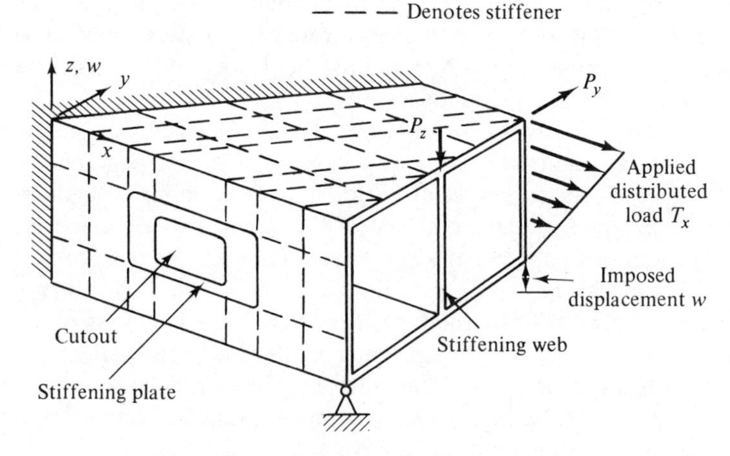

Fig. 3.8 General structural design situation.

lem. To these factors, which we discuss further in the following paragraphs, is added that of material property representation.

The most apparent advantage of finite element analysis, as implied by the comments above, is in the representation of many structural forms in a given analytical model—plates, solids, stiffeners, shell segments, and others. Thus there is wide scope *analytical representation*. There are in fact practical limitations on this capability and we shall explore such factors in the following chapters. When stiffening a plate, for example, the analytical attachment of the stiffeners to the plate occurs only at the nodal points. Thus, dependence is placed on the intrinsic nature of the formulation of the respective elements to ensure that continuity of structural behavior (displacement or stress) occurs on the juncture line between the node points. In general it is not possible to satisfy all aspects of these *interelement continuity* requirements, and a large share of theoretical developments in finite element analysis are devoted to the examination of this question and to the selection of those requirements that will be met in the element formulations.

Another limitation in analytical representation derives from the simplified functions chosen to represent conventional finite elements. In the vicinity of the corners of a cutout in a structure, for example, there is a sharply-varying stress intensity. Thus, considerable refinement of the grid is needed to describe this variation when an understanding of it is essential to design. In contrast to analytical methods that require regular grids, of course, the refinement can be accomplished rather easily but the extent of this refinement may be so great as to be economically infeasible for a desired level of solution accuracy.

In the cases described above it is possible to introduce special finite elements which are formulated on the basis of complicated functions which describe rapid variations in stress. A similar condition exists at the boundaries of structures at locations where concentrated loads are applied. The alternatives of grid refinement with simplified elements or of special elements in a coarser grid are again available.

The capability of *geometric representation*, i.e., of defining the grid lines for analysis in a nonrectangular and indeed rather irregular manner, has also long stood out as a special advantage of the finite element method. We have already encountered the idea of a triangular element for planar situations and in Chapter 5 and subsequent chapters we shall construct the force-displacement relationships for this element. The versatility in grid line definition furnished by the triangle is quite apparent. Less obvious but more powerful, however, is the convenience afforded in geometric representation by elements with curved boundaries. The particular case where the boundary curves are defined by polynomial expressions, known as *isoparametric* geometric representation, is developed in Section 8.8.

Both force (applied loading) and displacement *boundary conditions* are treated quite expeditiously in the procedure described in the previous sections. This prior discussion has implied that the applied loads are point forces. In practice, of course, many of the important loadings are distributed over the surface of the structure. Such conditions are treated through definition of statically equivalent joint loads and although the intuitively obvious process of prorating, or "lumping," will normally produce acceptable numerical results, we shall see in Chapter 6 that the theoretical basis of finite element analysis leads naturally to a more appropriate definition of joint loads equivalent to distributed loads that cannot be discerned intuitively.

A number of physical circumstances in practical design give rise to terms in the analysis that have the effect of loadings. The distribution of temperature in a structure may cause constrained thermal expansion and to deal with this problem numerically it is necessary to transform the thermal strains into fictitious loads or displacements. Chapter 6 also develops terms of the element force-displacement equations for these and other *initial strain* effects.

It is pertinent to note that the finite element method brings a number of special advantages to thermal stress analysis. A consistent methodology of finite element heat transfer analysis is available for the calculation of the temperature distribution in the structure. The basic ideas of steady-state finite element heat transfer analysis are outlined in Section 5.4. Refs. 3.7 and 3.8 describe more fully this nonstructural application of the finite element method, including procedures for the calculation of transient temperature distributions. It is possible to apply the same general-purpose finite element analysis program to the calculation of both temperatures due to a thermal input and to thermal stresses arising from these temperatures. Also in cases where the material properties are a function of the temperature, it is possible to assign properties to each element consistent with the temperature level of that element.

Displacement boundary conditions, in practice, will not always be described exclusively by suppressed d.o.f. (zero displacement). In some cases the displacement of a point will be a specified value; this can be introduced as such into the procedure of the previous sections. Elastic supports can be described either through elastic members (springs) introduced at the affected node points or in a special edge element that is formulated to account for elastic support on a portion of its boundary. In some cases the displacements of a number of points on the boundary of a structure are tied together by special *constraining* conditions. These and other facets of support definition are examined in detail in Section 3.5.

A more subtle attribute of the finite element method is its ability to deal with complex *material property* laws. Nearly all available classical solutions refer to structures composed of homogenous, *isotropic* materials. The restriction to homogeneity is difficult but not impossible to remove in finite element analysis and the treatment of nonhomogeneous situations is beyond the scope of this text. As demonstrated in the chapters concerned with element formulation, however, nonisotropic material properties are handled in the finite element method without any significant expansion of the cost or complexity of the numerical solution process. Indeed, the capability in this regard for the most part outdistances the availability of material property data that accurately characterize the degree of nonisotropy.

Our intent in the above was to refer to linear analysis situations. The scope of finite element analysis relative to classical solutions is even greater in the area of nonlinear analysis, as in the calculation of plastic deformation where even the simplest of geometric forms defy analytical treatment. We have excluded study of inelastic and other types of nonlinear analysis but to obtain a measure of progress in this direction the reader is advised to consult Refs. 3.9 and 3.10.

3.5 *Special Operations*

3.5.1 *Substructuring*

Many practical structures are so large and complex that the minimum-size finite element model of the complete structure places excessive demands on available equation-solving capabilities. It then becomes necessary to deal with the problem in stages in which major components of the structure, termed *substructures*, are first analyzed separately and then combined. Examples were given in Section 1.3. Also, the practical design process often begins with independent analyses of naturally occurring substructures and it is efficient to perform the final design analysis with the use of available substructure data. Furthermore, the substructure analysis approach enables one to keep in touch with the intermediate results for the component structures and is of value for repetitive analyses, as in the case of optimum design procedures and nonlinear analysis.

Figure 3.9 shows a complete structure, partitioned into three major substructures F, G, and H. Consider first the stiffness properties of the Gth substructure. The following subscripts are employed:

c – d.o.f. located on a substructure interface.
d – d.o.f. internal to substructure G, i.e, not associated with any other substructure.

The substructure stiffness equations are assumed to have been modified to account for the support conditions. The stiffness relationships for the Gth substructure can be written in the form (for simplicity we use no symbols to identify these with substructure G)

$$\begin{Bmatrix} \mathbf{F}_d \\ \mathbf{F}_c \end{Bmatrix} = \left[\begin{array}{c|c} \mathbf{k}_{dd} & \mathbf{k}_{dc} \\ \hline \mathbf{k}_{cd} & \mathbf{k}_{cc} \end{array}\right] \begin{Bmatrix} \mathbf{\Delta}_d \\ \mathbf{\Delta}_c \end{Bmatrix} \tag{3.20}$$

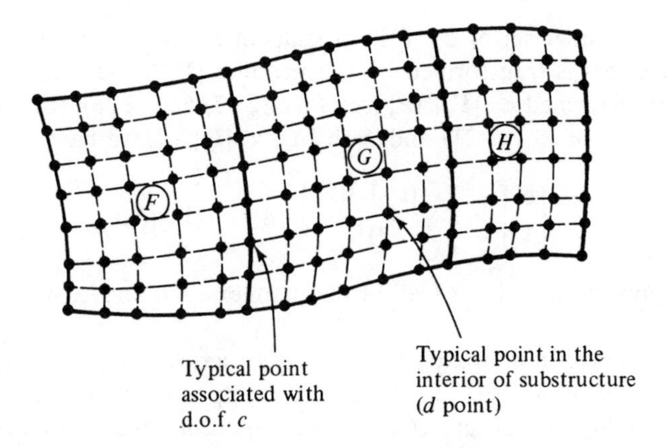

Fig. 3.9 Substructure analysis conditions (applied loads not shown).

One can first solve the upper part of Eq. (3.20) to express the displacements of the interior ($\{\mathbf{\Delta}_d\}$) in terms of the boundary displacements, noting also that the forces $\{\mathbf{F}_d\}$ are indeed the applied loads $\{\mathbf{P}_d\}$ at their d.o.f.

$$\{\mathbf{\Delta}_d\} = [\mathbf{k}_{dd}]^{-1}\{\mathbf{P}_d\} - [\mathbf{k}_{dd}]^{-1}[\mathbf{k}_{dc}]\{\mathbf{\Delta}_c\} \tag{3.21}$$

$$\{\mathbf{F}_c\} = [[\mathbf{k}_{cc}] - [\mathbf{k}_{cd}][\mathbf{k}_{dd}]^{-1}[\mathbf{k}_{dc}]]\{\mathbf{\Delta}_c\} + [\mathbf{k}_{cd}][\mathbf{k}_{dd}]^{-1}\{\mathbf{P}_d\} \tag{3.22}$$

For simplicity, we now define

$$\{\mathbf{R}_c\} = [\mathbf{k}_{cd}][\mathbf{k}_{dd}]^{-1}\{\mathbf{P}_d\} \tag{3.23}$$

$$[\hat{\mathbf{k}}_{cc}] = [[\mathbf{k}_{cc}] - [\mathbf{k}_{cd}][\mathbf{k}_{dd}]^{-1}[\mathbf{k}_{dc}]] \tag{3.24}$$

so that Eq. (3.22) becomes

$$\{\mathbf{F}_c\} = [\hat{\mathbf{k}}_{cc}]\{\mathbf{\Delta}_c\} + \{\mathbf{R}_c\} \tag{3.25}$$

This stiffness equation, as well as those for the other major substructures, can now be employed in the formation of the stiffness equations for the juncture d.o.f., i.e., for the complete structure (substructures F, G, and H):

$$\{\mathbf{P}_c\} = \{\mathbf{F}^F\} + \{\mathbf{F}^G\} + \{\mathbf{F}^H\} \tag{3.26}$$

where now the superscripts F, G, and H are used to identify the respective interface forces.

Solution of the resulting equations yields the interface displacements $\{\mathbf{\Delta}_c\}$. The displacements $\{\mathbf{\Delta}_c\}$ are then substituted back into the substructure equations [Eqs. (3.21) and (3.22)] to yield internal forces and the displacements of points on the interior of the substructure.

The desired condensation can also be achieved by a *coordinate transformation* process. It is recalled from Section 2.8 that if a set of d.o.f. is related to a smaller set of d.o.f. by a transformation matrix $[\mathbf{\Gamma}_0]$, then the original stiffness is transformed by the product $[\mathbf{\Gamma}_0]^T[\mathbf{K}][\mathbf{\Gamma}_0]$ and the force vector is transformed by $[\mathbf{\Gamma}_0]^T\{\mathbf{P}\}$ [see Eqs. (2.37) and (2.38)]. In the present case, by virtue of Eq. (3.21) and with $\{\mathbf{\Delta}_c\} = [\mathbf{I}]\{\mathbf{\Delta}_c\}$, we have

$$\begin{Bmatrix} \mathbf{\Delta}_d \\ \mathbf{\Delta}_c \end{Bmatrix} = \begin{bmatrix} -[\mathbf{k}_{dd}]^{-1}[\mathbf{k}_{dc}] \\ \hdashline [\mathbf{I}] \end{bmatrix} \{\mathbf{\Delta}_c\} = [\mathbf{\Gamma}_0]\{\mathbf{\Delta}_c\} \tag{3.27}$$

and, by application to Eq. (3.20) in the manner cited above, we obtain Eq. (3.25).

A more powerful approach to complex structure analysis is the *reduced* substructure technique.[3.11] To describe this approach analytically it is necessary to amplify the concept of constraint equations, which are taken up in Section 3.5.2.

3.5.2 Constraint Equations

Constraint equations are relationships between the d.o.f. that are supplemental to the relationships represented by the basic stiffness equations. The simple definition of a support condition, e.g., $\Delta_j = 0$, constitutes a constraint equation but, as we have already seen, it is a simple matter to account for this directly after the total global stiffness matrix is formed. More to the point of the present discussion is the case where a flexural member is connected to a solid as illustrated in Fig. 3.10. We see here that points 1–5 are constrained to form a linear variation of the w displacement, as dictated by the (straight-line) angular displacement of the shell element. Constraints arise in many other situations, including the reduced substructure scheme discussed in the next section, in certain approaches to the analysis of incompressible materials, in the treatment of special boundary conditions, and in attempts to impose specified patterns of displacement over certain portions of the structure. We shall encounter a number of these situations later in the text.

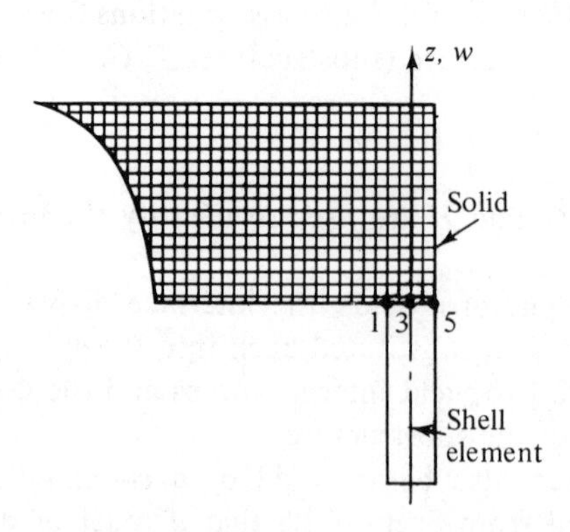

Fig. 3.10 Shell connected to solid.

Each constraint equation represents an opportunity to eliminate one of the d.o.f. in favor of the remainder. We shall now take advantage of this viewpoint to form a transformation matrix that can be used to condense as many d.o.f. as there are constraint equations from the stiffness equations; i.e., we shall adopt the general approach presented in Section 2.8.

Consider the case where r constraint equations are present in a system composed of n d.o.f. A general representation of linear constraint equations for this case would be of the form

$$[\mathbf{G}]_{r\times n}\{\mathbf{\Delta}\}_{n\times 1} = \{\mathbf{s}\}_{r\times 1} \tag{3.28}$$

where the terms of $[\mathbf{G}]$ are the coefficients of the constraint equations and $\{\mathbf{s}\}$ is a vector of known constants. For simplicity we shall deal only with the case $\{\mathbf{s}\} = \mathbf{0}$. Development of the more general case ($\{\mathbf{s}\} \neq \mathbf{0}$) is left as an exercise for the reader (see Prob. 3.18).

In order to develop the transformation matrix we once again divide the n d.o.f. into two groups, $\{\mathbf{\Delta}_e\}$ and $\{\mathbf{\Delta}_c\}$, where $\{\mathbf{\Delta}_e\}$ contains r d.o.f. and $\{\mathbf{\Delta}_c\}$ contains $(n - r)$ d.o.f.

$$[\mathbf{G}_{e_{r\times r}} \; \mathbf{G}_{c_{r\times n-r}}]\begin{Bmatrix}\mathbf{\Delta}_e\\ \mathbf{\Delta}_c\end{Bmatrix} = \mathbf{0} \tag{3.29}$$

Thus, the d.o.f. have been segregated so that the $\{\mathbf{\Delta}_e\}$ d.o.f. are selected in correspondence with the r number of constraint equations. The intent is to remove these d.o.f. from the potential energy functional by use of a condensation scheme. Although the choice of d.o.f. to be removed is often arbitrary, occasions do arise when great care must be exercised in their selection.[3.12]

Solving Eq. (3.29) for $\{\mathbf{\Delta}_e\}$,

$$\{\mathbf{\Delta}_e\} = -[\mathbf{G}_e]^{-1}[\mathbf{G}_c]\{\mathbf{\Delta}_c\} = [\mathbf{G}_{ec}]\{\mathbf{\Delta}_c\} \tag{3.30}$$

which in accordance with the procedures developed in Section 2.8 can be used to form the transformation of d.o.f.

$$\begin{Bmatrix}\mathbf{\Delta}_e\\ \mathbf{\Delta}_c\end{Bmatrix} = \begin{bmatrix}\mathbf{G}_{ec}\\ \hdashline \mathbf{I}\end{bmatrix}\{\mathbf{\Delta}_c\} = [\mathbf{\Gamma}_c]\{\mathbf{\Delta}_c\} \tag{3.31}$$

Applying this now to the global equations via $[\mathbf{\Gamma}_c]^{\mathrm{T}}[\mathbf{k}][\mathbf{\Gamma}_c]$ gives a reduced stiffness matrix referred to $\{\mathbf{\Delta}_c\}$ alone and also a reduced load vector

$$\{\hat{\mathbf{P}}_c\} = [\mathbf{\Gamma}_c]^{\mathrm{T}}\begin{Bmatrix}P_e\\ P_c\end{Bmatrix}$$

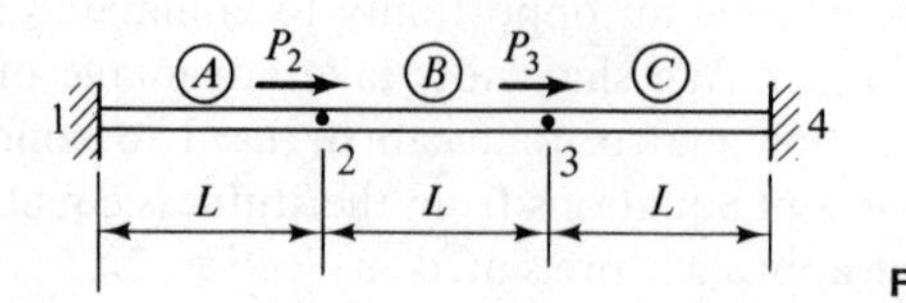

Fig. 3.11

Solution of the reduced stiffness equation yields $\{\mathbf{\Delta}_c\}$, which may then be substituted into Eq. (3.30) for calculation of $\{\mathbf{\Delta}_e\}$.

In illustration, consider the system of three axial members shown in Fig. 3.11. By direct stiffness analysis the stiffness of the system without supports is ($k_0 = AE/L$).

$$k_0 \begin{bmatrix} 1 & & \text{(Symmetric)} & \\ -1 & 2 & & \\ 0 & -1 & 2 & \\ 0 & 0 & -1 & 1 \end{bmatrix} \begin{Bmatrix} u_1 \\ u_2 \\ u_3 \\ u_4 \end{Bmatrix} = \begin{Bmatrix} P_1 \\ P_2 \\ P_3 \\ P_4 \end{Bmatrix}$$

and, with the support conditions $u_1 = u_4 = 0$,

$$k_0 \begin{bmatrix} 2 & -1 \\ -1 & 2 \end{bmatrix} \begin{Bmatrix} u_2 \\ u_3 \end{Bmatrix} = \begin{Bmatrix} P_2 \\ P_3 \end{Bmatrix}$$

Now, assume that a rigid link connects points 2 and 3 so that $u_2 = u_3$. In matrix format

$$\lfloor 1 \ -1 \rfloor \begin{Bmatrix} u_2 \\ u_3 \end{Bmatrix} = 0$$

Thus, the transformation relationship is

$$\begin{Bmatrix} u_2 \\ u_3 \end{Bmatrix} = \begin{bmatrix} 1 \\ 1 \end{bmatrix} \{u_3\}$$

and the reduced stiffness equations become

$$k_0 \lfloor 1 \quad 1 \rfloor \begin{bmatrix} 2 & -1 \\ -1 & 2 \end{bmatrix} \begin{Bmatrix} 1 \\ 1 \end{Bmatrix} u_3 = \lfloor 1 \quad 1 \rfloor \begin{Bmatrix} P_2 \\ P_3 \end{Bmatrix}$$

or $2k_0u_3 = P_2 + P_3$. This answer is in correspondence with the obvious solution to the problem. The constraint transforms member B into a rigid link between joints 2 and 3. Thus, the applied load is the sum of the forces at these joints ($P_2 + P_3$) and the coefficient of u_3 is the sum of the direct stiffness of the members A and C, which effectively meet at the same joint.

It should be noted that in many cases only a portion of the full set of d.o.f. may participate in the constraint equations. In Fig. 3.12, for example,

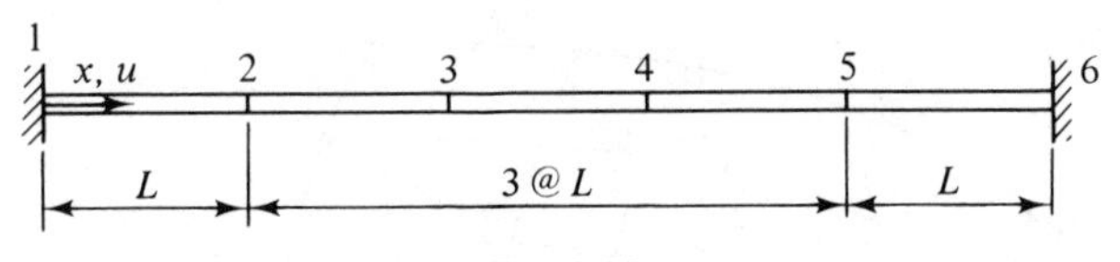

Fig. 3.12

if only u_2 and u_3 are coupled then u_4 and u_5 do not appear in the constraint equations. Suppose the full set of d.o.f. is now $\lfloor \mathbf{\Delta} \rfloor = \lfloor \lfloor \mathbf{\Delta}_e \rfloor \lfloor \mathbf{\Delta}_c \rfloor \lfloor \mathbf{\Delta}_g \rfloor \rfloor$ where the groups $\lfloor \mathbf{\Delta}_e \rfloor$ and $\lfloor \mathbf{\Delta}_c \rfloor$ participate in the constraint conditions as in Eq. (3.29) and the set $\lfloor \mathbf{\Delta}_g \rfloor$ is unaffected. The following transformation of d.o.f. can then be constructed:

$$\begin{Bmatrix} \mathbf{\Delta}_e \\ \mathbf{\Delta}_c \\ \hdashline \mathbf{\Delta}_g \end{Bmatrix} = \left[\begin{array}{c:c} \mathbf{\Gamma}_c & \mathbf{0} \\ \hdashline \mathbf{0} & \mathbf{I} \end{array} \right] \begin{Bmatrix} \mathbf{\Delta}_c \\ \hdashline \mathbf{\Delta}_g \end{Bmatrix} \tag{3.32}$$

where $[\mathbf{\Gamma}_c]$ is defined by Eq. (3.31). This transformation is applied directly to the global stiffness equations in the usual manner.

If there are relatively few d.o.f. affected by the constraints, it may be more efficient to incorporate the constraints into the global stiffness matrix by direct operations on the latter rather than by matrix transformation. The direct method would be similar to the approach described for special axes in Section 3.5.3.

We return now to the reduced substructure scheme where the concepts of condensation and constraint are incorporated into a single transformation matrix. Suppose that the points on the boundary are divided into two groups. As before, d.o.f. $\{\mathbf{\Delta}_c\}$ will be connected to adjacent substructures. The remaining boundary d.o.f. $\{\mathbf{\Delta}_e\}$ are *constrained* to displace in a form governed by the d.o.f. $\{\mathbf{\Delta}_c\}$; see Fig. 3.13. The d.o.f. $\{\mathbf{\Delta}_e\}$, for example, might be constrained to follow linear, quadratic, or higher-order edge displacement shapes. By partitioning of the substructure stiffness matrix, we have

$$\begin{Bmatrix} \mathbf{F}_d \\ \hdashline \mathbf{F}_e \\ \hdashline \mathbf{F}_c \end{Bmatrix} = \left[\begin{array}{c:c:c} \mathbf{k}_{dd} & \mathbf{k}_{de} & \mathbf{k}_{dc} \\ \hdashline \mathbf{k}_{ed} & \mathbf{k}_{ee} & \mathbf{k}_{ec} \\ \hdashline \mathbf{k}_{cd} & \mathbf{k}_{ce} & \mathbf{k}_{cc} \end{array} \right] \begin{Bmatrix} \mathbf{\Delta}_d \\ \hdashline \mathbf{\Delta}_e \\ \hdashline \mathbf{\Delta}_c \end{Bmatrix} \tag{3.33}$$

Also, we designate the constraining relationships between $\{\mathbf{\Delta}_e\}$ and $\{\mathbf{\Delta}_c\}$ as in Eq. (3.30), i.e., $\{\mathbf{\Delta}_e\} = [\mathbf{G}_{ec}]\{\mathbf{\Delta}_c\}$. Now, it is recalled from Section 2.8 that the desired transformation matrix is constructed by setting to zero the forces corresponding to the d.o.f. to be removed. Thus, solving the first partition for $\{\mathbf{\Delta}_d\}$ with $\{\mathbf{F}_d\} = 0$,

$$\{\mathbf{\Delta}_d\} = -[\mathbf{k}_{dd}]^{-1}[[\mathbf{k}_{de}]\{\mathbf{\Delta}_e\} + [\mathbf{k}_{dc}]\{\mathbf{\Delta}_c\}] \tag{3.34}$$

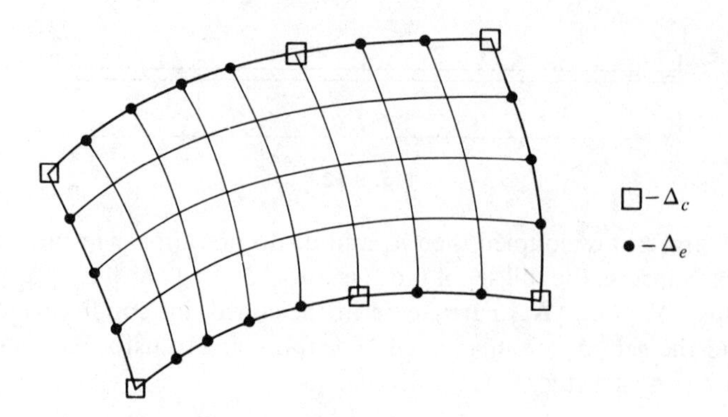

Fig. 3.13 Reduced-substructure analysis conditions.

and, by substitution of Eq. (3.30),

$$\{\mathbf{\Delta}_d\} = -[\mathbf{k}_{dd}]^{-1}[[\mathbf{k}_{de}][\mathbf{G}_{ec}] + [\mathbf{k}_{dc}]]\{\mathbf{\Delta}_c\} \tag{3.35}$$

Combining Eqs. (3.30), (3.35), and $\{\mathbf{\Delta}_c\} = [\mathbf{I}]\{\mathbf{\Delta}_c\}$, the desired transformation is

$$\begin{Bmatrix} \mathbf{\Delta}_d \\ \hline \mathbf{\Delta}_e \\ \hline \mathbf{\Delta}_c \end{Bmatrix} = \begin{bmatrix} -[\mathbf{k}_{dd}]^{-1}[[\mathbf{k}_{de}][\mathbf{G}_{ec}] + [\mathbf{k}_{dc}]] \\ \hline [\mathbf{G}_{ec}] \\ \hline [\mathbf{I}] \end{bmatrix} \{\mathbf{\Delta}_c\} = [\mathbf{\Gamma}]\{\mathbf{\Delta}_c\} \tag{3.36}$$

This may now be applied to Eq. (3.33), in the usual manner of a transformation, to yield a stiffness matrix referenced only to $\{\mathbf{\Delta}_c\}$ and a counterpart reduced load vector.

We can illustrate this scheme by reference to Fig. 3.14, where a structure is comprised of four planar rectangular elements, each of which has been formulated under the assumption of linear edge displacements (see Chapter 9 for details and pertinent element stiffness matrix). Suppose that a linear variation of displacement is imposed on the periphery of this structure. Then $u_2, u_3, u_4, u_5, v_2, v_3, v_4$, and v_5 comprise the d.o.f. $\{\mathbf{\Delta}_e\}$ and $u_6, u_7, u_8, u_9, v_6, v_7, v_8$, and v_9 comprise the d.o.f. $\{\mathbf{\Delta}_c\}$. The interior d.o.f. are u_1 and v_1 so

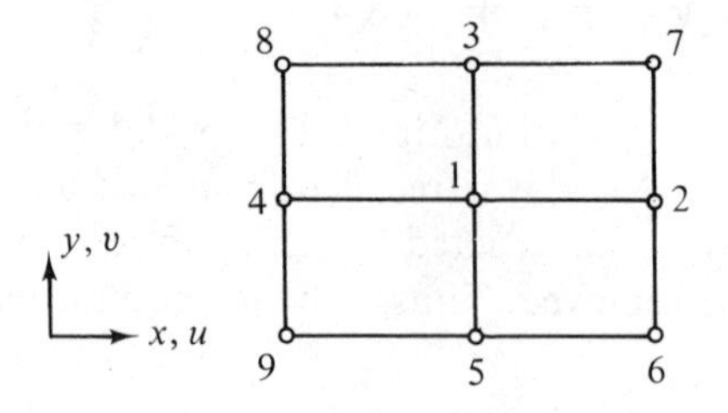

Fig. 3.14

that $\{\Delta_d\} = \lfloor u_1 \; v_1 \rfloor$. The matrix $[\mathbf{G}_{ec}]$ is constructed below for this problem in correspondence with Eq. (3.30). The other matrices required for the transformation matrix of Eq. (3.36), ($[\mathbf{k}_{dd}]$, $[\mathbf{k}_{de}]$, and $[\mathbf{k}_{dc}]$), are obtained from the stiffness matrix of the total structure. It is of interest to note that the stiffness matrix resulting from application of this transformation, which pertains only to the corner points, will be identical to that obtained by formulating the total region directly as a single element with linear edge displacements, using the element stiffness matrix given in Fig. 9.13.

$$\{\Delta_e\} = \begin{Bmatrix} u_2 \\ u_3 \\ u_4 \\ u_5 \\ v_2 \\ v_3 \\ v_4 \\ v_5 \end{Bmatrix} = \frac{1}{2}\begin{bmatrix} 1 & 1 & 0 & 0 & 0 & 0 & 0 & 0 \\ 0 & 1 & 1 & 0 & 0 & 0 & 0 & 0 \\ 0 & 0 & 1 & 1 & 0 & 0 & 0 & 0 \\ 1 & 0 & 0 & 1 & 0 & 0 & 0 & 0 \\ 0 & 0 & 0 & 0 & 1 & 1 & 0 & 0 \\ 0 & 0 & 0 & 0 & 0 & 1 & 1 & 0 \\ 0 & 0 & 0 & 0 & 0 & 0 & 1 & 1 \\ 0 & 0 & 0 & 0 & 1 & 0 & 0 & 1 \end{bmatrix} \begin{Bmatrix} u_6 \\ u_7 \\ u_8 \\ u_9 \\ v_6 \\ v_7 \\ v_8 \\ v_9 \end{Bmatrix} = [\mathbf{G}_{ec}]\{\Delta_c\}$$

3.5.3 Joint Coordinate Axes

It is occasionally necessary to describe a portion of the global equations in terms of *joint* coordinate axes, especially when support conditions are defined for directions different from the global axis directions or when performing shell analyses with flat plate elements.

A typical situation is described in Fig. 3.15(a). The displacement of point i is basically described by the global component displacements u_i and v_i. A simple specification of either or both of these components as zero will not, however, properly represent the constraint of the support in direction y''. The requirements can be met by expressing the behavior of point i in terms of coordinate displacements u_i'' and v_i'' after which the displacement v_i'' is stipulated as zero.

Figure 3.15(b) shows a shell structure that is idealized as a system of flat plate finite elements. Figures 3.15(c) and (d) examine the equilibrium of the moments at point i on Section A-A on a vectorial basis. Figure 3.15(c) shows that the net vectorial representations in global coordinates are substantial in both x and y axes. From Fig. 3.15(d), however, where we show the moment vectors $M_{x'}$ in the element axes and define joint axes $x'' - y''$ with x'' being the tangent direction to the real shell at point i, it is apparent that the net y'' component is extremely small in comparison with the x'' component. In fact, in the real structure, the y'' component is zero. This disproportionality of components in orthogonal directions produces severe

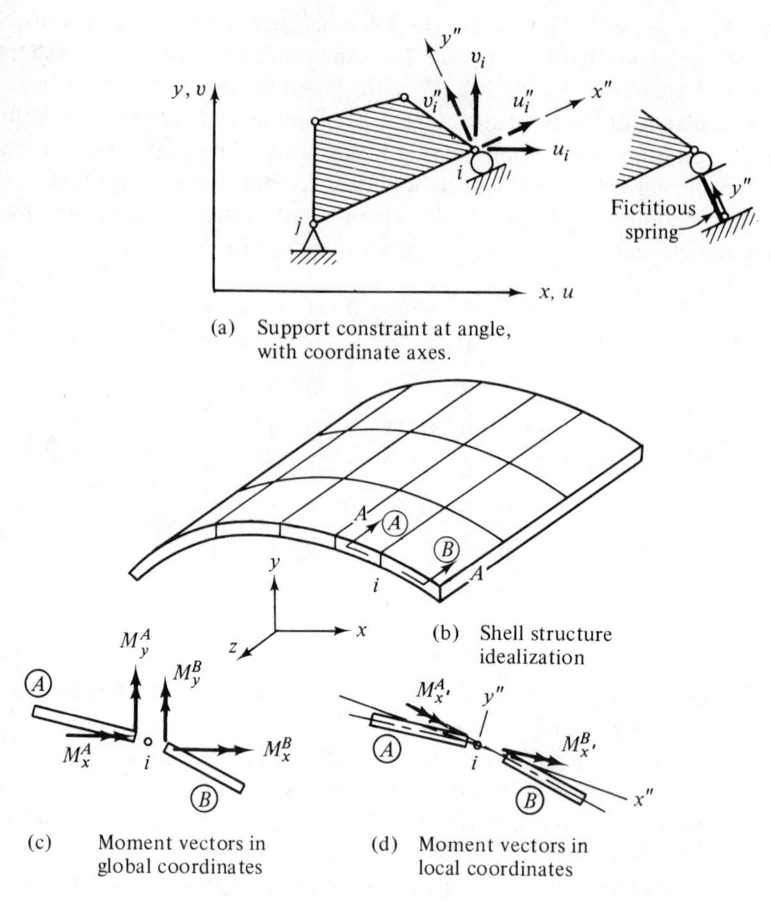

Fig. 3.15 Local coordinates in finte element analysis.

consequences in the solution of the global equations. One remedy is to establish $x'' - y''$ axes at each joint and to remove the small y'' component as if it were a supported d.o.f.

The manner in which the global stiffness matrix is modified to accommodate joint coordinates is now detailed, using as an example the problem of Fig. 3.15(a). On the basis of the coordinates of the points i and j, the direction cosines of the $x'' - y''$ axes with respect to the $x - y$ axes are $l_x = (x_i - x_j)/L$, $l_y = (y_i - y_j)/L$, with $L = \sqrt{(x_i - x_j)^2 + (y_i - y_j)^2}$.

Now, with use of these direction cosines, the u_i'', v_i'' displacement components can be expressed in terms of the u_i, v_i displacement components (see Section 2.7 for this type of transformation):

$$u_i'' = l_x u_i - l_y v_i \tag{3.37}$$

$$v_i'' = l_y u_i + l_x v_i \tag{3.38}$$

With respect to the effect on the global stiffness matrix this means that the column of the original global stiffness matrix multiplied by u_i should be multiplied by l_x and subtracted from the product of l_y and the column multiplied by v_i. The resulting column vector, which corresponds to u_i'', replaces the column vector corresponding to u_i. This operation is illustrated in Fig. 3.16. A similar operation, corresponding to Eq. (3.38), replaces v_i with v_i'', also as shown in Fig. 3.16.

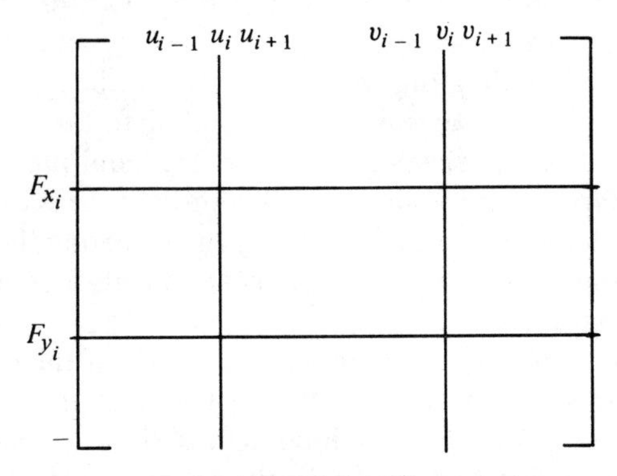

(a) Original global stiffness matrix

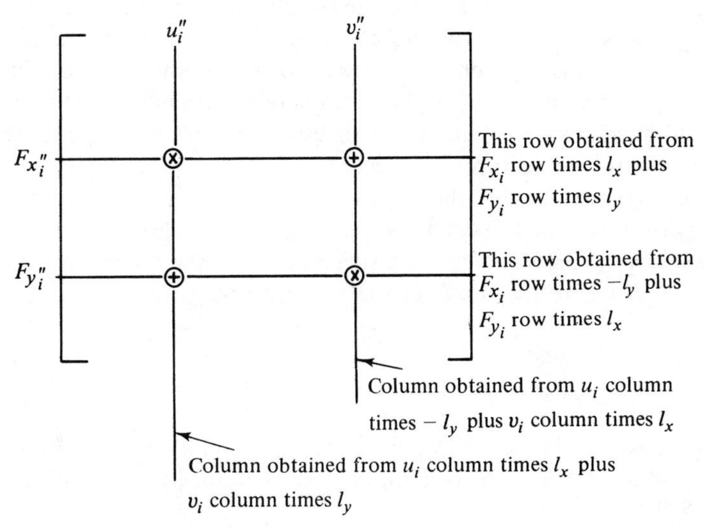

Note: Products of row and column multipliers at main diagonal locations (x) and at cross locations (+) gives squares and products of l_x and l_y.

(b) Modified global stiffness matrix

Fig. 3.16 Introduction of local coordinate displacements into global stiffness matrix.

The force equations (rows) of the global stiffness equations are governed by similar considerations. Thus, by the usual coordinate transformation,

$$F_{x_i''} = \quad l_x F_{x_i} + l_y F_{y_i} \tag{3.39}$$

$$F_{y_i''} = -l_y F_{x_i} + l_x F_{y_i} \tag{3.40}$$

In consequence of these equations, a new row in the F_{x_i} location is constructed by multiplying the row corresponding to F_{x_i} by l_x and the row corresponding to F_{y_i} by l_y and adding the two. A new row in the F_{y_i} location is also formed by performance of operations corresponding to Eq. (3.40).

The foregoing process must be repeated at every point requiring a transformation of axes. Boundary conditions in the new coordinate axes can then be applied to the revised global stiffness matrix ([**K**]). In evaluating the internal forces in the present approach, the calculated displacements first must be transformed back to global directions.

Alternative to the approach above is a procedure in which the element stiffness and stress matrices are formed directly in terms of the relevant local axes. This presents certain difficulties in input data assimilation because the support conditions are referenced to the joints rather than to the elements, but the procedure is nevertheless concise and efficient. Formation of the global stiffness matrix and all other operations then proceed in the usual manner. Still another procedure is to incorporate all the joint transformations above into a single global transformation matrix. This approach is algorithmically simple but will be effective only when a large share of the joints of the total structure are affected.

A scheme that avoids the algebraic complexity of all the approaches above is to introduce a special *boundary element*, as shown in the sketch to the right of Fig. 3.15(a). This is a special case of the procedure defined by Fig. 3.15(d), where the d.o.f. in the y'' direction is removed.

REFERENCES

3.1 Beaufait, F., W. H. Rowan, P. G. Hoadley, and R. M. Hackett, *Computer Methods of Structural Analysis*, Prentice-Hall, Inc., Englewood Cliffs, N.J., 1970.

3.2 Meek, J. L., *Matrix Structural Analysis*, McGraw-Hill Book Co., New York, N.Y., 1971.

3.3 Wang, C. K., *Matrix Methods of Structural Analysis*, 2nd ed., International Textbook Co., Scranton, Pa., 1970.

3.4 Willems, N. and W. Lucas, *Matrix Analysis for Structural Engineers*, Prentice-Hall, Inc., Englewood Cliffs, N.J., 1968.

3.5 Rose, D. J. and R. A. Willoughby (eds.), *Sparse Matrices and Their Applications*, Plenum Press, New York, N.Y., 1972.

3.6 FOX, R. and E. STANTON, "Developments in Structural Analysis by Direct Energy Minimization," *AIAA J.*, **6**, No. 6, June, 1968, pp. 1036–1042.

3.7 NICKELL, R. E. and E. L. WILSON, "Application of the Finite Element Method to Heat Conduction Analysis," *Nuc. Eng. Design*, **4**, 1966, pp. 276–286.

3.8 GALLAGHER, R. H., "Computational Methods in Nuclear Reactor Structural Design for High-Temperature Applications." Chapter 7, *Thermal Analysis*, Report ORNL-4756, July, 1972.

3.9 MARCAL, P. V., "Finite Element Analysis with Material Nonlinearities—Theory and Practice" in *Finite Element Method in Civil Engineering*, J. McCutcheon, M. S. Mirza, and A. Mufti, (eds.), McGill U., Montreal, Quebec, June, 1972, pp. 35–70.

3.10 GALLAGHER, R. H., "Geometrically Nonlinear Finite Element Analysis," in *Finite Element Method in Civil Engineering*, J. McCutcheon, M. S. Mirza, and A. Mufti (eds.), McGill U., Montreal, Quebec, June, 1972, pp. 3–34.

3.11 KAMEL, H., D. LIU, M. MCCABE, and V. PHILLIPOPOULOS, "Some Developments in the Analysis of Complex Ship Structures," in *Advances in Computational Methods in Structural Mechanics and Design*, J. T. Oden, et al. (eds.), U. of Alabama Press, University, Ala., 1972, pp. 703–726.

3.12 WALTON, W. C. and E. C. STEEVES, "A New Matrix Theorem and its Application for Establishing Independent Coordinates for Complex Dynamical Systems with Constraints," NASA TR R-326, Oct., 1969.

PROBLEMS

3.1 Form the stiffness matrix for the four-segment beam shown in Fig. P3.1, using the direct stiffness method.

$$EI = 20 \times 10^4 \text{ K in.}^2$$

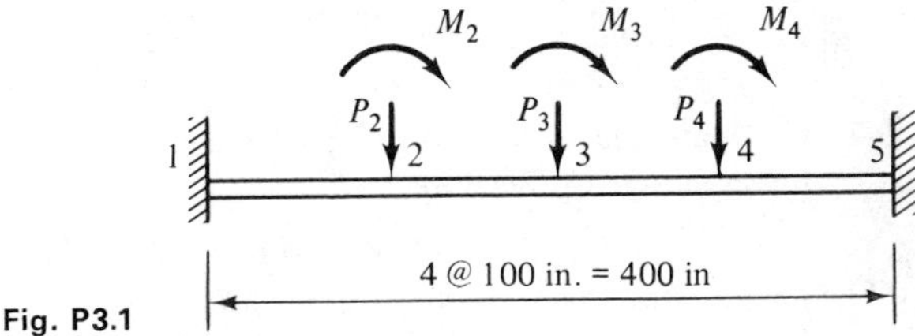

Fig. P3.1

3.2 Form the stiffness matrix for Prob. 3.1 by use of the congruent transformation approach.

3.3 Condense the stiffness matrix of Prob. 3.1 to order 3 × 3 by elimination of the angular displacements.

3.4 Apply the constraint $w_3 = w_4$ to the stiffness equations of Prob. 3.3 and construct the reduced stiffness matrix. Compare the solutions for w_4, with and without this constraint, for $P_4 = 4{,}800$ lb. (all other forces zero).

3.5 Solve for the member stresses and joint displacements of the structure given in Fig. P3.5, using the direct stiffness procedure. Present the computations in the sequence defined in Section 3.2. The stiffness matrix for the triangle in plane stress is given in Fig. 5.4.

$$E = 10^7 \text{ p.s.i.}, \qquad \mu = 0.3$$
$$A_{1-4} = A_{2-4} = A_{3-4} = 1.0 \text{ in.}^2$$

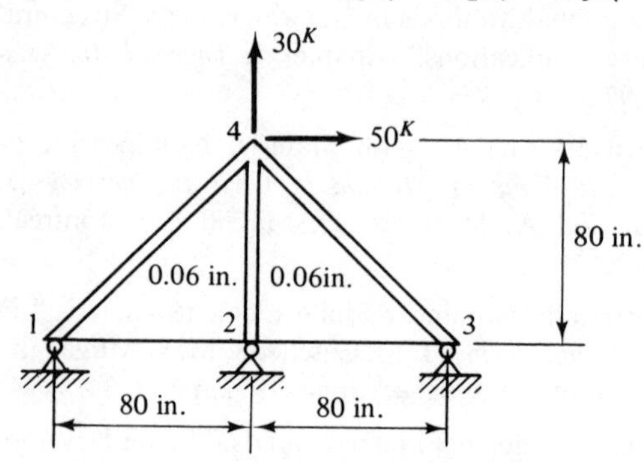

Fig. P3.5

3.6 Analyze the structure shown in Fig. P3.6 in the manner stipulated in Prob. 3.5. (Divide the rectangular elements into two triangular elements as indicated.)

$$E = 10^7 \text{ p.s.i.}, \qquad \mu = 0.3$$

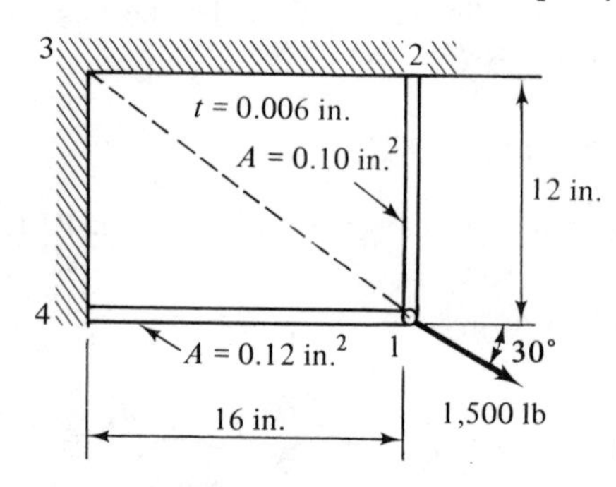

Fig. P3.6

3.7 Calculate the x-direction displacement at point A of Fig. P3.7 by use of direct stiffness procedures.

$$E = 10^7 \text{ p.s.i.}, \qquad \mu = 0.3$$

Divide the sheet material into four rectangular elements as indicated. The rectangular plate element stiffness matrix is given in Fig. 9.13. Invoke symmetry about the x axis.

3.8 The stiffness matrix for the structure shown in Fig. P3.8, in terms of x-y coordinates, is given. Establish the stiffness matrix in terms of P_{x_1}, P_{y_1}, and $P_{x_2''}$.

$$\begin{Bmatrix} P_{x_1} \\ P_{x_2} \\ P_{y_1} \\ P_{y_2} \end{Bmatrix} = 10^4 \begin{bmatrix} 10.0 & & \text{Sym.} & \\ -2.5 & 4.5 & & \\ 1.83 & 2.5 & 5.0 & \\ 2.5 & -2.5 & -2.5 & 2.5 \end{bmatrix} \begin{Bmatrix} u_1 \\ u_2 \\ v_1 \\ v_2 \end{Bmatrix}$$

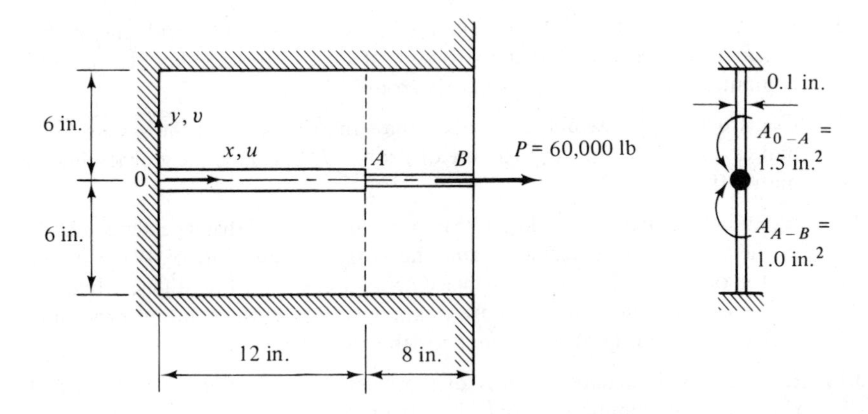

Fig. P3.7

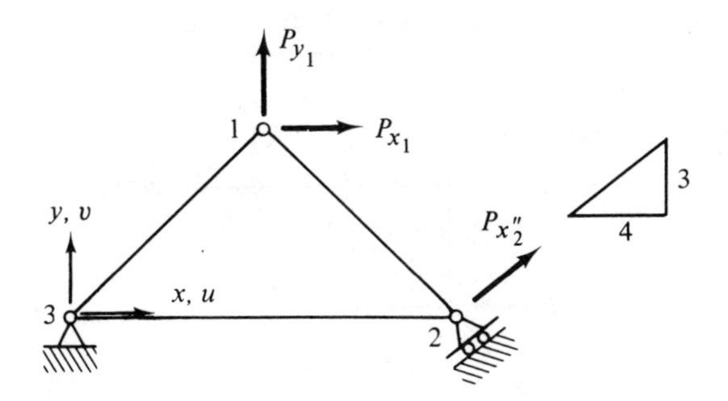

Fig. P3.8

3.9 Analyze the structure shown in Fig. P3.9 in the manner stipulated in Prob. 3.5. The trapezoidal plate is to be divided into a rectangular and a triangular plate. The relevant stiffness matrices are given in Fig. 5.4 and 9.13.

$$E = 20 \times 10^6 \text{ p.s.i.}, \qquad \mu = 0.2$$

Fig. P3.9

3.10 Construct the unassembled global stiffness matrix $\lceil \mathbf{k}^e \rfloor$ and the global kinematics matrix $[\mathcal{A}]$ for the structure of Prob. 3.5. Using the computer, calculate the global stiffness matrix from $[\mathcal{A}]^T \lceil \mathbf{k} \rfloor [\mathcal{A}]$.

3.11 Construct the unassembled global stiffness matrix $\lceil \mathbf{k}^e \rfloor$ and the global kinematics matrix $[\mathcal{A}]$ for the structure of Prob. 3.6. Calculate, by hand or computer, the global stiffness matrix from $[\mathcal{A}]^T \lceil \mathbf{k} \rfloor [\mathcal{A}]$.

3.12 Construct the unassembled global stiffness matrix $\lceil \mathbf{k}^e \rfloor$ and the system kinematics matrix $[\mathcal{A}]$ for the structure of Prob. 3.7. Calculate the global stiffness matrix $[\mathbf{k}]$.

3.13 Extend the statement of Eq. (3.5) to include initial displacements $\{\mathbf{\Delta}^{init.}\}$ in the vector of displacements and the corresponding forces $\{\mathbf{P}^{init.}\}$ in the vector of forces. The initial displacements are known quantities, while the forces corresponding to them are unknowns. Revise the solution procedure defined after Eq. (3.5) to account for this condition.

3.14 Examine the kinematic stability of the truss shown in Fig. P3.14 by use of Gauss-Jordan elimination applied to the three equations of overall equilibrium of the structure; i.e., ignore joints 5–9.

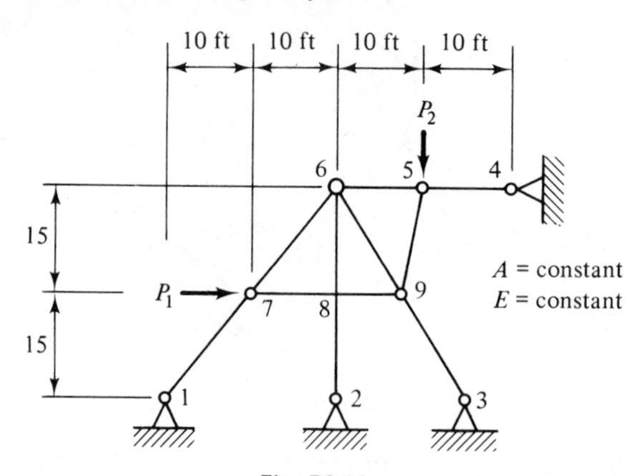

Fig. P3.14

3.15 Establish the matrix form of the equations of equilibrium for the truss shown in Fig. P3.15 and apply Gauss-Jordan elimination to identify redundant forces (all areas equal).

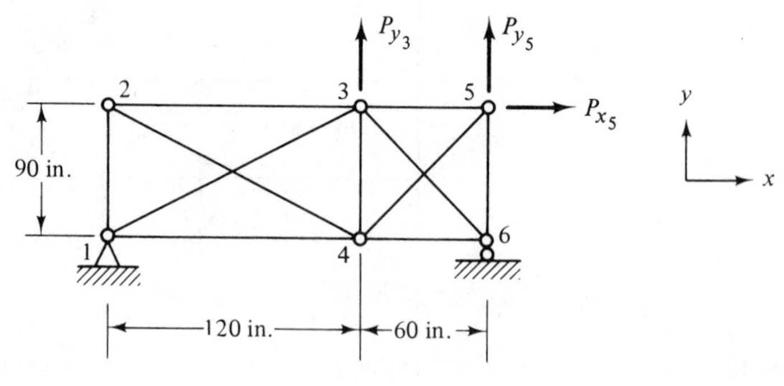

Fig. P3.15

3.16 Treating members 2-3 and 3-4 of the structure of Prob. 3.1 as a substructure, remove w_3 and θ_3 as internal d.o.f. and form the structural stiffness matrix with reference to w_2, θ_2, w_4, θ_4.

3.17 Members 1-2 and 2-3 shown in Fig. P3.17 only possess a torsional stiffness $k = GJ/L$. Calculate the angles of twist θ_x and θ_y due to the applied twisting moments M_{xx} and M_{yy}.

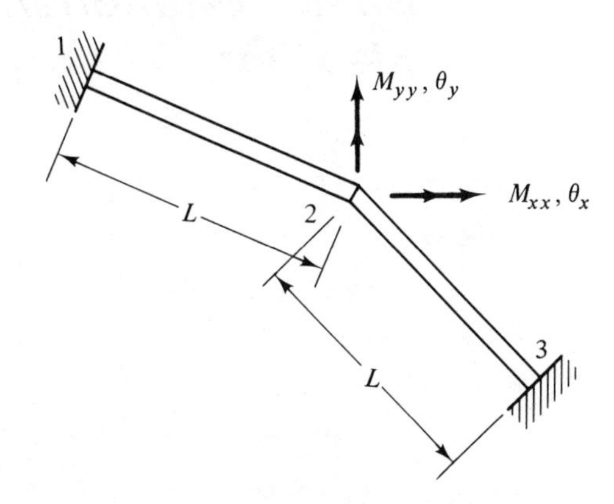

Fig. P3.17

3.18 Assume that the constraint conditions of a problem with n d.o.f. are of the form $[\mathbf{G}]\{\mathbf{\Delta}\} = \{\mathbf{s}\}$, where $[\mathbf{G}]$ is an $r \times n$ matrix. Divide the d.o.f. into two sets, $\{\mathbf{\Delta}_e\}_{r\times 1}$ and $\{\mathbf{\Delta}_c\}_{(n-r)\times 1}$, and form a transformation in the manner of Eq. (3.31) (but taking into account the vector $\{\mathbf{s}\}$). Develop the reduced form of the stiffness equations (i.e., the stiffness equations in terms of $\{\mathbf{\Delta}_c\}$).

4

Basic Relationships of Elasticity Theory

We develop in the following the basic differential relationships of linear elastic behavior. The detailed derivation of these expressions is performed for the two-dimensional case in cartesian coordinates since this case applies to most of the work covered in the chapters that formulate the fundamental theory of finite element analysis. The generalization of three-dimensional states of behavior is indicated but extensions to more special states and coordinate systems is reserved for those chapters where the related types of elements are examined.

It should also be noted that these developments are in the simplest of terms and with a minimum of rigor. The approach is at a level found in earlier texts on elasticity [4.1, 4.2] or more recent presentations characterized as *advanced strength of materials.* [4.3, 4.4] To obtain a more rigorous development, which includes representation of nonlinear phenomena and more general forms of behavior, the reader is advised to consult Refs. 4.5–4.7.

Three systems of relationships comprise elasticity theory: (1) the differential equations of equilibrium, (2) the strain-displacement and compatibility differential equations, and (3) the material constitutive laws. In any finite body, systems 1 and 2 are supplemented by the *boundary conditions.* Each of these is derived in the following. Then, their combination to form *governing* differential equations is outlined. Finally, some comments are given regarding the concept of uniqueness in the solution of elasticity problems and its significance to finite element analysis.

4.1 Equilibrium Differential Equations

For simplicity, we first study the equilibrium of a planar differential element, Fig. 4.1, subjected to normal stresses σ_x and σ_y, the shear stress τ_{xy}, and body forces (i.e., force per unit volume) X and Y. Body forces may arise from numerous sources but in this text they are used mainly to represent the inertia force effect in dynamics. The stresses are shown as constant with respect to the normal directions; i.e., although σ_x is shown to vary with respect to x, it is taken as constant across the face width dy. A detailed development which includes this variation will show that it produces terms which are of higher order than those taken into account in conventional linear elasticity theory.

For x-direction equilibrium we write (for a unit dimension in the direction normal to the $x - y$ plane)

$$\Sigma F_x = 0 = \left(\sigma_x + \frac{\partial \sigma_x}{\partial x} dx\right) dy - \sigma_x \, dy + X \, dx \, dy + \left(\tau_{yx} + \frac{\partial \tau_{yx}}{\partial y} dy\right) dx - \tau_{yx} \, dx \tag{4.1}$$

and, after clearing of terms,

$$\frac{\partial \sigma_x}{\partial x} + \frac{\partial \tau_{yx}}{\partial y} + X = 0 \tag{4.2a}$$

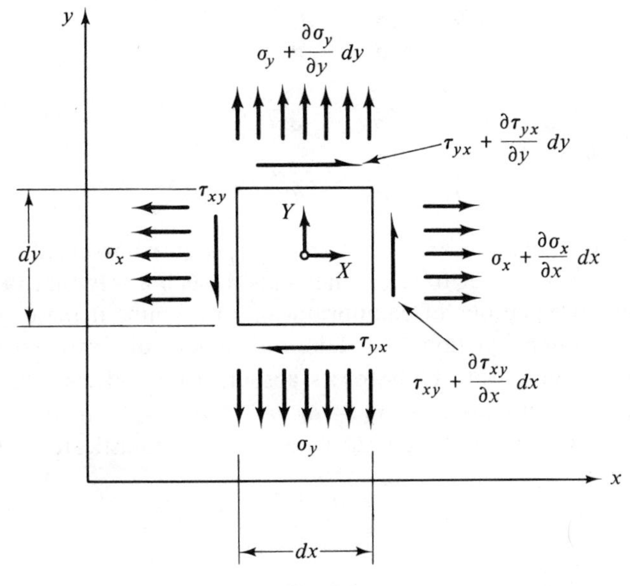

Fig. 4.1

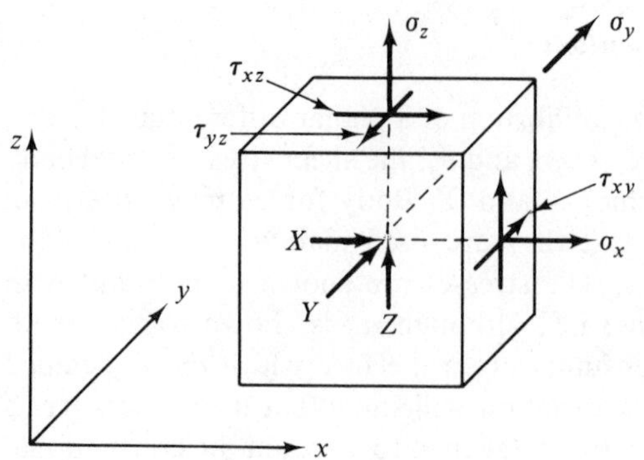

Fig. 4.2

A similar development in the y direction gives

$$\frac{\partial \sigma_y}{\partial y} + \frac{\partial \tau_{xy}}{\partial x} + Y = 0 \tag{4.2b}$$

In a plane, of course, three equilibrium conditions must be met, the third being the equilibrium of moments about an axis normal to the plane. Imposition of this condition simply proves that $\tau_{xy} = \tau_{yx}$. Thus Eqs. (4.2a) and (4.2b) comprise the desired equilibrium equations of plane elasticity. It is an easy matter to generalize these expressions to the three-dimensional case (with body forces X, Y, and Z). (See Fig. 4.2 for identification of stress and body force components.)

$$\begin{aligned}
\frac{\partial \sigma_x}{\partial x} + \frac{\partial \tau_{xy}}{\partial y} + \frac{\partial \tau_{xz}}{\partial z} + X = 0 \\
\frac{\partial \sigma_y}{\partial y} + \frac{\partial \tau_{xy}}{\partial x} + \frac{\partial \tau_{yz}}{\partial z} + Y = 0 \\
\frac{\partial \sigma_z}{\partial z} + \frac{\partial \tau_{xz}}{\partial x} + \frac{\partial \tau_{yz}}{\partial y} + Z = 0
\end{aligned} \tag{4.3}$$

Element formulations are in certain cases based on assumed stress fields and it is necessary either to select the fields in such a way that they satisfy the differential equations of equilibrium or to examine if these conditions are met by chosen functions that take no account of this circumstance *a priori*. For example, with body forces absent, if one selects a plane stress field in the form of constants, $\sigma_x = a_1$, $\sigma_y = a_2$, $\tau_{xy} = a_3$ then it is clear that Eqs. (4.2a) and (4.2b) are satisfied. A more sophisticated field is

$$\begin{aligned}
\sigma_x &= a_1 + a_2 y \\
\sigma_y &= a_3 + a_4 x \\
\tau_{xy} &= a_5
\end{aligned}$$

where $a_1, \ldots, a_5$ are constants, and this too satisfies Eqs. (4.2a) and (4.2b). On the other hand, the field

$$\sigma_x = a_1 + a_2 x$$
$$\sigma_y = a_3 + a_4 y$$
$$\tau_{xy} = a_5$$

does not satisfy the differential equations of equilibrium unless $a_2 = a_4 = 0$.

A convenient mode of selection of stress fields that satisfy the equilibrium differential equations is by use of *stress functions*. Stress functions are parameters which, when differentiated in accordance with certain rules, give stress components which automatically satisfy the differential equations of equilibrium. Plane stress is characterized by a single such quantity, Φ, termed the *Airy stress function*, defined by

$$\sigma_x = \frac{\partial^2 \Phi}{\partial y^2}, \qquad \sigma_y = \frac{\partial^2 \Phi}{\partial x^2}, \qquad \tau_{xy} = -\frac{\partial^2 \Phi}{\partial x\, \partial y} \tag{4.4}$$

Clearly, these stresses yield automatic satisfaction of Eq. (4.2) in the absence of body forces ($X = Y = 0$) for all choices of Φ. Consider, for example, $\Phi = a_6 + a_5 x + a_4 y + \frac{1}{2} a_2 x^2 + \frac{1}{2} a_1 y^2 - a_3 xy$. Then, $\sigma_x = a_1, \sigma_y = a_2$, $\tau_{xy} = a_3$, as in the example above.

Stress functions can also be constructed for three-dimensional elasticity, plate bending, and other specific forms of elastic representation. The plate bending parameters, termed *Southwell stress functions*, are especially useful in finite element analysis so that attention is given to them in Chapter 12. A difficulty with the stress function representation is its lack of physical meaning, which renders difficult the treatment of boundary conditions and other key aspects of any real problem.

4.2 Boundary Conditions on Stress

The differential equations of equilibrium apply to every point within the region of a body. Account must also be taken of equilibrium conditions on the boundary of the body (the *static boundary conditions*). Consider the boundary of a plane region, Fig. 4.3, subjected to *prescribed surface tractions* $\bar{T}_x$ and $\bar{T}_y$.† The usual manner of specification of these surface forces is such that they are expressed as x- and y-direction forces per unit area of the surface that is itself at an angle with the x and y directions. Figure 4.3 shows a length ds of the surface of the region in plane stress (since we are dealing with a plate of unit thickness, the length ds is numerically equal to the surface area).

†In this text all prescribed quantities (boundary tractions, displacements) are denoted by *barred symbols*.

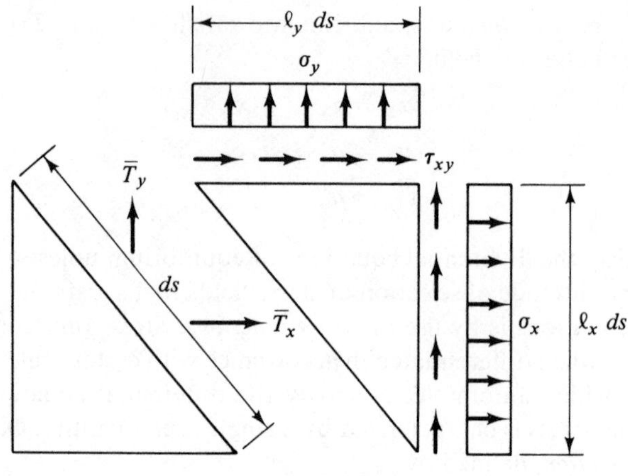

Fig. 4.3

The symbols l_x and l_y denote the direction cosines with respect to the x and y axes of the normal to the surface. For x-direction equilibrium we have

$$\bar{T}_x\, ds = \sigma_x(l_x\, ds) + \tau_{xy}(l_y\, ds)$$

or

$$\bar{T}_x = l_x\sigma_x + l_y\tau_{xy} \tag{4.5a}$$

and, for the y direction,

$$\bar{T}_y = l_y\sigma_y + l_x\tau_{xy} \tag{4.5b}$$

In finite element analysis one is concerned with the conditions of equilibrium not only in the interior of the structure and on the external surfaces but also on the interfaces between elements. A state of stress exists on the edge of each of the interfacing elements and Eq. (4.5) apply to each of these faces. Figure 4.4(a) shows two adjacent elements A and B whose interface is aligned with the global y axis. No applied loads are present at the interface. The elements are separated in Fig. 4.4(b) in order to clarify the interface equilibrium conditions. Due to the orientation of the interface, $l_y = 0$ and

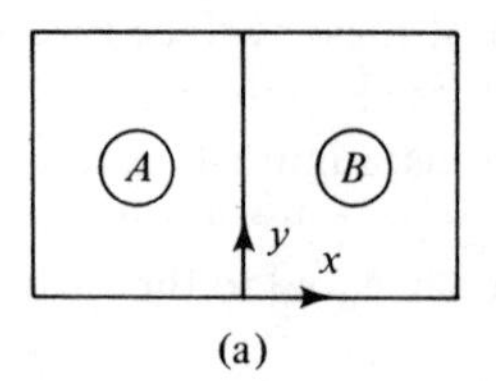

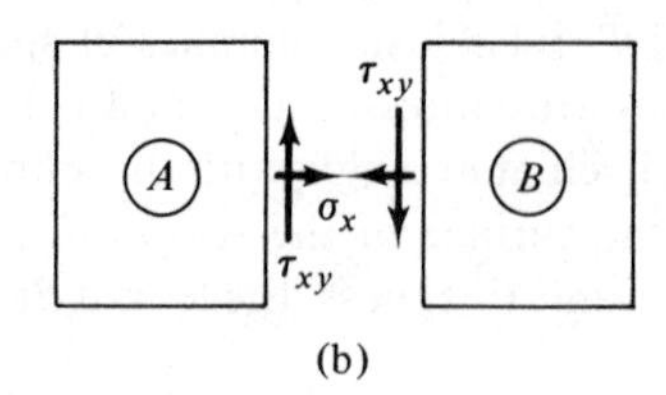

Fig. 4.4

$l_x = 1$ and Eqs. (4.5) reduce to

$$\bar{T}_x = \sigma_x, \qquad \bar{T}_y = \tau_{xy}$$

Consequently, with the element interface aligned with the y axis, for equilibrium to prevail it is only necessary that the normal stress σ_x and the shear stress τ_{xy} be continuous across the element boundary. It is permissible for a normal stress σ_y, if present, to be discontinuous in its variation in the x direction. When the orientation of the face is at an angle with the x and y axes, the components of the tractions normal and tangential to an unloaded interface ($\bar{T}_n$ and $\bar{T}_s$) must be continuous across the interface. On each side of the interface the two tractions are a function of the three stress components. Thus, the stresses in the coordinate directions may differ from one element to the next and equilibrium across the interface may still be preserved, depending on the values these stress states give to the traction components.

We again confront the necessity of designating a symbol which refers collectively to the components of a variable and which is distinct from the symbol for a vector that lists the values of these quantities at one or more specified points. A symbol in the first category is the stress tensor $\boldsymbol{\sigma}$ that describes the components $\sigma_x \ldots \tau_{zx}$; this is designated by boldface without braces. When it is necessary to array the components of $\boldsymbol{\sigma}$, we list them in a row or column vector in the order $\lfloor \sigma_x\ \sigma_y\ \sigma_z\ \tau_{xy}\ \tau_{yz}\ \tau_{zx} \rfloor$. Similarly, the field of prescribed surface tractions is collectively designated by $\bar{\mathbf{T}}$ and refers to the components $\lfloor \bar{T}_x\ \bar{T}_y\ \bar{T}_z \rfloor$, and the collective designation of body forces is $\mathbf{X} = \lfloor X\ Y Z \rfloor$. A symbol in the second category is the conventional row or column matrix $\lfloor\ \ \rfloor$ or $\{\ \ \}$. If, for example, the stresses in a plane stress situation are to be defined at two particular points, say points 1 and 2, then the designation is

$$\{\sigma\}^T = \lfloor \sigma_{x_1}\ \sigma_{y_1}\ \tau_{xy_1}\ \sigma_{x_2}\ \sigma_{y_2}\ \tau_{xy_2} \rfloor$$

4.3 Strain-Displacement and Compatibility Equations

The *kinematic* differential relationships between strain and displacement are of pivotal importance in the displacement-based approaches to finite element formulation. In contrast, the differential equations of equilibrium (*static* conditions), given in Section 4.1, play no explicit role in such approaches.

To develop the strain-displacement relationships of the differential element shown in Fig. 4.5, consider a small displacement from the undeformed state $ABCD$ to the deformed state $A'B'C'D'$. After straining, we have [for

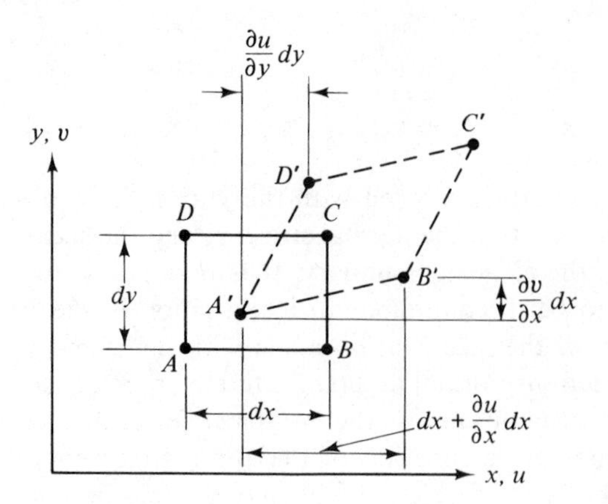

Fig. 4.5

small (linear) *strains*]

$$(A'B')^2 = \left(dx + \frac{\partial u}{\partial x}dx\right)^2 + \left(\frac{\partial v}{\partial x}dx\right)^2 \tag{4.6a}$$

By definition, the *engineering* strain (i.e., the change in length with respect to the original length) ϵ_x is

$$\epsilon_x = \frac{A'B' - AB}{AB}$$

or, with $AB = dx$,

$$A'B' = (1 + \epsilon_x)\,dx \tag{4.6b}$$

By squaring Eq. (4.6b), equating it to Eq. (4.6a), and dividing through by $(dx)^2$, we have

$$2\epsilon_x + \epsilon_x^2 = 2\frac{\partial u}{\partial x} + \left(\frac{\partial u}{\partial x}\right)^2 + \left(\frac{\partial v}{\partial x}\right)^2$$

Now, discarding the higher-order terms, consistent with our assumption of small strains, we have

$$\epsilon_x = \frac{\partial u}{\partial x} \tag{4.7a}$$

Similarly, for the y-direction strain,

$$\epsilon_y = \frac{\partial v}{\partial y} \tag{4.7b}$$

Shear strain (γ_{xy}) is defined as the change in value of an angle that is originally a right angle in the unstrained state. This mode of deformation is also shown in Fig. 4.5, which makes clear that the change in angle with respect to the x direction, due to the distortion of the line AB to $A'B'$, is

$$\frac{1}{dx}\left(\frac{\partial v}{\partial x}dx\right) = \frac{\partial v}{\partial x}$$

and likewise for the change in angle with respect to the y direction of the line AD as it deforms. Thus

$$\gamma_{xy} = \frac{\partial u}{\partial y} + \frac{\partial v}{\partial x} \tag{4.7c}$$

Equations (4.7a, b, c) are the *strain-displacement relationships* for planar behavior. For three-dimensional situations with w denoting the z-direction component of displacement, one need only add

$$\epsilon_z = \frac{\partial w}{\partial z} \tag{4.7d}$$

$$\gamma_{xz} = \frac{\partial w}{\partial x} + \frac{\partial u}{\partial z} \tag{4.7e}$$

$$\gamma_{yz} = \frac{\partial w}{\partial y} + \frac{\partial v}{\partial z} \tag{4.7f}$$

One factor that should be kept in mind in finite element analysis with reference to the strain-displacement equations is their association with *rigid body motions*. Strains exclude such motions but they appear in the displacements. Thus, in the determination of strains by differentiation of the displacements, we are eliminating the rigid body motions from the resulting expressions. For example, for an axial member, the horizontal displacement of a point may be described by the expression, Fig. 4.6,

$$u = a_1 + a_2 x$$

for which

$$\epsilon_x = \frac{du}{dx} = a_2$$

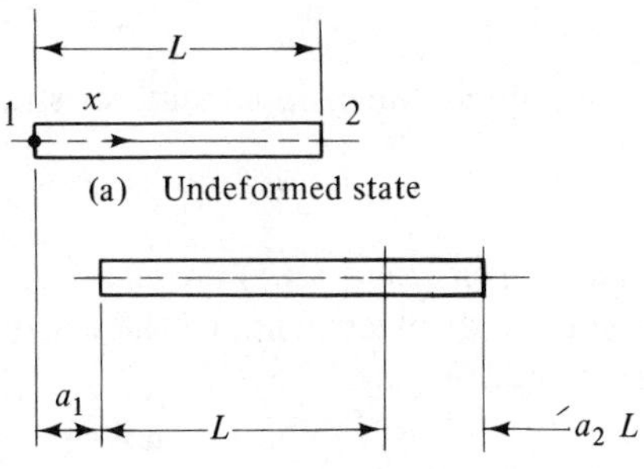

(a) Undeformed state

Fig. 4.6 (b) Displaced, deformed state

Consequently, the term a_1, which was removed in the differentiation operation, is the rigid body motion. This tells us that when an element is formulated on the basis of assumed displacements, the element strain state is described by independent parameters that are fewer in number than independent modes of the displacement field to the extent of the number of rigid body modes of the element.

Another factor intimately associated with and converse to the above enters into the basic relationships concerned with strain. In the planar case, three equations, [Eqs. (4.7a, b, c)] defining the three strain parameters are referenced to only two displacement components. In the three-dimensional case there are six strain parameters and three displacement parameters. Hence, in neither case will the equations possess a unique solution if the strains are arbitrarily prescribed. The necessary equations derive from the condition of *compatibility*, which imposes the requirement that the displacement components be single-valued continuous functions.

The compatibility condition is constructed here, in the most elementary way, through a series of differentiations. For planar elasticity we first differentiate γ_{xy} with respect to both x and y:

$$\frac{\partial^2 \gamma_{xy}}{\partial x\,\partial y} = \frac{\partial^2}{\partial x\,\partial y}\frac{\partial u}{\partial y} + \frac{\partial^2}{\partial x\,\partial y}\frac{\partial v}{\partial x} = \frac{\partial^2 \epsilon_x}{\partial y^2} + \frac{\partial^2 \epsilon_y}{\partial x^2} \tag{4.8}$$

where the second equation on the r.h.s. is obtained by noting that the condition of single-valued continuity of the displacements gives $\partial^2/\partial x\,\partial y = \partial^2/\partial y\,\partial x$. The extension of this condition to three dimensions gives a set of six equations.

As in the designation of stress quantities, it is necessary to distinguish between displacement fields *within the volume* of a body and the displacement fields of *boundary* (*surface*) points. The displacement field within the body is designated by $\mathbf{\Delta}$ and refers collectively to the coordinate displacements u, v, w. The boundary displacement field is symbolized by $\mathbf{u}$ and refers collectively to values u, v, and w of boundary points. Thus

$$\mathbf{\Delta} = \lfloor u\ v\ w \rfloor^{\mathrm{T}} \quad \text{(within body)}$$
$$\mathbf{u} = \lfloor u\ v\ w \rfloor^{\mathrm{T}} \quad \text{(on surface)}$$

Also, for a three-dimensional state of strain,

$$\boldsymbol{\epsilon} = \lfloor \epsilon_x\ \epsilon_y\ \epsilon_z\ \gamma_{xy}\ \gamma_{yz}\ \gamma_{zx} \rfloor^{\mathrm{T}}$$

Displacement (*kinematic*) *boundary conditions* simply demand adherence of the surface displacements of the elastic structure ($\mathbf{u}$) with the prescribed displacements ($\bar{\mathbf{u}}$); i.e.,

$$\bar{\mathbf{u}} - \mathbf{u} = 0 \tag{4.9}$$

4.4 Material Constitutive Relationships

The usual mode of introduction of the material *constitutive relationships*, which in the present discussion refer exclusively to material mechanical properties, is to postulate a full set of coefficients that connect each stress component to all strain components. Then, by consideration of symmetries and directional circumstances, these are reduced to forms appropriate to special cases, e.g., orthotropic plane strain. Here, in order to emphasize the physical and data measurement aspects of these properties, we advance in the opposite direction, from the simplest to more general forms.

The simplest mechanical properties derive from the test of a uniaxial specimen. The linear portion of the stress-strain diagram is represented algebraically by Hooke's law:

$$\sigma_x = E\epsilon_x \quad \text{or} \quad \epsilon_x = \frac{\sigma_x}{E}$$

This gives the strain due to stress. If a mechanism also exists to produce strain without impostition of stress, i.e., an initial strain $\epsilon_x^{\text{init.}}$, we have

$$\epsilon_x = \frac{\sigma_x}{E} + \epsilon_x^{\text{init.}} \quad \text{or} \quad \sigma_x = E\epsilon_x - E\epsilon_x^{\text{init.}} \tag{4.10}$$

To develop a two-dimensional case we first study an *isotropic* material and examine the response to imposed stress. An isotropic material is one whose stress-strain law is unaltered under orthogonal transformations of coordinates. Figure 4.7(a) shows the imposition of stress (σ_x) in the x direction and due to this a strain is produced in both the x and y directions. In the x direction the strain is simply σ_x/E. The specimen contracts in the y direction due to *Poisson's ratio* (μ) and here the strain is $-\mu\sigma_x/E$. Similarly, in Fig. 4.7(b) the imposition of σ_y produces x- and y-direction strains of $-\mu\sigma_y/E$ and σ_y/E, respectively. The response in normal strain is not influenced here by shearing deformation, Fig. 4.7(c), for which the stress-strain relationship

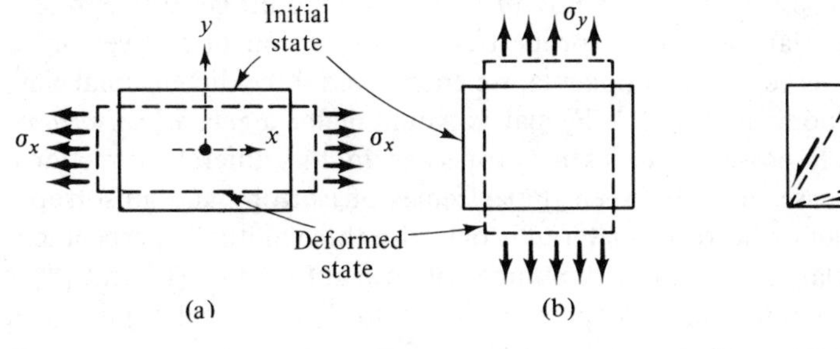

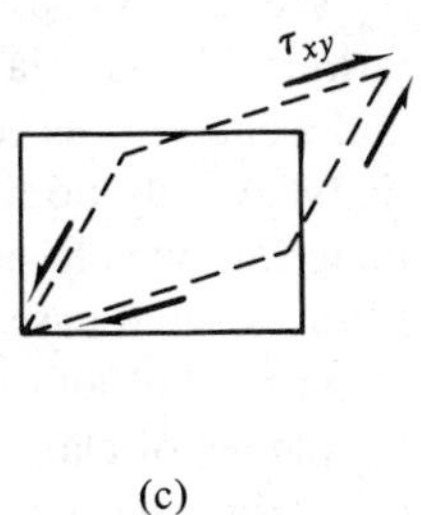

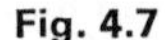

Fig. 4.7

is simply

$$\gamma_{xy} = \frac{2(1+\mu)}{E}\tau_{xy}$$

Superimposing the effects above and gathering them in matrix form (with $\boldsymbol{\sigma} = \lfloor \sigma_x \; \sigma_y \; \tau_{xy} \rfloor^T$, $\boldsymbol{\epsilon} = \lfloor \epsilon_x \; \epsilon_y \; \tau_{xy} \rfloor^T$) gives

$$\boldsymbol{\epsilon} = [\mathbf{E}]^{-1}\boldsymbol{\sigma} \tag{4.11}$$

where

$$[\mathbf{E}] = \frac{E}{(1-\mu^2)}\begin{bmatrix} 1 & \mu & 0 \\ \mu & 1 & 0 \\ 0 & 0 & \dfrac{(1-\mu)}{2} \end{bmatrix} \tag{4.12}$$

$$[\mathbf{E}]^{-1} = \frac{1}{E}\begin{bmatrix} 1 & -\mu & 0 \\ -\mu & 1 & 0 \\ 0 & 0 & 2(1+\mu) \end{bmatrix} \tag{4.13}$$

$[\mathbf{E}]$ is termed the *material stiffness matrix*, while $[\mathbf{E}]^{-1}$ is known as the *material flexibility matrix*. We immediately generalize the above to include initial strains $\epsilon^{\text{init.}} = \lfloor \epsilon_x^{\text{init.}} \; \epsilon_y^{\text{init.}} \; \tau_{xy}^{\text{init.}} \rfloor^T$, by analogy to Eq. (4.10), so that

$$\boldsymbol{\epsilon} = [\mathbf{E}]^{-1}\boldsymbol{\sigma} + \boldsymbol{\epsilon}^{\text{init.}} \tag{4.14}$$

Initial strains are of greatest interest in thermal expansion problems where $\epsilon_x^{\text{init.}} = \epsilon_y^{\text{init.}} = \alpha\Upsilon$, $\gamma_{xy}^{\text{init.}} = 0$ for an isotropic material with *coefficient of thermal expansion* α and a temperature rise Υ above the stress-free state.

The most popular approach to finite element formulation—the assumed displacement approach—requires an expression for stress in terms of strain. By inversion of Eq. (4.14), therefore, we have

$$\boldsymbol{\sigma} = [\mathbf{E}]\boldsymbol{\epsilon} - [\mathbf{E}]\boldsymbol{\epsilon}^{\text{init.}} \tag{4.15}$$

We now possess in the matrix Eqs. (4.14) and (4.15) quite general representations of material mechanical properties. By expansion of the vectors $\boldsymbol{\sigma}$ and $\boldsymbol{\epsilon}$ each to the full six components, we encompass three-dimensional elasticity. A fully populated 6×6 $[\mathbf{E}]$ matrix would define a *general anisotropic* material, which possesses different properties in the different directions. Many special cases exist between the extremes of isotropy and anisotropy. In particular, an orthotropic material is one with three mutually perpendicular planes of elastic symmetry. A number of special forms of $[\mathbf{E}]$ and $[\mathbf{E}]^{-1}$ are elaborated upon in later chapters dealing with elements that require such forms. An important property of all material flexibility and stiffness matrices for materials considered in this text is that of *symmetry* [note Eqs. (4.12) and (4.13)].

For the present it is sufficient to observe that the finite element method is especially suited to the solution of problems characterized by sophisticated forms of material stiffness properties. We shall see that the matrix [**E**] (or its inverse) is simply digested in an algorithm for numerical integration of the expression in which it appears. The limitations on sophistication in material stiffness representation are often quite practical; it is difficult to go beyond the measurement of orthotropic stress-strain data in two dimensions in most real design situations. Exceptions occur for laminated plates with orthotropic plies, where the ply properties are determined experimentally and the properties of the total laminated plate are calculated, and for composite materials (e.g., fiber glass). Because of the special role of composites as practical orthotropic materials, publications dealing with their design represent excellent sources for detailed development of quite general material stiffness relationships (cf., Ref. 4.8).

4.5 Equilibrium and Compatibility Differential Equations

The foregoing sets of equations can be combined to obtain alternative forms of differential equations whose exact solution satisfies all component requirements. These alternative forms are referred to as the *equilibrium* and *compatibility* differential equations, respectively.

A comment is in order regarding the motivation for development of such differential equations. Two sets of conditions, static and kinematic, have been independently formulated. The static conditions are expressed exclusively in terms of static variables (stresses, or stress functions). The kinematic conditions are written entirely in terms of kinematic variables (displacements or strains). For a unique solution a connection between the static and kinematic variables is needed. This is accomplished through the introduction of the constitutive relationships.

The equilibrium differential equations are developed first since the stiffness (or displacement) finite element analysis procedure represents an approach to the approximate solution of these equations. For simplicity we exclude consideration of body forces and initial strains ($X = Y = 0$, $\{\epsilon^{\text{init.}}\} = 0$). This development consists of the formation of stress-displacement equations and the substitution of these into the differential equations of equilibrium. For example, by substitution of the strain-displacement equations into the constitutive equation for σ_x, we have

$$\sigma_x = \frac{E}{(1-\mu^2)}\frac{\partial u}{\partial x} + \frac{\mu E}{(1-\mu^2)}\frac{\partial v}{\partial y} \tag{4.16}$$

and similarly for σ_y and τ_{xy}. Next, by substitution of these into the equilibrium differential relationships [Eqs. (4.2a) and (4.2b)], there results

$$\frac{E}{(1-\mu^2)}\left[\frac{\partial^2 u}{\partial x^2}+\frac{(1-\mu)}{2}\frac{\partial^2 u}{\partial y^2}\right]+\frac{E}{2(1-\mu)}\frac{\partial^2 v}{\partial x\,\partial y}=0$$
$$\frac{E}{(1-\mu^2)}\left[\frac{(1-\mu)}{2}\frac{\partial^2 v}{\partial x^2}+\frac{\partial^2 v}{\partial y^2}\right]+\frac{E}{2(1-\mu)}\frac{\partial^2 u}{\partial x\,\partial y}=0 \tag{4.17}$$

If single-valued displacement fields satisfying the conditions of continuity and the kinematic boundary conditions can be found to satisfy the equations above and the associated boundary conditions, then the exact solution will have been identified. This is termed the *uniqueness principle*.

Consider, for example, quadratic displacement fields for the planar case

$$u = a_1 + a_2x + a_3y + a_4x^2 + a_5y^2 + a_6xy$$
$$v = a_7 + a_8x + a_9y + a_{10}x^2 + a_{11}y^2 + a_{12}xy$$

where $a_1, a_2, \ldots, a_{12}$ are constants.

After substitution into Eq. (4.17),

$$\frac{2E}{(1-\mu^2)}\left[a_4+\frac{(1-\mu)}{2}a_5\right]+\frac{E}{2(1-\mu)}a_{12}=0$$
$$\frac{2E}{(1-\mu^2)}\left[\frac{(1-\mu)}{2}a_{10}+a_{11}\right]+\frac{E}{2(1-\mu)}a_6=0$$

Clearly this compatible displacement field does not represent an exact solution of an elasticity problem for a general choice of the constants a_i. For $a_5 = a_{10} = 0$, however, a solution may be represented by the above if

$$a_4 = -\frac{(1+\mu)}{4}a_{12} \quad \text{and} \quad a_{11} = -\frac{(1+\mu)}{4}a_6$$

These conditions, however, do not ensure that the displacement fields are the appropriate solution of the problem. The boundary conditions on displacement must be met by u and v, and the stress field represented by u and v (obtained by differentiating these components in accordance with the strain-displacement equations and introducing the stress-strain law) must satisfy the stress boundary conditions.

We now examine the formulation of the governing equations that are referenced to the compatibility conditions. The basic differential equation for compatibility in plane stress is Eq. (4.8). Substituting into this the constitutive relationships in the form of Eq. (4.11), there is obtained

$$\frac{\partial^2}{\partial y^2}(\sigma_x-\mu\sigma_y)+\frac{\partial^2}{\partial x^2}(\sigma_y-\mu\sigma_x)=2(1+\mu)\frac{\partial^2\tau_{xy}}{\partial x\,\partial y} \tag{4.18}$$

This equation involves three unknowns (σ_x, σ_y, τ_{xy}). We can transform this into an equation involving just one unknown by introducing the *Airy stress function*, Φ, which was defined in Section 4.1, Eq. (4.4). Substitution of these expressions into Eq. (4.18) gives

$$\frac{\partial^4 \Phi}{\partial x^4} + 2\frac{\partial^4 \Phi}{\partial x^2\, \partial y^2} + \frac{\partial^4 \Phi}{\partial y^4} = 0$$

or

$$\nabla^2 \nabla^2 \Phi = \nabla^4 \Phi = 0 \tag{4.19}$$

where

$$\nabla^2 = \frac{\partial^2}{\partial x^2} + \frac{\partial^2}{\partial y^2} \tag{4.20}$$

∇^2 is the *Laplacian* or *harmonic operator* and Eq. (4.19) is the *biharmonic equation.*

The considerations that applied in the choice of displacement fields in satisfaction of the equilibrium differential equations apply here in a similar way. The stress function, by its definition, satisfies the equations of equilibrium but the expression chosen to describe the stress function might very well not satisfy Eq. (4.19) which states the condition of compatibility. In this case the chosen expression would be an approximation to the exact solution of the problem. Also, the boundary conditions as well as Eq. (4.19) must be satisfied in an exact solution.

4.6 Concluding Remarks

The *uniqueness theorem* in the theory of elasticity, which was introduced in Section 4.5, formally states that *if, in addition to the body forces, either the surface forces or the surface displacements are specified, there exists only one form of distribution of stresses and strains in the body.* A solution meeting all conditions of equilibrium and compatibility, within the body and on its boundaries, is the unique solution.

In order for this condition to hold, the stress-strain relationship must be linearly elastic, the effects of deformation on the equilibrium conditions are ignored, and the analysis is limited to small displacements. In Chapter 13, for example, we shall take up the analysis of elastic instability, which is characterized by adjacent and, therefore, nonunique states of equilibrium. These are identified by taking into account the effects of deformation on equilibrium.

Awareness of the uniqueness theorem is of importance to the finite element analyst. If a finite element representation were to meet *all* equilibrium and compatibility conditions, then the *exact solution* would be in hand and *further grid refinement would produce no improvement in the answers.* All analysts accept, of course, the fact that a numerical solution procedure, whatever its basis—series representation, finite differences, finite elements—should produce improved results with grid refinement. This situation clearly denotes deficiencies in meeting some or all of the basic conditions on an exact solution by *any* available numerical method.

Thus, finite element analysis possesses no special deficiency because it fails to satisfy one or the other of the relevant equilibrium and compatibility conditions. Indeed, it will be shown that certain finite element approaches seek to group all deficiencies into a single category, e.g., lack of satisfaction of equilibrium, while meeting all requirements on displacement continuity. When this approach succeeds, it can be proved that the numerical solution will possess assurances regarding monotonicity of convergence and certain parameters of the solution, such as strain energy or individual influence coefficients, are guaranteed to be on a prespecified side of the exact solution. These *bounds* on the solution may prove invaluable in assessing the integrity of the solution.

REFERENCES

4.1 TIMOSHENKO, S. and J. GOODIER, *Theory of Elasticity*, 2nd ed., McGraw-Hill Book Co., New York, N.Y., 1951.

4.2 WANG, C. T., *Applied Elasticity*, McGraw-Hill Book Co., New York, N.Y., 1953.

4.3 ODEN, J. T., *Mechanics of Elastic Structures*, McGraw-Hill Book Co., New York, N.Y., 1967.

4.4 VOLTERRA, E. and J. GAINES, *Advanced Strength of Materials*, Prentice-Hall, Inc., Englewood Cliffs, N.J., 1971.

4.5 SOKOLNIKOFF, I. S., *Mathematical Theory of Elasticity*, 2nd ed., McGraw-Hill Book Co., New York, N.Y., 1956.

4.6 GREEN, A. and W. ZERNA, *Theoretical Elasticity*, 2nd ed., Oxford University Press, New York, N.Y., 1968.

4.7 NOVOZHILOV, V. V., *Theory of Elasticity*, Leningrad, 1958. Published by Office of Tech. Services, U.S. Dept. of Commerce, Washington, D.C., as OTS 61-11401.

4.8 ANONYMOUS, *Structural Design Guide for Advanced Composite Applications*, 2nd ed., U.S. Air Force Materials Laboratory, Wright Patterson AFB, Ohio, 1969.

PROBLEMS

4.1 Do the following stress distributions satisfy the conditions of equilibrium (zero body forces)? Sketch the state of edge stress consistent with these functions for the element shown in Fig. P4.1.

$$\sigma_x = a_1 + a_2 x$$
$$\sigma_y = a_3 + a_4 y$$
$$\tau_{xy} = a_5 - a_2 y - a_4 x$$

Fig. P4.1

4.2 The equilibrium and strain-displacement equations for plane stress in polar coordinates are given below. The constitutive relationships are identical to the rectangular coordinate case. Establish the governing equilibrium differential equations, including thermal strain.

$$\epsilon_r = \frac{\partial u}{\partial r}, \qquad \epsilon_\theta = \frac{u}{r} + \frac{1}{r}\frac{\partial v}{\partial \theta}, \qquad \gamma_{r\theta} = \frac{1}{r}\frac{\partial u}{\partial \theta} + \frac{\partial v}{\partial r} - \frac{v}{r}$$

$$\frac{\partial \sigma_r}{\partial r} + \frac{1}{r}\frac{\partial \tau_{r\theta}}{\partial \theta} + \frac{\sigma_r - \sigma_\theta}{r} = 0$$

$$\frac{1}{r}\frac{\partial \sigma_\theta}{\partial \theta} + \frac{\partial \tau_{r\theta}}{\partial r} + \frac{2\tau_{r\theta}}{r} = 0$$

Fig. P4.2

4.3 Formulate the linear equilibrium differential equations for three-dimensional elasticity, first taking into account the variations of stress with respect to all three coordinates and then discarding higher-order terms.

4.4 The displacement function given is to be employed to formulate a stiffness matrix for the illustrated parallelogram element in Fig. P4.4. Test the adequacy of this function with respect to
(a) equilibrium within the element,
(b) equilibrium across the element boundaries,
(c) interelement displacement continuity.

$$u = a_1 x + a_2 y + a_3\left(xy - \frac{x_4}{y_4}y^2\right) + a_4$$

$$v = a_5 x + a_6 y + a_7\left(xy - \frac{x_4}{y_4}y^2\right) + a_8$$

4.5 Form the stress-strain matrix **[E]** for the *plane stress* of an *orthotropic material*. In the final form this should connect $\boldsymbol{\sigma} = \lfloor \sigma_x \; \sigma_y \; \tau_{xy} \rfloor^T$ to $\boldsymbol{\epsilon} = \lfloor \epsilon_x \; \epsilon_y \; \gamma_{xy} \rfloor^T$.

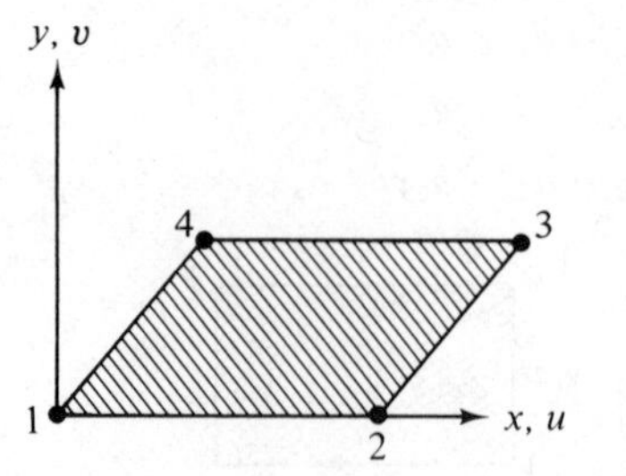

Fig. P4.4

For this case, $\tau_{xz} = \tau_{yz} = \sigma_x = 0$. The elastic moduli are E_x, E_y, G_{xy}. The Poisson ratio associated with y-direction strain caused by x-direction stress is μ_{yx}, etc.

4.6 Examine, for the element shown in Fig. P4.1, for the displacement field given below, the conditions of equilibrium. What is the physical significance of the deficiency in equilibrium evidenced by this function?

$$u = N_1u_1 + N_2u_2 + N_3u_3 + N_4u_4$$
$$v = N_1v_1 + N_2v_2 + N_3v_3 + N_4v_4$$

where

$$N_1 = \left(1 - \frac{x}{x_2}\right)\left(1 - \frac{y}{y_3}\right), \qquad N_2 = \frac{x}{x_2}\left(1 - \frac{y}{y_3}\right), \qquad N_3 = \frac{x}{x_2}\frac{y}{y_3},$$
$$N_4 = \left(1 - \frac{x}{x_2}\right)\left(\frac{y}{y_3}\right)$$

4.7 The displacements of the element boundary shown below are described by the functions

$$u = N_1u_1 + N_2u_2 + N_3u_3, \qquad v = N_1v_1 + N_2v_2 + N_3v_3$$

where

$$N_1 = \frac{(2s - a)(s - a)}{a^2}, \qquad N_2 = \frac{4s(a - s)}{a^2}, \qquad N_3 = \frac{s(2s - a)}{a^2}$$

Define the normal and tangential tractions $\bar{T}_n$ and $\bar{T}_s$ consistent with these displacements.

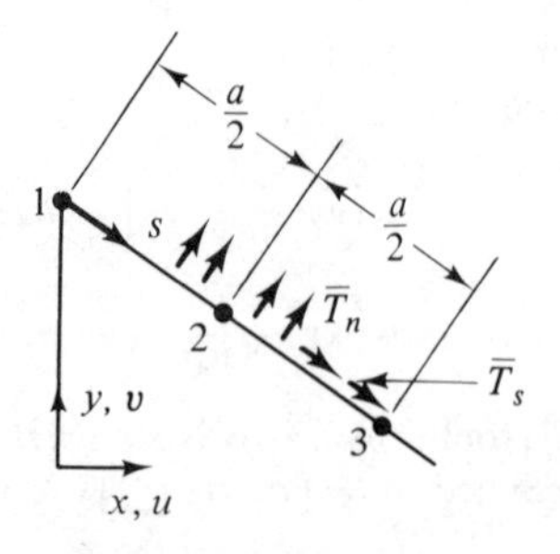

Fig. P4.7

4.8 Φ_1, Φ_2, and Φ_3 are three-dimensional stress functions defined as follows:

$$\sigma_x = \frac{\partial^2 \Phi_1}{\partial y\, \partial x}, \qquad \sigma_y = \frac{\partial^2 \Phi_2}{\partial z\, \partial x}, \qquad \sigma_z = \frac{\partial^2 \Phi_3}{\partial x\, \partial y}$$

$$\tau_{xy} = -\frac{1}{2}\frac{\partial}{\partial z}\left(\frac{\partial \Phi_1}{\partial x} + \frac{\partial \Phi_2}{\partial y} - \frac{\partial \Phi_3}{\partial z}\right)$$

$$\tau_{yz} = -\frac{1}{2}\frac{\partial}{\partial x}\left(-\frac{\partial \Phi_1}{\partial x} + \frac{\partial \Phi_2}{\partial y} + \frac{\partial \Phi_3}{\partial z}\right)$$

$$\tau_{zx} = -\frac{1}{2}\frac{\partial}{\partial y}\left(\frac{\partial \Phi_1}{\partial x} - \frac{\partial \Phi_2}{\partial y} + \frac{\partial \Phi_3}{\partial z}\right)$$

Prove that they satisfy the equilibrium differential equations and establish the compatibility equations corresponding to them.

5

Direct Methods of Element Formulation

This chapter initiates our treatment of formulative procedures for finite element force-displacement equations. Two approaches are examined: *the direct method* and *the method of weighted residuals.*

The direct method accomplishes the formulation of element relationships by direct combination of the three sets of elasticity equations given in the previous chapter: the equations of equilibrium, the strain-displacement equations, and the constitutive relationships. The method is especially useful in clarifying the fundamental relationships between the *finite element approximation* and *the real structure.* In this way we give theoretical justification of the idealization described in Sections 2.2 and 2.3. The direct method also introduces features common to all approaches to finite element formulation, especially the transformations from stress to force and from d.o.f. to strain. It comprises the basis used for element formulation in the early developmental phases of the method.[5.1, 5.2] The direct method is limited in scope, however, and is difficult or impossible to apply to the formulation of relationships for complicated elements and special phenomena.

The method of weighted residuals[5.3] is, conversely, of virtually unlimited scope and appears to possess advantages absent from alternative approaches. One form of this method yields formulations that are identical to those established by application of the variational principles described in Chapter 6. For certain classes of nonlinear problems the method of weighted residuals may be employed to produce relationships that cannot be obtained with use of conventional variational principles.[5.4, 5.5] The method also helps in understanding the physical basis of variational principles such as the potential and complementary energy principles.

Our development of structural analysis relationships in this chapter is devoted principally to the formulation of element stiffness matrices on the basis of assumed displacement fields. The methods to be described, however, are not limited in applicability to this type of relationship but are valid for formulation of any type of element relationship on the basis of assumed displacement and/or stress fields and indeed apply to a wide variety of nonstructural physical processes. We illustrate a few simple forms of the latter in this chapter.

5.1 Direct Method

The direct method of stiffness equation formulation consists of the following steps:

1. The element displacement field $\mathbf{\Delta}$ is expressed in terms of a finite number of parameters $\{\mathbf{a}\}$ that are preferably the element joint d.o.f. $\{\mathbf{\Delta}\}$. If parameters $\{\mathbf{a}\}$ lacking physical meaning are initially chosen, a transformation is established that connects them with the physical d.o.f. $\{\mathbf{\Delta}\}$.

2. The strain field $\boldsymbol{\epsilon}$ is expressed in terms of the d.o.f. $\{\mathbf{\Delta}\}$ by differentiation of the displacement field in accordance with the strain-displacement equations of elasticity [Eq. (4.7)].

3. The constitutive law [Eq. (4.15)] is introduced to establish the relationship between the stress field $\boldsymbol{\sigma}$ and the d.o.f. $\{\mathbf{\Delta}\}$.

4. Equations that describe the element joint forces $\{\mathbf{F}\}$ as functions of the stress field $\boldsymbol{\sigma}$ are constructed by defining these forces as the statical equivalents of the stresses acting on the element boundaries. Since equations for $\boldsymbol{\sigma}$ are available in terms of $\{\mathbf{\Delta}\}$ (step 3), it is now possible to relate $\{\mathbf{F}\}$ to $\{\mathbf{\Delta}\}$ and the resulting equations are, by definition, the element stiffness equations.

To illustrate the procedure above we formulate the stiffness matrices for three simple elements: an axial member, a beam element, and a triangle in plane stress.

Consider first the axial member (see Fig. 2.7 shown again here). We begin by defining the displacement field $\mathbf{\Delta} = u$ in terms of the *generalized displacements* $\{\mathbf{a}\}$. It is clear that two displacement d.o.f., the axial displacements at the end points 1 and 2, define the displaced state of this element

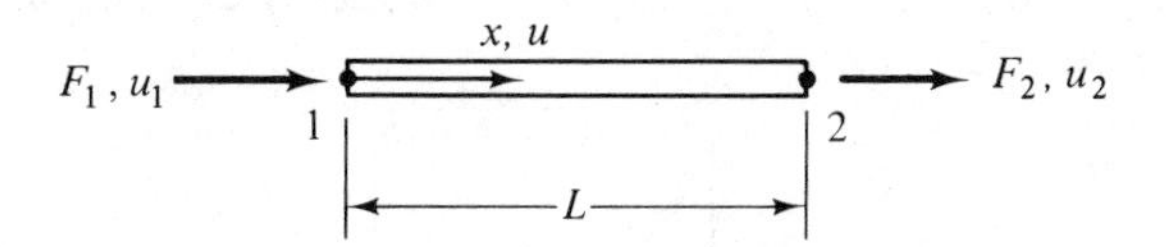

so that we choose two parameters in the set $\{\mathbf{a}\}$:

$$\{\mathbf{a}\} = \lfloor a_1 \; a_2 \rfloor^{\mathrm{T}} \tag{5.1}$$

In describing the one-dimensional variation of u between the end points we adopt a *polynomial representation.* This form of representation is consistent with our choice of such a form for most elements in two and three dimensions since the x, y, and z variables of the general polynomial fit in well with elements of all shapes. Additional theoretical justification for the choice of a polynomial will be given in Chapter 8. In the present case, with two available parameters, the logical choice is a linear polynomial in x:

$$u = a_1 + a_2 x = \lfloor 1 \; x \rfloor \begin{Bmatrix} a_1 \\ a_2 \end{Bmatrix} \tag{5.2}$$

Our symbolism for the general case is as follows:

$$\boldsymbol{\Delta} = [\mathbf{p}]\{\mathbf{a}\} \tag{5.2a}$$

The symbolism we adopt in Eq. (5.2a) deserves clarification. It was pointed out in Section 4.3 that the displacement of a point, symbolized by $\boldsymbol{\Delta}$, may consist of as many as three components, namely u, v, and w. Consequently, one may choose an independent polynomial expansion for each, in which case $[\mathbf{p}]$ would be a rectangular matrix of three rows. If, for example, the displacements in a three-dimensional situation were described as

$$u = a_1 + a_2 x$$
$$v = a_3 + a_4 y$$
$$w = a_5 + a_6 z$$

we would then have

$$\boldsymbol{\Delta} = \begin{Bmatrix} u \\ v \\ w \end{Bmatrix} = \begin{bmatrix} 1 & x & 0 & 0 & 0 & 0 \\ 0 & 0 & 1 & y & 0 & 0 \\ 0 & 0 & 0 & 0 & 1 & z \end{bmatrix} \begin{Bmatrix} a_1 \\ \cdot \\ \cdot \\ \cdot \\ a_6 \end{Bmatrix}$$

Now, returning to the problem of the axial member, in conformity with step 1 above, we seek to transform this representation into one that refers to the physical d.o.f. u_1 and u_2. This can be done by evaluation of Eq. (5.2) at $x = 0$ and $x = L$.

$$\begin{Bmatrix} u_1 \\ u_2 \end{Bmatrix} = \begin{bmatrix} 1 & 0 \\ 1 & L \end{bmatrix} \begin{Bmatrix} a_1 \\ a_2 \end{Bmatrix} \tag{5.3}$$

for which the general symbolism is

$$\{\mathbf{\Delta}\} = [\mathbf{B}]\{\mathbf{a}\} \tag{5.3a}$$

and, by inversion of $[\mathbf{B}]$,

$$\begin{Bmatrix} a_1 \\ a_2 \end{Bmatrix} = \frac{1}{L}\begin{bmatrix} L & 0 \\ -1 & 1 \end{bmatrix}\begin{Bmatrix} u_1 \\ u_2 \end{Bmatrix} \tag{5.4}$$

or, in the general case,

$$\{\mathbf{a}\} = [\mathbf{B}]^{-1}\{\mathbf{\Delta}\} \tag{5.4a}$$

Substituting this relationship into Eq. (5.2) we find

$$u = \left\lfloor \left(1 - \frac{x}{L}\right) \quad \frac{x}{L} \right\rfloor \begin{Bmatrix} u_1 \\ u_2 \end{Bmatrix} = \lfloor N_1 \quad N_2 \rfloor \begin{Bmatrix} u_1 \\ u_2 \end{Bmatrix} \tag{5.5}$$

or, in general symbolism,

$$\mathbf{\Delta} = [\mathbf{p}][\mathbf{B}]^{-1}\{\mathbf{\Delta}\} = [\mathbf{N}]\{\mathbf{\Delta}\} \tag{5.5a}$$

where the terms $[1 - (x/L)] = N_1$ and $x/L = N_2$ are termed the *shape functions* of the displacement field.

Now applying step 2 (introduction of the strain-displacement equation) we have $\boldsymbol{\epsilon} = \epsilon_x = u'$ where the prime denotes the derivative of u with respect to x. This step can be applied in either of two ways. For one, Eq. (5.2) can be differentiated and Eq. (5.4) can be invoked to produce the desired final relationship

$$u' = \lfloor 0 \quad 1 \rfloor \begin{Bmatrix} a_1 \\ a_2 \end{Bmatrix} \tag{5.6}$$

for which the general symbolism is

$$\boldsymbol{\epsilon} = [\mathbf{C}]\{\mathbf{a}\} \tag{5.6a}$$

and, by substitution of Eq. (5.4),

$$\boldsymbol{\epsilon} = u' = \frac{1}{L}\lfloor 0 \quad 1 \rfloor \begin{bmatrix} L & 0 \\ -1 & 1 \end{bmatrix}\begin{Bmatrix} u_1 \\ u_2 \end{Bmatrix} = \left\lfloor -\frac{1}{L} \quad \frac{1}{L} \right\rfloor \begin{Bmatrix} u_1 \\ u_2 \end{Bmatrix}$$

Alternatively we can differentiate Eq. (5.5) directly:

$$\boldsymbol{\epsilon} = u' = \left\lfloor -\frac{1}{L} \quad \frac{1}{L} \right\rfloor \begin{Bmatrix} u_1 \\ u_2 \end{Bmatrix} \tag{5.6b}$$

or

$$\boldsymbol{\epsilon} = [\mathbf{D}]\{\mathbf{\Delta}\} \tag{5.6c}$$

It can be observed, in Eq. (5.6), that the term a_1 has in fact been eliminated and that the equation might alternatively be written as $u' = a_2$. This contracted result is due to the fact, as noted in Section 4.3, that the differentiation of displacements to yield strains has the effect of eliminating rigid-body-motion terms. In this case the rigid-body-motion term is a_1. In the more general case we designate the rigid-body-displacement parameters as $\{\mathbf{a}_s\}$ and the remaining generalized parameters as $\{\mathbf{a}_f\}$. Then, the contracted form of the generalized displacement-to-strain transformation is written for the general case as

$$\boldsymbol{\epsilon} = [\mathbf{C}_f]\{\mathbf{a}_f\} \tag{5.6d}$$

In applying step 3 (introduction of the stress-strain law) to the case of the axial member we find that $[\mathbf{E}] = E$, and $\sigma = \sigma_x$ and, by use of Eq. (5.6b),

$$\sigma_x = E\begin{bmatrix} -\frac{1}{L} & \frac{1}{L} \end{bmatrix}\begin{Bmatrix} u_1 \\ u_2 \end{Bmatrix} \tag{5.7}$$

or, symbolically,

$$\boldsymbol{\sigma} = [\mathbf{E}][\mathbf{D}]\{\boldsymbol{\Delta}\} = [\mathbf{S}]\{\boldsymbol{\Delta}\} \tag{5.7a}$$

where $[\mathbf{S}] = [\mathbf{E}]\,[\mathbf{D}]$ is one form of the *element stress matrix.*

It will be recalled that the notion of an element stress matrix was introduced in Eq. (3.9) in the form $\{\boldsymbol{\sigma}\} = [\mathbb{S}]\{\boldsymbol{\Delta}\}$, which in accordance with our nomenclature indicated the evaluation of stress at *specific points*. For example, $\{\boldsymbol{\sigma}\} = \lfloor \sigma_{x_1}\ \sigma_{x_2} \rfloor^{\mathrm{T}}$ for the present element if we evaluate the stress at the end points. Thus, we use the symbol $[\mathbf{S}]$ to denote the transformation from the displacement vector $\{\boldsymbol{\Delta}\}$ to the field definition of stress $\boldsymbol{\sigma}$ and $[\mathbb{S}]$ to denote the transformation from $\{\boldsymbol{\Delta}\}$ to the vector of specified stress values $\{\boldsymbol{\sigma}\}$. The $\{\boldsymbol{\sigma}\} = [\mathbb{S}]\{\boldsymbol{\Delta}\}$ form of the element stress equations could be used in place of Eq. (5.7a); this form will in fact be used subsequently in the formulation of the flexural element.

As the final step, the transformation of stress into joint forces, it is noted that the joint forces are $\{\mathbf{F}\} = \lfloor F_1\ F_2 \rfloor^{\mathrm{T}}$ and that individual components are given by the stress multiplied by the element cross-sectional area A. We have (with F_1 acting in a direction opposite to positive σ_x)

$$\begin{Bmatrix} F_1 \\ F_2 \end{Bmatrix} = A\begin{bmatrix} -1 \\ 1 \end{bmatrix}\boldsymbol{\sigma}_x \tag{5.8}$$

or

$$\{\mathbf{F}\} = [\mathbf{A}]\{\boldsymbol{\sigma}\} \tag{5.8a}$$

and, by introduction of Eq. (5.7),

$$\begin{Bmatrix} F_1 \\ F_2 \end{Bmatrix} = AE\begin{bmatrix} -1 \\ 1 \end{bmatrix}\begin{bmatrix} -\frac{1}{L} & \frac{1}{L} \end{bmatrix}\begin{Bmatrix} u_1 \\ u_2 \end{Bmatrix} = \frac{AE}{L}\begin{bmatrix} 1 & -1 \\ -1 & 1 \end{bmatrix}\begin{Bmatrix} u_1 \\ u_2 \end{Bmatrix} \tag{5.9}$$

Thus, we have arrived at the expression of the element stiffness equations

$$\{\mathbf{F}\} = [\mathbf{k}]\{\mathbf{\Delta}\}$$

where

$$[\mathbf{k}] = [\mathbf{A}][\mathbf{E}][\mathbf{D}] \tag{5.10}$$

We find, therefore, that the stiffness matrix is formed of the product of three matrices:

[**D**]—*the d.o. f.-to-strain* transformation
[**E**]—the *material elastic stiffness* matrix
[**A**]—the *stress-to-force* transformation.

[**D**] can be broken down into more basic components. In the case where the displacement field is first written in terms of generalized d.o.f. we have, from Eqs. (5.4a) and (5.6a),

$$[\mathbf{D}] = [\mathbf{C}][\mathbf{B}]^{-1} \tag{5.11}$$

Two factors are noteworthy in the foregoing development. For one, no explicit consideration was given to the conditions of equilibrium within the element. We know, of course, that this element involves a state of constant stress and upon checking Eq. (5.7) it is seen that the chosen displacement field corresponds to this condition. In general, however, the equilibrium condition will not be met by the stress state corresponding to an assumed displacement field; but this will have no influence on our ability to formulate an element stiffness matrix via the sequence of operations above. Second, a continuity of displacements prevails within the element, due to the continuous nature of the chosen function, and across the boundaries from this element to those to which it will be joined. This is because the interface between the one-dimensional elements is totally comprised of the end points being joined. In the general case of two- and three-dimensional elements, however, the interface between elements is not totally comprised of the joints, and the displacement field for the element must be chosen with consideration of the properties of displacement continuity across the interfaces with adjacent elements. This point was discussed in Section 2.2 and will be examined again in Section 5.2. Since the present formulation meets all conditions of equilibrium and continuity of displacement, it gives the "exact" formulation of the element stiffness matrix.

In many analysis situations the applied loads are distributed as a continuous function of the x coordinate. The present approach implies that the distributed loads would be handled by forming statically equivalent joint forces. A more elegant way of dealing with this situation is taken up in Chapter 6.

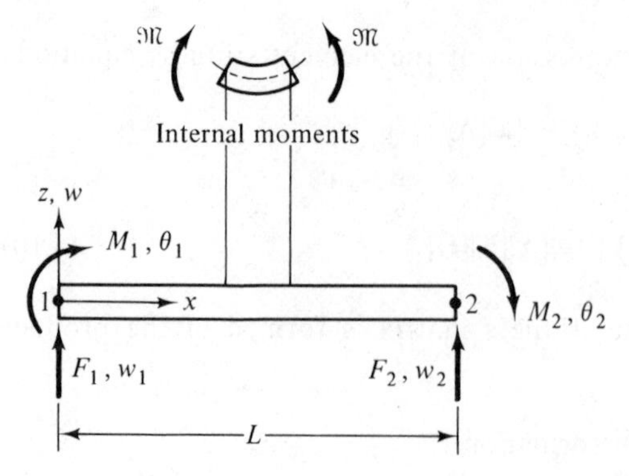

Fig. 5.1 Beam element.

We consider next the beam element, Fig. 5.1. The general features of this development are the same as for the axial member, but one significant detailed difference is in the type of joint d.o.f. to be specified. Another is that the strain field varies within the element. Consistent with the theory of flexure, which disregards transverse shear deformation, we must define not only transverse displacements at the end points (w_1 and w_2) but also angular displacements (θ_1 and θ_2). The latter are equal to the negative of the slope of the neutral axis since a positive (clockwise) rotation causes a negative transverse displacement.

$$\theta_1 = -\left.\frac{dw}{dx}\right|_{x=0}, \quad \theta_2 = -\left.\frac{dw}{dx}\right|_{x=L}$$

Thus

$$\{\Delta\} = \lfloor w_1\ \theta_1\ w_2\ \theta_2 \rfloor^{T} \tag{5.12}$$

As in the case of the axial member we will choose a polynomial to describe the displacement field $\mathbf{\Delta}$, which in this case is defined by w. We have four d.o.f. and if we are not to skip over any of the terms of the polynomial we must choose the cubic polynomial in order to have four terms.

$$w = a_1x^3 + a_2x^2 + a_3x + a_4 \tag{5.13}$$

and, by evaluation of w and $-dw/dx$ at points 1 and 2,

$$\begin{Bmatrix} w_1 \\ \theta_1 \\ w_2 \\ \theta_2 \end{Bmatrix} = \begin{bmatrix} 0 & 0 & 0 & 1 \\ 0 & 0 & -1 & 0 \\ L^3 & L^2 & L & 1 \\ -3L^2 & -2L & -1 & 0 \end{bmatrix} \begin{Bmatrix} a_1 \\ a_2 \\ a_3 \\ a_4 \end{Bmatrix}$$

for which the inverse is

$$\begin{Bmatrix} a_1 \\ a_2 \\ a_3 \\ a_4 \end{Bmatrix} = \frac{1}{L^3} \begin{bmatrix} 2 & -L & -2 & -L \\ -3L & 2L^2 & 3L & L^2 \\ 0 & -L^3 & 0 & 0 \\ L^3 & 0 & 0 & 0 \end{bmatrix} \begin{Bmatrix} w_1 \\ \theta_1 \\ w_2 \\ \theta_2 \end{Bmatrix}$$

and, by substitution into Eq. (5.13),

$$w = \lfloor \mathbf{N} \rfloor \{\mathbf{\Delta}\} \tag{5.14}$$

where

$$\lfloor \mathbf{N} \rfloor = \lfloor N_1 \, N_2 \, N_3 \, N_4 \rfloor$$

and

$$\begin{aligned} N_1 &= (1 + 2\xi^3 - 3\xi^2), \qquad & N_3 &= (3\xi^2 - 2\xi^3) \\ N_2 &= -x(\xi - 1)^2, & N_4 &= -x(\xi^2 - \xi) \end{aligned} \tag{5.14a}$$

with

$$\xi = \frac{x}{L}$$

Strains in the case of flexure equal the curvatures (second derivatives), w''. Hence

$$w'' = \lfloor \mathbf{N}'' \rfloor \{\mathbf{\Delta}\} \tag{5.15}$$

where

$$\begin{aligned} N_1'' &= -N_3'' = \frac{6}{L^2}(2\xi - 1) \\ N_2'' &= -\frac{2}{L}(3\xi - 2), \qquad N_4'' = -\frac{2}{L}(3\xi - 1) \end{aligned} \tag{5.15a}$$

Also, the *stresses* are the *internal* bending moments $\mathfrak{M}$ and for these the pertinent constitutive equation is of the form

$$\mathfrak{M} = EIw'' \tag{5.16}$$

Since the second derivatives [Eq. (5.15a)] vary linearly within the element, the curvature can be defined uniquely by the values of w'' at points 1 and 2. From Eq. (5.15),

$$\begin{Bmatrix} w_1'' \\ w_2'' \end{Bmatrix} = \frac{1}{L^2} \begin{bmatrix} -6 & 4L & 6 & 2L \\ 6 & -2L & -6 & -4L \end{bmatrix} \begin{Bmatrix} w_1 \\ \theta_1 \\ w_2 \\ \theta_2 \end{Bmatrix} = [\mathbf{D}]\{\mathbf{\Delta}\} \tag{5.15b}$$

In examining the equilibrium of forces we must first note that the internal moments at points 1 and 2, $\mathfrak{M}_1$ and $\mathfrak{M}_2$, are defined as positive when they produce a positive curvature as sketched in Fig. 5.1. Thus, $\mathfrak{M}_1 = M_1$ and $\mathfrak{M}_2 = -M_2$. The moment equilibrium relationships can be applied to give F_1 and F_2 in terms of $\mathfrak{M}_1$ and $\mathfrak{M}_2$, and collecting the full set of equations we write

$$\begin{Bmatrix} F_1 \\ M_1 \\ F_2 \\ M_2 \end{Bmatrix} = \frac{1}{L}\begin{bmatrix} -1 & 1 \\ L & 0 \\ 1 & -1 \\ 0 & -L \end{bmatrix}\begin{Bmatrix} \mathfrak{M}_1 \\ \mathfrak{M}_2 \end{Bmatrix} = [\mathbf{A}]\{\boldsymbol{\sigma}\} \tag{5.8b}$$

Also, because we wish to relate the two end point moments to the two end point curvatures, the flexural stress-strain law [Eq. (5.16)] must be written in the expanded form†

$$\{\mathfrak{M}\} = [\mathbf{E}]\{\mathbf{w}''\} \tag{5.16a}$$

where now

$$\begin{Bmatrix} \mathfrak{M}_1 \\ \mathfrak{M}_2 \end{Bmatrix} = EI\begin{bmatrix} 1 & 0 \\ 0 & 1 \end{bmatrix}\begin{Bmatrix} w_1'' \\ w_2'' \end{Bmatrix} = [\mathbf{E}]\{\boldsymbol{\epsilon}\} \tag{5.16a}$$

Finally, combining Eq. (5.15b), (5.16a), and (5.8b) in the product [**k**] = [**A**][**E**][**D**] we obtain

$$[\mathbf{k}] = \frac{2EI}{L^3}\begin{bmatrix} 6 & & \text{(symmetric)} & \\ -3L & 2L^2 & & \\ -6 & +3L & 6 & \\ -3L & L^2 & 3L & 2L^2 \end{bmatrix} \tag{5.17}$$

(columns labelled w_1, θ_1, w_2, θ_2)

Once again the stiffness matrix has been obtained without explicit consideration of internal equilibrium conditions, which for this element, in the absence of loads between the joints, are represented by the equation

$$\frac{d^4w}{dx^4} = 0 \tag{5.18}$$

Clearly, the fourth derivative of the cubic polynomial [Eq. (5.13)] is zero so that this condition is satisfied. Equilibrium is enforced in the global representation simply by summation of forces and moments at the joints as described in Chapter 3.

Intuitively, one expects that continuity of angular displacement should be maintained for a flexural element or else "kinks" would appear at the

†In the strictest sense, [A] and [E] relate to the field of stresses $\boldsymbol{\sigma}$ and the field of strains $\boldsymbol{\epsilon}$, respectively. Here, we are dealing with vectors of point stresses ($\{\boldsymbol{\sigma}\} = \lfloor \mathfrak{M}_1 \quad \mathfrak{M}_2 \rfloor^T$) and point strains ($\{\boldsymbol{\epsilon}\} = \lfloor w_1'' \quad w_2'' \rfloor^T$). To avoid introduction of new nomenclature, however, we employ the same symbols for both cases.

joints. The conditions on continuity of displacement across the joints of the global representation consequently require continuity of both w and θ. The present formulation satisfies these conditions. Since it satisfies all conditions of equilibrium when the loads are applied only at the joints the solution obtained is the exact one for such cases. One could construct an approximate displacement field (e.g., a linear field in terms only of w_1 and w_2, as in finite difference analysis) and this would produce an approximate solution if a finite number of segments were employed in the global representation.

The *shape function* representation of the displacement field [Eqs. (5.5) and (5.14)] played a central role in both of our illustrative formulations. Although the idea of a shape function is examined more closely in succeeding chapters, particularly Chapter 8, it is important to denote its basic properties at the outset of the study of element formulations. Consider first the case where the field (or trial function) of an independent variable ($\boldsymbol{\Delta}$) is expressed solely in terms of the values (Δ_i) of that variable at selected points. The axial member [Eq. (5.5)] exemplifies this case. For this circumstance the shape function N_i *must be of such form that it takes on the value* 1 *when evaluated at the geometric coordinates of the point at which* Δ_i *is defined and must be zero when evaluated at any point corresponding to the remaining* d.o.f. This must be so in order for $\Delta = \Delta_i$ at the point corresponding to Δ_i. The reason for designating N_i as a *shape function* can now be seen: It represents the shape of the variable Δ when plotted over the surface of the element for $\Delta_i = 1$ and all other d.o.f. held fixed. We illustrate this for the axial member in Fig. 5.2.

In some cases the description of the independent variable includes d.o.f. in the form of derivatives of the variable evaluated at specified points. The shape function representation of w for the flexural member [Eq. (5.14)] includes as d.o.f., for example, the derivatives of w (θ_1 and θ_2) at the end points. The shape functions which multiply these d.o.f. must possess units which cause the product of the shape function and corresponding d.o.f. to have the units of the displacement function. In the beam example, therefore, the multipliers of θ_1 and θ_2 in Eq. (5.14) have the units of length (displacement) since θ_1 and θ_2 are in radians.

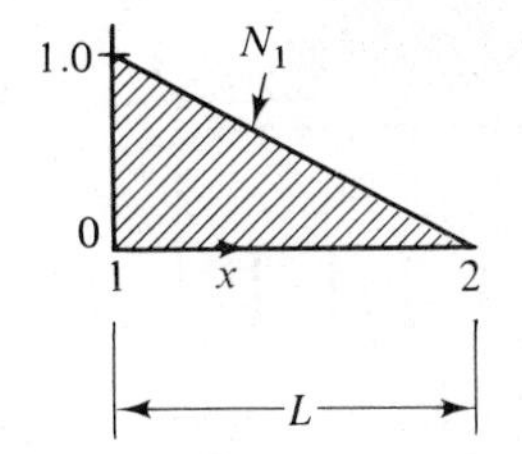

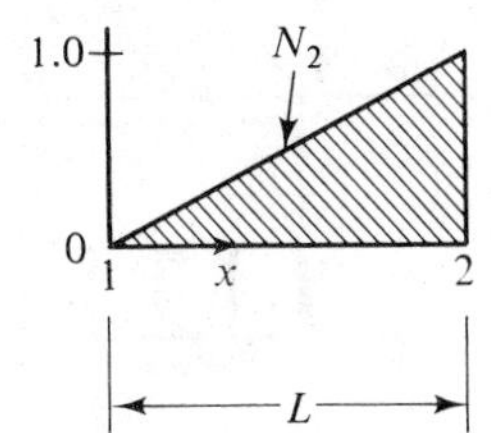

Fig. 5.2

5.2 Triangular Element in Plane Stress

We next illustrate the direct method in the formulation of element stiffness equations by reference to the triangle in plane stress, Figs. 5.3 and 5.4. The element is of constant thickness (t), it is composed of an isotropic material, and for convenience in development of the formulation we have positioned one edge along the element x axis. This illustrative example deserves separate attention because (1) it deals with a more general stress state (two dimensional), (2) the resulting stiffness equations lead to approximate solutions of the differential equations which govern the global analysis problems, and (3) this element is of primary importance in all areas of practical application.

The behavior of the element, as seen from Fig. 5.4, is described by six d.o.f.

$$\{\boldsymbol{\Delta}\} = \lfloor u_1\ u_2\ u_3\ v_1\ v_2\ v_3 \rfloor^{\mathrm{T}} \tag{5.19}$$

and since no bias should be shown to either the x or y directions, we choose three parameters to describe both u and v.

$$\begin{aligned} u &= a_1 + a_2 x + a_3 y \\ v &= a_4 + a_5 x + a_6 y \end{aligned} \tag{5.20}$$

Note that these expressions are complete *linear* polynomials. By evaluation

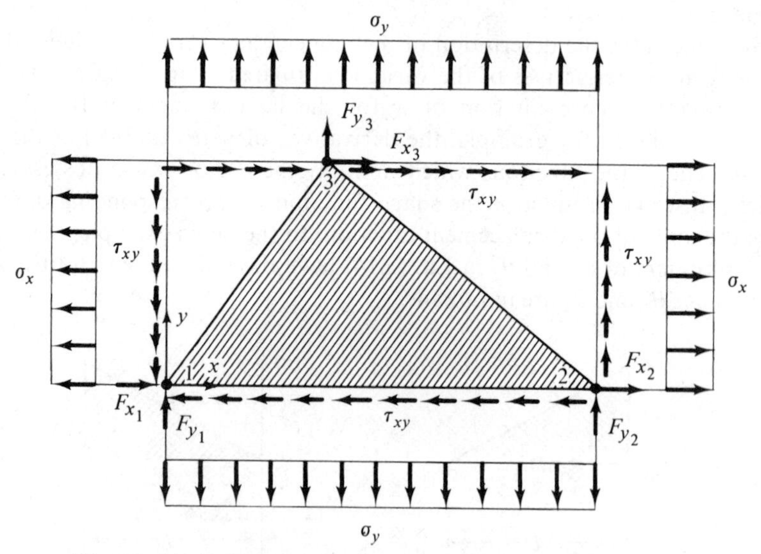

Fig. 5.3 Mode of representation of edge stesses on triangular element for purpose of calculating joint forces.

of u at points 1, 2, and 3,

$$\begin{Bmatrix} u_1 \\ u_2 \\ u_3 \end{Bmatrix} = \begin{bmatrix} 1 & 0 & 0 \\ 1 & x_2 & 0 \\ 1 & x_3 & y_3 \end{bmatrix} \begin{Bmatrix} a_1 \\ a_2 \\ a_3 \end{Bmatrix} = [\mathbf{B}_u]\{\mathbf{a}\}$$

and, by inversion and back-substitution into Eq. (5.20),

$$u = N_1 u_1 + N_2 u_2 + N_3 u_3 \tag{5.21a}$$

where

$$N_1 = \frac{1}{x_2 y_3}(x_2 y_3 - x y_3 - x_2 y + x_3 y)$$

$$N_2 = \frac{1}{x_2 y_3}(x y_3 - x_3 y)$$

$$N_3 = \frac{y}{y_3}$$

These same shape functions are obtained for the expression for v so that this displacement function is

$$v = N_1 v_1 + N_2 v_2 + N_3 v_3 \tag{5.21b}$$

Now, if the strain-displacement equations of plane elasticity [Eqs. (4.7 a, b, c)] are invoked and applied to Eqs. (5.21a, b), we obtain Eq. (5.6c), wherein

$$\boldsymbol{\epsilon} = \lfloor \epsilon_x \; \epsilon_y \; \gamma_{xy} \rfloor^{\mathrm{T}}$$

$$[\mathbf{D}] = \begin{bmatrix} N_{1,x} & N_{2,x} & N_{3,x} & 0 & 0 & 0 \\ 0 & 0 & 0 & N_{1,y} & N_{2,y} & N_{3,y} \\ N_{1,y} & N_{2,y} & N_{3,y} & N_{1,x} & N_{2,x} & N_{3,x} \end{bmatrix} \tag{5.22}$$

where $N_{1,x}$ represents the derivative of N_1 with respect to x, etc.

The matrix [E] for plane stress is presented in Eq. (4.12), so to complete the development of basic matrices for this element we need only construct the joint force-stress matrix, [A]. This is accomplished by a straightforward transformation of the edge stresses into joint forces. For force F_{x_2}, for example (see Fig. 5.3),

$$F_{x_2} = \frac{t}{2}[y_3 \sigma_x - x_2 \tau_{xy} + (x_2 - x_3)\tau_{xy}]$$

Applying this procedure to determine each joint force, we obtain Eq. (5.8a) in which

$$\{\mathbf{F}\} = \lfloor F_{x_1}\, F_{x_2}\, F_{x_3}\, F_{y_1}\, F_{y_2}\, F_{y_3} \rfloor^{\mathrm{T}}$$

$$\boldsymbol{\sigma} = \lfloor \sigma_x\, \sigma_y\, \tau_{xy} \rfloor^{\mathrm{T}}$$

$$[\mathbf{A}] = \frac{t}{2}\begin{bmatrix} -y_3 & 0 & x_3 - x_2 \\ y_3 & 0 & -x_3 \\ 0 & 0 & x_2 \\ 0 & x_3 - x_2 & -y_3 \\ 0 & -x_3 & y_3 \\ 0 & x_2 & 0 \end{bmatrix}$$

The stiffness matrix for this element, calculated from Eq. (5.10) by forming the product [A][E][D] with use of the relationships above, is presented in Fig. 5.4.

As in the preceding examples we now examine those aspects of an elasticity theory solution that were not covered explicitly in the formulation above. The element strains $\boldsymbol{\epsilon}$ are constant because they are obtained by differentiation of the linear displacement field. The stresses, being related to the strains by elastic constants, are also constant. Thus, the differential equations of equilibrium (Eq. 4.3), which involve differentiation of the stresses, are satisfied. $\boldsymbol{\sigma}$ is an *equilibrium stress field*, although no explicit attempt was made to construct it to meet that condition.

What of the equilibrium of stresses across the interface of adjacent elements? Figures 2.5(d) and (e), see page 122, illustrate such a juncture line between elements A and B. The stress matrix for the triangular element, constructed as a set of relationships between the stress field $\boldsymbol{\sigma}$ and the joint displacements $\{\boldsymbol{\Delta}\}$, in accordance with Eq. (5.7a), is presented in Fig. 5.5. From these relationships it is seen directly that each stress component is a function of all joints of an individual element. Consequently in the situation portrayed in Fig. 2.5(e), although both σ_n^A and σ_n^B are functions of u_1, u_2, v_1, and v_2, σ_n^A is a function of u_3 and v_3, while σ_n^B is a function of u_4 and v_4, and the normal and tangential element edge tractions (T_n and T_s) that these produce will not in general be equal. *Interelement equilibrium conditions are not met.*

Interelement continuity of the displacements u and v, on the other hand, is maintained. The displacement field is linear and this linearity describes the variation of displacements along the element edges. When the element edges are joined, the connection of the points 1 and 2 is sufficient to ensure continuity of displacement at all points between them. The satisfaction of this condition can be studied in another way by forming, from the displacement field [Eq. (5.21a)], expressions that describe the displacements of the edges. It will be found that the displacement of each edge is governed entirely

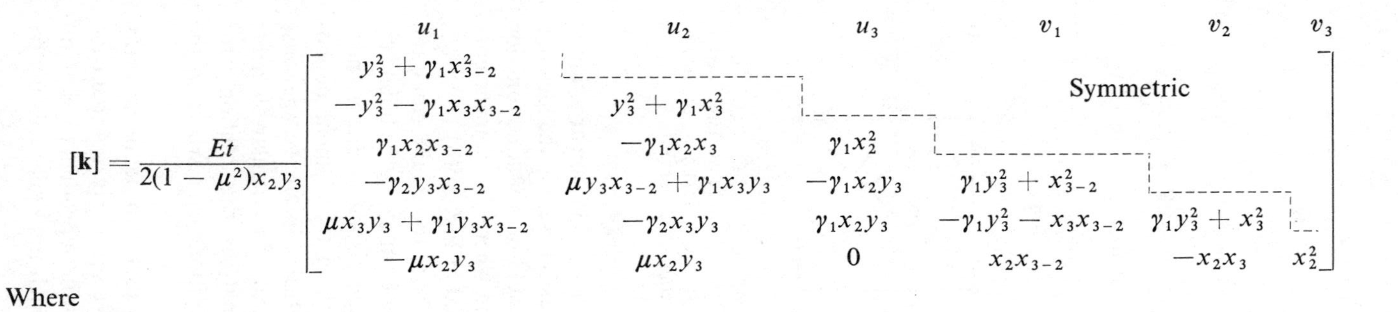

$$
[\mathbf{k}] = \frac{Et}{2(1-\mu^2)x_2y_3}
\begin{array}{c}
\begin{array}{cccccc} u_1 & u_2 & u_3 & v_1 & v_2 & v_3 \end{array} \\
\begin{bmatrix}
y_3^2+\gamma_1x_{3-2}^2 & & & & \text{Symmetric} & \\
-y_3^2-\gamma_1x_3x_{3-2} & y_3^2+\gamma_1x_3^2 & & & & \\
\gamma_1x_2x_{3-2} & -\gamma_1x_2x_3 & \gamma_1x_2^2 & & & \\
-\gamma_2y_3x_{3-2} & \mu y_3x_{3-2}+\gamma_1x_3y_3 & -\gamma_1x_2y_3 & \gamma_1y_3^2+x_{3-2}^2 & & \\
\mu x_3y_3+\gamma_1y_3x_{3-2} & -\gamma_2x_3y_3 & \gamma_1x_2y_3 & -\gamma_1y_3^2-x_3x_{3-2} & \gamma_1y_3^2+x_3^2 & \\
-\mu x_2y_3 & \mu x_2y_3 & 0 & x_2x_{3-2} & -x_2x_3 & x_2^2
\end{bmatrix}
\end{array}
$$

Where

$$\gamma_1 = \frac{1-\mu}{2}, \qquad \gamma_2 = \frac{1+\mu}{2}$$

$$x_{3-2} = x_3 - x_2$$

$$y_{3-2} = y_3 - y_2$$

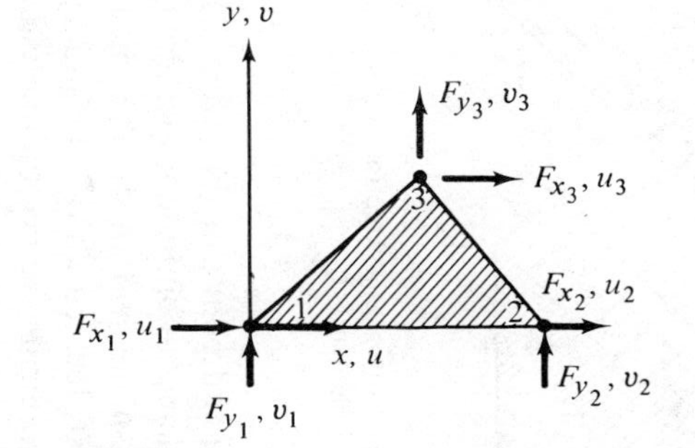

Fig. 5.4 Stiffness matrix for isotropic constant strain triangle in plane stress.

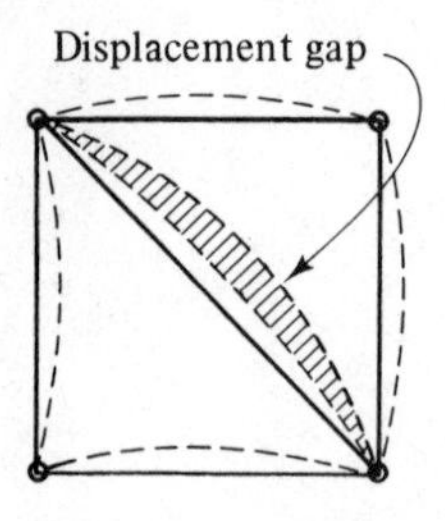

(b) Disparity of displacements along juncture line of connected elements

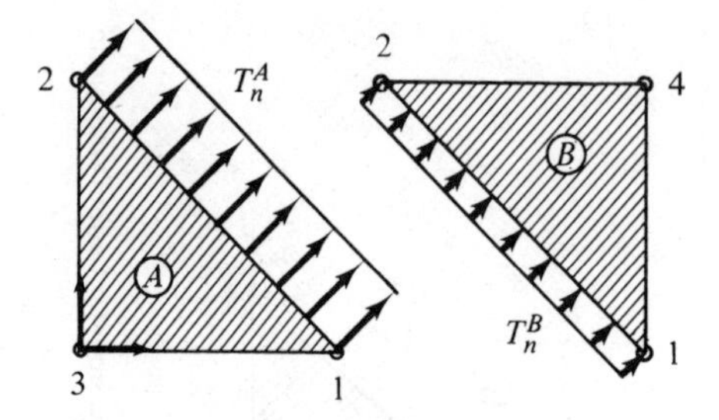

(d) Normal edge tractions on individual elements

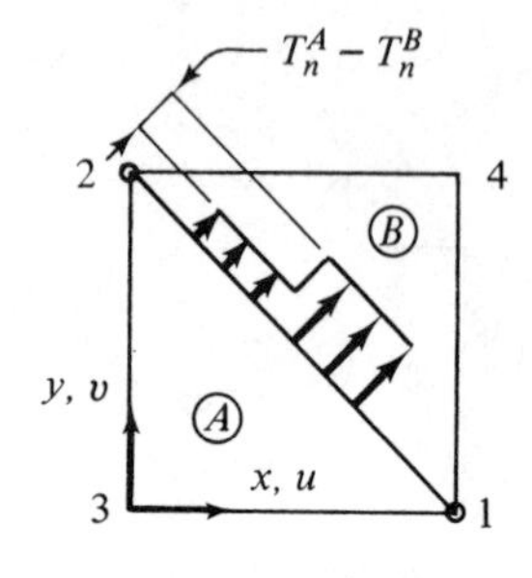

(e) Discontinuity of element edge tractions along juncture line of connected elements

Repeat of (b), (d), (e), Fig. 2.5.

by the end points of that edge. The general approach to the construction of interelement-continuous element displacement fields is based on the idea that the displacement of an element boundary line should be a unique function of d.o.f. located on that boundary line. A situation where the element edge displacements are not uniquely defined is portrayed in Fig. 2.5(b).

Thus, for the simple triangle in plane stress, both the equilibrium and compatibility conditions are met within the element but along the boundary only the condition of interelement continuity of the displacements u and v is met. Conditions of equilibrium are violated along the element boundaries but equilibrium of the boundary forces is satisfied in an average sense at the joints. The effect of a refinement of the grid of triangular elements is to contract, in the successively more refined analyses, the solution error due to the failure to satisfy the equilibrium conditions everywhere in the structure.

It was observed in Section 2.3 that it is often useful to define an array of stiffness coefficients in nondimensional form. We see in Fig. 5.4 that each term of the triangular element stiffness matrix contains a product (or square) of the element dimensions, while the constant which multiplies this matrix has such a product $(x_2 y_3)$ in the denominator. Hence, by factoring the latter into the matrix we obtain a set of nondimensional stiffness coefficients wherein individual terms involve ratios of the element overall dimensions, e.g. y_3/x_2.

$$
\begin{array}{c} \\ [\mathbf{S}] = \end{array}
\begin{array}{c}
\begin{matrix} \quad u_1 & \quad u_2 & \quad u_3 & \quad v_1 & \quad v_2 & \quad v_3 \end{matrix} \\
\begin{bmatrix} -y_3 & y_3 & 0 & \mu x_{3-2} & -\mu x_3 & \mu x_2 \\ -\mu y_3 & \mu y_3 & 0 & x_{3-2} & -x_3 & x_2 \\ \gamma_1 x_{3-2} & -\gamma_1 x_3 & \gamma_1 x_2 & -\gamma_1 y_3 & \gamma_1 y_3 & 0 \end{bmatrix}
\end{array}
\times \frac{E}{(1-\mu^2)x_2 y_3}
$$

$$
\gamma_1 = \frac{(1-\mu)}{2}, \qquad x_{3-2} = x_3 - x_2
$$

[$\boldsymbol{\sigma} = [\mathbf{S}]\{\boldsymbol{\Delta}\}$, where $\{\boldsymbol{\sigma}\} = \lfloor \sigma_x \; \sigma_y \; \tau_{xy} \rfloor^T$ and $\{\boldsymbol{\Delta}\}$ is defined by Eq. (5.19)].

Fig. 5.5 Stress matrix for isotropic constant stress element in plane stress.

5.3 Limitations of the Direct Method

The concept of an element stiffness matrix was introduced in Section 2.3, by definition and without reference to the procedures that must be applied to achieve formulation of its coefficients. Under the same conditions it was shown in Section 2.5 that it must possess the property of symmetry. The governing equation of the direct method [Eq. (5.10)], however, cannot ensure that the resulting stiffness matrix will be symmetric. The central matrix of the triple product ([**A**][**E**][**D**]), the material stiffness matrix [**E**], is by its intrinsic nature a symmetric matrix. The matrices [**A**] and [**D**], on the other hand, are independently formulated and are not necessarily congruent. A congruent transformation applied to the symmetric matrix [**E**] would ensure symmetry.

The difficulty in constructing a symmetric matrix is surmounted if one enforces congruence by supplanting [**A**] in Eq. (5.10) with the transpose of the strain-to-displacement transformation [**D**]. Then, $[\mathbf{k}] = [\mathbf{D}]^T[\mathbf{E}][\mathbf{D}]$. As shown in Section 6.4, a similar result is obtained by application of the *principle of minimum potential energy*. [The procedures differ somewhat for cases where the strains are functions of the spatial coordinates. In the direct method a discrete integration is employed (*vide* the flexural member), whereas the energy formula involves continuous integration.]

Alternatively, one might choose to deal only with the stress-to-force transformation [**A**]. In this case we must employ the inverse form of Eq. (5.10); i.e., $[\mathbf{f}] = [\mathbf{D}]^{-1}[\mathbf{E}]^{-1}[\mathbf{A}]^{-1}$ (with due account being taken of rigid body motion d.o.f.) and supplant $[\mathbf{D}]^{-1}$ with the transpose of $[\mathbf{A}]^{-1}$. The result obtained corresponds to use of the *principle of minimum complementary energy*. This principle will be studied in Section 6.6.

A second limitation in the direct method arises in the identification of the level of interelement displacement continuity, which must be satisfied by the chosen element shape functions. Consider, for example, the plane stress formulation of the prior section. From simple physical reasoning it would appear that the interelement continuity conditions are met fully by continuity of the displacements u and v. Is it necessary, however, to establish continuity of the displacement derivatives du/dx, dv/dy, etc., which are in fact strains? Do we require the continuity of even higher derivatives for plane stress? These questions, which are not answered on a theoretical basis by the direct method, are answered by *variational concepts*, which are also the subject matter of Chapter 6.

A third limitation of the direct method pertains to the treatment of distributed loads, initial strains, and other phenomena such as transients and elastic instability effects. It would appear that only a simple prorating or "lumping" process can be discerned for these effects in the context of the direct method. A more rational approach to formulation of terms for their representation is developed in the next and subsequent chapters on the basis of variational concepts.

Before concluding our remarks concerning the direct method in structural mechanics it is perhaps useful to summarize some of the transformations which we have introduced and that are employed continually in subsequent chapters.

Generalized Displacement-to-Displacement Field

$$\mathbf{\Delta} = [\mathbf{p}]\{\mathbf{a}\} \tag{5.2a}$$

Joint Displacement-to-Displacement Field

$$\mathbf{\Delta} = [\mathbf{N}]\{\mathbf{\Delta}\} \tag{5.5a}$$

Generalized Displacement-to-Joint Displacement

$$\{\mathbf{\Delta}\} = [\mathbf{B}]\{\mathbf{a}\} \tag{5.3a}$$

Generalized Displacement-to-Strain Field

$$\boldsymbol{\epsilon} = [\mathbf{C}]\{\mathbf{a}\} \tag{5.6a}$$

$$= [\mathbf{C}_f]\{\mathbf{a}_f\} \text{ (rigid body d.o.f. } \{\mathbf{a}_s\} = 0) \tag{5.6d}$$

Joint Displacement-to-Strain Field

$$\boldsymbol{\epsilon} = [\mathbf{D}]\{\mathbf{\Delta}\} \tag{5.6c}$$

Joint Displacement-to-Stress Field

$$\boldsymbol{\sigma} = [\mathbf{S}]\{\mathbf{\Delta}\} \tag{5.7a}$$

Joint Displacement-to-Point Stress

$$\{\boldsymbol{\sigma}\} = [\mathbb{S}]\{\boldsymbol{\Delta}\} \tag{3.9}$$

Point Stress-to-Force

$$\{\mathbf{F}\} = [\mathbf{A}]\{\boldsymbol{\sigma}\} \tag{5.8a}$$

5.4 The Direct Method in Nonstructural Problems

The direct method applies equally well to the formulation of finite element equations for physical processes other than elastic deformation. Consider, for example, the case of one-dimensional steady-state conduction heat transfer. This process is of considerable practical importance to structural designers who are concerned with problems of thermal stress analysis, where a consistency of the computational approaches to both the temperature and stress determinations is highly desirable.

The conditions under study are represented by the insulated rod shown in Fig. 5.6. We treat an "axial" thermal element of cross-sectional area A, length L, and thermal conductivity υ. We seek relationships between the temperatures (Υ_1, Υ_2) (°F) at points 1 and 2 and the thermal inputs at these points (H_1, H_2) (Btu).

The applicable constitutive relationship for the present case is Fourier's law of heat conduction:

$$h = -\upsilon \frac{d\Upsilon}{dx} \tag{5.23}$$

where h is the steady-state heat input per unit area (Btu/ft^2). The negative sign reflects the physical observation that heat flows in the direction in which temperature is decreasing. Consistent with prior developments for the axial structural member, we describe Υ as follows:

$$\Upsilon = \lfloor \left(1 - \frac{x}{L}\right) \quad \frac{x}{L} \rfloor \begin{Bmatrix} \Upsilon_1 \\ \Upsilon_2 \end{Bmatrix}$$

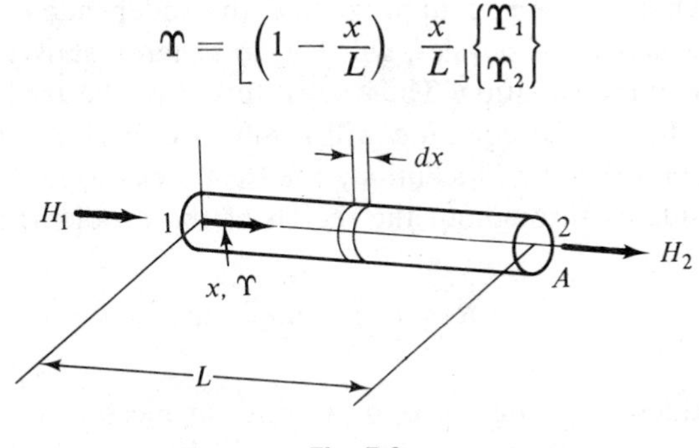

Fig. 5.6

Hence,

$$\frac{d\Upsilon}{dx} = \left\lfloor -\frac{1}{L} \quad \frac{1}{L} \right\rfloor \begin{Bmatrix} \Upsilon_1 \\ \Upsilon_2 \end{Bmatrix} \tag{5.24}$$

The total heat input at joint 1 is $H_1 = hA$, and at point 2 it is $H_2 = -hA$ (the negative sign accounts for the positive direction of H_2 being an *output* at that joint). Combining these relationships with Eqs. (5.23) and (5.24), we find

$$\begin{Bmatrix} H_1 \\ H_2 \end{Bmatrix} = \frac{A\upsilon}{L} \begin{bmatrix} 1 & -1 \\ -1 & 1 \end{bmatrix} \begin{Bmatrix} \Upsilon_1 \\ \Upsilon_2 \end{Bmatrix} \tag{5.25}$$

The set of coefficients on the right side of this equation is termed the *element conductivity matrix*.

It is also possible, by use of the direct method, to form thermal conductivity matrices for planar triangular elements and other simple forms. It is likewise possible to construct the relevant matrices for finite element representations of such processes as seepage, potential flow, and electromagnetism. As we have already noted, however, to deal with more complicated elements and more complicated aspects of these processes it is necessary to employ more sophisticated theoretical concepts. One such concept is explored in the next section.

5.5 Method of Weighted Residuals

With the expansion of the finite element method beyond structural mechanics there is a need for a more general approach to the formulation of element relationships. Such an approach is given by the *method of weighted residuals*, abbreviated here as MWR.[5.3] The method of weighted residuals presumes that a "trial function" (e.g., the polynomial series employed in Sections 5.1 and 5.2), which is chosen to approximate the independent variable in a problem of mathematical physics, will not, in general, satisfy the pertinent governing differential equation. Thus, substitution of the trial function into the governing differential equation will result in a *residual*, denoted by R. In order to obtain the "best" solution, one then seeks to minimize the integral of the residuals throughout the region of the problem; i.e.,

$$\int_{\text{vol}} R \cdot d(\text{vol}) = \text{minimum}$$

One can widen the range of opportunities in meeting this objective by requiring that a *weighted* value of the residual be a minimum throughout the region. The weighting function enables the weighted integral of residuals to achieve a value of zero. Denoting the weighting function by ϕ, the desired, more general form of the above becomes

$$\int_{\text{vol}} R \cdot \phi \, d(\text{vol}) = 0 \tag{5.26}$$

Thus Eq. (5.26) is the general statement of MWR.

The weighting functions may be chosen in a variety of ways and each choice corresponds to a different *criterion* in MWR. Our interest is in the *Galerkin criterion* since this approach yields equations that are identical to those obtained by the commonly used energy, or variational, equations.[5.6, 5.7, 5.8]

To describe the MWR with Galerkin criterion, consider the governing differential equation

$$\mathfrak{D}(\mathbf{\Delta}) = 0 \tag{5.27}$$

where $\mathfrak{D}(\ )$ is the differential operator and $\mathbf{\Delta}$ is the independent variable. The variable is to be approximated by $\bar{\mathbf{\Delta}}$ in the form of Eq. (5.5), i.e., by the sum of n shape functions N_i times their corresponding d.o.f. $\mathbf{\Delta}_i$. Substitution of $\bar{\mathbf{\Delta}}$ into Eq. (5.27) yields the residual

$$R = \mathfrak{D}(\bar{\mathbf{\Delta}}) \neq 0 \tag{5.28}$$

In accordance with the Galerkin criterion the shape functions N_i are chosen to play the role of weighting functions. Thus, for each i we have

$$\int_{\text{vol}} N_i \mathfrak{D}(\mathbf{\Delta}) \, d(\text{vol}) = 0 \quad (i = 1, \ldots, n) \tag{5.29}$$

and this leads to a total of n equations.

Equation (5.29) refers to points within a region, without reference to boundary conditions such as specified applied loads or displacements. To produce boundary conditions we apply integration by parts to Eq. (5.29), yielding integrals on both the region and its boundary.

We illustrate this method by first applying it to the formulation of stiffness equations of the axial member discussed earlier in this chapter but with a distributed load (q) applied as in Fig. 5.7. We seek a formulation in terms of displacements, where $\mathbf{\Delta} = u$. For this case the governing differential equation is obtained by substituting the stress-displacement relationship

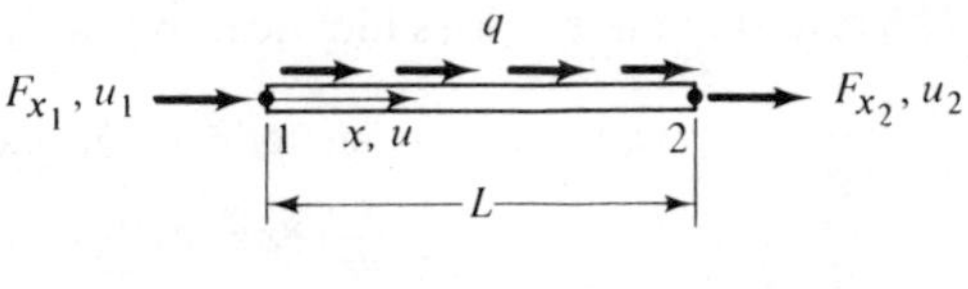

q

F_{x_1}, u_1 1 x, u 2 F_{x_2}, u_2

L

Fig. 5.7

$[\sigma_x = E\,(du/dx)]$ into the equilibrium equation $A\,(d\sigma_x/dx) + q = 0$. [Note that the latter is the one-dimensional case of Eq. (4.3) with $X = q/A$]. We have

$$EA\frac{d^2u}{dx^2} + q = 0 \tag{5.30}$$

The left side of this equation is $\mathfrak{D}(\Delta)$ for this case. The approximating function, $\bar{u}$, is given by Eq. (5.5) and, by substitution of this into Eq. (5.30) and introduction of the MWR concept with Galerkin criterion, we find

$$\int_0^L N_i\left(EA\frac{d^2\bar{u}}{dx^2}\right)dx = -\int_0^L N_i q\,dx \quad (i = 1, 2) \tag{5.31}$$

We now apply integration by parts† to the left-hand side, resulting in

$$\int_0^L \left(\frac{dN_i}{dx}\right)\frac{d\bar{u}}{dx}EA\,dx = \int_0^L N_i q\,dx + N_i EA\frac{d\bar{u}}{dx}\bigg|_0^L \tag{5.31a}$$

Since the parameters u_i are not functions of the coordinates,

$$\frac{d\bar{u}}{dx} = \Sigma\frac{dN_i}{dx}u_i = \left\lfloor\frac{\mathbf{dN}}{\mathbf{dx}}\right\rfloor\{\mathbf{u}\}$$

where $\{\mathbf{u}\}$ is the vector of element joint displacements. We now substitute this relationship into the left side of Eq. (5.31a)

$$EA\int_0^L\left(\frac{dN_i}{dx}\right)\left\lfloor\frac{\mathbf{dN}}{\mathbf{dx}}\right\rfloor dx\,\{\mathbf{u}\} = \int_0^L N_i q\,dx + N_i EA\frac{d\bar{u}}{dx}\bigg|_0^L \tag{5.31b}$$

and the complete set is, with $F_1 = EA\,(d\bar{u}/dx)$ and $N_1 = 1$ at $x = 0$, $N_1 = 0$ at $x = L$, and similarly for F_2 and N_2

$$[\mathbf{k}]\{\mathbf{u}\} = \{\mathbf{F}\} + \{\mathbf{F}^d\} \tag{5.32}$$

where

$$[\mathbf{k}] = \left[EA\int_0^L\left\{\frac{\mathbf{dN}}{\mathbf{dx}}\right\}\left\lfloor\frac{\mathbf{dN}}{\mathbf{dx}}\right\rfloor dx\right] \tag{5.33}$$

$$\{\mathbf{F}\} = \begin{Bmatrix}F_1\\F_2\end{Bmatrix}, \qquad \{\mathbf{F}^d\} = \begin{Bmatrix}\int_0^L N_1 q\,dx\\ \int_0^L N_2 q\,dx\end{Bmatrix} \tag{5.34), (5.35}$$

The resulting matrix $[\mathbf{k}]$ is identical to that formulated in Section 5.1.

†For this case, the appropriate formula for integration by parts is

$$\int_0^L N_i\frac{d^2\bar{u}}{dx^2}dx = N_i\frac{d\bar{u}}{dx}\bigg|_0^L - \int_0^L\frac{dN_i}{dx}\frac{d\bar{u}}{dx}dx$$

We see in the above that the governing differential equation in terms of displacement variables, the equilibrium differential equation, can be transformed via the method of weighted residuals into algebraic equations in terms of displacement parameters and with coefficients in the form of integrals. This result is identified in Chapter 6 as a potential energy formulation of the finite element equations. Correspondingly, the governing differential equations in terms of stress variables can be transformed into finite element equations by either the method of weighted residuals or the complementary energy approach; the results in either case are identical.

The governing differential equations result from the combination of the subsidiary equations of elasticity. In the development above the governing differential equation is obtained by substituting the stress-displacement equation into the equilibrium differential equation. What would happen if we were to apply the method of weighted residuals directly to the subsidiary equations? We have

$$\text{Equilibrium:}\ \frac{d\sigma_x}{dx} + \frac{q}{A} = 0 \tag{5.36}$$

$$\text{Stress-displacement:}\ \frac{du}{dx} - \frac{\sigma_x}{E} = 0 \tag{5.37}$$

In developing algebraic equations for this formulation we apply a weighting factor ψ to the equilibrium differential equation and a weighting factor ϕ to the stress-displacement differential equation

$$\int_0^L \left(\frac{d\bar{\sigma}_x}{dx} + \frac{q}{A}\right) \psi\, A\, dx = 0 \tag{5.38}$$

$$\int_0^L \left(\frac{d\bar{u}}{dx} - \frac{\bar{\sigma}_x}{E}\right) \phi\, A\, dx = 0 \tag{5.39}$$

The approximations to stress and displacement are chosen as

$$\bar{u} = \lfloor \mathbf{N} \rfloor \{\mathbf{u}\} \tag{5.5b}$$

$$\bar{\sigma}_x = \lfloor \boldsymbol{\Xi} \rfloor \{\boldsymbol{\sigma}\} \tag{5.40}$$

where the terms of $\lfloor \mathbf{N} \rfloor$ are the displacement shape functions and $\lfloor \boldsymbol{\Xi} \rfloor$ is employed to designate the stress shape functions. In selecting weighting factors from these terms we employ the displacement shape functions N_i for ψ and the stress shape functions Ξ_i for ϕ. Consider first the weighted integral of the equilibrium differential equation. By introduction of $\psi = N_i$ we have

$$\int_0^L \left(\frac{d\bar{\sigma}_x}{dx} + \frac{q}{A}\right) N_i A\, dx = 0 \tag{5.38a}$$

and, after integration by parts of the first term and rearrangement,

$$\int_0^L \frac{dN_i}{dx}\bar{\sigma}_x A\, dx = N_i A\bar{\sigma}_x \Big|_0^L + \int_0^L qN_i\, dx \tag{5.41}$$

Introducing Eq. (5.40) for $\bar{\sigma}_x$ and noting that $N_i A\bar{\sigma}_x = F_i$, we have for all values of N_i

$$[\mathbf{\Omega}_{21}]\{\boldsymbol{\sigma}\} = \{\mathbf{F}\} + \{\mathbf{F}^d\} \tag{5.42}$$

where

$$[\mathbf{\Omega}_{21}] = \left[\int_0^L \left\{\frac{\mathbf{dN}}{\mathbf{dx}}\right\}\lfloor \Xi \rfloor A\, dx\right] \tag{5.43}$$

$$\{\mathbf{F}^d\} = \left\{\int_0^L qN_i\, dx\right\} \tag{5.44}$$

and $\{\mathbf{F}\}$ is the conventional listing of joint forces.

Considering next the weighted integral of the stress-displacement equation we have, upon introduction of $\phi = \Xi_i$,

$$\int_0^L \left(\frac{d\bar{u}}{dx} - \frac{\bar{\sigma}_x}{E}\right)\Xi_i\, A\, dx = 0 \tag{5.39a}$$

and, after introduction of Eq. (5.5b) and (5.40) for $\bar{u}$ and $\bar{\sigma}_x$, we have for all Ξ_i

$$[\mathbf{\Omega}_{21}]^{\mathrm{T}}\{\mathbf{u}\} + [\mathbf{\Omega}_{11}]\{\boldsymbol{\sigma}\} = 0 \tag{5.45}$$

where $[\mathbf{\Omega}_{21}]$ is as defined by Eq. (5.43) and

$$[\mathbf{\Omega}_{11}] = -\left[\int_0^L \{\Xi\}\frac{\lfloor \Xi \rfloor}{E} A\, dx\right] \tag{5.46}$$

We see that this relationship connects $\{\mathbf{u}\}$ and $\{\boldsymbol{\sigma}\}$ so it is convenient to group both Eqs. (5.42) and (5.45) into a single matrix equation:

$$\begin{bmatrix} \mathbf{\Omega}_{11} & \mathbf{\Omega}_{21}^{\mathrm{T}} \\ \mathbf{\Omega}_{21} & 0 \end{bmatrix}\begin{Bmatrix} \boldsymbol{\sigma} \\ \mathbf{u} \end{Bmatrix} = \begin{Bmatrix} 0 \\ \mathbf{F} + \mathbf{F}^d \end{Bmatrix} \tag{5.47}$$

Thus, the equations we have derived are of the *mixed format* [see Section 2.3 and Eq. (2.3)]. It is shown in Chapter 6 that when variational principles are used in the formulation of element relationships the same result derives from a quantity known as the *Reissner energy*. Since combinations of the basic equations of elasticity, other than those employed above, are possible, it is clear that other mixed formats of the element force-displacement equations can be established.

Let us now examine the method of weighted residuals in a two-dimensional problem. We choose for this purpose the differential equation

$$\frac{\partial^2 \Phi}{\partial x^2} + \frac{\partial^2 \Phi}{\partial y^2} = C \tag{5.48}$$

or, in a more concise symbolism,

$$\nabla^2 \Phi = C \tag{5.49}$$

$$\nabla^2 = \frac{\partial^2}{\partial x^2} + \frac{\partial^2}{\partial y^2} \tag{4.20}$$

where ∇^2 is the *Laplacian operator*. Eq. (5.49), which is *Poisson's differential equation*, governs a very wide variety of physical problems. In structural mechanics it may describe a stretched membrane under transverse pressure, in which case Φ is the transverse displacement and C is a function of the pressure/tension ratio, or it may be employed to describe the torsion of noncircular sections, where Φ is the relevant stress function. The equation governs potential flow, in which Φ is the stream function or potential function, and conduction heat transfer, wherein Φ is the temperature.

In accordance with the method of weighted residuals, we approximate Φ as follows:

$$\bar{\Phi} = \lfloor \mathbf{N} \rfloor \{\mathbf{\Phi}\} \tag{5.50}$$

where $\lfloor \mathbf{N} \rfloor$, as before, comprises a set of shape functions that define the shape of $\bar{\Phi}$ in x and y coordinates and $\{\mathbf{\Phi}\}$ lists the values of $\bar{\Phi}$ at discrete points. We have, from the weighted residual concept,

$$\int_A (\nabla^2 \bar{\Phi} - C) N_i \, dA = 0 \tag{5.51}$$

As before, we apply integration by parts to the differential portion of this equation. This is given by the first form of Green's Theorem[5.9], which states

$$\int_A \nabla^2 \bar{\Phi} \cdot N_i \, dA = \int_S N_i \left(l_x \frac{\partial \bar{\Phi}}{\partial x} + l_y \frac{\partial \bar{\Phi}}{\partial y} \right) dS - \int_A \left(\frac{\partial \bar{\Phi}}{\partial x} \frac{\partial N_i}{\partial x} + \frac{\partial \bar{\Phi}}{\partial y} \frac{\partial N_i}{\partial y} \right) dA \tag{5.52}$$

After introducing Eq. (5.50) for $\bar{\Phi}$, we have for all $\bar{\Phi}_i$

$$[\mathbf{k}^n]\{\mathbf{\Phi}\} = \{\mathbf{F}^n\} + \{\mathbf{F}^c\} \tag{5.53}$$

where

$$[\mathbf{k}^n] = \left[\int_A \left[\left\{\frac{\partial \mathbf{N}}{\partial \mathbf{x}}\right\} \left\lfloor \frac{\partial \mathbf{N}}{\partial \mathbf{x}} \right\rfloor + \left\{\frac{\partial \mathbf{N}}{\partial \mathbf{y}}\right\} \left\lfloor \frac{\partial \mathbf{N}}{\partial \mathbf{y}} \right\rfloor\right] dA\right] \tag{5.54}$$

$$\{\mathbf{F}^n\} = \left\{\int_S \{\mathbf{N}\}\left(l_x \frac{\partial \bar{\Phi}}{\partial x} + l_y \frac{\partial \bar{\Phi}}{\partial y}\right) dS\right\} \tag{5.55}$$

$$\{\mathbf{F}^c\} = \left\{\int_A \{\mathbf{N}\} C \, dA\right\} \tag{5.56}$$

Eq. (5.55) indicates that the specified problem data will include the values of the derivatives of $\bar{\Phi}$ along the boundary. Problems governed by Eq. (5.48) are characterized however by boundary-condition equations that yield an expression alternative to Eq. (5.55).

Certain features of the method of weighted residuals, as it has been applied here, are noteworthy. First, it is clear that an integral whose value properly characterizes a certain type of behavior can be readily formed from a weighted integral of the approximated solution of the pertinent differential equation. A number of alternative forms of differential equations exist in structural mechanics and consequently there are alternative integrals.

The second feature of note is the role of integration by parts. The MWR formulation, prior to application of this theorem, considers only the interior of the element. The integration by parts, by generating surface integrals, gives a means of accounting for boundary conditions. Repeated application of the theorem enables definition of alternative formats for representation of the boundary integrals.

The method of weighted residuals with Galerkin criterion and the energy (variational) methods yield identical finite element formulations for the class of structural analysis problems covered in this text. Since the energy methods are more familiar to the structural engineer and are the basis for the preponderance of literature relating to element formulation, we restrict our attention to them in the next and following chapters.

REFERENCES

5.1 Turner, M., R. Clough, H. Martin, and L. Topp, "Stiffness and Deflection Analysis of Complex Structures," *J. Aero. Sci.*, **23**, No. 9, Sept., 1956, pp. 805–823, 854.

5.2 Gallagher, R. H., *Correlation Study of Methods of Matrix Structural Analysis*, Pergamon Press, New York, N.Y., 1964.

5.3 Finlayson, B., *The Method of Weighted Residuals and Variational Principles*, Academic Press, New York, N.Y., 1972.

5.4 Hutton, S. G. and D. L. Anderson, "Finite Element Method: A Galerkin

Approach," Proc. ASCE, *J. Engr. Mech. Div.*, **97**, No. EM5, Oct., 1971, pp. 1503–1520.

5.5 ARAL, M., P. MAYER, and C. V. SMITH, "Finite Element Galerkin Method Solutions to Selected Elliptic and Parabolic Differential Equations." Proc. of Third Air Force Conf. on Matrix Methods in Struct. Mech., Oct., 1971.

5.6 ZIENKIEWICZ, O. C. and C. J. PAREKH, "Transient Field Problems: Two-Dimensional and Three-Dimensional Analysis by Isoparametric Elements," *Int. J. Numerical Meth. Engr.*, **2**, No. 1, 1970, pp. 61–72.

5.7 SZABO, B. and G. C. LEE, "Derivation of Stiffness Equations for Problems in Elasticity by Galerkin's Method," *Internatl. J. Numerical Meth. Engr.*, **1**, No. 3, 1969, pp. 301–310.

5.8 SZABO, B. A. and G. C. LEE, "Stiffness Matrix for Plates by Galerkin's Method," Proc. ASCE, *J. Engr. Mech. Div.*, **95**, No. EM 3, June, 1969, pp. 571–585.

5.9 SOKOLNIKOFF, I. and R. REDHEFFER, *Mathematics of Physics and Modern Engineering*, 2nd ed., McGraw-Hill Book Co., New York, N.Y., 1966, pp. 370–375.

PROBLEMS

5.1 Construct the stiffness matrix for a simple torsional element, using the *direct* method.

5.2 Form the stiffness matrix for the rectangle in plane stress on the basis of linear edge displacements, using the *direct* method. Establish this result numerically by first forming the [D] and [A] matrices in general terms, evaluating them, and forming the product [A][E][D] on the computer. Use $t = 0.1$ in., $x_2 = 16$ in., $y_3 = 12$ in., $E = 10^7$, $\mu = 0.3$. Check your result with Fig. 9.13.

$$u = N_1 u_1 + N_2 u_2 + N_3 u_3 + N_4 u_4$$

$$v = N_1 v_1 + N_2 v_2 + N_3 v_3 + N_4 v_4$$

$$N_1 = (1 - \xi)(1 - \eta), \qquad N_2 = \eta(1 - \xi), \qquad N_3 = \xi\eta, \qquad N_4 = (1 - \xi)\eta$$

where $\xi = \dfrac{x}{x_2}$, $\eta = \dfrac{y}{y_3}$

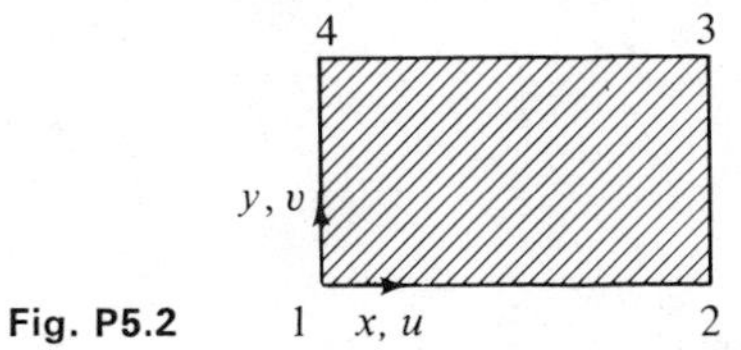

Fig. P5.2

5.3 Construct the stress matrix [S] for the rectangular element in plane stress whose formulation is based upon the shape functions given in Prob. 5.2. Retain x and y as variables.

5.4 Construct the stress matrix [S] for the triangle in plane stress formulated in Section 5.3, for orthotropic material properties.

5.5 Develop the procedure for formulation of the stiffness matrix for the beam element on the basis of the weighted residual-Galerkin approach and illustrate by formulating the first row (F_1 versus $w_1, \theta_1, w_2, \theta_2$).

5.6 Establish, using the weighted-residual-Galerkin approach, the necessary integral relationships for the direct construction of plane stress element stiffness matrices based on assumed displacement fields.

5.7 Define the *direct* approach to the direct formulation of element flexibility matrices and illustrate by construction of the flexibility matrix of a beam element supported as a cantilever.

5.8 Define the *direct* approach to the direct formulation of element force-displacement matrices in the mixed format (see Chapter 2) and illustrate by application to the beam element.

5.9 Construct the thermal conductivity matrix for the planar triangular element using the *direct* method and a linear temperature field $\Upsilon = N_1\Upsilon_1 + N_2\Upsilon_2 + N_3\Upsilon_3$, where Υ_1, Υ_2, and Υ_3 are the vertex temperatures.

5.10 The differential equation governing the buckling of beams is of the form

$$EI\frac{d^4w}{dx^4} + F_x\frac{d^2w}{dx^2} = 0$$

Using the weighted-residual-Galerkin approach, construct the integral form necessary for the construction of the pertinent element stiffness equations. Assume that the displacement field will be approximated by $w = N_1w_1 + N_2\theta_1 + N_3w_2 + N_4\theta_2$, where $N_1, \ldots, N_4$ are given by Eq. (5.14a).

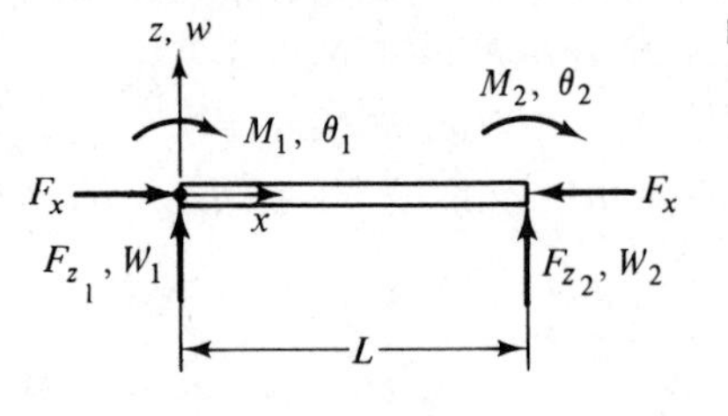

Fig. P5.10

6

Variational Methods of Element Formulation

The variational, or energy, methods of structural mechanics constitute a powerful and widely used approach to the formulation of element relationships. Rudimentary forms of these methods have been tools for the analysis of engineering structures for more than a century. Certain sophisticated forms of these methods are as recent as finite element analysis itself, however, and it can be conjectured that in some cases their emergence was motivated by an interest in establishing a new formulative basis for finite element analysis. Whether or not this has indeed been the case, modern developments have produced a comprehensive identification of the possible forms of variational principles in structural mechanics as well as their scope and limitations.

This chapter defines the respective variational principles in structural mechanics in terms of formulative procedures for *element* equations. Application to the construction of the equations for the *complete structure* is undertaken in Chapter 7. Thus, we adopt the view that force-displacement equations can be formulated for the individual finite element and that the construction of global equations is a separate operation. This is consistent with the *framework analysis* perspective on finite element analysis, first described in Section 2.2 and employed subsequently in Chapters 3 and 5. Energy concepts, however, permit a different view of finite element analysis, in which the global equations derive from a summation of element energies. We make a transition from one view to the other in this chapter and in Chapter 7.

We begin this chapter with a detailed derivation of the principle of virtual displacements. A brief outline is then given of basic concepts in variational calculus, followed by detailed developments of the minimum *potential* and *complementary* energy principles, respectively. These are special forms of the more basic principles of virtual displacements and virtual forces, respectively. Finally, an outline is given of the *mixed* variational principles, as well as of *hybrid* and *generalized variational* methods of element formulation that derive from either the potential or the complementary energy principles.

We are exclusively concerned with the individual finite element in this chapter. All expressions are therefore written as if the complete structure were the individual element and no subscripts or superscripts are marked to distinguish between element and global terms.

6.1 Principle of Virtual Work

6.1.1 Statement and Proof of the Principle

The *principle of virtual work* underlies the variational principles to be described subsequently—the conventional principles of stationary potential and complementary energy, respectively, as well as the less familiar *two-field* energy principles. The principle of virtual work is in fact an independent formulative procedure for finite element equations. There are two commonly used forms of the general principle, those of *virtual displacements* and *virtual forces*. These lead, respectively, to the conventional principles of stationary potential and complementary energy.

In the *virtual displacements* form of the principle of virtual work, it is assumed that a body in equilibrium under body forces and applied loads is subjected to a virtual (imaginary) displacement state described by components δu, δv, δw at each point. The virtual displacements must be *kinematically admissible*; i.e., they are continuous functions of the spatial coordinates and satisfy the kinematic boundary conditions on the portion of the surface on which such conditions are prescribed.

In Fig. 6.1(a), for example, a variety of admissible displacement shapes for the beam are shown with dotted lines. Each of these evidences the pinned condition of the end points and has a continuous variation of slope (a re-

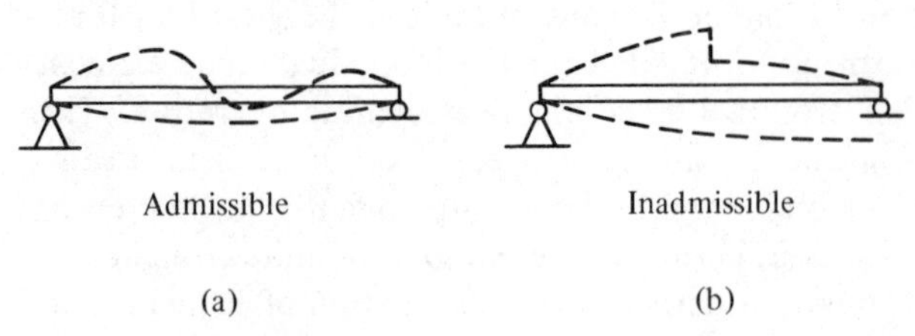

Fig. 6.1 Admissible versus inadmissible virtual displacements.

quirement for bending) between these points. The displacement fields of Fig. 6.1(b) either lack adherence to the support conditions or are discontinuous in slope between the end points and are, therefore, inadmissible. The meaning of the admissibility requirement with regard to the selection of displacement fields in finite element analysis will be given close scrutiny in subsequent sections.

Under the conditions above, the principle of virtual displacements stipulates that the sum of the potential of the applied loads (δV) and the stored strain energy (δU) during virtual displacements $\delta\mathbf{\Delta}$ is equal to zero. Thus,

$$\delta U + \delta V = 0 \tag{6.1}$$

To prove this statement we refer to a state of plane stress ($\sigma_z = \tau_{xz} = \tau_{yz} = 0$) for a plate of unit thickness with zero body forces and initial strains. (Extension to the general case presents no difficulty.) The state of stress in the body in equilibrium is represented by $\sigma_x, \sigma_y, \tau_{xy}$. The components of the virtual displacement field $\delta\mathbf{\Delta}$, are designated as δu and δv and corresponding to these are the variations in strain where, in consequence of the strain-displacement equations,

$$\delta\epsilon_x = \frac{\partial(\delta u)}{\partial x}, \qquad \delta\epsilon_y = \frac{\partial(\delta v)}{\partial y}, \qquad \delta\gamma_{xy} = \frac{\partial(\delta v)}{\partial x} + \frac{\partial(\delta u)}{\partial y} \tag{6.2}$$

We have, for the strain energy under the virtual displacements,

$$\delta U = \int_A \boldsymbol{\sigma}\cdot\delta\boldsymbol{\epsilon}\, dA = \int_A [\sigma_x(\delta\epsilon_x) + \sigma_y(\delta\epsilon_y) + \tau_{xy}(\delta\gamma_{xy})]\, dA \tag{6.3}$$

Note that the work due to change of stress in consequence of the virtual displacements is disregarded. Now, substituting the strain-displacement equations [Eq. (6.2)], we obtain

$$\delta U = \int_A \left\{\sigma_x \frac{\partial}{\partial x}(\delta u) + \sigma_y \frac{\partial}{\partial y}(\delta v) + \tau_{xy}\left[\frac{\partial}{\partial y}(\delta u) + \frac{\partial}{\delta x}(\delta v)\right]\right\} dA \tag{6.3a}$$

Next we observe that

$$\int_A \frac{\partial(\sigma_x\,\delta u)}{\partial x}\, dA = \int_A \delta u \frac{\partial\sigma_x}{\partial x}\, dA + \int_A \sigma_x \frac{\partial(\delta u)}{\partial x}\, dA$$

or

$$\int_A \sigma_x \frac{\partial(\delta u)}{\partial x}\, dA = \int_A \frac{\partial(\sigma_x\,\delta u)}{\partial x}\, dA - \int_A \delta u \frac{\partial\sigma_x}{\partial x}\, dA \tag{6.4}$$

and similarly for integrals involving

$$\sigma_y \frac{\partial}{\partial y}(\delta v), \qquad \tau_{xy}\frac{\partial}{\partial y}(\delta u), \qquad \tau_{xy}\frac{\partial}{\partial x}(\delta v)$$

Hence, by substitution,

$$\begin{aligned}\delta U = &-\int_A \left[\left(\frac{\partial \sigma_x}{\partial x} + \frac{\partial \tau_{xy}}{\partial y}\right)\delta u + \left(\frac{\partial \sigma_y}{\partial y} + \frac{\partial \tau_{xy}}{\partial x}\right)\delta v\right] dA \\ &+ \int_A \left[\frac{\partial}{\partial x}(\sigma_x\,\delta u) + \frac{\partial}{\partial x}(\tau_{xy}\,\delta v) + \frac{\partial}{\partial y}(\sigma_y\,\delta v) + \frac{\partial}{\partial y}(\tau_{xy}\,\delta u)\right] dA\end{aligned} \tag{6.3b}$$

The terms within parentheses in the first integral on the right side of this equation

$$\left(\frac{\partial \sigma_x}{\partial x} + \frac{\partial \tau_{xy}}{\partial x}\right), \qquad \left(\frac{\partial \sigma_y}{\partial y} + \frac{\partial \tau_{xy}}{\partial x}\right)$$

equal zero in accordance with the differential equations of equilibrium [Eq. (4.2)]. Thus, the expression for δU simplifies to

$$\delta U = \int_A \left[\frac{\partial}{\partial x}[\sigma_x\,\delta u + \tau_{xy}\,\delta v] + \frac{\partial}{\partial y}[\sigma_y\,\delta v + \tau_{xy}\,\delta u]\right] dA \tag{6.3c}$$

We now introduce *Gauss' theorem* (integration by parts in the plane), in order to transform the expression above, which applies to a point within the structure, into an expression that deals with both the interior and the boundaries of the structure. Physically, the theorem balances inflow and outflow with the quantities retained in the region.

In accordance with this theorem,

$$\begin{aligned}&\int_A \left[\frac{\partial}{\partial x}(\sigma_x\,\delta u + \tau_{xy}\,\delta v) + \frac{\partial}{\partial y}(\sigma_y\,\delta v + \tau_{xy}\,\delta u)\right] dA \\ &\qquad = \int_S [(\sigma_x\,\delta u + \tau_{xy}\,\delta v)l_x + (\sigma_y\,\delta v + \tau_{xy}\,\delta u)l_y]\, dS\end{aligned} \tag{6.5}$$

where l_x and l_y are direction cosines of the normal to the surface. Hence, Eq. (6.3c) becomes

$$\delta U = \int_S [(\sigma_x l_x + \tau_{xy} l_y)\,\delta u + (\sigma_y l_y + \tau_{xy} l_x)\,\delta v]\, dS \tag{6.6}$$

Now, we observe from Eq. (4.5) that $\bar{T}_x = \sigma_x l_x + \tau_{xy} l_y$ and $\bar{T}_y = \sigma_y l_y + \tau_{xy} l_x$, where $\bar{T}_x$ and $\bar{T}_y$ are the prescribed boundary tractions. Also, it is useful to distinguish between the portion of the boundary where tractions are prescribed, which we shall designate as S_σ, and the portion on which displacements are prescribed, which is designated as S_u. The virtual displacements are zero on the portion of the boundary where the displacements are prescribed. Thus,

$$\delta U = \int_{S_\sigma} (\bar{T}_x \, \delta u + \bar{T}_y \, \delta v) \, dS = -\delta V$$

since the potential of the applied loads due to the virtual displacement δV is clearly equal to the indicated integral. [We assign a minus sign to δV because it represents (the variation of) the potential energy of the applied loads, which is reduced as an elastic structure deflects.] The validity of the principle of virtual work is thereby proved.

Very often considerable emphasis is placed on the fact that no constitutive relationships enter into the above so that the stress-strain law in application of the principle is not restricted to be linear. In finite element analysis of material nonlinearities, however, the usual practice is to linearize the process within successive increments of load. The generality of the concept, however, is of importance in the establishment of formulations that account for the incrementation.

We shall not examine the principle of virtual forces in this text but rather give preference to the derivative principle of stationary complementary energy that is discussed in Section 6.6. If the virtual force concept were applied, consideration would be given to a virtual stress state that satisfies equilibrium conditions.

6.1.2 Finite Element Discretization of Virtual Work.

With the validity of the virtual work principle established, we turn to the establishment of a general procedure for finite element stiffness matrix formulation. We consider first the choice of an assumed displacement field $\mathbf{\Delta}$. As previously noted, we print $\mathbf{\Delta}$ in boldface to indicate that it may represent the full set of coordinate displacements u, v, and w. Consistent with prior terminology, we write the expression for this field in the form

$$\mathbf{\Delta} = [\mathbf{N}]\{\mathbf{\Delta}\} \tag{5.5a}$$

By application of the pertinent strain-displacement equations, we have

$$\boldsymbol{\epsilon} = [\mathbf{D}]\{\mathbf{\Delta}\} \tag{5.6c}$$

We take the distribution of virtual displacements ($\delta\mathbf{\Delta}$) and virtual strains ($\delta\boldsymbol{\epsilon}$) in the same forms as Eqs. (5.5a) and (5.6c). Hence

$$\delta\mathbf{\Delta} = [\mathbf{N}]\{\delta\mathbf{\Delta}\} \tag{6.7}$$

$$\delta\boldsymbol{\epsilon} = [\mathbf{D}]\{\delta\mathbf{\Delta}\} \tag{6.8}$$

Also, we apply the virtual displacement principle to a rather general case, one in which body forces $\mathbf{X}$ (denoting coordinate direction body forces X, Y, and Z) and initial strains $\boldsymbol{\epsilon}^{\text{init.}}$ are present. In the presence of the latter the constitutive equations are

$$\boldsymbol{\sigma} = [\mathbf{E}]\boldsymbol{\epsilon} - [\mathbf{E}]\boldsymbol{\epsilon}^{\text{init.}} \tag{4.15}$$

The virtual work expression can now be constructed. At this point we consider only joint forces F_i. Distributed loads will be dealt with subsequently. The potential of the applied loads of the joint forces $\{\mathbf{F}\}$ in consequence of the virtual joint displacements $\{\delta\boldsymbol{\Delta}\}$ is†

$$\delta V = -\lfloor \delta\boldsymbol{\Delta} \rfloor \{\mathbf{F}\} \tag{6.9}$$

The work of the internal forces arises from the action of the internal stresses $\boldsymbol{\sigma}$ through the strains $\delta\boldsymbol{\epsilon}$ associated with the virtual displacements. Generalizing Eq. (6.3) to a volume integral

$$\delta U = \int_{\text{vol}} \boldsymbol{\sigma}\, \delta\boldsymbol{\epsilon}\, d(\text{vol}) \tag{6.10}$$

and, with the stress-strain law [Eq. (4.15)] substituted for $\boldsymbol{\sigma}$,

$$\delta U = \int_{\text{vol}} \boldsymbol{\epsilon}[\mathbf{E}]\, \delta\boldsymbol{\epsilon}\, d(\text{vol}) - \int_{\text{vol}} \boldsymbol{\epsilon}^{\text{init.}}[\mathbf{E}]\, \delta\boldsymbol{\epsilon}\, d(\text{vol}) \tag{6.11}$$

Next, to obtain the discretized forms of the virtual work and energy expressions, we substitute Eq. (5.6c) for $\boldsymbol{\epsilon}$, and Eq. (6.8) for $\delta\boldsymbol{\epsilon}$:

$$\delta U = \lfloor \delta\boldsymbol{\Delta} \rfloor \{[\mathbf{k}]\{\boldsymbol{\Delta}\} - \{\mathbf{F}^{\text{init.}}\}\} \tag{6.12}$$

where

$$[\mathbf{k}] = \left[\int_{\text{vol}} [\mathbf{D}]^T[\mathbf{E}][\mathbf{D}]\, d(\text{vol})\right] \quad \text{(element stiffness matrix)} \tag{6.12a}$$

$$\{\mathbf{F}^{\text{init.}}\} = \left\{\int_{\text{vol}} [\mathbf{D}]^T[\mathbf{E}]\, \boldsymbol{\epsilon}^{\text{init.}}\, d(\text{vol})\right\} \quad \text{(element intial force vector)} \tag{6.12b}$$

To account for the body forces the potential of the applied loads δV must be supplemented by the integral $-\int_{\text{vol}} \delta\boldsymbol{\Delta} \cdot \mathbf{X}\, d(\text{vol})$. By substitution of $\delta\boldsymbol{\Delta} =$

†The indicated dot and vector products are consistent with the definition of work as the product of force and the displacement in the direction of the force. See Section 2.4. The work is performed while the force is effectively at its full intensity so that the factor of $\frac{1}{2}$ (for application of load with intensity increasing from zero to the final value) does not apply here.

$[\mathbf{N}]\{\delta\mathbf{\Delta}\}$ we obtain $-\lfloor\delta\mathbf{\Delta}\rfloor\{\mathbf{F}^b\}$, where

$$\{\mathbf{F}^b\} = \left\{\int_{\text{vol}} [\mathbf{N}]^{\mathrm{T}}\mathbf{X}\, d(\text{vol})\right\} \quad \text{(element body force vector)} \tag{6.12c}$$

Furthermore, we shall specifically characterize the body force as being due to dynamic behavior, in which case, by D'Alembert's principle, it is the reversed effective force (inertia force)

$$\mathbf{X} = -[\boldsymbol{\rho}]\ddot{\mathbf{\Delta}} \tag{6.13}$$

where $[\boldsymbol{\rho}]$ is the mass density per unit volume tensor, written here in matrix form. From Eq. (5.5a), with the assumption that the time variation of $\{\mathbf{\Delta}\}$ describes the motion fully, we have

$$\ddot{\mathbf{\Delta}} = [\mathbf{N}]\{\ddot{\mathbf{\Delta}}\} \tag{6.14}$$

so that

$$\{\mathbf{F}^b\} = -[\mathbf{m}]\{\ddot{\mathbf{\Delta}}\} \tag{6.12d}$$

where

$$[\mathbf{m}] = \int_{\text{vol}} [\mathbf{N}]^{\mathrm{T}}[\boldsymbol{\rho}][\mathbf{N}]\, d(\text{vol}) \quad \text{(element consistent mass matrix)} \tag{6.12e}$$

Equating δU to $-\delta V$ in accordance with the virtual work principle [Eq. (6.1)], one obtains

$$\lfloor\delta\mathbf{\Delta}\rfloor\{-[\mathbf{m}]\{\ddot{\mathbf{\Delta}}\} + \{\mathbf{F}\}\} = \lfloor\delta\mathbf{\Delta}\rfloor[[\mathbf{k}]\{\mathbf{\Delta}\} - \{\mathbf{F}^{\text{init.}}\}] \tag{6.15}$$

Finally, noting that this expression holds for any values of the virtual joint displacements $\{\boldsymbol{\delta}\mathbf{\Delta}\}$, we find

$$\{\mathbf{F}\} = [\mathbf{k}]\{\mathbf{\Delta}\} - \{\mathbf{F}^{\text{init.}}\} + [\mathbf{m}]\{\ddot{\mathbf{\Delta}}\} \tag{6.16}$$

i.e., the element stiffness equations written to account for initial strain effects and inertia forces. These are equations of equilibrium for the element. The formulative basis for the coefficients of these equations is given by Eqs. (6.12a) to (6.12e).

If *distributed loads* are applied to the surface, the potential of the applied loads δV [Eq. (6.9)] must be supplemented by an integral representing the work of these loads acting through the surface displacements. In order to distinguish the field of surface displacements from the field of displacements of points within the structure, we designate the former by $\mathbf{u}$. $\mathbf{u}$ is established by evaluation of $\mathbf{\Delta}$ on the surface and, since the latter is expressed in terms of joint displacements via Eq. (5.5a), $\mathbf{u}$ is likewise expressed in terms of the joint displacements:

$$\mathbf{u} = [\mathfrak{Y}]\{\mathbf{\Delta}\} \tag{6.17}$$

Designating the distributed applied loads (tractions) by $\bar{\mathbf{T}}$, we have the following contribution to the potential of the applied loads δV:

$$-\int_{S_\sigma} \delta\mathbf{u}\cdot\bar{\mathbf{T}}\, dS$$

where S_σ denotes the portion of the surface upon which the tractions $\bar{\mathbf{T}}$ are prescribed and $\delta\mathbf{u}$ are the virtual displacements of the surface. Again choosing the distribution of virtual displacements in the same form as the actual displacements [Eq. (6.17)] we have

$$-\delta\mathbf{u} = [\mathfrak{Y}]\{\delta\mathbf{\Delta}\} \tag{6.17a}$$

and, by substitution into the expression above for external virtual work,

$$\lfloor\delta\mathbf{\Delta}\rfloor\{\mathbf{F}^d\} = \lfloor\delta\mathbf{\Delta}\rfloor \int_{S_\sigma} [\mathfrak{Y}]^{\mathrm{T}}\cdot\bar{\mathbf{T}}\, dS \tag{6.18}$$

so that

$$\{\mathbf{F}^d\} = \int_{S_\sigma} [\mathfrak{Y}]^{\mathrm{T}}\cdot\bar{\mathbf{T}}\, dS \tag{6.12f}$$

The left-hand side of Eq. (6.18) should be added to the left side of the virtual work expression [Eq. (6.15)] and it follows that $\{\mathbf{F}^d\}$ must then be added to the left side of the stiffness equation [Eq. (6.16)]. The components of $\{\mathbf{F}^d\}$ are termed *work-equivalent loads*, for reasons discussed below.

We should note that as in Chapter 5 the assumed displacement field may also be represented in terms of generalized displacement parameters; i.e.,

$$\mathbf{\Delta} = [\mathbf{p}]\{\mathbf{a}\} \tag{5.2a}$$

and through application of processes described in Section 5.1 this can be transformed into an expression in terms of the joint displacements. The resulting formula was shown to be

$$\mathbf{\Delta} = [\mathbf{p}][\mathbf{B}]^{-1}\{\mathbf{\Delta}\} = [\mathbf{N}]\{\mathbf{\Delta}\} \tag{5.5a}$$

Chapter 8 is devoted, in part, to examination of the alternatives of expressing the field directly in terms of joint parameters or in terms of generalized parameters.

When the displacement field is expressed in terms of generalized parameters it is sometimes convenient to construct the element matrices with reference to these parameters. Consider, in particular, the element stiffness matrix, in which case the displacements are differentiated to give the strains, resulting in $\boldsymbol{\epsilon} = [\mathbf{C}]\{\mathbf{a}\}$ (see Eq. (5.6a)) and the strains due to virtual displacement are $\delta\boldsymbol{\epsilon} = [\mathbf{C}]\{\delta\mathbf{a}\}$. By substitution into δU (Eq. (6.11) one obtains (excluding, for simplicity, initial strains)

$$\delta U = \lfloor \delta\mathbf{a} \rfloor [\mathbf{k}^a]\{\mathbf{a}\}$$

$$[\mathbf{k}^a] = \left[\int_{\text{vol}} [\mathbf{C}]^T[\mathbf{E}][\mathbf{C}]\, d(\text{vol}) \right] \tag{6.12g}$$

where $[\mathbf{k}^a]$ will be referred to as the *kernel stiffness matrix* in this text. Similar kernel relationships can be established for the initial force vector, mass matrix, etc.

Note that the expressions for the stiffness and mass matrices [Eqs. (6.12a) and (6.12e)] are in the form of *congruent transformations*, which preserve symmetry in the product when the central matrix is symmetric. Since the elasticity matrix [E] and mass density $[\boldsymbol{\rho}]$ matrices are symmetric, symmetry of the resulting matrices is assured. This approach differs from the *direct* method of Section 5.1 in that the transformation from joint d.o.f. to strain is employed as the basis for the transformation from the joint forces to the stresses.

6.1.3 Consistency in Formulation of Virtual Work Terms

The development above typifies the principle of *consistency* in finite element formulation. It is apparent that the shape functions of the assumed displacement field are employed in each of the component matrices (basic stiffness, mass, and distributed load matrix) and that one and the same set of shape functions are used in each. Thus, the mass matrix is *consistent* with the basic stiffness matrix and such matrices formulated on this basis are termed *consistent mass matrices*.

The alternative to the above—*inconsistent matrices*—arises quite naturally in common practice. In dynamic analysis, for example, the global stiffness and mass matrices are often regarded as separate entities. It is assumed that the mass characteristics are inert and one therefore merely prorates, or *lumps*, mass quantities at each d.o.f. Mass matrices established in this manner are termed *lumped mass matrices*.

A physical view may likewise be taken of the distributed load situation. The mathematical model consists of fictitious force parameters $\{\mathbf{F}^d\}$ that account for the distributed loads. These are defined by an integrated product of the distributed loads and corresponding displacements in a manner such that an equality of work is established with the joint forces and corresponding displacements. Hence, $\{\mathbf{F}^d\}$ is a vector of *work-equivalent* loads.

The expressions above are developed once again, in Section 6.4, by use of the principle of stationary potential energy. Dual formulations are then constructed for the stationary complementary energy principle; after that, other (mixed) stationary principles are treated. First, however, it is necessary to review a number of fundamental considerations in the determination of stationary values for functions of many variables.

6.2 Calculus of Variations

6.2.1 Minimization Without Constraints

The principle of virtual work is characterized by variation of the strain energy and the potential of the applied loads. If we examine the quantities being varied, U and V, it is possible to establish a number of useful properties that they possess. This examination also discloses that the problem of structural analysis, when cast as the variation of $U + V$, is one that falls within the confines of the well-developed branch of mathematics known as the *calculus of variations*.[6.1-6.4] A number of the important results of this branch of mathematics can be applied directly to the finite element structural analysis problem.

In this section we develop some of the more elementary results of the calculus of variations. The continuous (integral or differential) forms of these results are preserved; we examine their discretization in subsequent sections. Consider first a one-dimensional problem described by the single independent variable $\Delta(x)$, where x is the spatial coordinate. The fundamental problem in the calculus of variations is to determine the value of $\Delta(x)$, which results in a *stationary value* of the integral

$$\Pi = \int f(x, \Delta, \Delta')\, dx \tag{6.19}$$

where $\Delta' = d\Delta/dx$. f is a function that takes such forms as the potential and complementary energy densities in structural mechanics and Π is a *functional*, i.e., the function of a function (in this case, a function of f). A stationary value might be either a maximum or a minimum or it might occur at a neutral point. These situations are portrayed schematically in Fig. 6.2. The function f must of course be twice-differentiable. A once-differentiable form of f would be linear and a linear relationship does not possess a minimum.

In order to establish the expressions that permit us to determine the point at which the stationary value occurs, and to distinguish between the situations shown in Fig. 6.2, we first consider the case of a *function* $\Pi(\Delta)$, where Δ is the variable. From elementary calculus we can expand such a function about the point Δ_0 by use of Taylor series, resulting in the expression

$$\Pi(\Delta) = \Pi(\Delta_0) + \frac{d\Pi(\Delta_0)}{d\Delta}(\Delta - \Delta_0) + \frac{1}{2}\frac{d^2\Pi(\Delta_0)}{d\Delta^2}(\Delta - \Delta_0)^2 + \cdots \tag{6.20}$$

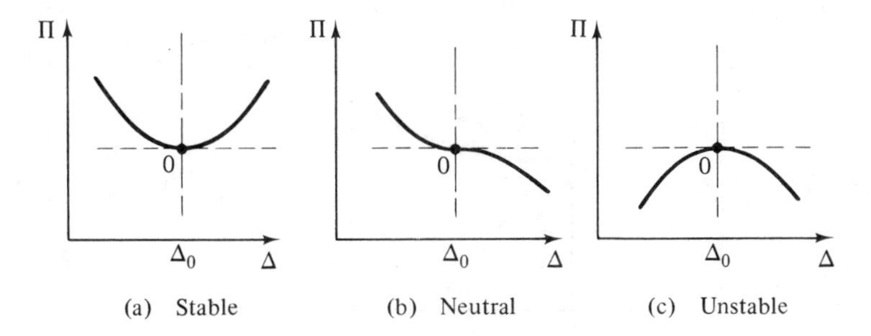

Fig. 6.2 Types of stationary points.

The test to determine the location of the extreme point derives from this expression when we designate the extreme point as Δ_0. As this point is approached, the distance $\Delta - \Delta_0$ becomes very small and the third term in the series above becomes insignificant in comparison with the second term. For a minimum condition all moves away from Δ_0 should cause an increase of $\Pi(\Delta)$, and the second term of the series must always be positive. Unless $d\Pi(\Delta_0)/d\Delta$ is zero, however, the second term might be positive or negative, depending on the sign of the move $d\Delta$. A similar reasoning holds true for a maximum point. Thus, we conclude that at a *stationary point* the following condition prevails:

$$\frac{d\Pi(\Delta_0)}{d\Delta} = 0 \tag{6.21}$$

This is of course the familiar requirement that the slope of a function be zero at a stationary point. In the calculus of variations it is known as *the first necessary condition.* A stationary point must meet this condition but the condition is not sufficient to ascertain if we are at a maximum, minimum, or neutral point. To make this determination, as Fig. 6.2 discloses, we must establish the sign of the curvature (the second derivative) of $\Pi(\Delta_0)$. For a minimum this curvature is positive, for a maximum it is negative, and for a neutral point it is zero. Symbolically, we have

$$\frac{d^2\Pi(\Delta_0)}{d\Delta^2} > 0 \quad \text{minimum} \tag{6.22a}$$

$$\frac{d^2\Pi(\Delta_0)}{d\Delta^2} < 0 \quad \text{maximum} \tag{6.22b}$$

$$\frac{d^2\Pi(\Delta_0)}{d\Delta^2} = 0 \quad \text{neutral} \tag{6.22c}$$

We now return to the matter of defining a stationary value for a *functional* $\Pi(\Delta)$. Fig. 6.3 shows Δ versus the spatial coordinate x. Suppose that the problem is defined within the interval from x_1 to x_2 and that Δ must meet certain conditions at the end points of the interval; these end point values

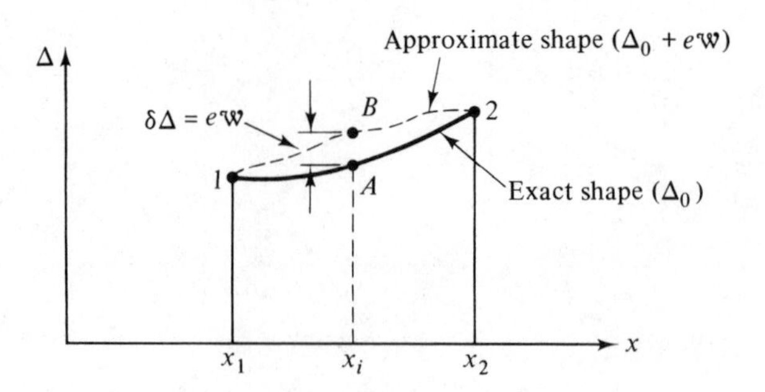

Fig. 6.3 Exact versus approximate shapes.

are designated as Δ_1 and Δ_2. The shape of Δ that causes $\Pi(\Delta)$ to achieve a stationary value is symbolized by Δ_0 and is shown by the solid line in Fig. 6.3. To determine Δ_0, the strategy pursued involves the selection of a shape that differs from Δ_0 by the amount $e\mathcal{W}$ where $\mathcal{W}$ is a shape of undefined amplitude that meets the conditions on Δ at x_1 and x_2 and e is the amplitude of the shape. Hence, the approximate shape is described by the expression

$$\Delta = \Delta_0 + e\mathcal{W} \tag{6.23a}$$

and the slope of this approximate shape is

$$\frac{d\Delta}{dx} = \Delta' = \Delta_0' + e\mathcal{W}' \tag{6.23b}$$

We further note that $e\mathcal{W}$ represents a small *variation* in the amplitude of Δ, which we call $\delta\Delta$. Thus

$$\delta\Delta = e\mathcal{W}, \qquad \delta\Delta' = e\mathcal{W}' \tag{6.24a, b}$$

The variation $\delta\Delta$ gives rise to a small change in the functional, denoted as $\delta\Pi$. $\delta\Pi$ is the *first variation of the functional.*

The symbol δ, or delta operator, represents a small arbitrary change in the dependent variable Δ for a *fixed* value of the independent variable x. As we see in Fig. 6.3 at the specified point x_i, $\delta\Delta$ is of an amplitude $B - A$. The distinction of the delta operator, (δ) with the differential calculus operator dy, is that the latter associates a dx with a dy; i.e., it is the vertical distance between points on a *given curve* at locations dx apart. The properties of the delta operator that are important to the development of variational relationships are that it is *commutative* with the differential and integral operators; i.e.,

$$\delta\left(\frac{d\Delta}{dx}\right) = \frac{d}{dx}(\delta\Delta)$$

$$\delta\left(\int \Delta\, dx\right) = \int (\delta\Delta)\, dx$$

With the considerations above in mind we proceed to the development of relationships that govern the stationary value of the functional Π. First we construct the expression for the functional associated with the approximate shape $\Delta_0 + e\mathcal{W}$. Eq. (6.19) then becomes

$$\Pi(e) = \int_{x_1}^{x_2} f(x, \Delta_0 + e\mathcal{W}, \Delta_0' + e\mathcal{W}')\, dx \tag{6.25}$$

Next we expand f in this expression about Δ_0 and Δ_0' (x fixed) resulting in

$$f(x, \Delta_0 + e\mathcal{W}, \Delta_0' + e\mathcal{W}') - f(x, \Delta_0, \Delta_0') = \left[\frac{\partial f}{\partial \Delta_0}(\delta\Delta) + \frac{\partial f}{\partial \Delta_0'}(\delta\Delta')\right] + \text{higher-order terms} \tag{6.26}$$

Now the left side of this equation is the change in f due to the variation $\delta\Delta = e\mathcal{W}$, i.e., it is δf. Thus, with the neglect of higher-order terms we can write the first variation of the functional as

$$\delta\Pi = \int_{x_1}^{x_2} \delta f\, dx = \int_{x_1}^{x_2} \left(\frac{\partial f}{\partial \Delta_0}\delta\Delta + \frac{\partial f}{\partial \Delta_0'}\delta\Delta'\right) dx = 0 \tag{6.27}$$

where we have set the expression to zero in accordance with the condition upon a stationary value.

To obtain a useful expression for $\delta\Pi$ we must integrate this expression by parts. As shown in Chapter 5, this operation is motivated partly by the need to produce boundary conditions. Here the result also allows us to factor out $\delta\Delta$. Only the second term need be integrated, as follows:

$$\int_{x_1}^{x_2} \frac{\partial f}{\partial \Delta_0'}(\delta\Delta')\, dx = \delta\Delta \cdot \frac{\partial f}{\partial \Delta_0'}\Bigg|_{x_1}^{x_2} - \int_{x_1}^{x_2} \delta\Delta \cdot \frac{d}{dx}\left(\frac{\partial f}{\partial \Delta_0'}\right) dx \tag{6.28}$$

Since $\delta\Delta$ must be zero at the limits x_1 and x_2, the first term on the right side must be zero. Equation (6.27) therefore becomes

$$\delta\Pi = \int_{x_1}^{x_2} \delta\Delta \left[\frac{\partial f}{\partial \Delta_0} - \frac{d}{dx}\left(\frac{\partial f}{\partial \Delta_0'}\right)\right] dx = 0 \tag{6.27a}$$

Clearly, since we have admitted the possibility of arbitrary shapes $\delta\Delta$, this integral can be zero only if

$$\frac{\partial f}{\partial \Delta_0} - \frac{d}{dx}\left(\frac{\partial f}{\partial \Delta_0'}\right) = 0 \tag{6.29}$$

This expression is known as the *Euler equation* (or *Euler-Lagrange equation*) of the functional Π. The function Δ which causes Π to have an extreme value is the function which will satisfy the associated Euler equation. In practical terms the Euler equation gives us a means for determining the governing differential equation of a physical circumstance that is described by a functional at the outset.

6.2.2 Example

To illustrate how the procedure above works, let us examine a functional associated with an axial member. We recall that virtual work stipulated that $\delta(U + V) = 0$. This represents the imposition of the first necessary condition on the functional, $U + V$. In the case of an axial member (see Section 5.5) we have

$$U + V = \frac{1}{2}\int_0^L \epsilon^2 EA\,dx - \int_0^L q \cdot u\,dx$$

$$= \int_0^L \left[\frac{1}{2}\left(\frac{du}{dx}\right)^2 EA - q \cdot u\right] dx$$

Hence, by comparison with Eq. (6.19),

$$f = \left[\frac{1}{2}\left(\frac{du}{dx}\right)^2 EA - q \cdot u\right]$$

Now applying the Euler differential equation [Eq. (6.29)], we find (since now $\Delta_0 = u$)

$$\frac{\partial f}{\partial u} = -q, \qquad \frac{\partial f}{\partial u'} = AE\frac{du}{dx}$$

$$\frac{\partial f}{\partial u} - \frac{d}{dx}\left(\frac{\partial f}{\partial u'}\right) = -q - AE\frac{d^2u}{dx^2} = 0$$

or

$$AE\frac{d^2u}{dx^2} + q = 0$$

which is the governing (equilibrium) differential equation for this problem.

6.2.3 Boundary Conditions and Constraints

The requirements that the independent variable, or derivatives of the independent variable, be of certain value at the end points are known as the *forced boundary conditions*. If the extremizing function does not meet these conditions, the first term on the right side of Eq. (6.28) can still be met if

$$\frac{\partial f}{\partial \Delta_0'} = 0 \tag{6.30}$$

The condition represented by Eq. (6.30) is known as a *natural boundary condition.* To illustrate this for the case of the axial member, we see from the previous work that $\partial f/\delta\Delta_0' = \partial f/\partial u' = AE\,(du/dx)$. We know, however, that $du/dx = \epsilon_x$, $E\epsilon = \sigma_x$, and $F = A\sigma_x$ so we can have $F = 0$ at this point. Thus, the natural boundary condition, that the force at the free end of the axial member be zero, can be met. This type of condition is represented directly in the energy functional by the term representing the work of the applied loads and is often not only approximately met in finite element analysis.

Before concluding this section it is important to examine a manner in which *constraining* conditions are taken into account in a variational procedure. One device for their inclusion is the *Lagrangian multiplier method.* Consider the problem of minimizing the functional $\Pi(\Delta)$ and let the constraining condition be represented by the equation

$$\mathcal{G}(\Delta) = 0 \tag{6.31}$$

Then we construct a new functional, Π^a, termed the *augmented functional,* by multiplying $\mathcal{G}$ by a constant λ and adding this product to the original functional; i.e.,

$$\Pi^a = \Pi + \lambda\mathcal{G} \tag{6.32}$$

where λ is termed the *Lagrangian multiplier*. Now, if Π has an extreme value at Δ_0 subject to $\mathcal{G}(\Delta) = 0$, the partial derivatives of Π^a, with respect to Δ and λ, respectively give conditions for the determination of Δ_0 and λ, in the manner of the first necessary condition, as follows:

$$\frac{d\Pi^a}{d\Delta} = \frac{\partial\Pi}{\partial\Delta} + \lambda\frac{\partial\mathcal{G}}{\partial\Delta} = 0 \tag{6.33a}$$

$$\frac{d\Pi^a}{d\lambda} = \mathcal{G} = 0 \tag{6.33b}$$

Note that one of the resulting relationships is the constraint itself, $\mathcal{G} = 0$. Rigorous proof of the above is given in standard texts on variational calculus (see, for example, Refs. 6.1–6.4).

The Lagrange multipliers may have important physical meaning in a given problem. In some cases it is possible to identify this meaning only by detailed investigation of the theoretical properties of Lagrange multipliers. In other cases it can be identified simply by examination of Π. For example, in the use to which we shall put the concepts above—the energy principles of structural mechanics—Π is an energy parameter with units of force times displacement. In certain energy formulations the constraint equations are specified relationships between displacements. For consistency of units, therefore, λ must have the units of force, and one can identify the Lagrange multipliers as *generalized forces.*

6.3 The Discretized Variational Problem

6.3.1 Unconstrained Minimization

We now proceed to an examination of *discretized* functionals, where a single independent variable $\mathbf{\Delta}$ is approximated by a finite sum of terms. Since we are concerned with the finite element concept, we take this approximation to be of the form of Eq. (5.5a); i.e., $\mathbf{\Delta} = \lfloor \mathbf{N} \rfloor \{\mathbf{\Delta}\}$. For simplicity we deal with the single variable $\mathbf{\Delta}$. The case of a field of variables (e.g., $\mathbf{\Delta} = \lfloor \mathbf{u\ v\ w} \rfloor$) follows directly the same development. In examining the properties of a discretized functional it is useful to visualize it as a surface in an $(n + 1)$-dimensional space where n of the orthogonal axes represent the n d.o.f. $\Delta_1, \Delta_2, \ldots, \Delta_n$ and the $(n + 1)$th axis defines values of the functional Π $(\{\mathbf{\Delta}\})$. Each point on such a surface is a value of Π $(\{\mathbf{\Delta}\})$. The surface for a problem with two d.o.f. (Δ_1, Δ_2) is sketched in Fig. 6.4. It is not possible to sketch the situation for more than two d.o.f. but the algebraic properties of the situation shown apply directly to the general n-dimensional problem. Since we are concerned with the properties of the extreme value of Π $(\{\mathbf{\Delta}\})$, we immediately generalize the statement of the Taylor series expansion, given by Eq. (6.20) for the continuous problem about the extreme point $\{\mathbf{\Delta}_0\}$:

$$\Pi(\{\Delta_i\}) = \Pi(\{\mathbf{\Delta}_0\}) + \sum_{i=1}^{n} \left.\frac{\partial \Pi}{\partial \Delta_i}\right|_{\{\mathbf{\Delta}_0\}} (d\Delta_i) + \frac{1}{2} \sum_{i=1}^{n} \sum_{j=1}^{n} \left.\frac{\partial^2 \Pi}{\partial \Delta_i\, \partial \Delta_j}\right|_{\{\mathbf{\Delta}_0\}} (d\Delta_i)(d\Delta_j) + \cdots \tag{6.34}$$

where $(d\Delta_i)$ is the difference between the ith component of $\{\mathbf{\Delta}\}$ and the corresponding component of $\{\mathbf{\Delta}_0\}$, and similarly for $(d\Delta_j)$. Alternatively, we may write this expansion in matrix format

$$\Pi(\{\mathbf{\Delta}\}) = \Pi_0 + \frac{\partial \Pi}{\lfloor \partial \mathbf{\Delta} \rfloor} \{d\mathbf{\Delta}\} + \frac{\lfloor d\mathbf{\Delta} \rfloor}{2} [\mathbf{\kappa}] \{d\mathbf{\Delta}\} + \cdots \tag{6.34a}$$

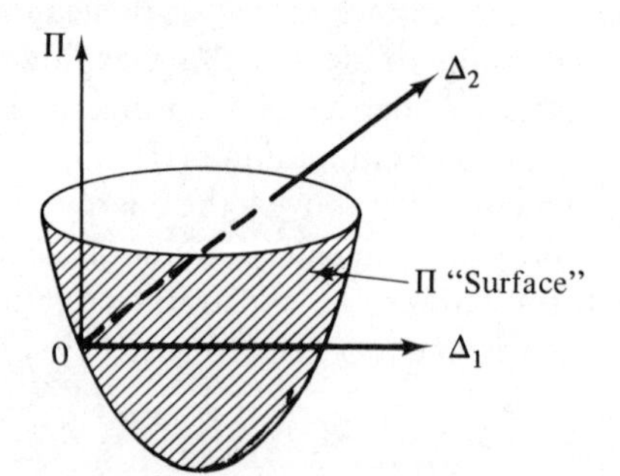

Fig. 6.4 Functional surface for two-d.o.f. system.

where $\{\mathbf{d\Delta}\} = \{\mathbf{\Delta}\} - \{\mathbf{\Delta}_0\}$ and the individual terms of $\{\boldsymbol{\kappa}\}$, the *Hessian* matrix, are $\kappa_{ij} = \partial^2\Pi/\partial\Delta_i\,\partial\Delta_j$. Both $\lfloor\partial\mathbf{\Pi}/\partial\mathbf{\Delta}\rfloor$ and $[\boldsymbol{\kappa}]$, as Eq. (6.34) discloses, are to be evaluated at $\{\mathbf{\Delta}_0\}$. We do not go beyond the third term in the expansion above because each of the functionals in linear structural mechanics with which we shall be concerned is of *quadratic* form. Thus, third- and higher-order differentiation appearing in subsequent terms produces no contribution to $\mathbf{\Pi}(\{\mathbf{\Delta}\})$.

Now, if $\mathbf{\Pi}(\Delta)$ possesses a *stationary point*, the tangent to such a point meets, by definition, the condition that an infinitesimal variation of the coordinates $(d\Delta_i)$ will cause no change in the functional to a first order of approximation. This requirement, the first necessary condition, was defined for the continuous case by Eq. (6.21) and in the present case we write

$$\delta\Pi\,(\{\mathbf{\Delta}\}) = 0 \tag{6.35}$$

To transform this expression into an operational procedure for constructing algebraic equations whose solution gives the coordinates of the stationary point, we employ δ as a differential operator. Hence

$$\delta\Pi(\{\Delta\}) = \frac{\partial\Pi}{\partial\Delta_1}\,\delta\Delta_1 + \frac{\partial\Pi}{\partial\Delta_2}\,\delta\Delta_2 + \cdots + \frac{\partial\Pi}{\partial\Delta_n}\,\delta\Delta_n = \frac{\partial\mathbf{\Pi}}{\lfloor\partial\mathbf{\Delta}\rfloor}\,\{\delta\mathbf{\Delta}\} = 0 \tag{6.35a}$$

and, since the variations $\delta\Delta_i$ are independent,

$$\left\{\frac{\partial\mathbf{\Pi}}{\partial\mathbf{\Delta}}\right\} = 0 \tag{6.35b}$$

This condition is applied to the $i = 1, \ldots, n$ d.o.f. Δ_i, resulting in a system of n simultaneous equations.

In certain cases the stationary point of the discretized functional possesses the additional property of being at an extreme point (maximum or minimum). When such a point is a minimum, then any move away from the point increases the value of $\mathbf{\Pi}$. Since $\lfloor\partial\mathbf{\Pi}/\partial\mathbf{\Delta}\rfloor\,\{d\mathbf{\Delta}\}$ is zero at the solution point, the condition for a minimum requires that

$$\lfloor\delta\mathbf{\Delta}\rfloor[\boldsymbol{\kappa}]\{\delta\mathbf{\Delta}\} > 0 \tag{6.36a}$$

Since $\{\delta\mathbf{\Delta}\}$ is arbitrary, this condition demands the property of positive definiteness for the Hessian matrix $[\boldsymbol{\kappa}]$. A positive definite matrix, by definition, is one that yields a positive value for an arbitrary vector $\{d\mathbf{\Delta}\} \neq 0$ when the quadratic product $\lfloor d\mathbf{\Delta}\rfloor[\boldsymbol{\kappa}]\{d\mathbf{\Delta}\}$ is formed.

Conversely, for a maximum point,

$$\lfloor\delta\mathbf{\Delta}\rfloor[\boldsymbol{\kappa}]\{\delta\mathbf{\Delta}\} < 0 \tag{6.36b}$$

so that $[\boldsymbol{\kappa}]$ is *negative definite*.

The variational formulation gives an opportunity for inspection of the

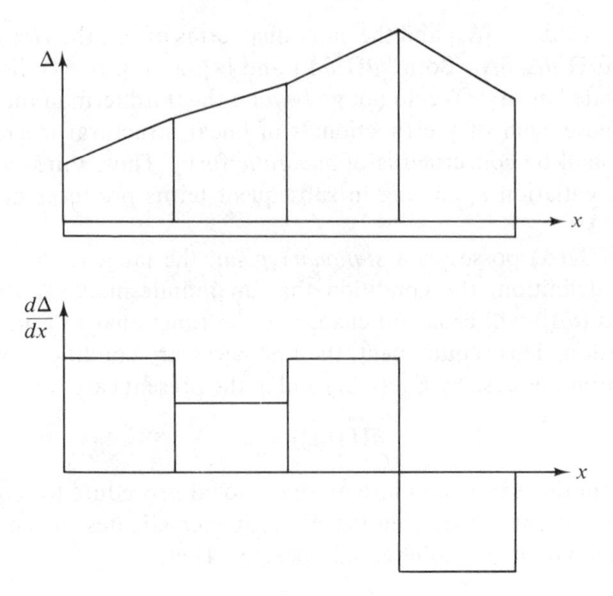

Fig. 6.5 Piecewise variation of $d\Delta/dx$.

question of what constitutes *admissibility* in the case of a finite element discretization of a physical problem. We have already seen that within a single region a function must be differentiable to the same order as the pertinent Euler equation (i.e., for an axial member the Euler equation is of second order no less than a quadratic function). In finite element analysis, the functional for the complete system is comprised of the sum of functionals of p individual regions (elements), Π^j,

$$\Pi = \sum_{j=1}^{p} \Pi^j \quad (j = 1, \ldots, p) \tag{6.37}$$

What, then, is admissibility condition across the element boundaries? We can obtain an idea of this condition by studying the variation of fields for a one-dimensional string of elements, as shown in Fig. 6.5. If, for this case, the functional consists of an integration of the first derivative ($d\Delta/dx$) over the system, it is seen that continuity of Δ itself permits a unique evaluation of Π. This situation can be generalized as follows: Unique evaluation of the functional can be achieved if *continuity of derivatives of one order less than the highest-order derivative appearing in the functional is maintained.*

6.3.2 Lagrange Multiplier Method for Constraints

The Lagrange multiplier concept is of value when constraints are present in the discretized problem. If there are r constraint conditions of the form

$$\mathcal{G}^k(\Delta_1, \ldots, \Delta_n) = 0 \quad (k = 1, \ldots, r) \tag{6.38}$$

one forms an augmented functional

$$\Pi^a(\{\Delta, \lambda\}) = \Pi(\{\Delta\}) + \sum_{k=1}^{r} \lambda_k \mathcal{G}^k \tag{6.39}$$

where the second term on the right side is the sum of the products of $\mathcal{G}_j$ and their associated Lagrange multipliers λ_k. The first necessary condition is applied to each of the d.o.f. Δ_i and then to each of the Lagrange multipliers λ_k. From the first series of differentiations we have

$$\frac{\partial \Pi}{\partial \Delta_i} + \sum_{k=1}^{r} \lambda_k \frac{\partial \mathcal{G}^k}{\partial \Delta_i} = 0 \tag{6.35c}$$

and from the differentiation with respect to λ_k we obtain the constraint relationships, Eq. (6.38).

We note, as we did in Section 6.2, that consistency of units enables the definition of the units of the Lagrange multiplier and identifies the physical meaning of the multiplier. If $\mathcal{G}^k = 0$ represents a condition among the displacements, then λ_k is an associated force. We shall have the opportunity to illustrate this point in Chapter 7.

6.4 Minimum Potential Energy

6.4.1 Properties of Potential Energy

The principle of minimum potential energy furnishes a variational basis for the direct formulation of element stiffness equations. The potential energy (Π_p) of a structure is given by the strain energy (U) plus the potential of the applied loads (V); i.e.,

$$\Pi_p = U + V \tag{6.40}$$

The principle can be stated as follows: *Among all displacements of an admissible form, those that satisfy the equilibrium conditions make the potential energy assume a stationary value. Thus*

$$\delta\Pi_p = \delta U + \delta V = 0 \tag{6.41}$$

For stable equilibrium, Π_p is a minimum. Hence

$$\delta^2\Pi_p = \delta^2 U + \delta^2 V > 0 \tag{6.42}$$

In establishing the principle above we shall exclude, for simplicity, the consideration of body forces. We designate the strain energy per unit volume or strain energy density (see Section 2.4 for the fundamental definition of strain energy) as dU. Then the change in strain energy density due to the change in strain caused by the virtual displacement ($\delta\boldsymbol{\epsilon}$) is given by

$$\delta(dU) = \boldsymbol{\sigma}\,\delta\boldsymbol{\epsilon} \tag{6.43}$$

where $\boldsymbol{\sigma}$ represents the equilibrium stress state prior to the imposition of the virtual displacement. The portion of the strain energy due to virtual strains acting over the associated changes in stress is disregarded from order-of-magnitude considerations. Introducing the stress-strain law [Eq. (4.15)], the expression for the change in strain energy becomes

$$\delta\,(dU) = \boldsymbol{\epsilon}[\mathbf{E}]\,\delta\boldsymbol{\epsilon} - \boldsymbol{\epsilon}^{\text{init.}}[\mathbf{E}]\,\delta\boldsymbol{\epsilon} \tag{6.44}$$

Then, by integration between zero and the strain $\boldsymbol{\epsilon}$, corresponding to $\boldsymbol{\sigma}$, there is obtained (with the transpose of terms of the second integral)

$$dU = \tfrac{1}{2}\boldsymbol{\varepsilon}[\mathbf{E}]\boldsymbol{\varepsilon} - \int_{\text{vol}} \boldsymbol{\varepsilon}[\mathbf{E}]\boldsymbol{\varepsilon}^{\text{init.}}\,d(\text{vol}) \tag{6.45}$$

and, for the complete finite element by integration of dU over the volume (designating the third term on the right side as $\mathbf{C}(\boldsymbol{\varepsilon}^{\text{init.}})$),

$$U = \frac{1}{2}\int_{\text{vol}} \boldsymbol{\varepsilon}[\mathbf{E}]\boldsymbol{\varepsilon}\,d(\text{vol}) - \int_{\text{vol}} \boldsymbol{\varepsilon}[\mathbf{E}]\boldsymbol{\varepsilon}^{\text{init.}}\,d(\text{vol}) + \mathbf{C}(\boldsymbol{\varepsilon}^{\text{init.}}) \tag{6.46}$$

We note also that the first variation of U, with δ applied in the manner of a differential, gives

$$\delta U = \int_{\text{vol}} \boldsymbol{\epsilon}[\mathbf{E}]\,\delta\boldsymbol{\epsilon}\,d(\text{vol}) - \int_{\text{vol}} \boldsymbol{\epsilon}^{\text{init.}}[\mathbf{E}]\,\delta\boldsymbol{\epsilon}\,d(\text{vol}) \tag{6.47}$$

The potential of the applied loads is given by

$$V = -\sum_{i=1} F_i\Delta_i - \int_{S_\sigma} \bar{\mathbf{T}}\cdot\mathbf{u}\,dS \tag{6.48}$$

where all symbols are as previously defined. Note again that S_u, the portion of the surface upon which the displacements are prescribed, is not represented by a surface integral because of the *kinematic admissibility* condition on the chosen displacement field; i.e., such geometric (forced) boundary conditions must be met exactly.

The first variation of V is given by

$$\delta V = -\sum F_i\,\delta\Delta_i - \int_{S_\sigma} \bar{\mathbf{T}}\cdot\delta\mathbf{u}\,dS \tag{6.49}$$

Referring once again to the principle of virtual work, represented by Eq. (6.1), we see that in accordance with Eq. (6.41)

$$\delta U + \delta V = \delta\Pi_p = 0$$

and it is demonstrated that the first variation of a properly constructed potential function Π_p is zero; i.e., Π_p is stationary at the solution point.

6.4.2 *Finite Element Discretization*

We base our development upon displacement fields expressed in terms of the joint d.o.f. so that, from Eq. (5.6c), $\boldsymbol{\epsilon} = [\mathbf{D}]\{\boldsymbol{\Delta}\}$. Thus, by substitution of this expression for $\boldsymbol{\epsilon}$ into Eq. (6.46), we have

$$U = \frac{\lfloor \boldsymbol{\Delta} \rfloor}{2}[\mathbf{k}]\{\boldsymbol{\Delta}\} - \lfloor \boldsymbol{\Delta} \rfloor\{\mathbf{F}^{\text{init.}}\} + \mathbf{C}(\boldsymbol{\varepsilon}^{\text{init.}}) \tag{6.50}$$

where $[\mathbf{k}]$ and $\{\mathbf{F}^{\text{init.}}\}$ are therefore exactly as defined by the expressions, derived from the virtual work principle, Eqs. (6.12a) and (6.12b).

Also, we have the discretized form of V [using from Eq. (6.17a) $\delta \mathbf{u} = [\boldsymbol{\mathcal{Y}}]\{\boldsymbol{\Delta}\}$]

$$V = -\lfloor \boldsymbol{\Delta} \rfloor\{\mathbf{F}\} - \lfloor \boldsymbol{\Delta} \rfloor\{\mathbf{F}^d\} \tag{6.51}$$

where $\{\mathbf{F}^d\}$ is defined by Eq. (6.12f).

Now, from Eqs. (6.50) and (6.51) the complete discretized expression for potential energy is given by

$$\boldsymbol{\Pi}_p = \frac{\lfloor \boldsymbol{\Delta} \rfloor}{2}[\mathbf{k}]\{\boldsymbol{\Delta}\} - \lfloor \boldsymbol{\Delta} \rfloor\{\{\mathbf{F}\} + \{\mathbf{F}^{\text{init.}}\} + \{\mathbf{F}^d\}\} \tag{6.39a}$$

which is a general quadratic form. Applying next the condition for a stationary value [i.e., $\{\partial \boldsymbol{\Pi}/\partial \boldsymbol{\Delta}\} = 0$, see Eq. (6.35b)], we have

$$[\mathbf{k}]\{\boldsymbol{\Delta}\} = \{\mathbf{F}\} + \{\mathbf{F}^{\text{init.}}\} + \{\mathbf{F}^d\} \tag{6.52}$$

To test if the potential energy is a maximum or a minimum, we apply the second variation. For cases of conservative loading, where $\{\mathbf{F}\}$ is a vector of constants,

$$\delta^2 \boldsymbol{\Pi}_p = \lfloor \delta \boldsymbol{\Delta} \rfloor[\mathbf{k}]\{\delta \boldsymbol{\Delta}\} \tag{6.53}$$

It is observed from physical considerations that strain energy must be a positive quantity. With the strain energy given by $U = \frac{1}{2}\lfloor \boldsymbol{\Delta} \rfloor[\mathbf{k}]\{\boldsymbol{\Delta}\}$, it is clear that $[\mathbf{k}]$ has the property of *positive definiteness* in view of the arbitrary nature of the vector $\{\boldsymbol{\Delta}\}$. Consequently, $\delta^2 \boldsymbol{\Pi}_p$ is nonnegative and the potential energy is a minimum.

The minimum character of a potential energy solution is a useful property since it permits the analyst to identify certain solution parameters as *bounds* on the exact solution. This property is explored further in Chapter 7 where we deal with the solution of the complete structure. We also observe that the positive definite character of $[\mathbf{k}]$ enabled the identification of the minimum property. In the case of certain mixed variational principles, to be described subsequently in this chapter, the basic matrix of coefficients in a finite element formulation do not possess this property and no bounds can be assigned the solution parameters.

It must be emphasized that the potential energy principle can be applied in the formulation of the stiffness matrix of an element as a *structural entity* without any reference to conditions to be met across element boundaries when the element is incorporated in the global structural representation. If these conditions are violated in forming the global representation, the analytical model is termed *interelement-incompatible* and there is no assurance of a lower bound solution. Elements which will not result in interelement-compatible global representations are formulated because they employ displacement fields which are simpler than those required to meet the interelement compatibility conditions. It is possible to test such formulations to determine if they will yield the correct solution in limit of grid refinement.[6.5] Examples of these elements are encountered in later chapters.

We have seen that the formulas for the element matrices for linear elastic behavior from the perspectives of either the virtual work or minimum potential energy principles are identical. The virtual work principle is more fundamental and its generalization enables formulation of finite element representations for nonstructural problems and for these reasons is preferred by many. Alternatively, expressions for strain energy are either very familiar or easily constructed for most situations in structural mechanics. The energy approach also enables the visualization of the minimum properties of a solution and suggests alternative solution algorithms based on these properties, as we shall see in Chapter 7.

6.4.3 Examples

It is of interest to apply the approach above to the construction of stiffness and other matrices of the elements studied in Chapter 5. For simplicity, in choosing the relevant element displacement fields, we adopt expressions in which the fields are written directly in terms of the element joint displacements rather than in terms of generalized parameters. Thus, for the axial member, discussed in Sections 5.1 and 5.5 (see Fig. 6.6), we have from Eq. (5.3)

$$u = \left(1 - \frac{x}{L}\right)u_1 + \frac{x}{L}u_2$$

so that

$$\lfloor \mathbf{N} \rfloor = \left\lfloor \left(1 - \frac{x}{L}\right) \quad \frac{x}{L} \right\rfloor$$

$$\lfloor \mathbf{N}' \rfloor = \left\lfloor -\frac{1}{L} \quad \frac{1}{L} \right\rfloor$$

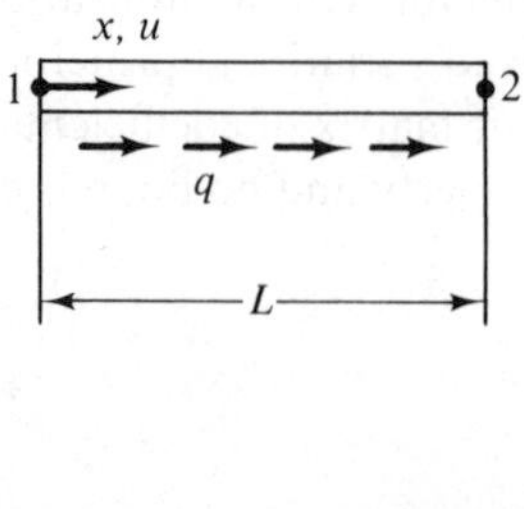

Fig. 6.6

By substitution into Eqs. (6.12a) and (6.12e) we have the element stiffness and mass matrices

$$[\mathbf{k}] = \int_0^L \begin{Bmatrix} -\frac{1}{L} \\ \frac{1}{L} \end{Bmatrix} E \left\lfloor -\frac{1}{L} \quad \frac{1}{L} \right\rfloor A\,dx = \frac{AE}{L}\begin{bmatrix} 1 & -1 \\ -1 & 1 \end{bmatrix}$$

$$[\mathbf{m}] = \int_0^L \begin{Bmatrix} \left(1 - \frac{x}{L}\right) \\ \frac{x}{L} \end{Bmatrix} \rho \left\lfloor \left(1 - \frac{x}{L}\right) \quad \frac{x}{L} \right\rfloor A\,dx = \rho AL \begin{bmatrix} \frac{1}{3} & \frac{1}{6} \\ \frac{1}{6} & \frac{1}{3} \end{bmatrix}$$

and, for the case in which the initial strain is due to temperature expansion ($e^{\text{init.}} = \alpha \Upsilon$), we have from Eq. (6.12b)

$$\{\mathbf{F}^{\text{init.}}\} = \int_0^L \begin{Bmatrix} -\frac{1}{L} \\ \frac{1}{L} \end{Bmatrix} E\alpha\Upsilon A\,dx = AE\alpha\Upsilon \begin{Bmatrix} -1 \\ 1 \end{Bmatrix}$$

Also, for an axially distributed load q (lb/in.) of constant value, $\mathbf{X} = q/A$. From Eq. (6.12c) we have

$$\{\mathbf{F}^b\} = \int_0^L \begin{Bmatrix} \left(1 - \frac{x}{L}\right) \\ \frac{x}{L} \end{Bmatrix} q\,dx = q\frac{L}{2}\begin{Bmatrix} 1 \\ 1 \end{Bmatrix}$$

The result for $[\mathbf{k}]$ is identical to that obtained by the direct procedure. Because of the simple form of the displacement field for this element, the prorating (or "lumping") of forces implied by the transpose of the strain-displacement matrix is the same as that performed by direct reasoning. With respect to the thermal forces we find, as must be the case, that the components of the vector $\{\mathbf{F}^{\text{init.}}\}$ indeed represent the forces necessary to suppress displacement of the element due to the temperature change Υ. Also, the distributed forces are those that would be obtained in a simple prorating operation.

Consider now the triangular element (Fig. 5.3). From Eq. (5.21a) we have

$$\lfloor \mathbf{N} \rfloor = \frac{1}{x_2 y_3} \lfloor (x_2 y_3 - x y_3 - x_2 y + x_3 y) \quad (x y_3 - x_3 y) \quad (x_2 y) \rfloor$$

and, from Eq. (5.22),

$$[\mathbf{D}] = \frac{1}{x_2 y_3}\begin{bmatrix} -y_3 & y_3 & 0 & 0 & 0 & 0 \\ 0 & 0 & 0 & x_3 - x_2 & -x_3 & x_2 \\ x_3 - x_2 & -x_3 & x_2 & -y_3 & y_3 & 0 \end{bmatrix}$$

The stiffness matrix obtained by use of **[D]** in the virtual work expression [Eq. (6.12a)] is again the same as that presented in Fig. 5.4, due to the simplicity of the linear field represented by $\lfloor \mathbf{N} \rfloor$. Derivation of **[m]** and $\{\mathbf{F}^{\text{init.}}\}$ matrices are left as exercises for the reader (see Prob. 6.4 and 6.7).

In plane stress the distributed loads are usually applied to the edges of a structure, rather than as a distributed load over the surface of the element. Consequently there are implications related to the distribution of loads on the overall edges in the calculation of $\{\mathbf{F}^d\}$ and it is appropriate to reserve discussion of this calculation until Chapter 9, which deals with global aspects of plane stress analysis.

Both the axial member and the linear-displacement triangle are deceptive representatives of the potential energy (or virtual work) concept in element formulation because the assumed displacement fields correspond to stress fields that satisfy the differential equations of equilibrium. For example, in the case of the triangle, it is found that the differential equation of equilibrium

$$\frac{\partial \sigma_x}{\partial x} + \frac{\partial \tau_{xy}}{\partial y} = 0$$

when applied to the expressions for stress obtained from $[\boldsymbol{\sigma}] = [\mathbf{E}][\mathbf{D}]\{\boldsymbol{\Delta}\}$ [Eq. (5.7a)] is identically satisfied. The selection of a kinematically admissible displacement field, however, is usually accomplished without reference to the conditions of equilibrium and therefore will not, in general, satisfy these conditions. This is demonstrated in the formulation of more complex elements in later chapters.

6.4.4 Approximation of Geometry

It was pointed out in Section 3.4 that one of the advantages of the finite element method is in the analysis of structures possessing geometric complexity. It must, however, be recognized that the geometry of the real structure is nearly always approximated somewhat in analysis and that this is also a source of analysis error. Approximation of behavior (e.g., displacement) is given the most emphasis, but questions relating to geometric approximation may be of equal or greater importance. As demonstrated now, the variational approach facilitates the incorporation of more realistic approximations of structural geometry.

In discussing this topic it is useful to distinguish between solid (three-dimensional) structures and plate and prismatic structures. In solids we are concerned mainly with curved surfaces, while in plates and prismatic members the parameters are thickness and area variations. Some basic considerations in the approximation of the latter is taken up here. The approximation of geometry of solid elements is taken up in later chapters.

The varying section axial member shown in Fig. 6.7 illustrates the key

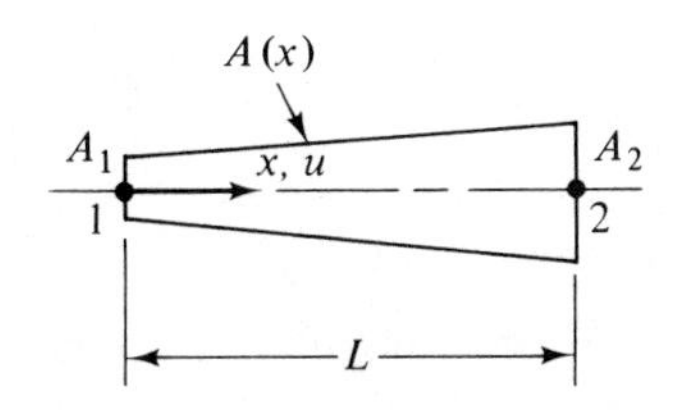

Fig. 6.7 Tapered axial member.

factors in the approximation of geometry of tapered prismatic members and plates. The usual approach in analysis practice is to represent a tapered member in stepped fashion, using constant-depth elements. This is a good approximation if a sufficient number of elements is employed, but numerical evidence shows that the solution error may be more significant than the type of error associated with approximate displacement fields.

An alternative to the stepped representation is one in which a simple approximation of the variation of $A(x)$ is applied either to the member as a whole or to segments into which it will be divided. An *approximation* is desired because, if we seek the development of an explicit stiffness matrix for the element, it is easy to see that no single representation of $A(x)$ can cover exactly all possible specified variations.

With the factors above in mind, the strain energy of the member is written as

$$U = \frac{E}{2}\int_0^L \left(\frac{du}{dx}\right)^2 [A(x)]\,dx$$

The displacement function used previously for the case of a uniform-section member [Eq. (5.5)] is not the exact function for the present case since it represents a condition of constant strain, which no longer prevails along the axis of the member. It is a convenient approximation, however, and we adopt it here.

In this example we assume that $A(x)$ (Fig. 6.7) varies linearly between points 1 and 2. Thus, we write

$$A(x) = \left\lfloor \left(1 - \frac{x}{L}\right) \quad \frac{x}{L} \right\rfloor \begin{Bmatrix} A_1 \\ A_2 \end{Bmatrix}$$

where A_1 and A_2 are the cross sectional areas at points 1 and 2. With these representations for displacement and geometry we have for the strain energy

$$U = E\frac{\lfloor u_1 \quad u_2 \rfloor}{2L^2}\int_0^L \begin{Bmatrix} -1 \\ 1 \end{Bmatrix}\left[\left\lfloor \left(1 - \frac{x}{L}\right) \quad \frac{x}{L} \right\rfloor \begin{Bmatrix} A_1 \\ A_2 \end{Bmatrix}\right]\lfloor -1 \quad 1 \rfloor\,dx \begin{Bmatrix} u_1 \\ u_2 \end{Bmatrix}$$

or, upon integration,

$$U = \lfloor u_1 \; u_2 \rfloor [\mathbf{k}] \begin{Bmatrix} u_1 \\ u_2 \end{Bmatrix}$$

where

$$[\mathbf{k}] = \frac{(A_1 + A_2)}{2L} E \begin{bmatrix} 1 & -1 \\ -1 & 1 \end{bmatrix}$$

By comparison, it is possible to construct the "exact" stiffness matrix for an axial member with linear taper and this gives

$$[\mathbf{k}] = \frac{E}{L} \cdot \frac{(A_2 - A_1)}{ln \dfrac{A_2}{A_1}} \begin{bmatrix} 1 & -1 \\ -1 & 1 \end{bmatrix}$$

where *ln* denotes the natural logarithm. Comparing the alternative forms of stiffness matrix it is seen that the exact form requires the evaluation of a logarithmic function.

The procedure for the formulation of tapered beam, plate and shell elements on the basis of simple approximations of geometry parallels that described above for the case of the tapered axial member. The displacement approximation for the untapered element is adopted and also furnishes the basis, in the shape functions, for the approximation of geometry. This concept is known as *isoparametric representation*—i.e., the same (*iso-*) parameters are employed for both displacement and geometric approximation. The level of interelement continuity of the displacement field embodied in the shape functions then transfers over to the geometric representation. In the example above the function itself (the area) is made continuous from one element to the next. We will present the general theory of isoparametric representation in Section 8.8.

Design practice has not favored the construction of even simple descriptions of taper within prismatic and plate elements. Practitioners prefer to employ uniform-depth elements in a stepped representation. Generally, sufficient computer resources are available for these classes of structure to describe taper accurately with use of large numbers of elements. Thus, concepts of isoparametric representation have not been exploited to a significant extent for analysis of tapered finite elements. This is not the case in the analysis of solids, however, because there even relatively crude finite element analysis gridworks demand very large computational resources.

An alternative approach to the treatment of tapered members is by application of numerical integration directly to the strain energy integral. Tabulations or functional representations of the geometric properties must be available for all sampling points of the numerical integration and of course this approach applies equally well to the integral stemming from the concept (isoparametric) above. Indeed, in the application of isoparametric concepts as described in Section 8.8, the resulting strain energy integral is usually too complicated for explicit integration and numerical integration is required.

6.5 *Displacement Hybrid and Generalized Potential Energy Formulations*

6.5.1 Displacement Hybrid-I Formulation

The assumed displacement-hybrid procedures and the generalized potential energy approach represent alternatives to the single-field, interelement-compatible formulations. Both the hybrid and generalized potential energy schemes involve *multifield* representations of behavior in which one form of displacement field is described within the element and independent displacement and/or stress fields are defined on the element boundaries. The hybrid procedures establish the element equations through the elimination of generalized parameters, while the generalized potential energy approach "patches up" displacement disparities along the boundaries of elements that have been formulated with use of interelement-incompatible displacement fields.

In this section we examine two hybrid formulations that derive from a potential energy functional. In the first of these (*Hybrid I*) the element interior displacement field is expressed in terms of generalized displacements and the boundary stress field is independently described in terms of joint forces. This yields an element flexibility matrix. The second hybrid formulation (*Hybrid II*) is an extension of the concept above in that both the interior displacements and the edge (boundary) stresses are described in terms of generalized parameters, while the boundary displacements are independently described in terms of joint displacements. This leads to an element stiffness matrix.

In order to deal with independent fields a *modification* of the potential energy is needed. In describing the modification applicable to the hybrid-I scheme we consider only *interior* elements, i.e., elements with no boundaries on the edges of the structure, and exclude body or initial forces. The pertinent *boundary* of the element is the complete boundary (S_n) and this is loaded by the interelement tractions $\bar{\mathbf{T}}$. Thus, in view of Eqs. (6.40) and (6.49) we have the modified potential energy

$$\Pi^{m}_{p}l = U - \int_{S_n} \bar{\mathbf{T}}\cdot\mathbf{u}\, dS \tag{6.54}$$

where $\mathbf{u}$ represents the boundary displacements consistent with the chosen interior displacement field $\mathbf{\Delta}$. The generalization of the conventional potential energy consists of the fact that $\bar{\mathbf{T}}$ will be written in terms of joint force parameters so that both the *displacement* parameters of $\mathbf{u}$ (and $\mathbf{\Delta}$) and the joint *force* parameters will appear as unknowns in $\Pi^{m}_{p}l$. A conventional potential energy principle involves only displacement parameters. In order to clarify the descriptions of interior and boundary fields that follow, we have shown, in Fig. 6.8, a hypothetical element with the assumed displacement and stress fields sketched thereon.

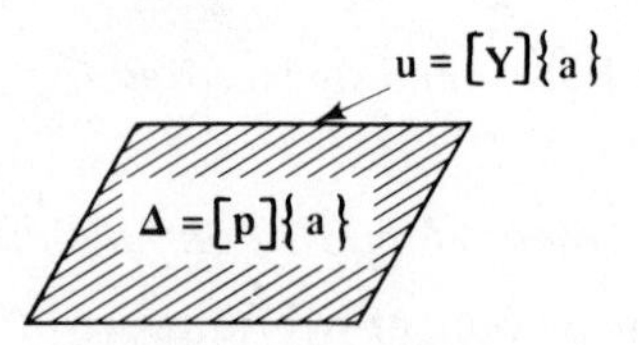

(a) Description of Displacements (interior and surface displacements expressed in terms of same generalized parameters $\{a\}$)

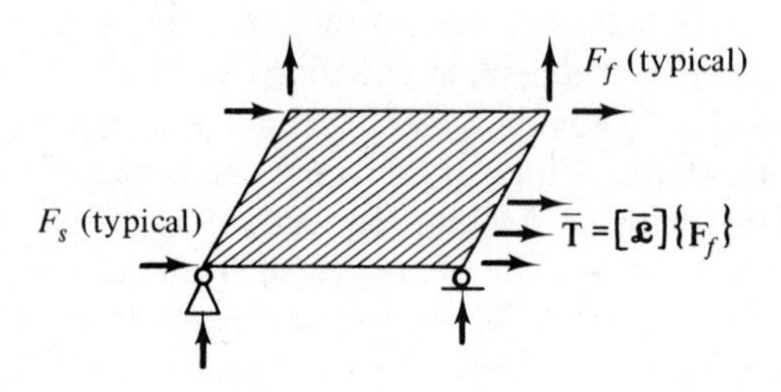

(b) Description of stresses (surface forces expressed in terms of forces at joints free to displace)

Fig. 6.8 Stress and displacement assumptions employed in first displacement hybrid method.

Consistent with the terminology employed in Chapter 5, we designate the generalized parameters of the interior displacement field as $\{\mathbf{a}\}$. We have, for the usual polynomial representation,

$$\mathbf{\Delta} = [\mathbf{p}]\{\mathbf{a}\} \tag{5.2a}$$

and, by application of the strain-displacement equations,

$$\boldsymbol{\epsilon} = [\mathbf{C}_f]\{\mathbf{a}_f\} \tag{5.6d}$$

where $\{\mathbf{a}_f\}$ refers to the d.o.f. remaining after the rigid-body-motion d.o.f. $\{\mathbf{a}_s\}$ have been eliminated through the differentiation associated with the strain-displacement equations. We also require the boundary values $\mathbf{u}$ of this field. These are obtained simply by evaluating $\mathbf{\Delta}$ along the element boundaries

$$\mathbf{u} = [\mathbf{Y}]\{\mathbf{a}\} = [\mathbf{Y}_f \ \mathbf{Y}_s]\begin{Bmatrix} \mathbf{a}_f \\ \mathbf{a}_s \end{Bmatrix} \tag{6.55}$$

where, for purposes of the development to follow, we have preserved the distinction between $\{\mathbf{a}_f\}$ and $\{\mathbf{a}_s\}$.

The final ingredient of the subject displacement-hybrid method is the description of the boundary tractions $\bar{\mathbf{T}}$ in terms of joint forces $\{\mathbf{F}_f\}$. The subscript f designates a system of joint forces exclusive of those necessary to furnish statically determinate support of the element. This is a consequence of the requirement that the vector $\bar{\mathbf{T}}$ represent a system of self-equilibrating forces for cases of zero body force. We symbolize these relationships as follows:

$$\bar{\mathbf{T}} = [\bar{\mathfrak{L}}]\{\mathbf{F}_f\} \tag{6.56}$$

$\bar{\mathbf{T}}$ must represent tractions in equilibrium with the tractions of the adjacent element (with proper account being taken of any applied loads along such interelement boundaries). It should be emphasized that it will in general be difficult, or impossible, to construct equations of the form of Eq. (6.56) to meet these conditions. A more convenient procedure, detailed in Section 6.6.4 and in Chapter 7, is to employ stress functions in place of the stress field and joint values of the stress functions in place of $\{\mathbf{F}_f\}$. The use of joint forces $\{\mathbf{F}_f\}$ is motivated, however, by our use of beam elements in exemplification of various formulative approaches. The appropriate beam element joint parameters are the forces $\{\mathbf{F}_f\}$.

We may now construct the discretized form of the modified potential energy [Eq. (6.54)]. First we observe that in constructing the work of the boundary tractions (the integral over S_n) the contribution of these self-equilibrated forces acting through rigid body displacements is zero. Since the rigid body displacements $\mathbf{u}_s$ are equal to $[\mathbf{Y}_s]\{\mathbf{a}_s\}$ in Eq. (6.55), we consider only the product $[\mathbf{Y}_f]\{\mathbf{a}_f\}$ in treating $\mathbf{u}$, and term this $\mathbf{u}_f$. With

$$U = \frac{1}{2}\int_{\text{vol}} \boldsymbol{\epsilon}\,[\mathbf{E}]\,\boldsymbol{\epsilon}\,d(\text{vol})$$

and, by substitution into Eq. (6.54) of expressions for $\boldsymbol{\epsilon}$, $\mathbf{u}_f$, and $\bar{\mathbf{T}}$, using Eq. (5.6d), the left partition of Eq. (6.55), and Eq. (6.56), respectively, we have

$$\Pi_p^{m_1} = \frac{\lfloor \mathbf{a}_f \rfloor}{2}[\mathbf{H}]\{\mathbf{a}_f\} - \lfloor \mathbf{a}_f \rfloor[\mathbf{J}]\{\mathbf{F}_f\} \tag{6.54a}$$

where

$$[\mathbf{H}] = \left[\int_{\text{vol}} [\mathbf{C}_f]^{\mathrm{T}}[\mathbf{E}][\mathbf{C}_f]\,d(\text{vol})\right] \tag{6.57}$$

$$[\mathbf{J}] = \left[\int_{S_n} [\mathbf{Y}_f]^{\mathrm{T}}[\bar{\mathfrak{L}}]\,dS\right] \tag{6.58}$$

By variation of Eq. (6.54a) with respect to $\lfloor \mathbf{a}_f \rfloor$

$$[\mathbf{H}]\{\mathbf{a}_f\} - [\mathbf{J}]\{\mathbf{F}_f\} = 0$$

or

$$\{\mathbf{a}_f\} = [\mathbf{H}]^{-1}[\mathbf{J}]\{\mathbf{F}_f\}$$

By substitution into Eq. (6.54a)

$$\Pi_p^{m_1} = -\frac{\lfloor \mathbf{F}_f \rfloor}{2}[\mathbf{f}]\{\mathbf{F}_f\} \tag{6.54b}$$

where the derived flexibility matrix is given by

$$[\mathbf{f}] = [\mathbf{J}]^{\mathrm{T}}[\mathbf{H}]^{-1}[\mathbf{J}] \tag{6.59}$$

6.5.2 Example of Displacement Hybrid-I Method

We illustrate this approach through construction of the flexibility matrix of the cantilever beam element (see Fig. 6.9). For this case $\boldsymbol{\Delta} = w$ and since the "surface" or boundary consists of discrete points, the boundary integral in Eq. (6.54) is supplanted by a finite sum. We seek the conventional flexibility matrix for this element and therefore choose a cubic polynomial in description of w, as in Chapter 5, Eq. (5.13)

$$w = x^3a_1 + x^2a_2 + xa_3 + a_4 = [\mathbf{p}_f\ \mathbf{p}_s]\begin{Bmatrix} \mathbf{a}_f \\ \mathbf{a}_s \end{Bmatrix}$$

with

$$[\mathbf{p}_f] = \lfloor x^3 \quad x^2 \rfloor, \qquad [\mathbf{p}_s] = \lfloor x \quad 1 \rfloor$$
$$\{\mathbf{a}_f\} = \lfloor a_1 \quad a_2 \rfloor^{\mathrm{T}}, \qquad [\mathbf{a}_s] = \lfloor a_3 \quad a_4 \rfloor^{\mathrm{T}}$$

Also,

$$w' = \lfloor 3x^2 \quad 2x \quad 1 \rfloor \begin{Bmatrix} a_1 \\ a_2 \\ a_3 \end{Bmatrix} = -\theta$$

$$\boldsymbol{\epsilon} = w'' = \lfloor 6x \quad 2 \rfloor \begin{Bmatrix} a_1 \\ a_2 \end{Bmatrix} = [\mathbf{C}_f]\{\mathbf{a}_f\}$$

We obtain boundary values of this field by evaluation of the expressions

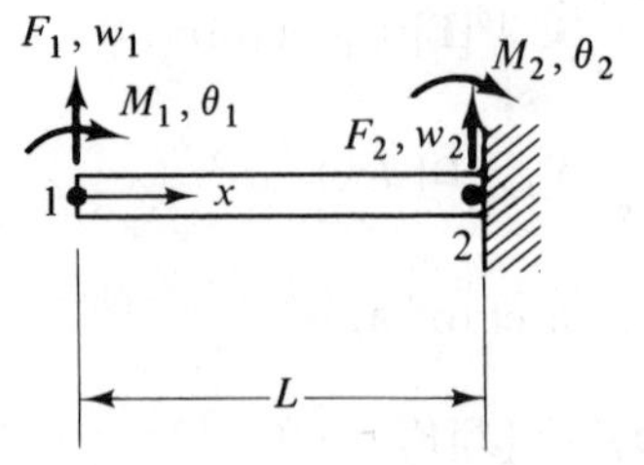

Fig. 6.9

above for w and w' at the points 1 and 2. Thus

$$\mathbf{u} \equiv \begin{Bmatrix} w_1 \\ \theta_1 \\ w_2 \\ \theta_2 \end{Bmatrix} = \left[\begin{array}{cc:cc} 0 & 0 & 0 & 1 \\ 0 & 0 & -1 & 0 \\ L^3 & L^2 & L & 1 \\ -3L^2 & -2L & -1 & 0 \end{array}\right] \begin{Bmatrix} a_1 \\ a_2 \\ \hdashline a_3 \\ a_4 \end{Bmatrix} = [\mathbf{Y}_f \ \mathbf{Y}_s] \begin{Bmatrix} \mathbf{a}_f \\ \mathbf{a}_s \end{Bmatrix}$$

The boundary tractions for this case are simply the joint forces; i.e.,

$$\bar{\mathbf{T}} = \lfloor F_1 \ M_1 \ F_2 \ M_2 \rfloor^{\mathrm{T}}$$

As noted above, however, relationships exist among F_1, M_1, F_2, M_2 due to the conditions of static equilibrium. In particular, $F_2 = -F_1$ and $M_2 = -F_1L - M_1$. Hence

$$\bar{\mathbf{T}} = \begin{Bmatrix} F_1 \\ M_1 \\ F_2 \\ M_2 \end{Bmatrix} = \begin{bmatrix} 1 & 0 \\ 0 & 1 \\ -1 & 0 \\ -L & -1 \end{bmatrix} \begin{Bmatrix} F_1 \\ M_1 \end{Bmatrix} = [\bar{\mathfrak{L}}]\{\mathbf{F}_f\}$$

In the case of the beam element the strain energy is

$$\frac{EI}{2} \int_L (w'') \, dx$$

so we have, from Eq. (6.57)

$$[\mathbf{H}] = \left[EI \int_L [\mathbf{C}_f]^{\mathrm{T}} [\mathbf{C}_f] \, dx \right]$$

and by substitution of the expressions above for $[\mathbf{C}_f]$, $[\mathbf{Y}_f]$, and $[\bar{\mathfrak{L}}]$ into this form of $[\mathbf{H}]$ and into Eq. (6.58) we obtain

$$[\mathbf{H}] = EIL \begin{bmatrix} 12L^2 & 6L \\ 6L & 4 \end{bmatrix}$$

$$[\mathbf{J}] = \begin{bmatrix} 2L^3 & 3L^2 \\ L^2 & 2L \end{bmatrix}$$

so that, from Eq. (6.59)

$$[\mathbf{f}] = [\mathbf{J}]^{\mathrm{T}} [\mathbf{H}]^{-1} [\mathbf{J}] = \frac{L}{6EI} \begin{bmatrix} 2L^2 & 3L \\ 3L & 6 \end{bmatrix}$$

which is the correct representation for the beam element.

6.5.3 Displacement Hybrid-II Method

The second assumed displacement hybrid method[6.6] extends further the concept above to produce directly an element stiffness matrix. We choose here a system of interelement-compatible boundary displacements $\bar{\mathbf{u}}$, expressed in terms of joint displacements $\{\boldsymbol{\Delta}\}$. These are chosen independently of the field $\boldsymbol{\Delta}$ that describes displacements within the element in terms of parameters $\{\mathbf{a}\}$ (see Fig. 6.10). Thus, in the general case, there is a disparity of these displacements along the element boundaries, as given by $(\bar{\mathbf{u}} - \mathbf{u})$, where, as before, $\mathbf{u}$ are the boundary displacements consistent with $\{\mathbf{a}\}$.

It is recalled from Section 6.2 that in the rigorous form of the potential energy principle the displacement boundary conditions are met exactly and comprise the *forced* boundary conditions. Since they are not met exactly in this development, they are treated as *natural* boundary conditions. It is further recalled that the natural boundary conditions are represented directly in the energy functional by a work term. Thus, we form the integral representing the work done by the displacement discrepancy $(\bar{\mathbf{u}} - \mathbf{u})$ under the action of the edge tractions $\mathbf{T}$, i.e.,

$$\int_{S_n} \mathbf{T}(\bar{\mathbf{u}} - \mathbf{u})\, dS$$

$\bar{u} = [\mathcal{Y}]\{\Delta\}$

$u = [Y]\{a_f\}$

Δ_i (typical)

$\Delta = [P]\{a\}$

(a) Description of displacements (interior and associated surface displacements (u) expressed in terms of generalized parameters $\{a_f\}$ Prescribed surface displacements ($\bar{u}$) expressed in terms of joint displacements $\{\Delta\}$).

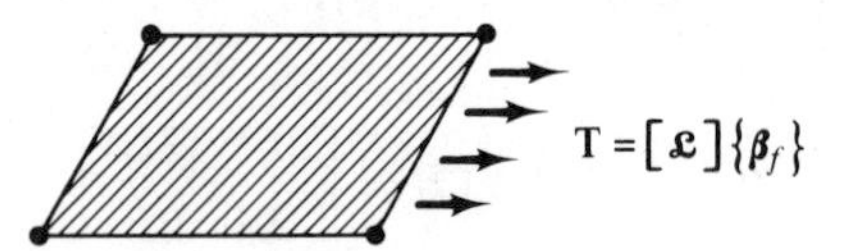

(b) Description of stresses (surface forces expressed in terms of generalized parameters $\{\beta_f\}$).

Fig. 6.10 Stress and displacement assumptions employed in second displacement hybrid method.

and modify the potential energy accordingly†

$$\Pi_p^{m_2} = U - \int_{S_\sigma} \bar{\mathbf{T}} \cdot \mathbf{u}\, dS - \int_{S_n} \mathbf{T}(\bar{\mathbf{u}} - \mathbf{u})\, dS \tag{6.60}$$

In this development the boundary S_σ is present to accommodate element edges that are on the exterior boundary of the structure. Since we restrict our attention to the formulation of interior elements, we disregard the integral on S_σ in what follows. Thus, the modified potential energy for this development becomes

$$\Pi_p^{m_2} = U - \int_{S_n} \mathbf{T}(\bar{\mathbf{u}} - \mathbf{u})\, dS \tag{6.60a}$$

The discretization of Eq. (6.60a) requires field representations of $\boldsymbol{\epsilon}$, $\mathbf{u}$, $\bar{\mathbf{u}}$, and $\mathbf{T}$. The representations of $\boldsymbol{\epsilon}$ and $\mathbf{u}$ are already in hand via Eqs. (5.6d) and (6.55). We must now establish appropriate representations of $\bar{\mathbf{u}}$ and $\mathbf{T}$.

It is required that $\bar{\mathbf{u}}$ be expressed in terms of joint displacements $\{\boldsymbol{\Delta}\}$ and we have already symbolized such relationships as [see Eq. (6.17)]

$$\bar{\mathbf{u}} = [\bar{\mathfrak{Y}}]\{\boldsymbol{\Delta}\}$$

where now the overline is inserted to identify prescribed quantities. Also, the boundary tractions $\mathbf{T}$ are to be written in terms of generalized parameters $\{\boldsymbol{\beta}_f\}$. We invoke the designation of parameters that exclude rigid body terms (the subscript f) consistent with our prior discussion of the definition of boundary tractions [see comments above Eq. (6.56)]. Hence, we denote these relationships as

$$\mathbf{T} = [\mathbf{L}]\{\boldsymbol{\beta}_f\} \tag{6.61}$$

†An alternative view of the term

$$\int_{S_n} \mathbf{T}(\bar{\mathbf{u}} - \mathbf{u})\, dS$$

is obtained if we assume that $\Pi_p^{m_2}$ is a functional that seeks to eliminate the displacement disparity $(\bar{\mathbf{u}} - \mathbf{u})$. Thus, we wish to introduce the constraint condition $(\bar{\mathbf{u}} - \mathbf{u}) = 0$. To do so, we first employ the Lagrange multiplier concept (Section 6.3), which requires that we add the term

$$\int_{S_n} \lambda(\bar{\mathbf{u}} - \mathbf{u})\, dS$$

to the potential energy. As observed earlier, however, the Lagrange multiplier has here the units of a load parameter and in this case it is the boundary traction $\mathbf{T}$ associated with the gap $(\bar{\mathbf{u}} - \mathbf{u})$. Thus,

$$\int_{S_n} \mathbf{T}(\bar{\mathbf{u}} - \mathbf{u})\, dS$$

supplements the basic potential energy.

A discretization of $\Pi_p^{m_2}$ can now be formed by substitution into Eq. (6.60a) of $\boldsymbol{\epsilon}$, $\mathbf{u}_f$, $\bar{\mathbf{u}}$, and $\mathbf{T}$, from Eq. (5.6d), the left portion of Eq. (6.55), and Eqs. (6.17) and (6.61), respectively. We then have.

$$\Pi_p^{m_2} = \frac{\lfloor \mathbf{a}_f \rfloor}{2}[\mathbf{H}]\{\mathbf{a}_f\} - \lfloor \beta_f \rfloor[\mathcal{J}]\{\boldsymbol{\Delta}\} + \lfloor \mathbf{a}_f \rfloor[\mathcal{Q}]\{\beta_f\} \tag{6.60b}$$

where $[\mathbf{H}]$ is as defined by Eq. (6.57) and

$$[\mathcal{J}] = \left[\int_{S_n} [\mathbf{L}]^{\mathrm{T}}[\overline{\mathcal{Y}}]\, dS\right] \tag{6.62}$$

$$[\mathcal{Q}] = \left[\int_{S_n} [Y_f]^{\mathrm{T}}[\mathbf{L}]\, dS\right] \tag{6.63}$$

To develop the desired stiffness matrix we construct algebraic equations by first varying $\Pi_p^{m_2}$ with respect to $\{\mathbf{a}_f\}$ and then with respect to $\{\boldsymbol{\beta}_f\}$, yielding

$$[\mathbf{H}]\{\mathbf{a}_f\} + [\mathcal{Q}]\{\beta_f\} = 0 \tag{6.64a}$$

$$[\mathcal{J}]\{\boldsymbol{\Delta}\} - [\mathcal{Q}]^{\mathrm{T}}\{\mathbf{a}_f\} = 0 \tag{6.64b}$$

By solution of these for $\{\mathbf{a}_f\}$ and $\{\beta_f\}$ in terms of $\{\boldsymbol{\Delta}\}$ and by back-substitution into Eq. (6.60b), there is obtained

$$\Pi_p^{m_2} = \frac{\lfloor \boldsymbol{\Delta} \rfloor}{2}[\mathbf{k}]\{\boldsymbol{\Delta}\} \tag{6.65}$$

where

$$[\mathbf{k}] = [\mathcal{J}]^{\mathrm{T}}[[\mathcal{Q}]^{\mathrm{T}}[\mathbf{H}]^{-1}[\mathcal{Q}]]^{-1}[\mathcal{J}] \tag{6.66}$$

6.5.4 *Example of Hybrid-II Method*

We again examine the beam element: for purposes of illustration see Fig. 6.11. The definition of $\boldsymbol{\epsilon}$ and $\mathbf{u}$ is as in the previous illustration and the matrix $[\mathbf{H}]$ is also identical. The boundary displacements $\bar{\mathbf{u}}$ equal the joint displacements

$$\bar{\mathbf{u}} = \lfloor w_1\ \theta_1\ w_2\ \theta_2 \rfloor^{\mathrm{T}} = \lfloor \boldsymbol{\Delta} \rfloor^{\mathrm{T}}$$

so it is apparent that $[\overline{\mathcal{Y}}] = [\mathbf{I}]$ (unit matrix). Since the present approach requires that the vector of boundary tractions $\mathbf{T} \equiv \lfloor F_1\ M_1\ F_2\ M_2 \rfloor^{\mathrm{T}}$ be defined

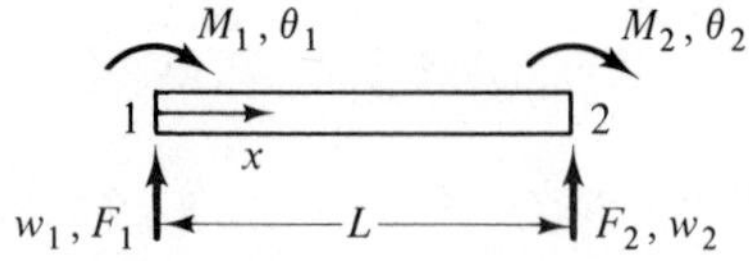

M_1, θ_1 M_2, θ_2 1 x 2 w_1, F_1 L F_2, w_2

Fig. 6.11

in terms of generalized parameters, we choose one generalized parameter for each joint force; i.e., $\lfloor \beta_1 \ \beta_2 \ \beta_3 \ \beta_4 \rfloor^T$. As before, however, $\mathbf{T}$ is a system of self-equilibrating forces and since from equilibrium $F_2 = -F_1$ and $M_2 = -F_1 L - M_1$, it follows that $\beta_3 = -\beta_1$ and $\beta_4 = -\beta_1 L - \beta_2$. Thus,

$$\mathbf{T} \equiv \lfloor F_1 \ M_1 \ F_2 \ M_2 \rfloor^T = [\mathbf{L}]\{\boldsymbol{\beta}_f\}$$

where $[\mathbf{L}]$ is identical to the matrix $[\bar{\mathfrak{L}}]$ constructed for the illustration of the first hybrid method and $\{\boldsymbol{\beta}_f\} = \lfloor \beta_1 \ \beta_2 \rfloor^T$.

We have, from the illustrative example for $\Pi_p^{m_1}$, the matrix $[\mathbf{Y}_f]$ for this case. Using this, together with the expression for $[\mathbf{L}]$ above and $[\overline{\mathfrak{Y}}] = [\mathbf{I}]$ in Eqs. (6.62) and (6.63), we find

$$[\mathfrak{G}] = [\mathbf{L}]^T = \begin{bmatrix} 1 & 0 & -1 & -L \\ 0 & 1 & 0 & -1 \end{bmatrix}$$

$$[\mathfrak{Q}] = \begin{bmatrix} 2L^3 & 3L^2 \\ L^2 & 2L \end{bmatrix}$$

Substituting these and the previously derived matrix $[\mathbf{H}]$ into Eq. (6.66) we obtain the conventional beam element stiffness matrix.

6.5.5 Generalized Potential Energy

The *generalized potential energy* approach can be identified by reinterpretation of Eq. (6.60a). Consider the evaluation of the strain energy (U) and of the surface integrals as separate operations. The strain energy is solely a function of the interior displacements, $\boldsymbol{\Delta}$. In the particular form of generalized potential energy method we are discussing here[6.6–6.8], the interior displacements are written in terms of joint displacements; i.e., $\boldsymbol{\Delta} = [\mathbf{N}]\{\boldsymbol{\Delta}\}$, but these will not meet interelement displacement continuity requirements. Thus, a stiffness matrix, which we term the *basic stiffness* $[\mathbf{k}_0]$ is calculated by substitution of $\boldsymbol{\Delta}$ into U. It will not be interelement-compatible when joined to adjacent elements.

Consider now the surface integral on S_n in Eq. (6.60a). (We shall again discuss only interior elements and therefore exclude examination of the surface integral on S_σ.) Recall from the prior discussion that this integral reconstitutes, or enforces, satisfaction of the required displacement continuity along the element boundaries. As before, we describe the boundary displacements ($\bar{\mathbf{u}}$) independently of the interior displacements, in a manner that is interelement-compatible, but also in terms of the joint displacements $\{\boldsymbol{\Delta}\}$. Now, however, the boundary tractions $\mathbf{T}$ are first written in terms of the displacement derivatives by use of the appropriate equations of elasticity [Eq. (4.5), followed by substitution of the stress-strain law and the strain-displacement equation]. These displacements are also approximated by $\bar{\mathbf{u}}$. The result is an integral which is a quadratic function of the joint displacements $\{\mathbf{\Delta}\}$ and which contains as its Hessian matrix a *corrective* stiffness matrix $[\mathbf{k}_c]$. In consequence, the total stiffness matrix becomes

$$[\mathbf{k}] = [\mathbf{k}_0] + [\mathbf{k}_c] \tag{6.67}$$

An alternative[6.9, 6.10] to this form of the generalized potential energy approach is one in which the basic element stiffnesses $[\mathbf{k}_0]$ are numerically evaluated and summed to form the global stiffness matrix without applying any correction for interelement displacement discontinuity to the individual element relationships. Then constraint equations are formed that represent the average satisfaction of these discontinuities across the element interfaces and these equations are appended to the global equations through application of the Lagrange multiplier procedure. We shall discuss this approach once again in Chapter 7 since it is more properly categorized as a global analysis procedure.

We have chosen to introduce the ideas of hybrid and generalized potential energy element formulations above principally through simple applications. Consequently, our development is lacking in generality to a large measure, as seen by reference to the comments associated with the construction of certain displacement and boundary traction fields [see lines below Eq. (6.56)]. We shall return to these types of formulations again in the chapters on plane stress and bending, however, and at that point a more general perspective will emerge. Developments that are still more general are presented in Refs. 6.5–6.8, 6.11, and 6.12.

6.6 Minimum Complementary Energy

6.6.1 Properties of Complementary Energy

The principle of minimum complementary energy furnishes a variational basis for the direct formulation of element flexibilities, i.e., expressions for element displacement parameters in terms of force parameters. The complementary energy (Π_c) of a structure is given by the sum of the complementary strain energy (U^*) and the potential of boundary forces acting through prescribed displacements (V^*); i.e.,

$$\Pi_c = U^* + V^* \tag{6.68}$$

The principle can be stated as follows: *Among all states of stress which satisfy the equilibrium conditions in the interior of the body and the prescribed surface stress conditions, the state of stress which also satisfies the stress-displacement relations in the interior and all prescribed displacement boundary conditions makes the complementary energy assume a stationary value. Thus*

$$\delta \Pi_c = \delta U^* + \delta V^* = 0 \tag{6.69}$$

For linear theory of elasticity equilibrium, Π_c is a minimum.

$$\delta^2 \Pi_c = \delta^2 U^* + \delta^2 V^* > 0 \tag{6.70}$$

We can establish the validity of the statements above through a procedure that parallels the one used in Section 6.4 to justify the principle of potential energy. The virtual displacement must be replaced by a virtual stress state imposed on the existing true displacement state. Noting that the boundary conditions on stress must be satisfied in the choice of a virtual stress state, the procedure yields Eq. (6.69), i.e., $\delta \Pi_c = 0$, in which the complementary energy is of the form

$$\Pi_c = \frac{1}{2} \int_{\text{vol}} \boldsymbol{\sigma} [\mathbf{E}]^{-1} \boldsymbol{\sigma} \, d(\text{vol}) - \int_{S_u} \mathbf{T} \cdot \bar{\mathbf{u}} \, dS \tag{6.68a}$$

where the first integral on the r.h.s. is U^* and the second integral is $-V^*$. S_u is the surface upon which displacements $\bar{\mathbf{u}}$ are prescribed and $\mathbf{T}$ are the associated boundary tractions.

6.6.2 Finite Element Discretization Using Joint Forces

Consider now the discretization of Π_c for the purpose of constructing finite element representations. The most direct and familiar mode of discretization involves the description of the element stress state in terms of joint forces. This description can be written in the form

$$\boldsymbol{\sigma} = [\mathbf{Z}]\{\mathbf{F}_f\} \tag{6.71}$$

where $\{\mathbf{F}_f\}$ is a set of joint forces excluding the reaction forces associated with a statically determinate element support system. A description of the boundary tractions is obtained by evaluation of Eq. (6.71) on these boundaries and we symbolize the result, as before, by Eq. (6.56)

$$\mathbf{T} = [\mathbf{\mathcal{L}}]\{\mathbf{F}_f\}$$

We note again, as we did in Section 6.5.1, that in the general case it will be difficult or impossible to write $\mathbf{T}$ as a function of the joint forces $\{\mathbf{F}_f\}$. It is feasible to define this transformation, however, for beam and axial elements.

By substitution into Eq. (6.68a) of the expressions for $\boldsymbol{\sigma}$ and $\mathbf{T}$, obtained from Eqs. (6.71) and (6.56), we find

$$\Pi_c = \frac{\lfloor \mathbf{F}_f \rfloor}{2}[\mathbf{f}]\{\mathbf{F}_f\} - \lfloor \mathbf{F}_f \rfloor\{\bar{\mathbf{\Delta}}\} \tag{6.68b}$$

$$[\mathbf{f}] = \left[\int_{\text{vol}} [\mathbf{Z}]^T[\mathbf{E}]^{-1}[\mathbf{Z}]\, d(\text{vol})\right] \quad \text{(element } \textit{flexibility} \text{ matrix)} \tag{6.72}$$

$$\{\bar{\mathbf{\Delta}}\} = \left\{\int_{S_u} [\mathcal{L}]^T\bar{\mathbf{u}}\, dS\right\} \quad \text{(element } \textit{prescribed} \text{ displacement vector)} \tag{6.73}$$

As in the development of the potential energy in discretized form, it is possible by use of the above to demonstrate the minimum character of Π_c.

6.6.3 Example

The application of the ideas above can once again be illustrated by formulation of the beam element flexibility matrix for a cantilever support condition, see Fig 6.12. The basic complementary energy expression for this case is

$$\Pi_c = \frac{1}{2EI}\int_L (\mathfrak{M})^2\, dx - \lfloor F_1 \; M_1 \rfloor \begin{Bmatrix} w_1 \\ \theta_1 \end{Bmatrix}$$

Note that the *stress* for this case is the bending moment $\mathfrak{M}$, that joint displacements w_1 and θ_1 play the role of prescribed displacements, and $[\mathcal{L}]$ is an identity matrix and need not be represented explicitly. It is clear from the figure that the moment varies linearly.

$$\mathfrak{M} = xF_1 + M_1 = \lfloor x \; 1 \rfloor \begin{Bmatrix} F_1 \\ M_1 \end{Bmatrix} = [\mathbf{Z}]\{\mathbf{F}_f\}$$

Thus

$$\Pi_c = \frac{\lfloor F_1 \; M_1 \rfloor}{2}[\mathbf{f}]\begin{Bmatrix} F_1 \\ M_1 \end{Bmatrix} - \lfloor F_1 \; M_1 \rfloor \begin{Bmatrix} w_1 \\ \theta_1 \end{Bmatrix}$$

where

$$[\mathbf{f}] = \frac{1}{EI}\int_L \begin{Bmatrix} x \\ 1 \end{Bmatrix} \lfloor x \quad 1 \rfloor\, dx = \frac{L}{6EI}\begin{bmatrix} 2L^2 & 3L \\ 3L & 6 \end{bmatrix}$$

Element flexibility matrices derived in the manner above can be trans-

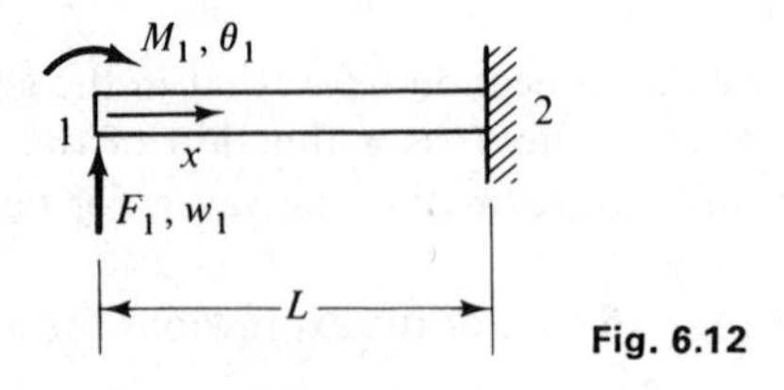

Fig. 6.12

formed into element stiffness matrices using the procedure described in Section 2.6 or they can be employed directly in a global flexibility analysis. When the element flexibilities are written in terms of member forces, the associated global analysis is a *matrix force method*, which involves the definition of *redundant*, self-equilibrating force systems as basic unknowns of the global analysis. As shown in Chapter 7, this type of global analysis is awkward both from the view of selecting the redundant force systems and from the standpoint of the required matrix operations.

6.6.4 Finite Element Discretization Using Stress Functions

The complexities of redundant force analysis may, in large measure, be avoided if the parameters of the element flexibility analysis are either stresses or stress functions rather than forces. In plane stress, for example, the complementary strain energy is

$$U^* = \frac{1}{2}\int_A \lfloor \sigma_x \; \sigma_y \; \tau_{xy} \rfloor [\mathbf{E}]^{-1} \begin{Bmatrix} \sigma_x \\ \sigma_y \\ \tau_{xy} \end{Bmatrix} t\,dA \tag{6.74}$$

Now it will be recalled from Section 4.1 that the stresses σ_x, σ_y, and τ_{xy} may be written as derivatives of the Airy stress function Φ as follows [see Eq. (4.4)]:

$$\sigma_x = \frac{\partial^2\Phi}{\partial y^2} = \Phi,_{yy}, \qquad \sigma_y = \frac{\partial^2\Phi}{\partial x^2} = \Phi,_{xx}, \qquad \tau_{xy} = \frac{-\partial^2\Phi}{\partial x \partial y} = -\Phi,_{xy}$$

Hence

$$U^* = \frac{1}{2}\int_A \lfloor \Phi,_{yy} \; \Phi,_{xx} \; - \Phi,_{xy} \rfloor [\mathbf{E}]^{-1} \begin{Bmatrix} \Phi,_{yy} \\ \Phi,_{xx} \\ -\Phi,_{xy} \end{Bmatrix} t\,dA \tag{6.74a}$$

If the stress function field is written in shape function form,

$$\mathbf{\Phi} = \lfloor \mathbf{N} \rfloor \{\mathbf{\Phi}\} \tag{6.75}$$

where $\{\mathbf{\Phi}\}$ designates a vector of stress function parameters at the element joints. We designate the vector of second derivatives as

$$\lfloor \Phi,_{yy} \; \Phi,_{xx} \; - \Phi,_{xy} \rfloor^{\mathrm{T}} = [\mathbf{N}'']\{\mathbf{\Phi}\} \tag{6.76}$$

The discretized form of U^* becomes

$$U^* = \frac{\lfloor \mathbf{\Phi} \rfloor}{2}[\mathbf{f}]\{\mathbf{\Phi}\} \tag{6.74b}$$

where now, instead of Eq. (6.72), the flexibility matrix is given by

$$[\mathbf{f}] = \left[\int_A [\mathbf{N}'']^T [\mathbf{E}]^{-1} [\mathbf{N}''] t \, dA \right] \tag{6.72a}$$

The significant advantages of this format of the element flexibility matrix derive from two factors. For one, the joint d.o.f. are connectable to joint d.o.f. of adjacent elements as in stiffness analysis so that the manner of construction of the associated global flexibility matrix can be exactly the same as in direct stiffness analysis, as described in Section 3.2. Thus, a *direct flexibility* method is established.[6.13]

The second factor relates to certain dual characteristics of stress functions and displacements. The homogeneous differential equation of the Airy stress function is identically that of plate flexure in the transverse displacement w for zero distributed load. Thus, in Eq. (6.74a), if the stress function is supplanted by w and if $[\mathbf{E}]^{-1}$ is replaced by $[\mathbf{E}]$, the integral defines the strain energy of thin plate flexure. It follows, therefore, that the appropriate choice for a stress function field (Φ field) for a given form of element in plane stress is identical to the choice of the transverse displacement field (w field) for plate flexure and the respective flexibility and stiffness matrices differ only in regard to the difference in elastic coefficients, $[\mathbf{E}]^{-1}$ versus $[\mathbf{E}]$. Since *dual* stress functions can be defined for other situations (e.g., Southwell stress functions for plate flexure, which are dual to in-plane displacements for plane stress), this means that much of the element stiffness matrix technology, established first in terms of assumed, compatible displacement fields, is transferable to flexibility analysis. We shall return to this point in later chapters.

6.7 Assumed Stress Hybrid Approach[6.14, 6.15]

6.7.1 Basic Theory

The assumed stress-hybrid method is an approach to the formulation of element stiffness matrices that is based upon a generalization of the complementary energy principle. As in the discussion of the assumed displacement-hybrid methods, we restrict our attention to the formulative procedure for an element bounded entirely by other elements with no loads applied to the element surfaces or along the interelement boundaries between joints. For the present case the formula for Π_c [Eq. (6.68a)] need only be reinterpreted in the boundary integral in order to yield the desired modified functional Π_c^m.

The philosophy of the stress-hybrid approach is to assume an *equilibrium stress field* $\boldsymbol{\sigma}$ within the element in terms of generalized parameters $\{\boldsymbol{\beta}_f\}$ and to assign to the boundary, simultaneously, an *interelement-compatible dis-*

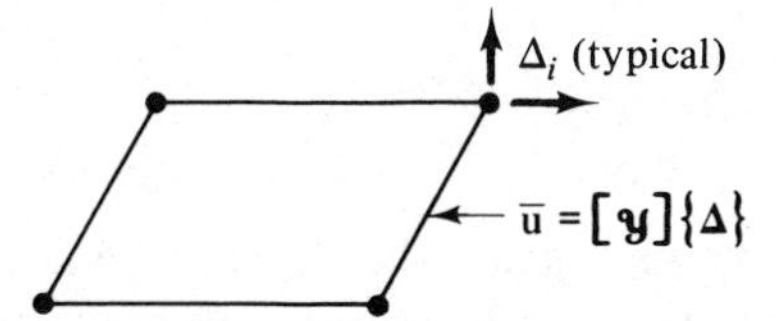

(a) Description of displacements (surface displacements expressed in terms of joint displacements $\{\Delta\}$).

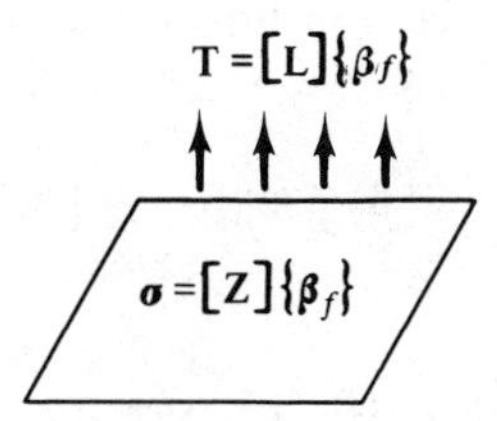

(b) Description of stresses (interior and surface stresses expressed in terms of generalized parameters $\{\beta_f\}$).

Fig. 6.13 Stress and displacement assumptions employed in assumed stress-hybrid method.

placement field $\bar{\mathbf{u}}$ expressed in terms of the element joint displacements $\{\boldsymbol{\Delta}\}$. The system of boundary tractions $\mathbf{T}$ is defined consistent with $\boldsymbol{\sigma}$ so that it also is expressed in terms of $\{\boldsymbol{\beta}_f\}$ (see Fig. 6.13). The modified complementary energy is

$$\Pi_c^m = \frac{1}{2}\int_{\text{vol}} \boldsymbol{\sigma}[\mathbf{E}]^{-1}\boldsymbol{\sigma}\, d(\text{vol}) - \int_{S_n} \mathbf{T}\cdot\bar{\mathbf{u}}\, dS \tag{6.68c}$$

To discretize Π_c^m we first select an assumed stress field in terms of generalized parameters $\{\boldsymbol{\beta}_f\}$.

$$\boldsymbol{\sigma} = [\mathbf{Z}]\{\boldsymbol{\beta}_f\} \tag{6.77}$$

It is again noted that the stresses are described by parameters that exclude the rigid-body-motion (support) terms. This stems from the requirement that $\boldsymbol{\sigma}$ be an equilibrium stress field. It can be seen in another way by noting that if the stresses were basically defined by a stress function field [see Eq. (4.4) and (6.76)], this field would encompass the *support* parameters $\{\boldsymbol{\beta}_s\}$ and these would disappear upon differentiation in accordance with Eq. (4.4).

The element boundary tractions $\mathbf{T}$ can readily be expressed in terms of $\{\boldsymbol{\beta}_f\}$ by appropriate evaluation of Eq. (6.77) along the boundaries. This operation also appeared in the second potential energy hybrid method so we have from Eq. (6.61)

$$\mathbf{T} = [\mathbf{L}]\{\boldsymbol{\beta}_f\}$$

Finally, we note once again that the independently prescribed boundary displacements $\bar{\mathbf{u}}$ are related to the joint displacements by Eq. (6.17)

$$\bar{\mathbf{u}} = [\overline{\mathcal{Y}}]\{\boldsymbol{\Delta}\}$$

Now, with use of the relationships for $\boldsymbol{\sigma}$, $\mathbf{T}$, and $\bar{\mathbf{u}}$, given by Eq. (6.77), (6.61), and (6.17), respectively, we write the discretized form of Eqs. (6.68c):

$$\Pi_c^m = \tfrac{1}{2}\lfloor\boldsymbol{\beta}_f\rfloor[\mathcal{H}]\{\boldsymbol{\beta}_f\} - \lfloor\boldsymbol{\beta}_f\rfloor[\mathcal{G}]\{\boldsymbol{\Delta}\} \tag{6.68d}$$

where

$$[\mathcal{H}] = \left[\int_{\text{vol}} [\mathbf{Z}]^{\mathrm{T}}[\mathbf{E}]^{-1}[\mathbf{Z}]\, d(\text{vol})\right] \tag{6.78}$$

and, as in Eq. (6.62)

$$[\mathcal{G}] = \left[\int_{S_n} [\mathbf{L}]^{\mathrm{T}}[\overline{\mathcal{Y}}]\, dS\right]$$

By variation of Eq. (6.68d) with respect to $\{\boldsymbol{\beta}_f\}$

$$[\mathcal{H}]\{\boldsymbol{\beta}_f\} - [\mathcal{G}]\{\boldsymbol{\Delta}\} = 0$$

or

$$\{\boldsymbol{\beta}_f\} = [\mathcal{H}]^{-1}[\mathcal{G}]\{\boldsymbol{\Delta}\}$$

and, by substitution into Eq. (6.68d),

$$\Pi_c^m = -\frac{\lfloor\boldsymbol{\Delta}\rfloor}{2}[\mathbf{k}]\{\boldsymbol{\Delta}\}$$

with†

$$[\mathbf{k}] = [\mathcal{G}]^{\mathrm{T}}[\mathcal{H}]^{-1}[\mathcal{G}] \tag{6.79}$$

6.7.2 Example

In illustrating this approach for the beam element we note that the "stress field" is the bending moment distribution ($\mathfrak{M}$) and

†It should be noted that the stiffness matrix $[\mathbf{k}]$ will be deficient in rank if the number of parameters in $\{\boldsymbol{\Delta}\}$ exceeds the number of terms in $\{\boldsymbol{\beta}_f\}$ by more than the number of rigid body motions. Eq. (6.79) implies that there is no limit to the relationship between the numbers of terms in these vectors.

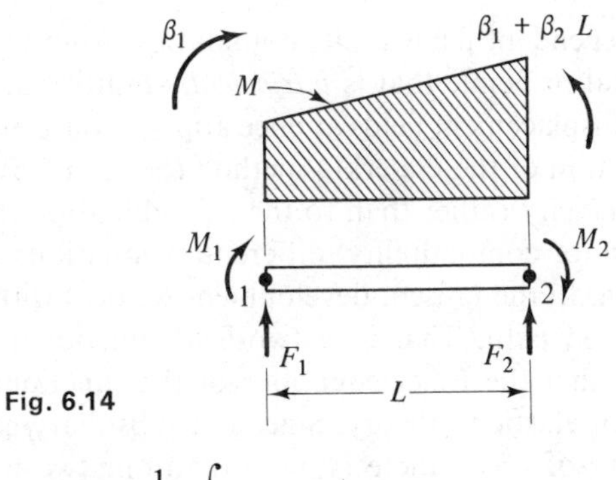

Fig. 6.14

$$U^* = \frac{1}{2EI}\int_L (\mathfrak{M})^2\,dx$$

$$[\mathcal{H}] = \left[\frac{1}{EI}\int_L [\mathbf{Z}]^{\mathrm{T}}[\mathbf{Z}]\,dx\right]$$

Since the bending moment varies linearly for this case,

$$\mathfrak{M} = \lfloor 1\ x \rfloor \begin{Bmatrix}\beta_1\\ \beta_2\end{Bmatrix} = [\mathbf{Z}]\{\boldsymbol{\beta}_f\}$$

In accordance with this distribution, see Fig. 6.14, we find that $M_1 = \beta_1$, $-M_2 = \beta_1 + \beta_2 L$, $F_1 = \beta_2$, $F_2 = -\beta_2$. Hence

$$\mathbf{T} \equiv \lfloor F_1\ M_1\ F_2\ M_2 \rfloor^{\mathrm{T}} = \begin{bmatrix} 0 & 1\\ 1 & 0\\ 0 & -1\\ -1 & -L \end{bmatrix}\begin{Bmatrix}\beta_1\\ \beta_2\end{Bmatrix} = [\mathbf{L}]\{\boldsymbol{\beta}_f\}$$

and $\bar{\mathbf{u}} = \lfloor w_1\ \theta_1\ w_2\ \theta_2 \rfloor$ ($[\overline{\mathcal{Y}}]$ is an identity matrix).

Substituting the above into the appropriate form of Π_c^m, we obtain

$$[\mathcal{H}] = \frac{1}{6EI}\begin{bmatrix} 6L & 3L^2\\ 3L^2 & 2L^3 \end{bmatrix}$$

$$[\mathcal{G}] = [\mathbf{L}]^{\mathrm{T}}$$

It is readily verified that the conventional beam element stiffness matrix results from the use of the matrices $[\mathcal{H}]$ and $[\mathcal{G}]$ above in Eq. (6.79).

6.8 Reissner Energy and Alternative Functionals

6.8.1 Basic Theory

The hybrid approach applied the *multifield* concept to modified potential and complementary principles, using one field within the element and one or two independent fields on its boundaries. Another approach is to employ a variational principle that is *intrinsically* multifield, in which the respective stress and displacement fields at once apply to the element as a whole.

Application of the Galerkin method (Section 5.5) to the subsidiary equations of elasticity rather than to their combination in the form of combined (equilibrium or compatibility) differential equations will produce a two-field formulation. In the present development we deal with the same formulations via a reversed path. That is, a two-field functional is stated and it can be established that the Euler equations of the functional are certain subsidiary equations of elasticity theory. Since the subsidiary equations can be written in a number of ways, there is more than one two-field functional. Here we deal with the Reissner functional (Π_R),[6.16] which has attracted the most attention in finite element analysis.

We can develop this functional using the potential energy as a basis. Excluding once again initial strains and body forces we first note that, by definition,

$$U^* + U = \int_{\text{vol}} \boldsymbol{\sigma} \cdot \boldsymbol{\epsilon} \, d(\text{vol}) \tag{6.80}$$

or

$$U = \int_{\text{vol}} \boldsymbol{\sigma\epsilon} \, d(\text{vol}) - \mathbf{U}^* \tag{6.80a}$$

where U^*, the complementary strain energy is given by the first integral on the r.h.s. of Eq. (6.68a). Now, by substitution into Π_p [Eq. (6.40)] we have the *Reissner functional*, Π_R:

$$\Pi_R = \int_{\text{vol}} \boldsymbol{\sigma} \, \mathbf{D\Delta} d(\text{vol}) - U^* + V \tag{6.81}$$

where $\mathbf{D\Delta}$ represents the displacement derivatives in replacement of the strains, in accordance with Eqs. (4.7). We therefore have a functional expressed in terms of both stresses and displacements. We see that if independent choices are made for both stress-related and displacement fields, then the surface integral (V) must account for *both* prescribed boundary tractions and prescribed displacements; i.e.,

$$V = -\int_{S_\sigma} \bar{\mathbf{T}} \cdot \mathbf{u} \, dS - \int_{S_u} \mathbf{T} \cdot (\mathbf{u} - \bar{\mathbf{u}}) \, dS \tag{6.82}$$

where now S_u is the portion upon which the displacements $\bar{u}$ are prescribed.

It can be shown, by variation of Eq. (6.81) and integration by parts, that the Euler equations of Π_R are the differential equations of equilibrium [Eq. (4.3)] and the stress-displacement differential equations; i.e., the equations obtained by substituting the strain-displacement equations [Eq. (4.7)] into the constitutive equations [Eq. (4.15)]. The converse was proved in Section 5.5 by use of the method of weighted residuals.

In discretizing the above for finite element analysis we consider here only fields that have been written in terms of physical d.o.f. Hence, discretizing $\boldsymbol{\sigma}$, $\boldsymbol{\epsilon}$, $\mathbf{T}$, and $\mathbf{u}$ by use of Eqs. (6.71), (5.6c), (6.56), and (6.17), we find

$$\Pi_R = \lfloor \mathbf{F}_f \rfloor [\boldsymbol{\Omega}_{12}]\{\boldsymbol{\Delta}\} - \frac{\lfloor \mathbf{F}_f \rfloor}{2}[\boldsymbol{\Omega}_{11}]\{\mathbf{F}_f\} - \lfloor \boldsymbol{\Delta} \rfloor \{\bar{\mathbf{F}}\} + \lfloor \mathbf{F}_f \rfloor \{\bar{\boldsymbol{\Delta}}_f\} \tag{6.81a}$$

where

$$[\boldsymbol{\Omega}_{12}] = \left[\int_{\text{vol}} [\mathbf{Z}]^{\mathrm{T}}[\mathbf{D}]\, d(\text{vol})\right] - \left[\int_{S_u} [\mathbf{L}]^{\mathrm{T}}[\mathbf{Y}]\, dS\right] \tag{6.83a}$$

$$[\boldsymbol{\Omega}_{11}] = \left[\int_{\text{vol}} [\mathbf{Z}]^{\mathrm{T}}[\mathbf{E}]^{-1}[\mathbf{Z}]\, d(\text{vol})\right] \tag{6.83b}$$

$$\{\bar{\mathbf{F}}\} = \left\{\int_{S_\sigma} [\mathbf{Y}]^{\mathrm{T}} \cdot \bar{\mathbf{T}}\, dS\right\} \tag{6.83c}$$

$$\{\bar{\boldsymbol{\Delta}}_f\} = \left\{\int_{S_u} [\mathbf{L}]^{\mathrm{T}} \bar{\mathbf{u}}\, dS\right\} \tag{6.83d}$$

and $[\mathbf{Z}]$, $[\mathbf{D}]$, $[\mathbf{L}]$, and $[\mathbf{Y}]$ are the stress-joint force, strain-joint displacement, boundary traction-joint force, and boundary displacement-joint displacement transformation matrices.

Variation of Π_R, in the form of Eq. (6.81a), with respect to both $\{\mathbf{F}_f\}$ and $\{\boldsymbol{\Delta}\}$, yields the following mixed force-displacement matrix:

$$\begin{bmatrix} -\boldsymbol{\Omega}_{11} & \boldsymbol{\Omega}_{12} \\ \boldsymbol{\Omega}_{12}^{\mathrm{T}} & 0 \end{bmatrix} \begin{Bmatrix} \mathbf{F}_f \\ \boldsymbol{\Delta} \end{Bmatrix} = \begin{Bmatrix} \bar{\boldsymbol{\Delta}}_f \\ \bar{\mathbf{F}} \end{Bmatrix} \tag{6.84}$$

This same element matrix format was presented earlier, as Eq. 2.3, and a similar representation appeared in Eq. (5.47).

6.8.2 Example

We again consider, for the purpose of illustration, the simple flexural element, see Fig. 6.15. In applying this approach to the construction of an element that is bounded entirely by elements of the same type, it is assumed that the chosen displacement field will permit satisfaction of the relevant

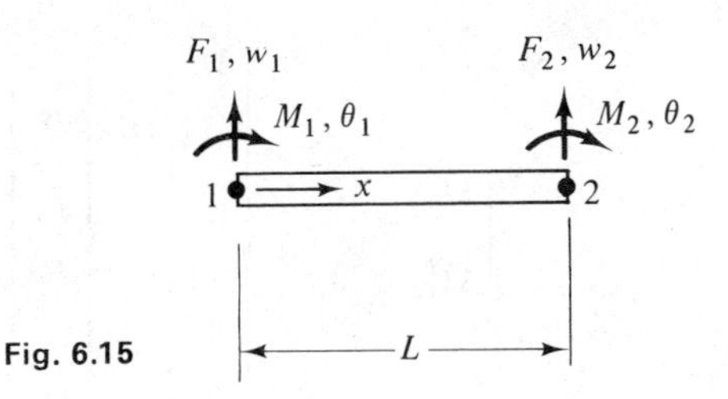

Fig. 6.15

interelement displacement continuity conditions. Thus, $\mathbf{u} - \bar{\mathbf{u}}$ is zero in the surface integral on S_u in Eq. (6.82) and it follows that the surface integral in $[\boldsymbol{\Omega}_{12}]$ and $\{\bar{\boldsymbol{\Delta}}_f\}$ are zero in the discretized functional [see Eqs. (6.83a) and (6.83d)].

For the beam element the *stress field* $\boldsymbol{\sigma}$ is the moment $\mathfrak{M}$, the *displacement field* $\boldsymbol{\Delta}$ is the transverse displacement w, and the *strain field* is the curvature w''. The surface integral on S_σ is a summation of discrete quantities, $F_1 w_1 + F_2 w_2 + M_1\theta_1 + M_2\theta_2$. Hence, the Reissner functional can be written as

$$\Pi_R = \int_L \mathfrak{M} w''\, dx - \frac{1}{2}\int_L \frac{\mathfrak{M}^2}{EI}\, dx - (F_1 w_1 + F_2 w_2 + M_1\theta_1 + M_2\theta_2)$$

The curvature expression has been defined previously; from Eq. (5.15),

$$w'' = \lfloor \mathbf{N}'' \rfloor \{\boldsymbol{\Delta}\}$$

where

$$\lfloor \mathbf{N}'' \rfloor = \frac{2}{L^2}\lfloor 3(2\xi - 1) \mid -3(2\xi - 1) \mid -L(3\xi - 2) \mid -L(3\xi - 1) \rfloor$$

$$\{\boldsymbol{\Delta}\} = \lfloor w_1\; w_2\; \theta_1\; \theta_2 \rfloor^{\mathrm{T}}$$

and $\xi = x/L$.

The moment field is described by a linear variation

$$\mathfrak{M} = \lfloor (1 - \xi) \quad \xi \rfloor \begin{Bmatrix} M_1 \\ M_2 \end{Bmatrix}$$

Substitution of these expressions into Π_R yields

$$\Pi_R = -\frac{\lfloor M_1\; M_2 \rfloor}{2EI} \int_0^L \begin{Bmatrix} (1 - \xi) \\ \xi \end{Bmatrix} \lfloor (1 - \xi)\; \xi \rfloor\, dx \begin{Bmatrix} M_1 \\ M_2 \end{Bmatrix}$$

$$+ \lfloor M_1\; M_2 \rfloor \int_0^L \begin{Bmatrix} (1 - \xi) \\ \xi \end{Bmatrix} \lfloor \mathbf{N}'' \rfloor\, dx\, \{\boldsymbol{\Delta}\} - F_1 w_1 - F_2 w_2 - M_1\theta_1 - M_2\theta_2$$

After performance of the indicated integration and variation of Π_R with respect to M_1, M_2, w_1, w_2, θ_1, and θ_2, we obtain

$$\begin{bmatrix} \boldsymbol{\Omega}_{11} & \boldsymbol{\Omega}_{12} \\ \boldsymbol{\Omega}_{12}^{\mathrm{T}} & \mathbf{0} \end{bmatrix} \begin{Bmatrix} M_1 \\ M_2 \\ \hline w_1 \\ w_2 \\ \theta_1 \\ \theta_2 \end{Bmatrix} = \begin{Bmatrix} 0 \\ \hline F_1 \\ F_2 \\ M_1 \\ M_2 \end{Bmatrix}$$

where

$$[\Omega_{11}] = -\frac{L}{6EI}\begin{bmatrix} 2 & 1 \\ 1 & 2 \end{bmatrix}$$

$$[\Omega_{12}] = \frac{1}{L}\begin{bmatrix} -1 & 1 & -L & 0 \\ 1 & -1 & 0 & L \end{bmatrix}$$

This set of equations can be employed directly in the formation of a global system of equations that will have both forces (M_1, M_2) and displacements (w_1, w_2, θ_1, θ_2) as unknowns. In this special case it is possible to operate upon the element equations to establish the familiar stiffness matrix for this element. To do so, one first solves the upper partition:

$$\begin{Bmatrix} M_1 \\ M_2 \end{Bmatrix} = -[\mathbf{\Omega}_{11}]^{-1}[\mathbf{\Omega}_{12}]\{\mathbf{\Delta}\}$$

and, by substitution into the lower partition,

$$-[\mathbf{\Omega}_{12}]^{\mathrm{T}}[\mathbf{\Omega}_{11}]^{-1}[\mathbf{\Omega}_{12}]\{\mathbf{\Delta}\} = \begin{Bmatrix} F_1 \\ F_2 \\ M_1 \\ M_2 \end{Bmatrix}$$

and it will be verified that $[\mathbf{k}] = -[\mathbf{\Omega}_{12}]^{\mathrm{T}}[\mathbf{\Omega}_{11}]^{-1}[\mathbf{\Omega}_{12}]$ is the stiffness matrix for the beam element.

Fraeijs de Veubeke,[6.17] Prager,[6.18] Washizu,[6.4] Sewell,[6.19] Kikuchi and Ando,[6.8] and others describe in detail more general variational principles of which the potential, complementary, and Reissner principles are special cases. One alternative form derives from the solution of Eq. (6.80) for U^*, substitution of this into Eq. (6.68), and the simultaneous relaxation of the boundary conditions in the manner of Eq. (6.82). The alternatives in element formulation embodied in these more general functionals have been exploited, in some measure, in Sections 6.5 and 6.7.

6.9 An Overview and Summary

We have seen in this chapter that there are numerous independent approaches to the formulation of element flexibility, stiffness, and mixed-variable equations. These alternatives stem basically from potential, complementary, and mixed-energy principles; within each of these categories there are alternatives predicated upon the choices made of the behavior fields in conjunction with appropriate *relaxation* of certain conditions of the basic forms of the energy principles.

Although the *potential energy* (or virtual displacements) procedure is the predominant approach to the formulation of element force-displacement equations, it is not universally the most convenient approach. In many practical circumstances it is difficult to choose an element displacement field that will satisfy all conditions of interelement displacement continuity pertinent to the connection of the element to adjacent elements. Bending elements illustrate this situation. They require continuity not only of the displacement function itself but also of the derivatives of this function (the angular displacements) across the element boundaries. No simple displacements are capable of satisfying these requirements.

For the reason above, it is not uncommon for plate bending element formulation to be established on the basis of a displacement field that is continuous within the element but not across the element boundary when it is joined to other elements. The potential energy principle applies to the individual element formulation but the solution of the global representation will not represent a rigorous application of the minimum potential energy principle since there will be displacement discontinuities along the interelement boundaries.

Similar difficulties are confronted in the development of element formulations on the basis of the *complementary energy* principle. This is particularly the case when the element stress field is described by stress-function fields. Then, a direct correspondence with assumed displacement-potential energy formulations can be constructed and any difficulties associated with the latter accrue to the counterpart complementary formulations.

Alternatives to the potential or complementary energy formulations, with or without discontinuities in the behavior functions across element boundaries, are the generalized potential or complementary energy principles, the hybrid procedures, and multifield functionals. The *generalized potential energy* approach, when applied at the element formulation level, produces a *corrective* element stiffness matrix. Chapter 7 shows that corrective equations can also be applied after a global potential energy representation with discontinuities of the displacement fields across element boundaries has been assembled.

Hybrid formulations not only involve the generalization of the conventional energy principles but also introduce multifield representations of element behavior. For example one form of stress or displacement field is described *within* the element and independent stress and/or displacement fields are described on the *boundary* of the element. All but one of the fields are described in terms of generalized parameters; that field is given in terms of physical d.o.f. The appropriate energy expression (a modification of either potential or complementary energy) is first formed in terms of both classes of parameters and then the stationary condition is applied to the generalized parameter set. This yields a system of equations for the generalized parameters in terms of the physical d.o.f. These equations are used to eliminate the generalized parameters from the energy expression. The resulting energy expression then contains an identifiable stiffness or flexibility matrix of conventional form.

Multifield variational principles lead directly to mixed formats of the element force-displacement equations. Because the Euler equations of these functionals are the more basic equations of elasticity, with lower-order derivatives, the continuity requirements on the assumed fields are of lower order than for the conventional variational principles.

We summarize, as in Section 5.4, the transformations employed in the foregoing element formulation procedures. These transformations have been designated by *barred* superscripts when the quantities being described (boundary tractions or displacements) are regarded as prescribed in a given formulative approach. Also, the subscripts f and s are assigned to the forces and displacements and the transformation matrix is appropriately partitioned when a distinction between *free* and *supported* d.o.f. is necessary.

Generalized Force Parameters-to-Stress Field

$$\boldsymbol{\sigma} = [\mathbf{Z}]\{\boldsymbol{\beta}_f\} \tag{6.77}$$

Joint Forces-to-Stress Field

$$\boldsymbol{\sigma} = [\boldsymbol{\mathfrak{Z}}]\{\mathbf{F}_f\} \tag{6.71}$$

Generalized Force Parameters-to-Edge Traction

$$\mathbf{T} = [\mathbf{L}]\{\boldsymbol{\beta}_f\} \tag{6.61}$$

Joint Forces-to-Edge Traction

$$\mathbf{T} = [\boldsymbol{\mathfrak{L}}]\{\mathbf{F}_f\} \tag{6.56}$$

Generalized Displacement Parameters-to-Boundary Displacements

$$\mathbf{u} = [\mathbf{Y}]\{\mathbf{a}\} \tag{6.55}$$

Joint Displacements-to-Boundary Displacements

$$\mathbf{u} = [\boldsymbol{\mathfrak{Y}}]\{\boldsymbol{\Delta}\} \tag{6.17}$$

REFERENCES

6.1 Mikhlin, S. G., *Variational Methods in Mathematical Physics*, Pergamon Press, Oxford, 1964.

6.2 SCHECTER, R., *The Variational Method in Engineering*, McGraw-Hill Book Co., New York, N.Y., 1967.

6.3 LANGHAAR, H. L., *Energy Methods in Applied Mechanics*, John Wiley & Sons, Inc., New York, N.Y., 1962.

6.4 WASHIZU K., *Variational Methods in Elasticity and Plasticity*, Pergamon Press, Oxford, 1968.

6.5 STRANG, G. and G. FIX., *An Analysis of the Finite Element Method*, Prentice-Hall, Inc., Englewood Cliffs, N.J., 1973.

6.6 TONG, PIN., "New Displacement Hybrid Finite Element Model for Solid Continua," *Internat'l. J. Num. Meth. Eng.*, **2**, 1970, pp. 73–83.

6.7 MCLAY, R. W., "A Special Variational Principle for the Finite Element Method," *AIAA J.*, **7**, No. 3., Mar., 1969, pp. 533–534.

6.8 KIKUCHI, F. and Y. ANDO, "New Variational Functional for the Finite Element Method and Its Application to Plate and Shell Problems," *Nuc. Eng. Design*, **21**, 1972, pp. 95–113.

6.9 GREENE, R. E., R. E. JONES, R. W. MCLAY, and D. R. STROME, "Generalized Variational Principles in the Finite-Element Method," *AIAA J.*, **7**, No. 7, July, 1969, pp. 1254–1260.

6.10 HARVEY, J. W. and S. KELSEY, "Triangular Plate Bending Element with Enforced Compatibility," *AIAA J.*, **9**, No. 6, June, 1971, pp. 1023–1026.

6.11 PIAN, T. H. H., and PIN TONG, "Basis of Finite Element Methods for Solid Continua", *Internat'l. J. Num. Meth. Eng.*, **1**, No. 1, 1969, pp. 3–29.

6.12 PIAN, T. H. H., "Hybrid Models" in *Numerical and Computer Methods in Structural Mechanics*, S. J. Fenves, et al (ed.), Academic Press, New York, N.Y., 1973.

6.13 GALLAGHER, R. H. and A. DHALLA, "Direct Flexibility Finite Element Analysis," Proc. of First Int. Conf. on Struct. Mech. in Nuclear Reactor Technology, Berlin, 1971.

6.14 PIAN, T. H. H., "Derivation of Element Stiffness Matrices by Assumed Stress Distributions," *AIAA J.*, **2**, pp. 1333–1336, 1964.

6.15 PIAN, T. H. H., "Element Stiffness Matrices for Boundary Compatibility and Prescribed Boundary Stresses," Proc. of Conf. on Matrix Methods in Struct. Mechanics, AFFDL TR 66–80, 1965, pp. 457–477.

6.16 REISSNER, E., "On a Variational Theorem in Elasticity," *J. Math. Phys.*, **29**, 1950, p. 90.

6.17 FRAEIJS DE VEUBEKE, B., "Displacement and Equilibrium Models in the Finite Element Method," Chapter 9, *Stress Analysis*, O. C. Zienkiewicz and G. Holister (ed.), John Wiley, Ltd., London, 1965.

6.18 PRAGER, W., "Variational Principles of Linear Elastostatics for Discontinuous Displacements, Strains, and Stresses," *Recent Progress in Applied Mechanics:*

The F. Odqvist Volume, John Wiley & Sons, Inc., New York, 1967, pp. 463–474.

6.19 SEWELL, M. J., "On Dual Approximation Principles and Optimization in Continuum Mechanics," Phil. Trans., Royal Soc. of London, **265**, No. 1162, 13 Nov. 1969, pp. 319–351.

PROBLEMS

6.1 Verify the principle of virtual forces, $\delta U^* = -\delta V$, where U^* is the complementary strain energy. The principle requires the imposition of a virtual stress state $\delta\sigma$, which meets all stress boundary conditions.

6.2 Find the work-equivalent loads at the joints of an axial member for a distribution of load described by

$$q = q_0\left(1 - \left(\frac{x}{L}\right)^2\right)$$

6.3 Derive the *consistent* mass matrix for elementary beam flexure.

6.4 Construct the consistent mass matrix **[m]** for the isotropic triangle in plane stress (Fig. 5.3), ρ = mass density per unit volume. Geometric properties of the triangle need only be symbolized by the symbol ℓ_{nm}, where

$$\ell_{nm} = \int_A x^n y^m \, dA$$

6.5 The distribution of loads shown in Fig. P6.5 acts upon the face of an element whose displacements vary linearly:

$$v = \left(1 - \frac{x}{a}\right)v_1 + \left(\frac{x}{a}\right)v_2$$

Calculate the work-equivalent loads at points 1 and 2.

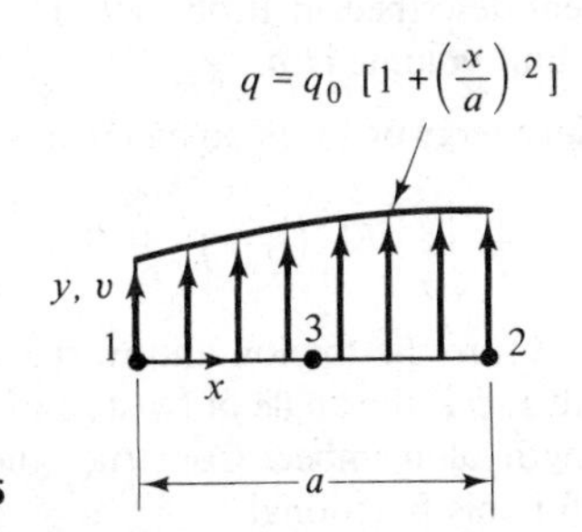

Fig. P6.5

6.6 The distribution of loads defined in Prob. 6.5 acts upon the face of an element whose displacements vary quadratically

$$v = \frac{(2x - a)(x - a)}{a^2}v_1 + 4x\frac{(a - x)}{a^2}v_3 + x\frac{(2x - a)}{a^2}v_2$$

Calculate the work-equivalent loads at points 1, 2, and 3.

6.7 Establish the vector of x-direction thermal forces for the triangular element of Fig. 5.3 when the element is subjected to a temperature change described by

$$\Upsilon = \sum_{i=1}^{3} N_i \Upsilon_i$$

where N_i are the element shape functions and Υ_i are the joint temperatures.

6.8 Find the work-equivalent joint forces and moments for a beam element of length a subjected to a transverse load q distributed in the manner of the loading of Prob. 6.5.

6.9 Construct the work-equivalent load vector for the uniformly loaded six-node triangular element shown in Fig. P6.9 whose displacement field is described by

$$w = \frac{1}{(x_2 y_3)^2} \sum_{i=1}^{6} N_i w_i$$

where

$$N_1 = (x_2^2 y_3^2 + 2y_3^2 x^2 + \tfrac{1}{2} x_2^2 y^2 - 3x_2 y_3^2 x + 2x_2 y_3 xy - \tfrac{3}{2} x_2^2 y_3 y)$$
$$N_2 = (2y_3^2 x^2 - x_2 y_3^2 x + \tfrac{1}{2} x_2^2 y^2 - 2x_2 y_3 xy + \tfrac{1}{2} x_2^2 y_3 y)$$
$$N_3 = (2x_2^2 y^2 - x_2^2 y_3 y)$$
$$N_4 = 4(xx_2 y_3^2 - y_3^2 x^2 - \tfrac{1}{2} x_2^2 y_3 y + \tfrac{1}{4} x_2^2 y^2)$$
$$N_5 = 4(x_2 y_3 xy - \tfrac{1}{2} x_2^2 y^2)$$
$$N_6 = 4(x_2^2 y_3 y - x_2 y_3 xy - \tfrac{1}{2} x_2^2 y^2)$$

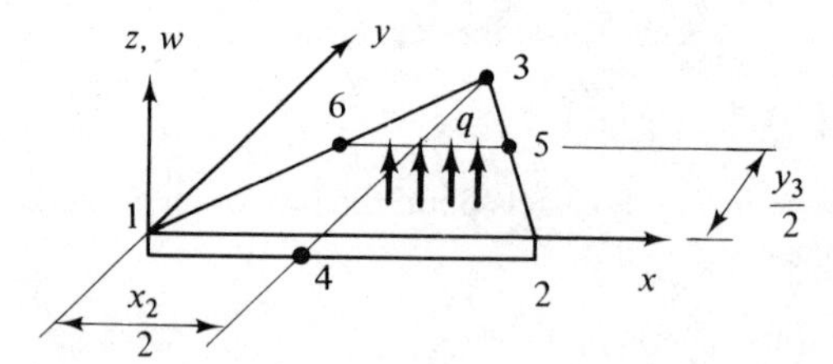

Fig. P6.9

6.10 Formulate the term that connects F_{z_3} and w_5 in the *consistent* mass matrix of the element described in Prob. 6.9. The element thickness is t, the mass density per unit volume is ρ.

6.11 The potential energy of a torsion member is given by the following expression:

$$\Pi_p = \frac{1}{2} \int_L \left[E\Gamma \left(\frac{d^2\phi}{dx^2} \right)^2 + GJ \left(\frac{d\phi}{dx} \right)^2 \right] dx - \int_L \bar{M} \cdot \phi \, dx$$

where J and Γ are the torsion and warping constants, respectively; G is the shear modulus; ϕ is the angle of twist; and $\bar{M}$ is the applied twisting moment per unit length of member. Construct the Euler equations and boundary conditions for this functional.

6.12 The potential energy of a system described by the parameters (Δ_1, Δ_2) and subjected to the load P is given by $\Pi_p = (6 - 3P)\Delta_1^2 - 5(1 - P)\Delta_1 \Delta_2 + (4 - P)\Delta_2^2$. Calculate the value of P at which neutral equilibrium prevails.

6.13 Formulate the 3×3 stiffness matrix for the three-jointed axial member

$$u = \frac{1}{L_2}[(2x - L)(x - L)u_1 + 4(L - x)xu_2 + x(2x - L)u_3]$$

Reduce to the conventional 2×2 axial stiffness matrix.

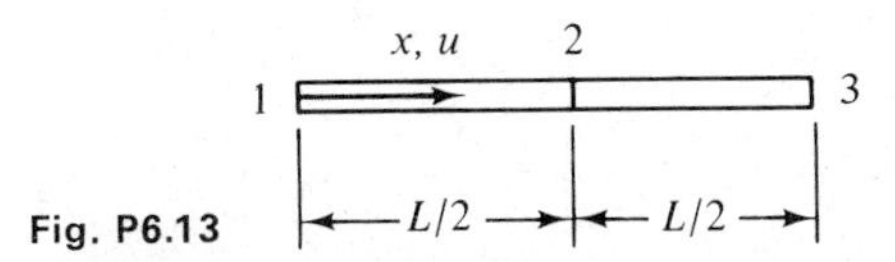

Fig. P6.13

6.14 Formulate an approximate stiffness matrix for the haunched axial member shown in Fig. P6.14 using a linear displacement field

$$u = \left(1 - \frac{x}{L}\right)u_1 + \frac{x}{L}u_2$$

and the principle of minimum potential energy. The member is of constant width b.

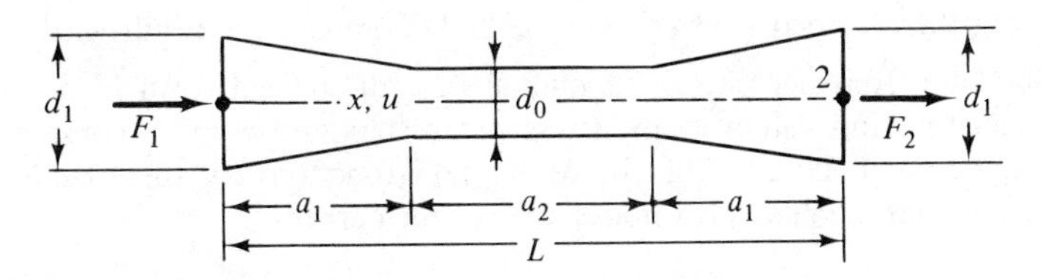

Fig. P.6.14

6.15 Derive the stiffness matrix for the rectangular element in plane stress defined in Prob. 5.2, using the displacement field given therein and the principle of minimum potential energy. Compare your result with the solution given in Fig. 9.13.

6.16 Construct the flexibility matrix for the triangle in plane stress (Fig. 5.3) by use of the displacement hybrid method. Impose the support conditions $u_1 = v_1 = v_2 = 0$. Compare your result with the matrix given in Prob. 2.3.

6.17 Formulate the *exact* flexibility matrix for the tapered beam element shown in Fig. P6.17, using the complementary energy principle. Invert the flexibility to form the corresponding stiffness matrix.

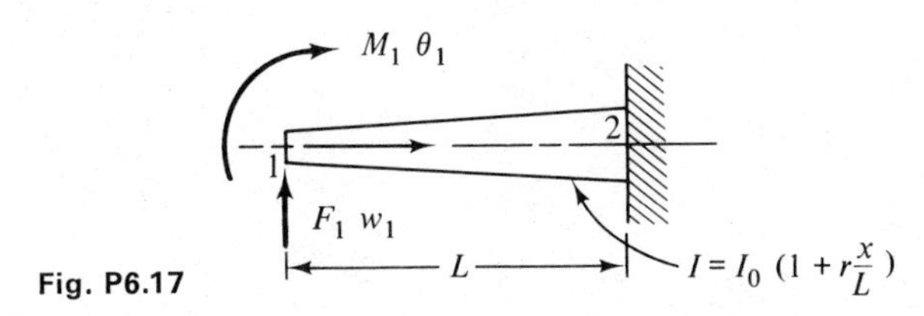

Fig. P6.17

6.18 Formulate the flexibility matrix for the three-node triangular element in plane stress, using the complementary energy principle. Use the support conditions $u_2 = v_2 = v_3 = 0$ (see Fig. 5.3). Compare the result with the flexibility matrix given in Prob. 2.4.

6.19 Formulate, via the complementary energy principle, the flexibility matrix of the curved ring element shown in Fig. P6.19. Compare with the matrix given in Prob. 2.6. (Support at point 2.)

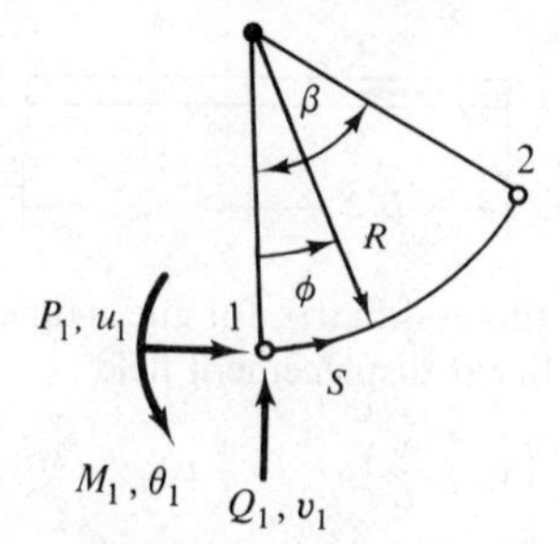

Fig. P6.19

6.20 Formulate the stiffness matrix for the triangular element in plane stress (Fig. 5.3) based upon a constant stress field and linear edge displacements, using the assumed stress-hybrid method. Compare the result with Fig. 5.4.

6.21 State the Reissner energy functional in a discretized form involving Airy stress function values as the stress parameters and u and v components of displacement. Discuss the choice of shape functions for these fields for the rectangular element with nodes at the four corners.

7

Variational Principles in Global Analysis

The application of variational principles to the formulation of element relationships, described in Chapter 6, produced element stiffness, flexibility, or mixed relationships. Such element stiffness relationships may be employed directly in the formation of the equations describing the behavior of the complete structure, using the procedures given in Chapter 3. Thus it might appear that no further reference need be made to variational principles beyond their role in the element formulative process. Actually the variational principles are highly advantageous in certain aspects of the analysis of the complete structure and are essential in others.

For one, the variational principles suggest alternative approaches to the construction of the global equations. This role in global analysis is largely one of furnishing an alternative view of algebraic operations that could be justified on different grounds. Special global operations can also be given the variational perspective; for the particular case of constraint conditions treated by means of the Lagrange multiplier approach, the variational concept is essential. Furthermore, variational principles enable mathematical proofs of convergence, including the direction of convergence for particular forms of these principles.

We examine the potential energy approach in some detail and consider also the complementary energy approach. Mixed methods are not treated because procedures for constructing the global equations parallel those of the conventional variational principles. No convergence properties exist for these methods that would permit the definition of a solution as being an upper or lower bound to the exact solution.

7.1 Minimum Potential Energy Principle

To explain the potential energy approach to global analysis, we observe again that strain energy is a scalar quantity. Hence, the strain energy (U) of a complete structure, consisting of p elements, is simply the sum of the p element strain energies

$$U = \sum_{i=1}^{p} U^i = \frac{1}{2} \sum_{i=1}^{p} \lfloor \mathbf{\Delta}^i \rfloor [\mathbf{k}^i] \{\mathbf{\Delta}^i\} \tag{7.1}$$

where $\lfloor \mathbf{\Delta}^i \rfloor$ and $[\mathbf{k}^i]$ are the joint displacement vector and stiffness matrix of the ith element, respectively. To make use of this fact in defining the algebra of constructing the global equations, we define the following arrays:

$\{\mathbf{\Delta}^e\} = \lfloor \lfloor \mathbf{\Delta}^1 \rfloor \lfloor \mathbf{\Delta}^2 \rfloor \ldots \lfloor \mathbf{\Delta}^p \rfloor \rfloor$. A vector containing all sets of element d.o.f.; defined previously as Eq. (3.12). (7.2)

$\lceil \mathbf{k}^e \rfloor$ *The unassembled global stiffness* matrix, a diagonal matrix of submatrices, each of which is an element stiffness matrix. All element stiffness matrices are included in this array. $\lceil \mathbf{k}^e \rfloor$ was introduced and defined in Section 3.3, Eq. (3.13).

On the basis of these definitions, we can rewrite Eq. (7.1) as

$$U = \tfrac{1}{2} \lfloor \mathbf{\Delta}^e \rfloor \lceil \mathbf{k}^e \rfloor \{\mathbf{\Delta}^e\} \tag{7.3}$$

We must now effect the connection of the elements. To do so, we again refer to Section 3.3 and invoke Eq. (3.14); i.e., $\{\mathbf{\Delta}^e\} = [\mathcal{C}]\{\mathbf{\Delta}\}$, where $\{\mathbf{\Delta}\}$ lists the *global joint displacements* and $[\mathcal{C}]$, as noted in the earlier development, is the global connectivity matrix. Applying this to U in the manner of a conventional transformation results in

$$U = \frac{\lfloor \mathbf{\Delta} \rfloor}{2} [\mathbf{K}] \{\mathbf{\Delta}\} \tag{7.4}$$

where

$$[\mathbf{K}] = [\mathcal{C}]^{\mathrm{T}} \lceil \mathbf{k}^e \rfloor [\mathcal{C}] \tag{7.5}$$

and $[\mathbf{K}]$ is identically the global stiffness matrix constructed in a direct way in Section 3.3. A numerical example of the procedure above was also given in that section.

In Section 3.3 it was stated that the element stiffness matrices that comprise $\lceil \mathbf{k}^e \rfloor$ need not include rigid-body-motion d.o.f. This can now be explained from an energy viewpoint: The $[\mathbf{k}^i]$ matrix of each element is constructed so as to achieve definition of the strain energy and, as disclosed in Section 2.4, the strain energy of an element is defined completely by the stiffness matrix referenced to the d.o.f. beyond those that are suppressed to give statically determinate support to the element. It can further be observed that the transformation represented by the r.h.s. of Eq. (7.5) "frees" the individual elements from their respective supports.

We now return to the general theory of the potential energy representation of the complete structure. The potential energy is completed by formation of the potential of the loads, V. The simplest situation exists when point loads $\{\mathbf{P}\}$ are applied in each d.o.f. Then

$$V = -\lfloor \mathbf{\Delta} \rfloor \{\mathbf{P}\} \tag{7.6}$$

If distributed loads are present, then the product represented by Eq. (7.6) is construed as integration of the vector product of the distributed load and the corresponding displacement state. The latter is given by the associated element displacement field evaluated on the surface of interest. As shown in Chapter 6, the development of these integrals is done on an element-by-element basis and the resulting vector products are appropriately summed to yield a global vector product of the form of Eq. (7.6).

It was shown in Section 6.4 that the potential energy Π_p is given by

$$\Pi_p = U + V \tag{7.7}$$

and, by insertion of the expressions for U and V from Eqs. (7.4) and (7.6), we have

$$\Pi_p = \tfrac{1}{2}\lfloor \mathbf{\Delta} \rfloor [\mathbf{K}] \{\mathbf{\Delta}\} - \lfloor \mathbf{\Delta} \rfloor \{\mathbf{P}\} \tag{7.8}$$

and, applying to this the first necessary condition for a minimum, i.e.,

$$\left\{ \frac{\partial \Pi_p}{\partial \mathbf{\Delta}} \right\} = 0 \tag{7.9}$$

there results

$$[\mathbf{K}]\{\mathbf{\Delta}\} = \{\mathbf{P}\} \tag{7.10}$$

One important conceptual distinction between the process above and direct stiffness analysis should be noted. The direct stiffness method produces each equation by explicit study of the equilibrium of joint forces in each d.o.f. The potential energy approach accomplishes the same result through an addition of component element energies, a key factor being the establishment of the correspondence of element and global d.o.f. (matrix $[\mathcal{A}]$). The latter philosophy is of value for situations where the force quantities associated with certain types of d.o.f. do not have well-defined physical meaning (e.g., d.o.f. that are of the form of higher derivatives of the displacements).

It was pointed out in the beginning of Chapter 6 that many descriptions of finite element analysis adopt energy concepts exclusively. With respect to potential energy this means that element strain energies are summed as in Eq. (7.1) and the potential of applied loads is formed directly from prescribed loads. The mechanics of formation of the global stiffness matrix is then identically a direct stiffness procedure, but the notion of element joint forces, their associated potential ($-\lfloor \mathbf{F}^e \rfloor \{\mathbf{\Delta}^e\}$) and the formation of stiffness equations by explicit study of the equilibrium of joint forces in each d.o.f. does not appear. The construction of global finite element equations for all of the conventional, mixed, and hybrid principles described in Chapter 6 can be accomplished through energy methods in the same manner.

It is also possible to employ a procedure[7.1] that preserves the consideration of the minimum character of the potential energy into the solution phase of analysis. Equation (7.8) illustrates that Π_p is a quadratic function of the variables $\Delta_1, \ldots, \Delta_n$ and the condition for a stable solution is that Π_p attain a minimum. Many reliable algorithms are available for calculation of the set of parameters that will render a minimum quadratic function of such parameters. Since mathematical algorithms are beyond the scope of this text, we do not review these alternatives here; the reader is advised to consult Refs. 7.1 and 7.2. We note, however, one feature of this approach. It is, in fact, possible to construct the element-to-global kinematics matrices on an element-by-element basis, i.e., in the form

$$\{\mathbf{\Delta}^i\} = [\mathcal{A}^i]\{\bar{\mathbf{\Delta}}^i\} \tag{7.11}$$

where now $\{\bar{\mathbf{\Delta}}^i\}$ is intended to signify a vector of *external* d.o.f. for the ith element. Equation (7.1) then becomes

$$U = \frac{1}{2}\sum_{i=1}^{p} \lfloor \bar{\mathbf{\Delta}}^i \rfloor [\mathcal{A}^i]^{\mathrm{T}}[\mathbf{k}^i][\mathcal{A}^i]\{\bar{\mathbf{\Delta}}^i\} \tag{7.12}$$

With this format the scalar value U is evaluated without formation of the global matrices $[\mathbf{K}]$ and $[\mathcal{A}]$; the actual effect is to eliminate all operations on zero matrices, as would be required by Eq. (7.5).

7.2 Lower Bound Character of Minimum Potential Energy Solution

A numerical solution that adheres to all the conditions of a minimum potential energy solution is termed a *lower bound* solution because the strain energy and also the main-diagonal flexibility coefficients of the numerical solution are smaller in value than the "exact" (i.e., infinite number of elements) solution.

The contention above can be demonstrated quite simply for the case of the main-diagonal flexibility coefficient, f_{ii}. As the load P_i is increased from zero value to its stipulated amplitude (with all other loads absent), the work done is $P_i \Delta_i/2$ and this is equal to the internal strain energy, U. The potential of the applied loads is $-P_i \Delta_i$ so that the exact minimum potential energy is

$$\Pi_{p_{\text{exact}}} = U + V = \frac{P_i \Delta_i}{2} - P_i \Delta_i = \frac{-P_i \Delta_i}{2} = \frac{-P_i^2 f_{ii_{\text{exact}}}}{2} \tag{7.13}$$

The approximate value of the potential energy for the same load is

$$\Pi_{p_{\text{approx.}}} = \frac{-P_i^2 f_{ii_{\text{approx.}}}}{2} \tag{7.14}$$

The approximation of Π_p is *algebraically* higher than the exact value since the exact value is the minimum. Noting that $\Pi_{p_{\text{exact}}}$ is a negative value, we see by comparison of Eqs. (7.13) and (7.14) that $f_{ii_{\text{approx.}}}$ must be smaller in value than $f_{ii_{\text{exact}}}$:

$$f_{ii_{\text{exact}}} \geq f_{ii_{\text{approx.}}} \tag{7.15}$$

and it can be concluded that a minimum potential energy solution yields a *lower bound* prediction of the main-diagonal flexibility coefficient.

To illustrate the points above, consider first the structure composed of two axial members (Fig. 7.1). The potential energy for this problem is

$$\Pi_p = \frac{3}{4}\frac{AE}{L}u_2^2 - P_2 u_2$$

Figure 7.2 is a plot of Π_p as a function of various "guessed at" values of u_2. If, for example, we estimate that

$$u_2 = \frac{1}{2}\frac{P_2 L}{AE}$$

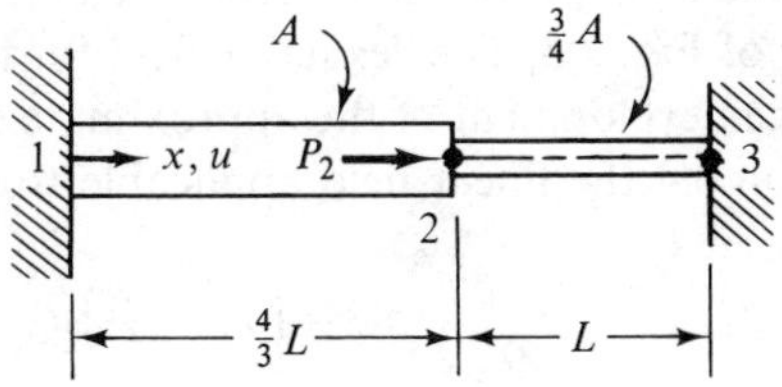

Fig. 7.1

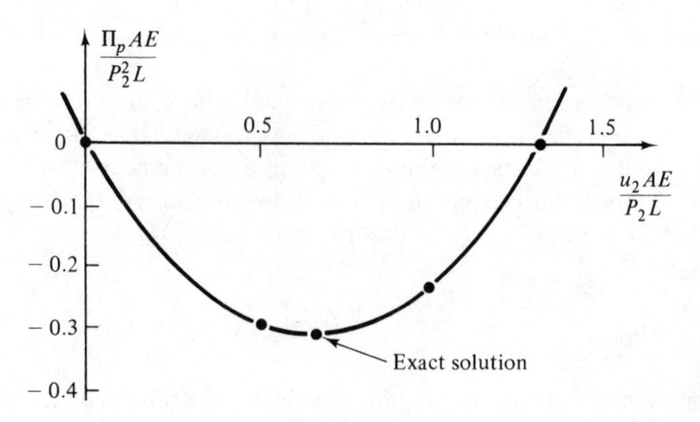

Fig. 7.2

then

$$\Pi_p = -\frac{5}{16}\frac{P_2^2 L}{AE}$$

while for

$$u_2 = \frac{P_2 L}{AE}$$

we have

$$\Pi_p = -\frac{1}{4}\frac{P_2^2 L}{AE}$$

The correct solution is, of course $\frac{2}{3}(P_2 L/AE)$, for which

$$\Pi_p = -\frac{1}{3}\frac{P_2^2 L}{AE}$$

If we were to describe the character of the potential energy for a more complicated problem, with many d.o.f., we would need one orthogonal coordinate axis for each d.o.f. and an additional axis for Π_p. It is not possible to portray this situation for a system of more than two d.o.f. but such geometric properties of this characterization of the problem are exploited in the analysis approach that was discussed briefly at the close of Section 7.1.

To describe the significance of the lower bound theorem in the choice of displacement fields for the individual elements we examine the tapered axial member of Fig. 7.3. The "exact" u field for this case is a logarithmic function but in the development of the approximate stiffness coefficients we employed in Section 6.4 the linear field applicable to uniform members; i.e.,

$$u = \left(1 - \frac{x}{L}\right)u_1 + \frac{x}{L}u_2$$

We examine the case where the area of the right end is twice the area of the left end. If the left end of the element is fixed against displacement, then the displacement u_2 due to a unit value of F_2 is, in the exact solution,

$$u_2 = \frac{L}{(A_2 - A_1)E} \cdot \ln\left(\frac{A_2}{A_1}\right) = 0.69315\frac{L}{A_1E}$$

The approximate solution, on the other hand, is

$$u_2 = \frac{2L}{(A_1 + A_2)E} = 0.66667\frac{L}{A_1E}$$

Thus, the approximate solution is nearly 4% less than the exact solution and is on the low side. By forming element representations of the member with various numbers of segments, the convergence to the correct solution is as shown in Fig. 7.4.

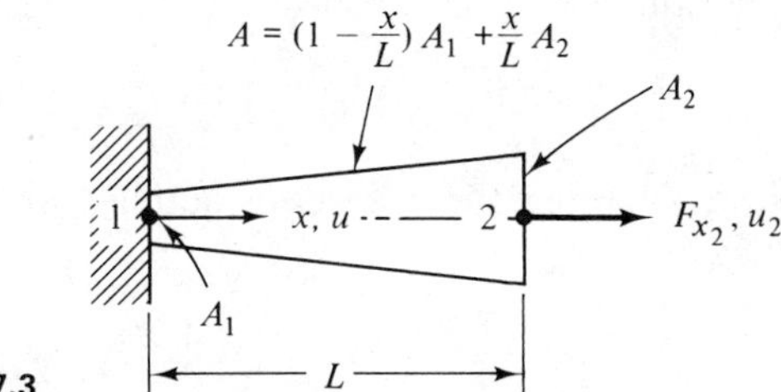

Fig. 7.3

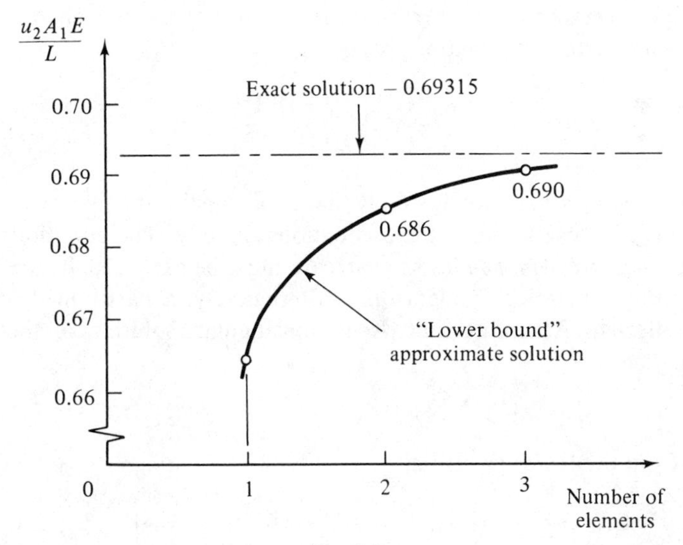

Fig. 7.4

7.3 *Constraint Equations by the Lagrange Multiplier Method*

As we have seen in Section 6.2, the Lagrange multiplier procedure is an approach to treating constraint conditions in the context of classical concepts in variational calculus. We now apply the procedure to multiple d.o.f. systems. In this form it represents an alternative to the *transformation* method of constraint incorporation described in Section 3.5.2.

In accordance with the Lagrange multiplier concept, the extreme value of a functional subject to constraints can be obtained simply by multiplying each constraint by a constant (λ_i—the Lagrange multiplier), adding the result to the original functional, and applying the variation to each d.o.f. and to each multiplier. As before (see Eq. 3.28), for r constraint conditions the constraint equations of an n-d.o.f. system will in general be of the form

$$[\mathbf{G}]_{r\times n}\{\mathbf{\Delta}\}_{n\times 1} = \{\mathbf{s}\}_{r\times 1} \tag{7.16}$$

Thus, r values of λ_i are defined

$$\lfloor \boldsymbol{\lambda} \rfloor = \lfloor \lambda_1 \ldots \lambda_i \ldots \lambda_r \rfloor \tag{7.17}$$

Now, adopting the procedure outlined above, we form the *augmented functional* Π_p^a:

$$\Pi_p^a = \frac{\lfloor \mathbf{\Delta} \rfloor}{2}[\mathbf{K}]\{\mathbf{\Delta}\} - \lfloor \mathbf{\Delta} \rfloor\{\mathbf{P}\} + \lfloor \boldsymbol{\lambda} \rfloor[\mathbf{G}]\{\mathbf{\Delta}\} - \lfloor \boldsymbol{\lambda} \rfloor\{\mathbf{s}\} \tag{7.18}$$

and, by performance of the first variation with respect to each Δ_i and λ_i, the following system of equations results:

$$\begin{bmatrix} \mathbf{K} & \mathbf{G}^{\mathrm{T}} \\ \mathbf{G} & \mathbf{O} \end{bmatrix} \begin{Bmatrix} \mathbf{\Delta} \\ \boldsymbol{\lambda} \end{Bmatrix} = \begin{Bmatrix} \mathbf{P} \\ \mathbf{s} \end{Bmatrix} \tag{7.19}$$

Note that the lower partition is simply the set of constraint equations.

One may choose to solve these equations directly. The equations to be solved are *positive semidefinite* so that care must be exercised in the choice of an appropriate solution algorithm. Alternatively, a partitioned solution may be effected. Assuming that [K] is nonsingular, solution of the upper partition yields

$$\{\mathbf{\Delta}\} = [\mathbf{K}]^{-1}\{\mathbf{P}\} - [\mathbf{K}]^{-1}[\mathbf{G}]^{\mathrm{T}}\{\boldsymbol{\lambda}\} \tag{7.20}$$

so that, from the lower partition,

$$\{\boldsymbol{\lambda}\} = ([\mathbf{G}][\mathbf{K}]^{-1}[\mathbf{G}]^{\mathrm{T}})^{-1}([\mathbf{G}][\mathbf{K}]^{-1}\{\mathbf{P}\} - \{\mathbf{s}\}) \tag{7.21}$$

and $\{\mathbf{\Delta}\}$ is obtained by back-substitution into Eq. (7.20).

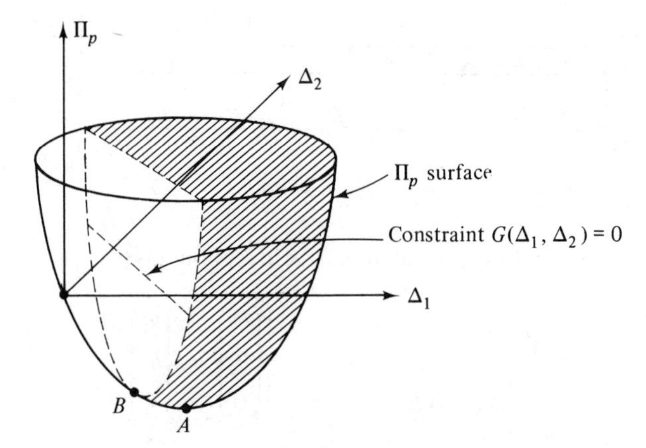

Fig. 7.5 Constrained functional surface for two-d.o.f. system.

It should be observed that Eq. (7.19) is in the format of a *mixed formulation*. (Compare with Eq. 2.3). This is also recognizable from the fact that the dimensional consistency of the augmented functional assigns the units of force parameters to the Lagrange multipliers. In view of the semidefinite nature of Eq. (7.19), lower bound characteristics cannot in general be proved for the potential energy solution constrained in this manner.

It should also be noted that in the transformation procedure of Section 3.5.2 the transformation matrix $[\mathbf{\Gamma}_c]$ is first applied to the potential energy functional and then the transformed Π_p is varied with respect to the retained d.o.f. This is illustrated in Fig. 7.5, in which the potential energy surface for a two-d.o.f. system (Δ_1 and Δ_2), portrayed earlier in Fig. 6.4, has been plotted. In the present case a linear constraint $G(\Delta_1, \Delta_2) = 0$ prevails. This defines a plane perpendicular to the Δ_1, Δ_2 plane which cuts off the portion of the energy surface which contains the previous minimum point A. The minimum now occurs at the point B, on the section through the energy surface made by the constraint plane.

The transformation method contracts the system of equations to be solved, while the Lagrange multiplier method expands it. It should be observed, however, that the transformation method involves extensive matrix manipulation.

To illustrate the Lagrange multiplier method, consider the system of axial members shown in Fig. 7.6 and the constraint condition $u_2 - u_3 = 0$. We have, in accordance with the Lagrange multiplier method, the system of equations (with $k_0 = AE/L$)

$$\begin{bmatrix} 2k_0 & -k_0 & 1 \\ -k_0 & 2k_0 & -1 \\ 1 & -1 & 0 \end{bmatrix} \begin{Bmatrix} u_2 \\ u_3 \\ \lambda \end{Bmatrix} = \begin{Bmatrix} P_2 \\ P_3 \\ 0 \end{Bmatrix}$$

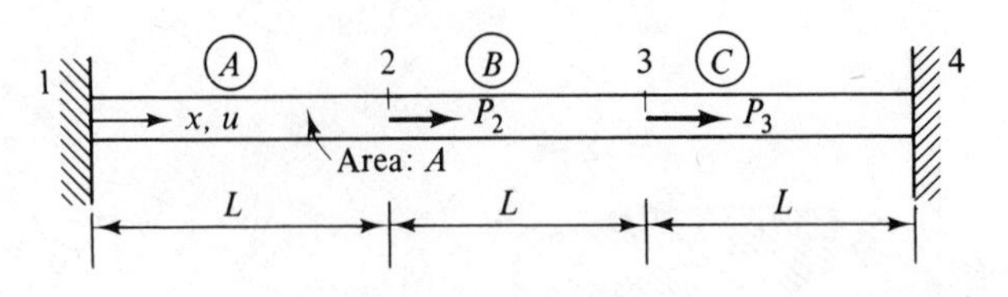

Fig. 7.6

Solution of these equations, through matrix inversion, gives

$$\frac{1}{2k_0}\begin{bmatrix} 1 & 1 & k_0 \\ 1 & 1 & -k_0 \\ k_0 & -k_0 & -3(k_0)^2 \end{bmatrix} \begin{Bmatrix} P_2 \\ P_3 \\ 0 \end{Bmatrix} = \begin{Bmatrix} u_2 \\ u_3 \\ \lambda \end{Bmatrix}$$

The condition $u_2 = u_3$ implies the insertion of a rigid link between points 2 and 3 so that only the members A and C participate in the elastic deformation. Thus, as the solution shows, the displacement of point 2 due to the load P_2 alone is $P_2/2k_0$ and is of the same value for P_3 alone. The Lagrange multiplier $\lambda = \frac{1}{2}(P_2 + P_3)$—a force quantity—and in this case represents the force transmitted through the rigid link. Note that the constraints are applied to a supported structure. Thus the solution procedure of Eqs. (7.20) and (7.21), in which the basic stiffness matrix is inverted, can be employed.

Support conditions, e.g., $\Delta_i = 0$, which are constraint equations as well, can also be handled by the Lagrange multiplier method. The usual practice (Section 3.2) is to apply these conditions directly to the stiffness matrix by striking out the affected columns and removing the corresponding rows. The global stiffness matrix can be left undisturbed in the Lagrange multiplier approach, however, in which case each support condition times a Lagrange multiplier is added to the potential energy. The procedure can be illustrated by reference to the axial member, Fig. 7.7., supported at the left end so that $u_1 = 0$. The algebraic equations of the Lagrange multiplier formulation are then

$$\begin{bmatrix} \dfrac{AE}{L} & \dfrac{-AE}{L} & 1 \\ \dfrac{-AE}{L} & \dfrac{AE}{L} & 0 \\ 1 & 0 & 0 \end{bmatrix} \begin{Bmatrix} u_1 \\ u_2 \\ \lambda \end{Bmatrix} = \begin{Bmatrix} F_1 \\ F_2 \\ 0 \end{Bmatrix}$$

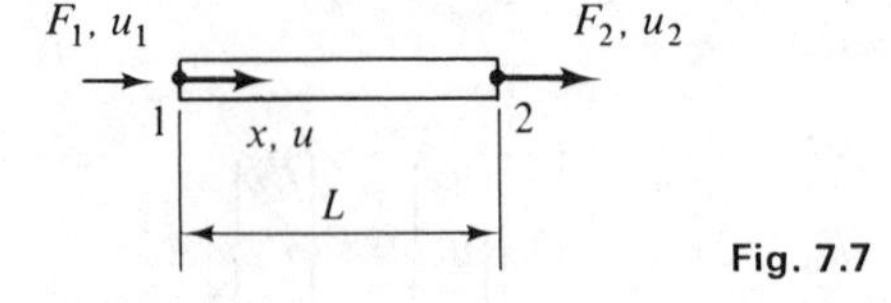

Fig. 7.7

If the first and third columns of this augmented stiffness matrix are interchanged, the equations are of a readily solvable form and we obtain

$$\begin{bmatrix} 1 & 1 & 0 \\ 0 & \dfrac{L}{AE} & 1 \\ 0 & 0 & 1 \end{bmatrix} \begin{Bmatrix} F_1 \\ F_2 \\ 0 \end{Bmatrix} = \begin{Bmatrix} \lambda \\ u_2 \\ u_1 \end{Bmatrix}$$

i.e., $u_2 = F_2L/AE$, $u_1 = 0$, and $\lambda = F_1 + F_2$. The Lagrange multiplier is seen to be the sum of the x-direction forces in this problem. In this case the constraints are applied to the unsupported structure and the basic stiffness matrix is singular. The procedure represented by Eqs. (7.20) and (7.21) cannot be applied.

7.4 Generalized Potential Energy Approach

Chapter 6 examined a number of alternatives to conventional potential and complementary energy procedures in which the requirements on interelement continuity of the assumed fields were relaxed. The examined procedures permitted the formulation of element relationships of a "self-contained" type that could be connected to adjacent elements without modification of global analysis procedures. The present section refers to one of a different class of alternative procedures in which interelement continuity is again relaxed but for which special global operations (specifically, the imposition of constraint conditions on the global stiffness equations) are required.

In the approach discussed here it is assumed that the element trial functions are described in terms of the element joint d.o.f., i.e., $\mathbf{\Delta} = \lfloor \mathbf{N} \rfloor \{\mathbf{\Delta}\}$, that the d.o.f. $\{\mathbf{\Delta}\}$ are to be connected to the corresponding d.o.f. of adjacent elements and that the trial functions are not fully compatible along interelement boundaries. Suppose, for example, that the u displacements along the edge 1-2 of elements A and B of Fig. 7.8 are described by the functions

$$u_{1-2}^A = N_1^A u_1 + N_2^A u_2 + N_3^A u_3 + N_4^A u_4 \tag{7.22a}$$

$$u_{1-2}^B = N_1^B u_1 + N_2^B u_2 + N_5^B u_5 + N_6^B u_6 \tag{7.22b}$$

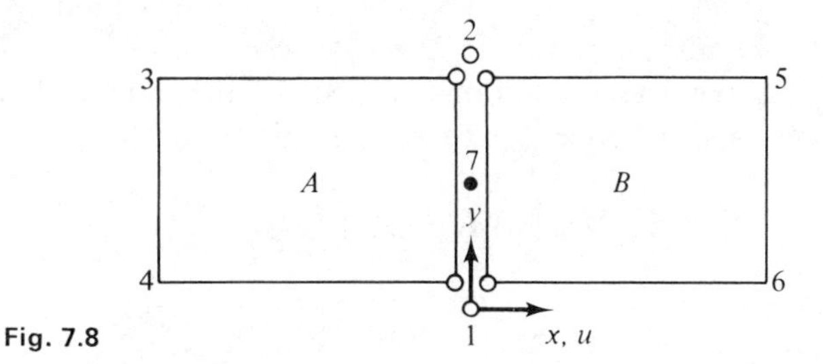

Fig. 7.8

where $N_1^A \ldots N_6^B$ are *quadratic* functions of y (such functions are, in general, functions of both x and y but here they are evaluated along a line where x is a constant). For neither element A nor element B is the displacement defined uniquely by the two end point displacements u_1 and u_2. u_{1-2}^A and u_{1-2}^B differ along the edge 1–2 and a discontinuity of displacement in the amount $u_{1-2}^A - u_{1-2}^B$ prevails. Interelement continuity can be restored, however, by imposition of the condition

$$\int_0^{y_2} (u_{1-2}^A - u_{1-2}^B)\, dy = 0 \tag{7.23}$$

and, by introduction of Eq. (7.22),

$$\int_0^{y_2} [(N_1^A - N_1^B)u_1 + (N_2^A - N_2^B)u_2 + N_3^A u_3 + N_4^A u_4 - N_5^B u_5 - N_6^B u_6]\, dy \tag{7.24}$$

which, after integration, yields a linear algebraic equation of the form

$$G_{11}u_1 + G_{12}u_2 + G_{13}u_3 + G_{14}u_4 + G_{15}u_5 + G_{16}u_6 = 0 \tag{7.25}$$

Using the Lagrange multiplier procedure, Eq. (7.25) can be accounted for in the solution process. In the present illustration one such constraint equation occurs for each displacement component on each element boundary.

In the present case the Lagrange multipliers represent the average internal forces along the lines where the displacement discontinuities are eliminated. Furthermore, the system of global equations has the form of Eq. (7.19), which is the form of the force-displacement equations for the mixed formulation [see Eq. (2.3)]. Thus mixed formulations can be interpreted as applications of the conventional energy methods with "relaxed" continuity requirements.

If the discontinuity of displacement along each interelement boundary is of higher degree, additional constraint equations are needed on each such boundary. One way of dealing with this is to require zero discontinuity everywhere along the interface. We can write the constraint times a Lagrange multiplier as $\lambda(u_{1-2}^A - u_{1-2}^B) = 0$ where λ is now a continuous function of the boundary coordinate. The potential energy is then augmented with $\int \lambda(u_{1-2}^A - u_{1-2}^B)\, dy$. To produce algebraic equations from this we expand λ in a polynomial series $\lambda = \lambda_0 + \lambda_1 y + \lambda_2 y^2 + \ldots$, choosing as many terms as there are conditions needed for the unique definition of displacements along that edge. We thereby obtain

$$\int \lambda(u_{1-2}^A - u_{1-2}^B)\, dy = \lambda_0 \int (u_{1-2}^A - u_{1-2}^B)\, dy + \lambda_1 \int (u_{1-2}^A - u_{1-2}^B) y\, dy + \lambda_2 \int (u_{1-2}^A - u_{1-2}^B) y^2\, dy + \ldots$$

so that the constraint equations are

$$\int (u_{1-2}^A - u_{1-2}^B)\,dy = 0, \qquad \int (u_{1-2}^A - u_{1-2}^B)y\,dy = 0,$$
$$\int (u_{1+2}^A - u_{1-2}^B)y^2\,dy = 0, \ldots \tag{7.26}$$

It should be noted that an alternative approach to the definition of constraint equations to restore boundary displacement continuity is to specify such continuity at discrete points along the boundary. In the example above there is a one-d.o.f. disparity between the edge displacements. Denoting a point midway between points 1 and 2 in Fig. 7.8 as point 7, we can write

$$u_7^A - u_7^B = 0$$

Then by evaluation of the respective (A and B) element displacement fields at point 7, the constraint equation will be produced in a form similar to Eq. 7.25. If we impose this constraint condition by use of the Lagrange multiplier technique, then the multiplier gives the corresponding force quantity at the point in question.

The generalized variational approach has proved especially advantageous in plate bending formulations and we shall have the opportunity to illustrate it in that regard in Chapter 12.

7.5 Minimum Complementary Energy Principle

When joint or edge forces are employed as the unknowns in a flexibility formulation of finite element analysis, it is not possible to construct the associated complementary energy formulation with the same ease as the potential energy representation in a stiffness formulation. This is because the transformation from element joint forces ($\{\mathbf{F}^e\}$) to applied loads ($\{\mathbf{P}\}$) cannot be formed directly for statically indeterminate structural idealizations. If, on the other hand, stress functions are employed as the basic unknowns, the formulation parallels that which is employed in stiffness representations. Both approaches are now described.

Considering first the case in which *forces* are employed as unknowns and once again excluding body forces and initial strains, we adapt Eq. (6.68b) to define the complementary strain energy (U^*) of a finite element model consisting of p elements:

$$U^* = \sum_{i=1}^{p} U^{i*} = \frac{1}{2}\sum_{i=1}^{p} \lfloor \mathbf{F}^i \rfloor [\mathbf{f}^i] \{\mathbf{F}^i\} \tag{7.27}$$

where $\{\mathbf{F}^i\}$ and $[\mathbf{f}^i]$ denote the force vector and flexibility matrix of the ith element, respectively. Alternatively, this may be written in the form

$$U^* = \tfrac{1}{2}\lfloor \mathbf{F}^e \rfloor \lceil \mathbf{f}^e \rfloor \{\mathbf{F}^e\} \tag{7.28}$$

where $\{\mathbf{F}^e\}$ is a vector containing all sets of element force vectors $\lfloor \mathbf{F}^i \rfloor$ *plus* all support reaction forces $\lfloor \mathbf{R}^s \rfloor$; i.e.,

$$\lfloor \mathbf{F}^e \rfloor = \lfloor \lfloor \mathbf{F}^1 \rfloor \lfloor \mathbf{F}^2 \rfloor \dots \lfloor \mathbf{F}^p \rfloor \lfloor \mathbf{R}^s \rfloor \rfloor$$

$\lceil \mathbf{f}^e \rfloor$ is the unassembled *global* flexibility matrix, a diagonal matrix of submatrices, each of which is an element flexibility. All element flexibility matrices $[\mathbf{f}^i]$, $i = 1, \dots, p$, are contained in this array; *in addition*, null rows and columns are included to correspond to the support reaction forces.

The logical next step would appear to be the transformation of the internal and reaction forces $\{\mathbf{F}^e\}$ into applied joint forces $\{\mathbf{P}\}$ but as noted above this cannot be accomplished directly in a statically indeterminate structure. Consequently, $\{\mathbf{F}^e\}$ is expressed as the sum of two force systems, $\{\mathbf{F}^0\}$ and $\{\mathbf{F}^r\}$. $\{\mathbf{F}^0\}$ can be any system of internal forces in equilibrium with $\{\mathbf{P}\}$ and, unless the choice of this system corresponds to the final solution of the problem, it will be associated with an incompatible state of deformation. The forces $\{\mathbf{F}^r\}$ are the amplitudes of self-equilibrating force distributions equal in number to the degree of indeterminacy of the problem.

As shown in Section 3.3, the selection of redundant forces and the construction of equations that relate the internal forces to the applied loads and redundant forces can be accomplished through operations on the global statics equations, $\{\mathbf{P}\} = [\mathfrak{B}]\{\mathbf{F}^e\}$ [Eq. (3.15)]. Alternatively, physical reasoning may be applied for this purpose. In any case the resulting equations are of the form

$$\{\mathbf{F}^e\} = [\mathfrak{D}_1 \,\vdots\, \mathfrak{D}_2]\begin{Bmatrix} \mathbf{P} \\ \hline \mathbf{F}^r \end{Bmatrix} \tag{3.15d}$$

Next the complementary potential (V^*) of the *prescribed displacements* must be formulated. For simplicity we exclude the case of nonzero prescribed displacements. A number of joint d.o.f., those corresponding to supports, will be prescribed to be zero so these make no contribution to V^*. Also, in common with conventional developments of *redundant force* analysis, we consider only concentrated applied (joint) forces, P_i. In treating these, however, we temporarily adopt the view that the corresponding d.o.f. Δ_i are prescribed and the immediately following solution procedure treats the P_i as parameters subject to variation. Hence

$$V^* = -\lfloor \mathbf{P} \rfloor \{\boldsymbol{\Delta}\} \tag{7.29}$$

By substitution into Eq. (7.28) and by expansion, and with $\Pi_c = U^* + V^*$ (see Eq. 6.68) we have

$$\begin{aligned}\Pi_c = {} & \frac{\lfloor \mathbf{P} \rfloor}{2}[\mathfrak{D}_1]^{\mathrm{T}}\lceil \mathbf{f}^e \rfloor[\mathfrak{D}_1]\{\mathbf{P}\} + \lfloor \mathbf{F}^r \rfloor[\mathfrak{D}_2]^{\mathrm{T}}\lceil \mathbf{f}^e \rfloor[\mathfrak{D}_1]\{\mathbf{P}\} \\ & + \frac{\lfloor \mathbf{F}^r \rfloor}{2}[\mathfrak{D}_2]^{\mathrm{T}}\lceil \mathbf{f}^e \rfloor[\mathfrak{D}_2]\{\mathbf{F}^r\} - \lfloor \mathbf{P} \rfloor\{\mathbf{\Delta}\}\end{aligned} \tag{7.30}$$

To establish the stationary value of Π_c, variations are performed with respect to all parameters, both $\{\mathbf{P}\}$ and $\{\mathbf{F}_r\}$, yielding

$$\left[\begin{array}{c|c}[\mathfrak{D}_1]^{\mathrm{T}}\lceil \mathbf{f}^e \rfloor[\mathfrak{D}_1] & [\mathfrak{D}_1]^{\mathrm{T}}\lceil \mathbf{f}^e \rfloor[\mathfrak{D}_2] \\ \hline [\mathfrak{D}_2]^{\mathrm{T}}\lceil \mathbf{f}^e \rfloor[\mathfrak{D}_1] & [\mathfrak{D}_2]^{\mathrm{T}}\lceil \mathbf{f}^e \rfloor[\mathfrak{D}_2]\end{array}\right]\left\{\begin{array}{c}\mathbf{P} \\ \hline \mathbf{F}^r\end{array}\right\} = \left\{\begin{array}{c}\mathbf{\Delta} \\ \hline \mathbf{0}\end{array}\right\} \tag{7.31}$$

Solution of the lower partition gives

$$\{\mathbf{F}^r\} = -[[\mathfrak{D}_2]^{\mathrm{T}}\lceil \mathbf{f}^e \rfloor[\mathfrak{D}_2]]^{-1}[[\mathfrak{D}_2]^{\mathrm{T}}\lceil \mathbf{f}^e \rfloor[\mathfrak{D}_1]]\{\mathbf{P}\} \tag{7.32}$$

and, by substitution into the upper partition,

$$[\mathfrak{F}]\{\mathbf{P}\} = \{\mathbf{\Delta}\} \tag{7.33}$$

where the *assembled global flexibility matrix* is

$$[\mathfrak{F}] = [[\mathfrak{D}_1]^{\mathrm{T}}\lceil \mathbf{f}^e \rfloor[\mathfrak{D}_1] - [\mathfrak{D}_1]^{\mathrm{T}}\lceil \mathbf{f}^e \rfloor[\mathfrak{D}_2][[\mathfrak{D}_2]^{\mathrm{T}}\lceil \mathbf{f}^e \rfloor[\mathfrak{D}_2]]^{-1}[\mathfrak{D}_2]^{\mathrm{T}}\lceil \mathbf{f}^e \rfloor[\mathfrak{D}_1]]\{\mathbf{P}\} \tag{7.33a}$$

while the internal force distribution (and support reaction forces, if they are included in $\{\mathbf{F}^e\}$) is obtained by insertion of Eq. (7.32) into Eq. (3.15d).

$$\{\mathbf{F}^e\} = [[\mathfrak{D}_1] - [\mathfrak{D}_2][[\mathfrak{D}_2]^{\mathrm{T}}\lceil \mathbf{f}^e \rfloor[\mathfrak{D}_2]]^{-1}[\mathfrak{D}_2]^{\mathrm{T}}\lceil \mathbf{f}^e \rfloor[\mathfrak{D}_1]]\{\mathbf{P}\} \tag{7.34}$$

The applied joint forces are now regarded as known values.

It will be observed that in comparison with stiffness analysis the approach to flexibility analysis above requires a very extensive sequence of matrix transformations. More significant is the labor involved in construction of Eq. (3.15d) from the matrix $[\mathcal{B}]$ by means of the procedure described in Section (3.3). The matrix $[\mathcal{B}]$ is formed of equilibrium equations in each d.o.f. Thus, there will be exactly the same number of rows in $[\mathcal{B}]$ as there are equations in a direct stiffness analysis. The method of Gauss-Jordan elimination is a method of matrix inversion so the effort expended in this operation corresponds to that required to establish the inverse of $[\mathbf{K}]$, the assembled global stiffness matrix.

A method of circumventing the difficulties above is to employ *stress functions* as the *stress field parameters*, as introduced in Section 6.6. In accordance with this scheme, the complementary strain energy of the ith element is of the form

$$U^{i*} = \frac{\lfloor \mathbf{\Phi}^i \rfloor}{2}[\mathbf{f}^i]\{\mathbf{\Phi}^i\} \tag{6.74b}$$

where for plane stress $\{\mathbf{\Phi}^i\}$ would contain, as parameters, the Airy stress function and appropriate derivatives at the element joints [see Eq. (6.75)]. Other stress function definitions apply for different types of stress states. The element flexibility $[\mathbf{f}^i]$ is in this case defined by Eq. (6.72a).

Continuity of the stress field across element boundaries, when the element stress fields are chosen to permit such continuity, is established by equating the stress function parameters at the element juncture points. (The element stress fields may be written in terms of joint stress function parameters but of a form that would violate continuity of the tractions across the element boundaries. Such elements would also be joined by equating the stress-function parameters at the juncture points but would not produce a valid global representation of complementary energy.) This procedure may be performed in a manner identical to that of direct stiffness analysis, as detailed in Section 3.2. In this way a *direct flexibility* procedure is established and we symbolize the resulting global complementary strain energy U^* as (for p elements)

$$U^* = \sum_{i=1}^{p} U^{i*} = \frac{\lfloor \mathbf{\Phi} \rfloor}{2}[\mathfrak{F}]\{\mathbf{\Phi}\} \tag{7.35}$$

where now $[\mathfrak{F}]$ is the *global flexibility matrix referenced to stress functions* (rather than to forces) and $\{\mathbf{\Phi}\}$ is the vector of global stress function parameters. Alternatively, U^* could be formed by means of the *congruent transformation* scheme of Section 3.3.

It is essential to take account of the fact that the element complementary strain U^{i*} is formed of the stress field $\boldsymbol{\sigma}$ that is given by the derivatives, to appropriate order, of the stress-function field $\mathbf{\Phi}$ [see Eqs. (6.74) and (4.4)]. The appropriate order of derivative, for example, is of second order for plane stress, Eq. (4.4). Consequently, *U^* is undefined to the extent of the terms that have vanished on account of imposed differentiation.* The situation is exactly the same as that which prevails in a displacement representation where rigid body motion is undefined because of the differentiation of the displacements $\mathbf{\Delta}$ to produce the strain field $\boldsymbol{\epsilon}$.

In order to meet the requirement above, one may suppress enough joint stress parameters to give a *statically determinate, stable support condition.* We have noted, for example, that the Airy stress function for plane stress is the *dual* of transverse displacement in thin plate flexure theory where three properly chosen d.o.f. must be suppressed to give the desired support. It follows that three stress-parameter d.o.f. would likewise be suppressed in the complementary formulation.

It is not necessary that a suppressed d.o.f., say $\mathbf{\Phi}_j$, be imposed directly on the complementary strain energy by substitution of $\mathbf{\Phi}_j = 0$ into U^*. Rather, this constraint can be treated by the Lagrange multiplier procedure noted in Section 7.3 where it was shown that if the constraint equations of a problem give it the property of definiteness and these constraints are handled through the Lagrange multiplier method, then it is not necessary to suppress *rigid-body-motion* d.o.f. in the basic matrix, in this case the "as constructed" matrix $[\mathfrak{F}]$. The treatment of the applied loads results in a system of constraint equations and if the system of applied loads is *self-equilibrated*, then these constraint equations are sufficient to the purpose above.

Consider now the V^* component of the complementary energy. Concentrated (joint) values of the stress function parameters are not given quantities in the analysis and, additionally, the associated deformation parameters do not have a physical meaning of practical interest. Thus, the device of treating the joint deformation parameters as prescribed values is of no value here and $V^* = 0$. This gives, from Eqs. (6.68) and (7.35)

$$\Pi_c = U^* = \frac{\lfloor \mathbf{\Phi} \rfloor}{2}[\mathfrak{F}]\{\mathbf{\Phi}\}$$

and apparently the variation of Π_c with respect to $\mathbf{\Phi}$ gives the indeterminate result $[\mathfrak{F}]\{\mathbf{\Phi}\} = 0$.† Clearly, however, we have not yet at all taken cognizance of the actually present applied loads. By so doing, we form constraint equations that give the solution its proper character.

To describe this, consider the case of plane stress in which an element with an edge parallel to the x axis is acted upon by distributed normal stresses σ_y, Fig. 7.9. From Eq. (4.4), at any point along this edge

$$\frac{\partial^2 \Phi}{\partial x^2} = \sigma_y$$

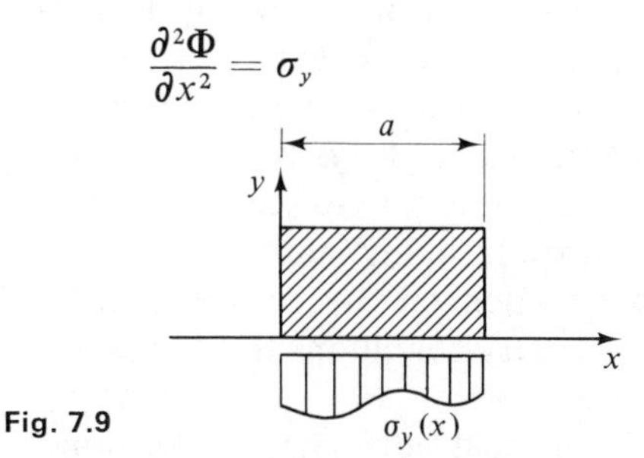

Fig. 7.9

†A nonzero r.h.s. will of course be present if initial strains or body forces exist but this factor does not alter the need to formulate constraint equations to account for surface loadings.

Integrating this expression twice and evaluating the constants of integration in terms of Φ and $\partial\Phi/\partial x = \Phi_x$ at the end points (i.e., $\Phi_i, \Phi_j, \Phi_{x_i}, \Phi_{x_j}$), there results

$$-\Phi_{x_i} + \Phi_{x_j} = \int_0^a \sigma_y \, dx \tag{7.36a}$$

$$-\Phi_i + \Phi_j - \Phi_{x_i} a = \int_0^a \int_0^x \sigma_y \, dx \tag{7.36b}$$

Since σ_y is given and is only a function of x, Eq. (7.36) can be integrated to yield two constraint equations. Operations with respect to all other relevant stress boundary conditions yield additional constraint equations. Preserving the notation of prior sections, the entire set of constraint equations is written in the form

$$[\mathbf{G}]\{\boldsymbol{\Phi}\} = \{\mathbf{s}\} \tag{7.37}$$

These constraints may be taken into account either through the Lagrange multiplier method (Section 7.3) or the condensation method (Section 3.5.2). Adopting the former and with $\lfloor \boldsymbol{\lambda} \rfloor$ defined as the vector of Lagrange multipliers, the *augmented complementary energy function* becomes

$$\Pi_c^a = \frac{\lfloor \boldsymbol{\Phi} \rfloor}{2}[\boldsymbol{\mathfrak{F}}]\{\boldsymbol{\Phi}\} + \lfloor \boldsymbol{\lambda} \rfloor[\mathbf{G}]\{\boldsymbol{\Phi}\} - \lfloor \boldsymbol{\lambda} \rfloor\{\mathbf{s}\} \tag{7.38}$$

and, by variation with respect to $\{\boldsymbol{\Phi}\}$ and $\lfloor \boldsymbol{\lambda} \rfloor$,

$$\begin{bmatrix} \boldsymbol{\mathfrak{F}} & \mathbf{G}^{\mathrm{T}} \\ \mathbf{G} & \mathbf{0} \end{bmatrix} \begin{Bmatrix} \boldsymbol{\Phi} \\ \boldsymbol{\lambda} \end{Bmatrix} = \begin{Bmatrix} \mathbf{0} \\ \mathbf{s} \end{Bmatrix} \tag{7.39}$$

which is now solvable provided the "rigid body constraints" referred to above are represented in this system. They will be represented in the constraint equations, Eq. (7.37), if these equations are written for a system of applied loads that is fully equilibrated. Since unknown support reactions are usually present and prevent a knowledge of the full self-equilibrating surface load system, it is generally necessary to account for this condition by direct modification of the global flexibility matrix.

Detailed information on the construction of constraint equations for applied stress boundary conditions for various types of stress states is contained in Ref. (7.5, 7.6).

One should note that adherence to the principle of minimum complementary energy in a global sense can be retained only if the applied stresses are distributed on the element boundaries in the same form as the assumed stress fields of the elements that comprise these boundaries. If they do not, then the stress constraint equations [e.g., Eq. (7.37)] represent an approximate satisfaction of conditions that are assumed to be met exactly in the minimum complementary energy principle.

7.6 Upper Bound Character of Minimum Complementary Energy Solution

A valid minimum complementary energy solution will, under certain specified conditions, ensure that the values of the calculated deflection influence coefficients are *upper bounds* to the values that would be obtained at the limit of grid refinement.

Consider the case where prescribed displacements are zero so that $V^* = 0$ and $\Pi_c = U^*$. Also, for a single applied load P_i and the associated displacement Δ_i,

$$U^* = \frac{P_i \Delta_i}{2} \tag{7.40}$$

We compare the exact and approximate values of the complementary strain energy, noting that the exact complementary strain energy is a minimum. Thus

$$U^*_{\text{exact}} < U^*_{\text{approx.}} \tag{7.41}$$

and by substitution of Eq. (7.40)

$$P_i(\Delta_i)_{\text{exact}} < P_i(\Delta_i)_{\text{approx.}}$$

or

$$\frac{(\Delta_i)_{\text{exact}}}{(P_i)} = (f_{ii})_{\text{exact}} < (f_{ii})_{\text{approx.}} = \frac{(\Delta_i)_{\text{approx.}}}{(P_i)} \tag{7.42}$$

i.e., the approximate f_{ii} is an upper bound.

Physical meaning can be perceived for both the upper bound solution discussed above and for the lower bound (minimum potential energy) solution. The approximate complementary energy solution features discontinuities of the displacement field and is therefore more flexible than the exact solution. The potential energy solution, with a continuous but approximate displacement field, implies constraints with respect to the exact field and therefore is a stiffer representation.

Mixed and hybrid formulations do not possess upper or lower bound properties. It can be reasoned, however, that they hold out the possibility of a solution that lies between those limits. Suppose, for example, that a stress hybrid formulation employs a stress field that satisfies not only equilibrium within the element but also across element boundaries so that a conventional complementary energy formulation on the basis of this same field would produce an upper bound (high flexibility) solution. The choice of a boundary displacement field in the hybrid formulation constrains the representation somewhat, reduces the flexibility, and moves the results in the direction of the exact solution. It is, of course, possible for the boundary displacement field to "overconstrain" the analytical representation and shift the results to the opposite side of the exact solution, toward the "lower bound" solution associated with the displacement field implied by the boundary displacement field.

The ideas above are meaningful only when referenced to known characteristics of upper or lower bound solutions and must operate within the limits defined by the fields upon which those solutions are based; i.e., the degree of sophistication of the boundary displacement field in a stress hybrid formulation must in some measure correspond to that of the assumed interior stress field. A higher degree of sophistication might very well prove counterproductive. An assessment of these considerations can be found in Ref. 7.7.

REFERENCES

7.1 Fox, R. and E. Stanton, "Developments in Structural Analysis by Direct Energy Minimization," *AIAA J.*, **6**, No. 6, pp. 1036–1042, June, 1968.

7.2 Fried, I., "More on Gradient Iterative Methods in Finite Element Analysis," *AIAA J.*, **7**, No. 3, pp. 565–567, Mar., 1969.

7.3 Greene, R. E., R. E. Jones, R. W. McLay, and D. R. Strome, "Generalized Variational Principles in the Finite-Element Method," *AIAA J.*, **7**, No. 7, pp. 1254–1260, July, 1969.

7.4 Harvey, J. and S. Kelsey, "Triangular Plate Bending Element with Enforced Compatibility," *AIAA J.*, **9**, No. 6, pp. 1023–1026, June, 1971.

7.5 Gallagher, R. H. and A. K. Dhalla, "Direct Flexibility Finite Element Elastoplastic Analysis," Proc. First Internat'l. Conf. on Struct. Mech. in Nuc. React. Tech., Berlin, 1971.

7.6 Morley, L. S. D., "The Triangular Equilibrium Element in the Solution of Plate Bending Problems," *Aero. Quarterly*, **XIX**, pp. 149–169, May, 1968.

7.7 Tong, P. and T. H. H. Pian, "Bounds to the Influence Coefficients by the Assumed Stress Method," *Internat'l. J. Solids Struct.*, **6**, pp. 1429–1432, 1970.

PROBLEMS

7.1 Through expansion of the relevant matrix products, verify that the condensation of a system of stiffness equations can be accomplished by setting to zero the loads associated with the d.o.f. to be removed and forming a transformation matrix on that basis (i.e., verify the footnote in Section 2.7).

7.2 Calculate the tip deflection of the end-loaded cantilever beam in Fig. P7.2. To do this, form a tapered-member stiffness matrix using the exact taper and the shape function for the uniform-section member. Perform calculations for one- and two-segment representations and verify the lower bound character of the solution.

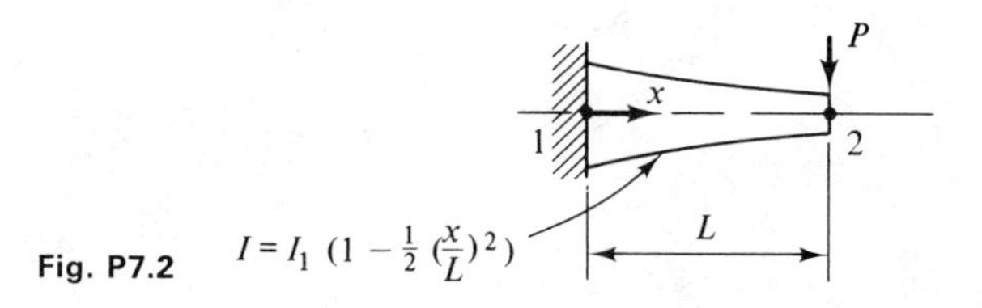

Fig. P7.2

7.3 Define the procedure and general matrix relationships to account for initial strains in the matrix force (complementary energy) method.

7.4 Perform a matrix force analysis of the problem shown in Prob. 3.5, but with the plate elements absent.

7.5 Solve the illustrative problem of Chapter 3 (Figs. 3.4 and 3.6), using the approach based on minimizing a quadratic function of the joint displacements [Eqs. (7.11) and (7.12)].

7.6 Two rectangular elements shown in Fig. P7.6 are to be joined. The v displacement of the respective elements along the juncture lines are as follows:

$$v^A = \left[\frac{(x - x_2)(x - 2x_2)}{2(x_2)^2}\right]v_1 + \left[\frac{x(2x_2 - x)}{(x_2)^2}\right]v_2 + \left[\frac{x(x - x_2)}{2(x_2)^2}\right]v_3$$

$$v^B = \left(\cos\frac{\pi x}{4x_2}\right)v_1 + \left(\sin\frac{\pi x}{4x_2}\right)v_3$$

Establish, in algebraic form, the constraint equation that enforces continuity of displacement at point 2.

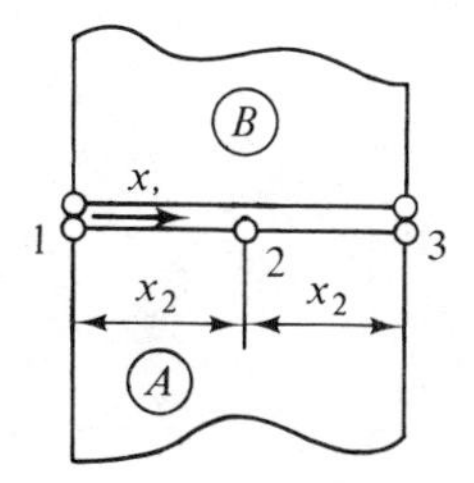

Fig. P7.6

7.7 Construct a stiffness matrix for the beam shown below in terms of the d.o.f. w_2, θ_2^A and θ_2^B. Then, impose the condition of continuity of angular displacement across the hinge using the Lagrange multiplier method and solve for the transverse displacement at point 2. Finally, set up and solve the problem directly for continuity of angular displacement at point 2, solve for the internal bending moment at 2, and compare the result with the calculated value of the Lagrange multiplier.

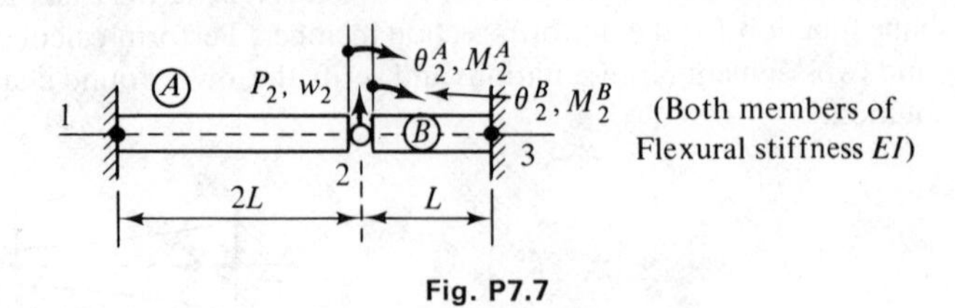

Fig. P7.7

8

Representation of Element Behavior Functions and Geometry

Our prior discussion of element formulation methods has been general enough to admit theories based upon assumed stress, stress function, strain, and displacement fields. We now turn to the problem of selecting such fields, or *behavior functions*, in an organized and rational manner and in forms that are most suited to finite element computational procedures. The subsequent discussion is largely in terms of assumed displacement functions, reflecting the predominance of element formulations based upon assumed displacements. Yet there is a continuing growth of activity and interest in assumed stress or stress function and hybrid formulations and nearly all the considerations developed herein for displacement functions are relevant to the other types of functions.

This chapter begins with an examination of the general conditions to be met by the form of the assumed behavior representations. Then the choice of such representations in terms of polynomial series is discussed, followed by the description of an organized approach to the establishment of representations directly in terms of the physical d.o.f., i.e., in *shape function* form. For triangular (two-dimensional) elements this approach takes the form of special *triangular coordinates*, while for tetrahedral (three-dimensional) elements a corresponding device is available in the form of *tetrahedronal coordinates*. Interpolation concepts are then described for generation of functions for two- and three-dimensional quadrilateral and hexahedronal elements.

A useful concept, stemming from the idea of functional representations for the behavior parameters, is that of representation of the element geometry in the same form, in order to define elements of more general shape such as arbitrary quadrilaterals or elements with curved boundaries. This concept, termed *isoparametric* representation, is also treated in this chapter.

8.1 Requirements on Representation of Element Behavior Functions

One advantage of variational principles in formulation of element relationships for finite element analysis is that they place guidelines on the choice of the *trial functions* or *fields* that describe element behavior. The guidelines that evolve from variational and other considerations can be stated as follows:

1. The chosen functions should possess continuity, of the type required by the variational basis of the formulation, within the element and across the element boundaries when the element is joined to the same type of element or to elements with the same shape functions along contiguous boundaries.
2. The force-displacement equations derived on the basis of the chosen functions should evidence zero strain energy when the element is subjected to rigid body motion.
3. The assumed functions should include representation of constant values of all pertinent stresses or strains.

With respect to condition 1, the trial functions must necessarily be differentiable to the highest order of derivative appearing in the variational functional or else the contributions of any such derivative to the evaluation of the functional will disappear or be otherwise compromised. Derivatives of nth order require that at least nth-order terms be present in polynomial series representation of the behavior function. Clearly, it is not difficult to choose a function that will satisfy this aspect of condition 1.

The satisfaction of condition 1 across element boundaries is loosely termed the *interelement continuity* condition. It was demonstrated in Section 6.3 that when a variational principle is employed in the formulation of an element force-displacement relationship, condition 1 arises from the requirement for a unique evaluation of the integral (functional) identified with that variational principle. In particular, continuity of all derivatives up to one order less than the maximum order appearing in the functional is mandated. The satisfaction of this aspect of condition 1 has presented some of the more formidable difficulties in the construction of behavior functions in finite element analysis. Organized procedures are available for accomplishing this objective for simple elements, however, and these procedures are developed later in this chapter.

A solution that represents rigorous adherence to the minimum potential energy principle requires interelement compatible *displacement* fields in the construction of Π_p. Functions describing equilibrium stress fields that meet *equilibrium* conditions across element boundaries are needed in the construction of Π_c if a valid minimum complementary energy solution is to be realized. As we have seen (Sections 7.2 and 7.6), such solutions have the advantage of furnishing bounds on certain solution parameters. The monotonic convergence of these parameters with grid refinement can also be proved.[8.1-8.2]

Because of these considerations we give close attention to the definition of functions that meet the requirements of conventional energy principles. It must be emphasized, however, that *some* level of interelement continuity is required of functions for the alternative principles (Reissner, hybrid, etc.) and even for *interelement-incompatible* fields that depend on the conventional energy principles in the element formulation stage. The enforcement of continuity of physical d.o.f. is inevitably required in the formation of the global equations when such d.o.f. are employed.

Guarantee of a minimum energy solution also requires satisfaction of condition 2. It was shown in Section 2.9 that the number of rigid body modes contained in a system of element stiffness equations can be ascertained by calculation of the eigenvalues of the matrix of stiffness coefficients. This condition is often specialized to the requirement that no strains due to elastic deformation arise in consequence of a rigid body motion. In the case of simpler elements it is not difficult to test this requirement. In the triangle of Fig. 5.4, for example, the strain ϵ_x is represented by [see Eqs. (5.21a) and (5.22)]

$$\epsilon_x = \frac{1}{x_2 y_3}(-y_3 u_1 + y_3 u_2)$$

Since $u_1 = u_2$ for rigid body motion, we see that $\epsilon_x = 0$. The same condition is satisfied for ϵ_y and γ_{xy}.

Many formulations consciously violate condition 2 if the representation of rigid body motion in the assumed displacement functions may require extremely complicated expressions and procedures in the formulation of element force-displacement equations. This is especially true with respect to formulations in curvilinear coordinates. Many formulations of curved elements accept this violation of the condition of zero strain under rigid body motion in order to produce a simplified formulation. Numerical evidence[8.3, 8.4] for certain elements disclosed that lack of adherence to this condition slowed, but did not preclude, convergence to the correct result.

More serious consequences accrue from violation of condition 3. The exclusion of constant strain states can result in convergence toward an incorrect solution, and in certain cases the error is significant. In the past, it has appeared desirable to construct an element displacement field to exclude a specific constant strain state; in other cases it has been excluded inadvertently. Convergence to an incorrect solution occurs because as the grid is refined, the strain state within an individual element should approach a constant value but cannot since the constant value lacks representation. An example of such an erroneous formulation is presented in Section 12.2.

We study two major related classes of behavior function representation in the sections to follow. These are the *polynomial series* and *shape function* representations. The requirements above are examined in both classes.

8.2 Polynomial Series

The simplest approach to the analytical description of an element behavior field is by means of a *polynomial series* whose coefficients are the *generalized parameters* a_i. Even when the description of the element field is in terms of shape functions, the shape functions can often be identified as transformations of a polynomial field.

In discussing polynomial series we deal with the two-dimensional case for simplicity and assume that the field $\mathbf{\Delta}$ is described by the single quantity Δ. We denote such polynomial series in the alternative forms

$$\Delta = \sum_{i=1}^{n} x^j y^k a_i$$

or

$$\Delta = \lfloor \mathbf{p}(m) \rfloor \{\mathbf{a}\} \tag{8.1}$$

where n is the total number of terms in the series and the superscripts j and k are integer exponents whose values are related to the value of the integer subscript i as follows:

$$i = \tfrac{1}{2}(j + k)(j + k + 1) + k + 1 \tag{8.2}$$

Also m, the *order* of the polynomial, is the highest power to which the variables are raised (the largest value of the sum of the integer exponents j and k). The polynomial is *complete* to a given order if it contains all terms of that order and below. The row matrix $\lfloor \mathbf{p}(m) \rfloor$ denotes, in particular, the vector of spatial variables for a polynomial that is complete to the degree m. The number of terms in a complete polynomial is given by the expression

$$n = \tfrac{1}{2}(m + 1)(m + 2) \tag{8.3}$$

As an example, we have for the linear polynomial

$$\Delta = a_1 + a_2 x + a_3 y = \lfloor \mathbf{p}(1) \rfloor \{\mathbf{a}\}$$

Here $n = 3$, a circumstance that is confirmed by Eq. (8.3). Also, in the second term, $j = 1$ and $k = 0$ and Eq. (8.2) confirms that $i = 2$.

We shall find it convenient in the study of many aspects of polynomial series to refer to the *Pascal triangle* scheme for arraying the terms of the series. This is given by

$$\begin{array}{ccccccccc} & & & & a_1 & & & & \\ & & & a_2x & & a_3y & & & \\ & & a_4x^2 & & a_5xy & & a_6y^2 & & \\ & a_7x^3 & & a_8x^2y & & a_9xy^2 & & a_{10}y^3 & \\ a_{11}x^4 & & a_{12}x^3y & & a_{13}x^2y^2 & & a_{14}xy^3 & & a_{15}y^4 \end{array} \qquad \begin{array}{l} \text{(constant—1 term)} \\ \text{(linear—2 terms)} \\ \text{(quadratic—3 terms)} \\ \text{(cubic—4 terms)} \\ \text{(quartic—5 terms)} \end{array}$$

and so on to any desired order of polynomial. The Pascal triangle shows at a glance how many terms are present in a complete polynomial of any specified order.

The generalized parameters are normally chosen *equal* in number to the element joint d.o.f.; although this case has already been covered in both Chapters 5 and 6, we examine it first in the following for the sake of completeness and to identify features not previously discussed.

The evaluation of the generalized parameters in terms of the joint d.o.f. is accomplished by evaluating the polynomial series at each d.o.f. This results in as many equations as there are d.o.f. Given a representation in the form of Eq. (8.1), this produces

$$\{\mathbf{\Delta}\} = [\mathbf{B}]\{\mathbf{a}\} \tag{5.3a}$$

The terms in [**B**] are constants, in the form of either integers or functions of the given element dimensions. The generalized displacements are sought in terms of the joint d.o.f. so that

$$\{\mathbf{a}\} = [\mathbf{B}]^{-1}\{\mathbf{\Delta}\} \tag{5.4a}$$

and we have, from Eq. (8.1),

$$\Delta = \lfloor \mathbf{p}(m) \rfloor [\mathbf{B}]^{-1}\{\mathbf{\Delta}\} = \lfloor \mathbf{N} \rfloor \{\mathbf{\Delta}\} \tag{8.2a}$$

or, for the more general case of a field $\mathbf{\Delta}$,

$$\mathbf{\Delta} = [\mathbf{p}(m)][\mathbf{B}]^{-1}\{\mathbf{\Delta}\} = [\mathbf{N}]\{\mathbf{\Delta}\} \tag{5.5a}$$

where rectangular matrices, with as many rows as there are components of the field $\mathbf{\Delta}$, supplant the row matrices. In certain cases, e.g., when the element geometry results in a dependence of one d.o.f. on another, the chosen field may result in a singular matrix [**B**]. This occurs when the element geometry results in a dependence of one d.o.f. on another or if a combination of the terms in the polynomial represents a shape function that gives zero value at all node points. Such a "null" shape function represents a degeneracy of rank one of the [**B**] matrix.

The point at which the polynomial series should be truncated, i.e., how many terms of {**a**} should be chosen, deserves careful attention. This matter first involves examination of the requirements discussed in Section 8.1. The requirements on rigid body motion and constant strain present no problem for the linear, flat, and solid elements discussed in this text since they are straightforwardly met by choice of a series that includes all constant and linear terms.

The interelement continuity condition is not so easily met. It can be recalled from prior chapters that if the description of a function along an interelement boundary is specified uniquely by d.o.f. along that boundary, then interelement continuity of that function is preserved in the joining of the element to a counterpart element along the same boundary. For example, if a cubic function is employed to describe the displacements of an element, four independent d.o.f. must be available on each edge for the purpose of describing the function on that edge.

By coupling the factor above with the idea of *geometric isotropy*[8.5] we can establish a criterion for the number of terms to be chosen in a polynomial series representation of an element behavior function. Geometric isotropy seeks to preserve, in any coordinate transformation between cartesian axes, the completeness of all terms in a given degree of polynomial.

According to Eq. (8.3), the total number of terms *available* in a complete mth-degree two-dimensional polynomial is $\frac{1}{2}(m+1)(m+2)$. Consider an $\mathfrak{N}$-sided plane polygonal element wherein the parameter Δ is to be defined by the polynomial series in terms of its values at specified locations on only the *boundary* of the element. If this polynomial is to be uniquely defined on each boundary segment, it must be defined by $(m+1)$ points, and a total of $\mathfrak{N}(m+1) - \mathfrak{N} = \mathfrak{N}m$ such points will be *required* on the boundary of the complete polygon (account must be taken of $\mathfrak{N}$ vertex points, which are common to two sides). Equating the available number of coefficients [given by Eq. (8.3)] to the required number, we have

$$\frac{1}{2}(m+1)(m+2) = \mathfrak{N}m \tag{8.4}$$

a condition that is met only by $\mathfrak{N} = 3$ and $m = 1$ or $m = 2$. These elements are the simple and six-jointed triangles in Fig. 8.1.

It is possible to produce other *triangular* elements that meet the conditions above if we relax the requirement that all node points lie on the periphery of the element. This is done in Section 8.5. One could also permit derivatives of the behavior function as node point d.o.f. Rectangular elements

Fig. 8.1 Triangular elements with nodes on periphery.

cannot meet the conditions on a "complete" polynomial as it is defined above, even with the relaxation of the condition of edge-only node points and d.o.f. in the form of the function itself. This point is amplified in Section 8.4.

The test above can be extended to three-dimensional elements in direct stress and also to plate bending elements. Two points should be noted with respect to the criterion above. First, one should add the formal mathematical definition of completeness of a series, which requires that if a function $\boldsymbol{\Delta}$ is represented by the series $\sum a_i\Delta_i$, then over the region of application

$$\Delta - \sum_{i\to\infty} a_i\Delta_i \longrightarrow 0 \tag{8.5}$$

This condition is met by polynomial series.

Second we emphasize once again that elements may be defined and polynomial series constructed for them in a form that violates interelement compatibility conditions. The resulting formulation may prove to be entirely satisfactory apart from the lack of assuredness of an upper bound or lower bound solution.

Polynomial expressions which are written in terms of generalized displacements can be employed directly in the construction of an element stiffness matrix referenced to such generalized displacements. This was done earlier in Chapters 5 and 6 in construction of basic element matrices, and in Chapter 7 in conjunction with the generalized variational principles. Such stiffnesses serve as *kernel* stiffness matrices which are amenable to transformation into matrices which are referenced to alternative physical d.o.f.

Consider, for example, the four-point axial member shown in Fig. 8.2, which is to be formulated on the basis of a cubic polynomial

$$u = \lfloor x^3\ x^2\ x\ 1 \rfloor \begin{Bmatrix} a_1 \\ a_2 \\ a_3 \\ a_4 \end{Bmatrix} = \lfloor \mathbf{p}(3) \rfloor \{\mathbf{a}\}$$

Fig. 8.2 Four-node axial segment.

The basic formula for the kernel stiffness matrix, given by Eq. (6.18) is

$$[\mathbf{k}^a] = \left[\int_{\text{vol}} [\mathbf{C}]^{\mathrm{T}}[\mathbf{E}][\mathbf{C}]d(\text{vol})\right] \tag{6.18}$$

In the present case $d(\text{vol}) = A\,dx$, $[\mathbf{E}] = E$, and $[\mathbf{C}]$, the transformation of generalized displacements-to-strains, is simply the row matrix $\lfloor \mathbf{p}' \rfloor$, where

$$\lfloor \mathbf{p}' \rfloor = \frac{d}{dx}\lfloor \mathbf{p}(3) \rfloor$$

so that

$$[\mathbf{k}^a] = \left[AE\int_0^{3L} \{\mathbf{p}'\}\lfloor \mathbf{p}' \rfloor\,dx\right]$$

If one seeks the element stiffness matrix with reference to the physical d.o.f. u_1, u_2, u_3, and u_4, then the transformation from generalized-to-joint displacements (Eq. 5.3(a)) is constructed: In the present case it is of the form

$$\begin{Bmatrix} u_1 \\ u_2 \\ u_3 \\ u_4 \end{Bmatrix} = \begin{bmatrix} 0 & 0 & 0 & 1 \\ L^3 & L^2 & L & 1 \\ (2L)^3 & (2L)^2 & 2L & 1 \\ (3L)^3 & (3L)^2 & 3L & 1 \end{bmatrix} \begin{Bmatrix} a_1 \\ a_2 \\ a_3 \\ a_4 \end{Bmatrix}$$

Inversion of the matrix on the r.h.s. produces a matrix that transforms $a_1, \ldots, a_4$ into $u_1, \ldots, u_4$. This transformation is then applied to the kernel stiffness matrix in the usual manner.

Alternatively, one may wish to reference the stiffness matrix to the end point displacements (u_1, u_4) and the derivatives of the displacement $(du_1/dx, du_4/dx)$ at these same points. The transformation in this case is exactly that which was constructed for the flexural member in Chapter 5 (except for the sign of the equations for the derivatives) and will not be repeated here.

8.3 Direct Construction of Shape Functions Through Interpolation

Although the polynomial form of an assumed displacement field is useful in assessment of the completeness of a function and of its adherence to certain required conditions and is sometimes essential to the particular approach taken to finite element analysis, most often it is preferable to be able to express the assumed field directly in terms of the joint d.o.f., i.e., in shape function form. This can be accomplished through the use of *inter-*

polation procedures. In this section we demonstrate the application of these procedures to shape functions in a single coordinate direction.

8.3.1 Lagrangian Interpolation

Lagrangian interpolation permits determination of the coefficients of a polynomial series representation of a function in terms of values of the function at points on a line.

Consider the line shown in Fig. 8.3(a), divided into equal-length segments by the $(m + 1)$ points $1, 2, \ldots, m + 1$. The locations of the points are defined by the *physical coordinates* $x_1, x_2, \ldots x_{m+1}$. We wish to define a function Δ that takes on the specified values $\Delta_1, \Delta_2, \ldots, \Delta_{m+1}$ at these points. This can be accomplished by fitting a polynomial of order m through these points. The resulting expression will be of the form

$$\Delta = \sum_{i=1}^{m+1} N_i \Delta_i = \lfloor \mathbf{N} \rfloor \{\boldsymbol{\Delta}\}$$

where, clearly, the terms N_i are shape functions because they have the property $N_i = 1$ at Δ_i and $N_i = 0$ at Δ_j, $j \neq i$. We have performed this operation previously in Section 5.1 and elsewhere by forming the relationships $\{\boldsymbol{\Delta}\} = [\mathbf{B}]\{\mathbf{a}\}$ and solving for the terms $\{\mathbf{a}\}$. Fortunately, in the one-dimensional case we have a formula available, as follows, due to Lagrange:

$$N_i = \frac{\prod\limits_{\substack{j=1 \\ j \neq i}}^{m+1} (x - x_j)}{\prod\limits_{\substack{j=1 \\ j \neq i}}^{m+1} (x_i - x_j)} \tag{8.6}$$

where the symbol Π denotes a product of the indicated binomials $[(x - x_j)$ or $(x_i - x_j)]$ over the indicated range of j.

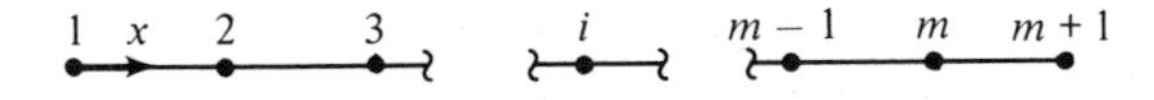

(a) Division into intervals for purposes of physical coordinates

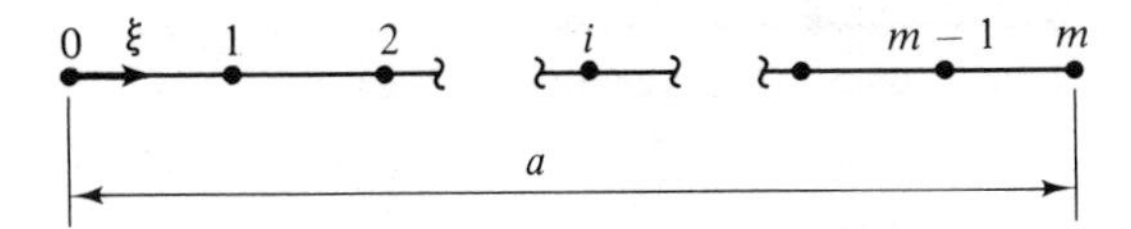

(b) Division into intervals for purposes of natural coordinates

Fig. 8.3 Intervals for Lagrange interpolation.

In expanded form, this gives the following typical expressions:

$$N_1 = \frac{(x - x_2)(x - x_3) \cdots (x - x_{m+1})}{(x_1 - x_2)(x_1 - x_3) \cdots (x_1 - x_{m+1})}$$

$$N_2 = \frac{(x - x_1)(x - x_3) \cdots (x - x_{m+1})}{(x_2 - x_1)(x_2 - x_3) \cdots (x_2 - x_{m+1})}$$

$$N_{m+1} = \frac{(x - x_1)(x - x_2) \cdots (x - x_m)}{(x_{m+1} - x_1)(x_{m+1} - x_2) \cdots (x_{m+1} - x_m)}$$

As an example, consider the simple axial member where, with two points, $m = 1$. In this case, $x_1 = 0$ and

$$\Delta = \frac{(x - x_2)}{-x_2}\Delta_1 + \frac{x}{x_2}\Delta_2 = \left(1 - \frac{x}{x_2}\right)\Delta_1 + \frac{x}{x_2}\Delta_2$$

which is the familiar representation for this type of element.

For three points in a line, Fig. 8.4(a), we have $m = 2$. Again, $x_1 = 0$ and with equal spacing $x_3 = 2x_2$. Thus,

$$\Delta = \frac{(x - x_2)(x - 2x_2)}{2x_2^2}\Delta_1 + x\frac{(2x_2 - x)}{x_2^2}\Delta_2 + \frac{x(x - x_2)}{2x_2^2}\Delta_3$$

An alternative mode of representation of the coordinates of points on a line is by means of *natural coordinates*. Such coordinates represent a *mapping* of the physical coordinates into a nondimensionalized system in which the natural coordinates take on values of one or zero at the node points. Thus, they are useful for definition of shape functions.

To describe natural coordinates in the one-dimensional case, consider the segment of length x_2 in Fig. 8.5. The distances from point i to points 1 and 2, designated as L_1 and L_2, are nondimensionalized so that

$$L_1 + L_2 = 1 \tag{8.7}$$

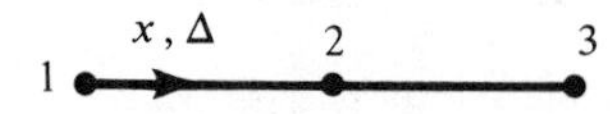

(a) Physical coordinate designations

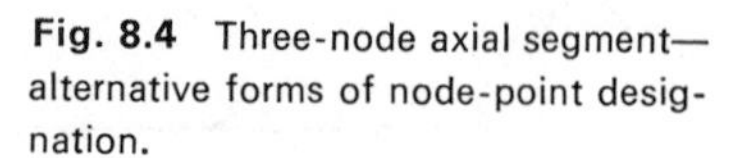

(b) Natural coordinate designations

Fig. 8.4 Three-node axial segment—alternative forms of node-point designation.

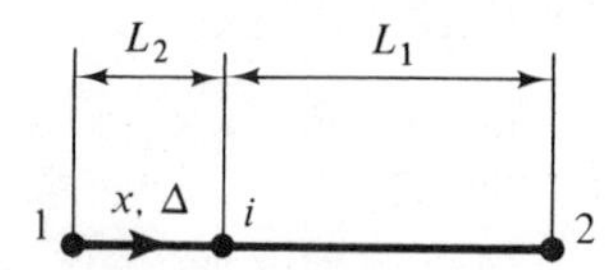

Fig. 8.5 One-dimensional natural coordinates.

Consequently, consistent with the above definition of natural coordinates, we see that $L_1 = 1$ at point 1 and $L_1 = 0$ at point 2, $L_2 = 1$ at point 2 and $L_2 = 0$ at point 1.

We can also describe the x coordinate of a point in the form:

$$x = L_1 x_1 + L_2 x_2 \tag{8.8}$$

Collecting Eq. (8.7) and (8.8), we can write

$$\begin{Bmatrix} 1 \\ x \end{Bmatrix} = \begin{bmatrix} 1 & 1 \\ x_1 & x_2 \end{bmatrix} \begin{Bmatrix} L_1 \\ L_2 \end{Bmatrix} \tag{8.9}$$

and by inversion we have the definition of L_1 and L_2 in terms of the physical coordinates x_1, x_2:

$$\begin{Bmatrix} L_1 \\ L_2 \end{Bmatrix} = \frac{1}{(x_2 - x_1)} \begin{bmatrix} x_2 & -1 \\ -x_1 & 1 \end{bmatrix} \begin{Bmatrix} 1 \\ x \end{Bmatrix} \tag{8.10}$$

We emphasize at this juncture that the designation of the end points of the line as points 1 and 2 is merely for the purpose of defining natural coordinates that refer to the total length of the line. As we subdivide the total length into a number of segments, the points at the ends and at the interior will be designated differently.

The natural coordinates enable a particularly simple mode of representation of shape functions for lines divided into any number of segments. To pursue this development it is convenient to shift the joint numbering scheme to that shown in Fig. 8.3(b). The leftmost point is designated as 0 and the rightmost point as m. Construction of the shape function will represent the fitting of an mth-order curve through these points.

Again designating a typical d.o.f. as i, the transformation to natural coordinates of the Lagrangian interpolation formula gives

$$\begin{aligned} N_i(L_1) &= \prod_{j=1}^{i} \left(\frac{mL_1 - j + 1}{j} \right) && \text{for } i \geq 1 \\ &= 1 && \text{for } i = 0 \end{aligned} \tag{8.11}$$

A similar formula holds for $N_i(L_2)$. To construct the total shape function in the terms above we first note that each point can be identified by its location from the two end points of the member. We use the subscripts p and q, where p identifies the number of points from the point in question to the right end point and q identifies the number of points from the left end point. In Fig. 8.4(b) for example, the three points are denoted as 20, 11, and 02 and the corresponding three shape functions are N_{20}, N_{11}, and N_{02}. We now describe the shape function as follows:

$$N_{pq} = N_p(L_1)N_q(L_2) \tag{8.12}$$

where $N_p(L_1)$ and $N_q(L_2)$ are given by Eq. (8.11), with p and q supplanting i, as appropriate.

For the element above, for example, we have (with $m = 2$) $N_2(L_1) = L_1(2L_1 - 1)$, $N_2(L_2) = L_2(2L_2 - 1)$, $N_1(L_1) = 2L_1$, $N_1(L_2) = 2L_2$, $N_0(L_1) = N_0(L_2) = 0$. The characterization of the displacement function in natural coordinates is then (we now designate the joint displacements with the same double subscripts as the shape functions)

$$\begin{aligned}\Delta &= N_{20}\Delta_{20} + N_{11}\Delta_{11} + N_{02}\Delta_{02}\\ &= L_1(2L_1 - 1)\Delta_{20} + 4L_1L_2\Delta_{11} + L_2(2L_2 - 1)\Delta_{02}\end{aligned}$$

As far as the definitions of L_1 and L_2 are concerned, the x_2 of Eq. (8.10) is x_{02} while x_1 is x_{20}. To verify the correspondence of the above result with the previous construction of this function in terms of physical coordinates it is necessary to define a further transformation from the coordinates of Fig. 8.4(b) to those of Fig. 8.4(a). The expression above is then found to be identical to the one previously given for this case, except that now the subscript on x is reduced by one (x_1 in the present case corresponds to x_2 of the previous construction).

Use of the shape functions in terms of L_1 and L_2 in the formula for element stiffness matrices (see Section 6.2) results in integral terms of the general form

$$\int_0^a L_1^b L_1^c \, dx$$

where a is the total length of the element. An advantage of the nondimensional coordinates L_1 and L_2 is that an explicit algebraic expression can be written for this integral.[8.6] To prove this we first note that $L_2 = 1 - L_1$ and $dx = a\, d\xi$. Thus

$$\int_0^a L_1^b L_1^c \, dx = \int_0^1 L_1^b(1 - L_1)^c a \, d\xi$$

The transformed integral is of the form,[8.7]

$$a\int_0^1 L_1^b(1 - L_1)^c \, d\xi = a\frac{\Gamma(b+1)\Gamma(c+1)}{\Gamma(b+c+2)}$$

where $\Gamma(b + 1)$, $\Gamma(c + 1)$, $\Gamma(b + c + 2)$ are the *gamma functions*, for which

$\Gamma(b+1) = b!$ and similarly for $\Gamma(c+1)$, $\Gamma(b+c+2)$. Hence

$$\int_0^a L_1^b L_1^c \, dx = \frac{ab!c!}{(b+c+1)!} \tag{8.13}$$

One should note that $0! = 1$.

We shall not have occasion to make use of the one-dimensional natural coordinate scheme in this text. It lays the groundwork, however, for its development in two and three dimensions, where it proves invaluable.

8.3.2 Hermitian Interpolation

Problems of flexure introduce the need to satisfy both the function and its first derivative. In other cases, where the representation of the first derivative is not essential, it may prove desirable to introduce the first derivative, and even higher derivatives, as d.o.f. This objective can be met by *Hermitian polynomial interpolation*, a process that is now described.

We examine a normalized interval connecting points 1 and 2, with the coordinate $\xi = x/L$, as shown in Fig. 8.6. We seek to construct a function Δ that will satisfy the conditions on itself and its derivatives up to $(m-1)$th order at the end points. This function can be written in the shape function form

$$\begin{aligned}\Delta = N_1\Delta_1 + N_2\Delta_1' + \cdots + N_m\Delta_1^{m-1} \\ + N_{m+1}\Delta_2 + N_{m+2}\Delta_2' + \cdots + N_{2m}\Delta_2^{m-1}\end{aligned} \tag{8.14}$$

where the superscripts on Δ_1 and Δ_2 (e.g., $m-1$) denote the order of differentiation with respect to x. Now $2m$ conditions are available for the construction of each of the shape functions N_i since each such function (or its appropriate derivative) must have a value of 1 when Δ (or its appropriate derivative) is evaluated at the d.o.f. associated with N_i and will have a value of zero when the evaluation is performed for the remaining $(2m-1)$ d.o.f. Following our practice of choosing polynomial descriptions of behavior, the availability of $2m$ conditions suggests that we describe each N_i by polynomials of order $2m-1$, i.e., polynomials with $2m$ coefficients:

$$N_i = a_1 + a_2\xi + a_3\xi^2 + \cdots + a_{2m}\xi^{2m-1} \tag{8.15}$$

We solve for the $2m$ values of a_i by use of the conditions cited above. The process is then repeated for each of the $(2m)$ shape functions N_i.

We can illustrate the above via the simple flexural member, see Fig. 8.6. Since the continuity of the transverse displacement w and of its first derivative is sought at each end of the element, $m = 2$. Here we have $\Delta = w$, and at

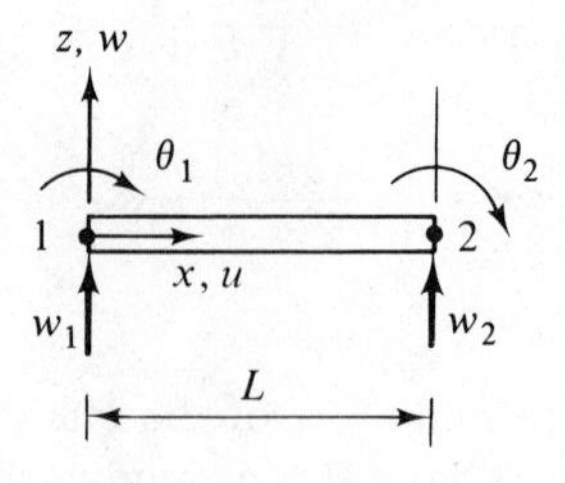

Fig. 8.6 D.o.f. for Hermitian interpolation of cubic polynomial in one dimension.

point 1

$$\Delta_1 = w_1, \qquad \Delta_1' = -\frac{dw}{dx}\bigg|_1 = \theta_1$$

and similarly for point 2. From Eq. (8.15), each shape function in this case is of the form

$$N_i = a_1 + a_2\xi + a_3\xi^2 + a_4\xi^3$$

Consider now the formation of N_1. This is first represented as $N_1 = a_1 + a_2\xi + a_3\xi^2 + a_4\xi^3$. By $N_1 = 1$ at $x = 0$, $N_1 = 0$ at $x = L$ and $N_1' = 0$ at both $x = 0$ and $x = L$ we have

$$1 = a_1 \qquad (N_1 = 1 \text{ @ } \xi = 0)$$

$$0 = \frac{a_2}{L} \qquad (N_1' = 0 \text{ @ } \xi = 0)$$

$$0 = (a_1 + a_2 + a_3 + a_4) \qquad (N_1 = 0 \text{ @ } \xi = 1)$$

$$0 = \frac{a_2}{L} + \frac{2a_3}{L} + \frac{3a_4}{L} \qquad (N_1' = 0 \text{ @ } \xi = 1)$$

and, by solution,

$$a_1 = 1, \qquad a_2 = 0, \qquad a_3 = -3, \qquad a_4 = 2$$

so that

$$N_1 = 1 - 3\xi^2 + 2\xi^3$$

In a similar way we find the remaining shape functions are as given originally by Eq. (5.14a),

$$N_3 = 3\xi^2 - 2\xi^3, \qquad N_2 = -x(\xi - 1)^2, \qquad N_4 = -x(\xi^2 - \xi)$$

8.4 Rectangular Elements

To generalize the interpolation concept to two dimensions and produce functions that are uniquely defined on each edge of a rectangle by the d.o.f. present on that edge and its corner points, we can employ a simple product of the one-dimensional shape functions for the x and y directions, respectively. The rectangle with joints only at the corners, Fig. 8.7(a), for which

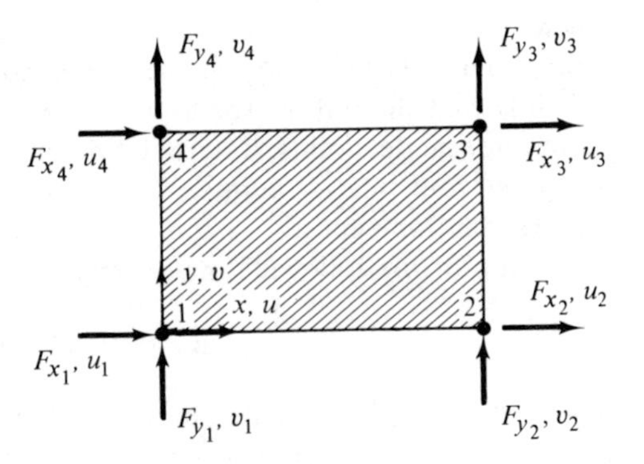

(a) Bilinear rectangle

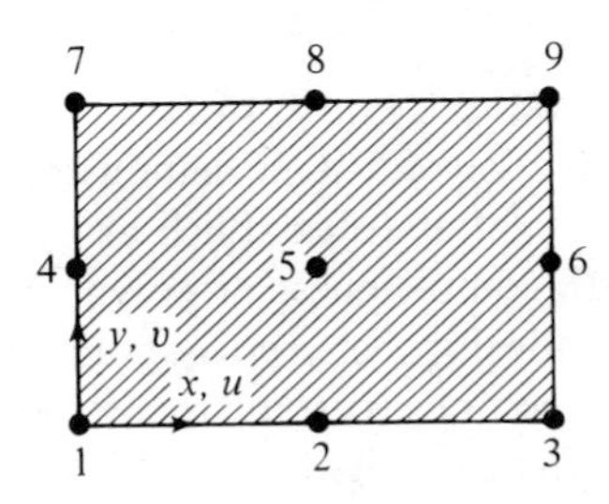

(b) Biquadratic rectangle

Fig. 8.7 Lagrangian interpolation in two dimensions.

we seek a linear displacement field, illustrates the procedure. We have for the displacement component Δ (which may be either u or v)

$$\Delta = (N_{1x}N_{1y})\Delta_1 + (N_{2x}N_{1y})\Delta_2 + (N_{2x}N_{2y})\Delta_3 + (N_{1x}N_{2y})\Delta_4 \tag{8.16}$$

where

$$N_{1x} = (1 - \xi), \qquad N_{2x} = \xi, \qquad N_{1y} = (1 - \eta), \qquad N_{2y} = \eta$$

with $\xi = x/x_2$, $\eta = y/y_3$. Equation (8.16) represents the *bilinear interpolation* formula.

One can proceed to the rectangle with mid-side joints and an interior node, Fig. 8.7(b)† (*biquadratic interpolation*)

$$\begin{aligned}\Delta = (N_{1x}N_{1y})\Delta_1 &+ (N_{2x}N_{1y})\Delta_2 + (N_{3x}N_{1y})\Delta_3 + (N_{1x}N_{2y})\Delta_4 \\ &+ (N_{2x}N_{2y})\Delta_5 + (N_{3x}N_{2y})\Delta_6 + (N_{1x}N_{3y})\Delta_7 + (N_{2x}N_{3y})\Delta_8 \\ &+ (N_{3x}N_{3y})\Delta_9\end{aligned} \tag{8.17}$$

†We depart here, and in other examples in this chapter, from the joint numbering convention described in Section 2.1.

where $N_{1x} = [(x - 2x_2)(x - x_2)]/2x_2^2$, etc., as defined by the quadratic Lagrangian interpolation scheme presented earlier.

Note that the full interpolation in a quadratic or higher function results in interior node points. Interpolation of a cubic function gives a 4×4 array and four interior points.

The development above can be accomplished quite simply through the formation of the following matrix triple product:

$$\Delta = \lfloor \mathbf{N}_\xi \rfloor [\mathbf{\mathfrak{R}}] \{\mathbf{N}_\eta\} \tag{8.18}$$

where $\lfloor \mathbf{N}_\xi \rfloor$, $\lfloor \mathbf{N}_\eta \rfloor$ denote the shape function vectors in the x and y directions, respectively, and $[\mathbf{\mathfrak{R}}]$ is a matrix of the joint displacements. For bilinear interpolation, for example, we have

$$\Delta = \lfloor (1 - \xi) \quad \xi \rfloor \begin{bmatrix} \Delta_1 & \Delta_4 \\ \Delta_2 & \Delta_3 \end{bmatrix} \begin{Bmatrix} (1 - \eta) \\ \eta \end{Bmatrix}$$

which produces Eq. (8.16).

In Section 8.5 it is demonstrated, using the Pascal triangle, that it is a simple matter to establish arrays of points for triangular elements to accomplish completeness of polynomials up to any desired order. To establish the relationship between Lagrangian interpolation for rectangular elements and the completeness of the associated polynomial, we again consider the Pascal triangle, Fig. 8.8.

First it must be noted that each order of one-dimensional polynomial corresponds exactly to the corresponding order of Lagrangian interpolation formula; e.g., $\Delta = a_1 + a_2 x$ corresponds to linear interpolation. Then bilinear interpolation, when defined in terms of generalized coordinates, can be described on the Pascal triangle by the product of linear functions. Figure 8.8(a) shows that this gives $\Delta = a_1 + a_2 x + a_3 y + a_5 xy$. Polynomial coefficients of biquadratic interpolation are now easily identified, Fig 8.8(b), as are any higher-order representations, by the appropriate products.

The Pascal triangle makes apparent that bivariate Lagrangian interpolation encompasses the complete order n in the polynomial expansion and encompasses individual terms to order $2n$. The linear (first-order) interpolation is complete in the first-order terms ($a_2 x$, $a_3 y$) and is incomplete in the second-order terms (x^2 and y^2 terms are missing; only xy is represented). Since convergence rates are governed by the highest order of complete polynomial,[8.8] a full expansion in two dimensions on the rectangle is inefficient. This is one motivation for elimination of certain d.o.f. The presence of interior points, as in biquadratic or higher-order representation, is also generally a nuisance from the view of data handling, and a function is desired that can be expressed in terms of edge and corner node points only.

A simple approach to the accomplishment of the latter objective is illus-

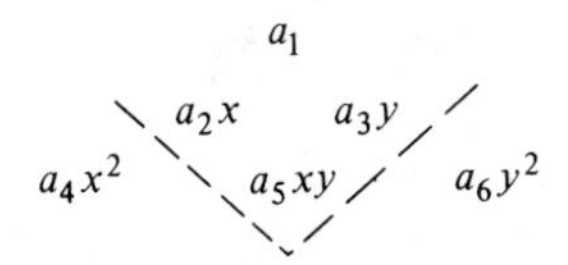

(a) Bilinear interpolation

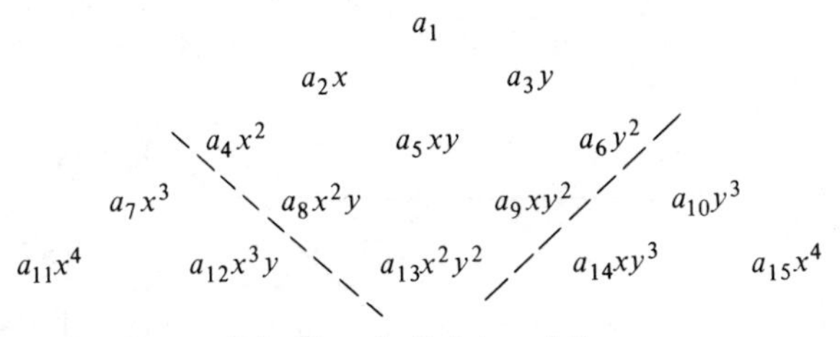

(b) Biquadratic interpolation

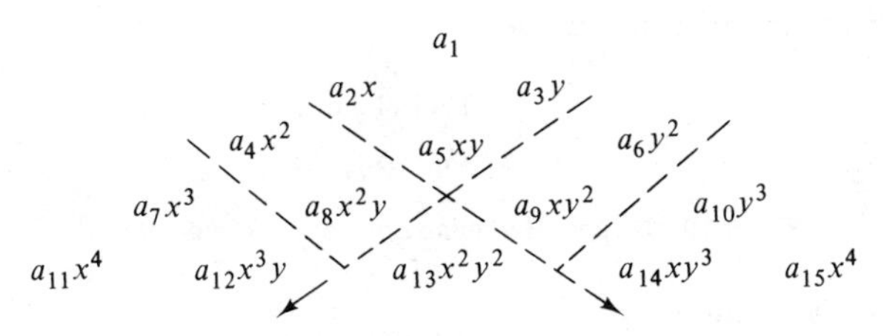

(c) Polynomial basis for eight-node rectangle

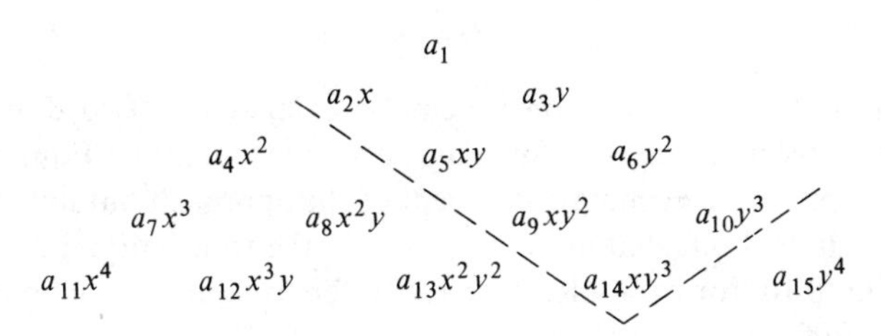

(d) Polynomial basis for rectangle with linear expansion in x and quadratic expansion in y (six-nodes)

Fig. 8.8 Interpretation of shape functions in terms of polynomial series.

trated in Fig. 8.9 and 8.8(d). The rectangle of Fig. 8.9 features six node points arranged so that linear interpolation is appropriate in the x direction and quadratic interpolation applies in the y direction. The polynomial terms associated with this expansion are shown in Fig. 8.8(d). It is apparent that the Lagrangian interpolation can be applied in this case to give the displacement field directly in shape function form. Using the scheme of Eq. (8.18), we have linear interpolation functions in $\lfloor \mathbf{N}_\xi \rfloor$ and quadratic interpolation

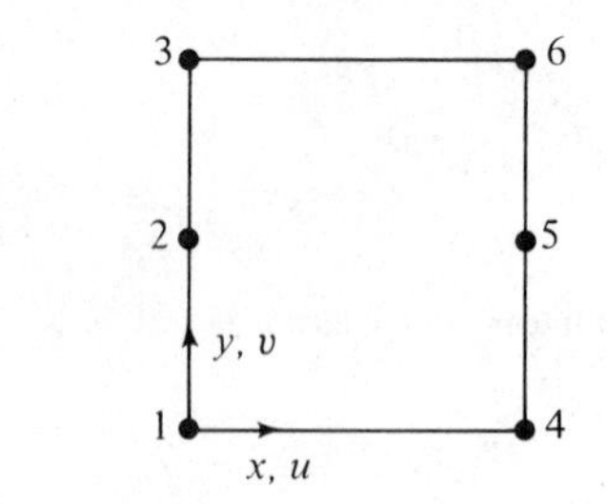

Fig. 8.9 Unequal interpolation in x and y directions.

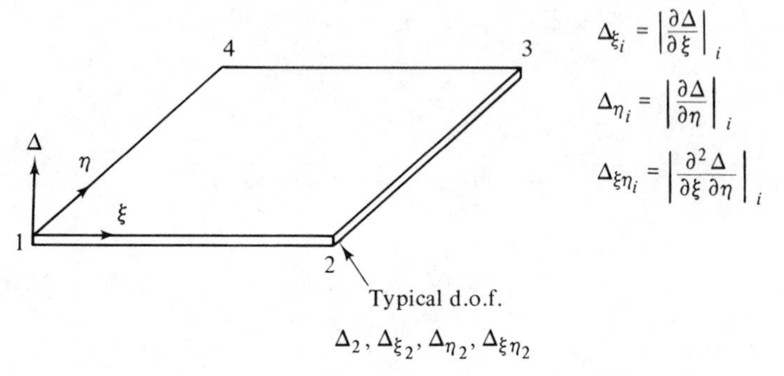

Fig. 8.10 Bicubic Hermitian polynomial interpolation.

functions in $\lfloor \mathbf{N}_\eta \rfloor$ and

$$[\mathcal{R}] = \begin{bmatrix} \Delta_1 & \Delta_2 & \Delta_3 \\ \Delta_4 & \Delta_5 & \Delta_6 \end{bmatrix}$$

The Hermitian interpolation scheme is extended to two dimensions in the manner discussed above for Lagrange interpolation. Equation (8.18) once again applies as a general statement of the approach but now the matrix $[\mathcal{R}]$ must include d.o.f. that are derivatives of the translational d.o.f. For the plate of Fig. 8.10, for example, we have (refer to figure for definition of the following d.o.f.)

$$[\mathcal{R}] = \begin{bmatrix} \Delta_1 & \Delta_{\eta_1} & \Delta_4 & \Delta_{\eta_4} \\ \Delta_{\xi_1} & \Delta_{\xi\eta_1} & \Delta_{\xi_4} & \Delta_{\xi\eta_4} \\ \Delta_2 & \Delta_{\eta_2} & \Delta_3 & \Delta_{\eta_3} \\ \Delta_{\xi_2} & \Delta_{\xi\eta_2} & \Delta_{\xi_3} & \Delta_{\xi\eta_3} \end{bmatrix}$$

and now

$$\lfloor \mathbf{N}_\xi \rfloor = \lfloor N_{1\xi} \; N_{2\xi} \; N_{3\xi} \; N_{4\xi} \rfloor$$

where $N_{1\xi} \ldots N_{4\xi}$ are the x-direction Hermitian shape functions, as defined at the close of Section 8.3.2. $\lfloor \mathbf{N}_\eta \rfloor$ is similarly constructed.

To determine the relationship of the above to a polynomial expansion we need only refer to Fig. 8.11, which demonstrates that 16 terms are en-

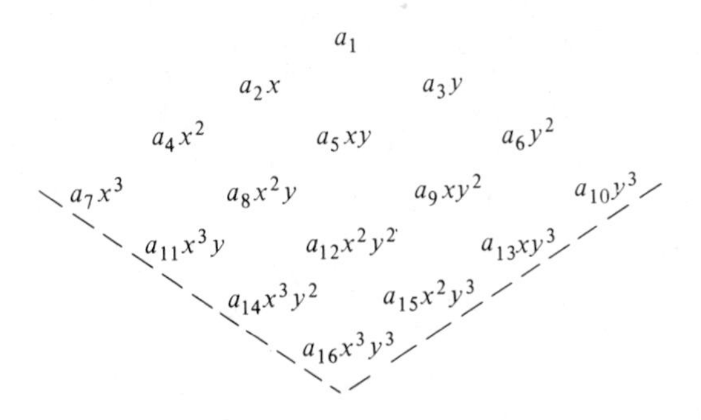

Fig. 8.11 Bicubic expansion.

compassed, as would be expected from the form of the [$\mathfrak{R}$] matrix above. The present shape function is examined once again, in greater detail, in Section 12.2.

On occasion it is necessary to define different numbers of nodes on each of the four edges or to have the same number of nodes on each edge but remove the interior nodes. If one can identify the terms of the polynomial expansion pertinent to any such case, it is a simple matter to construct the transformation between the generalized parameters of the polynomial $\{\mathbf{a}\}$ and the joint displacements $\{\mathbf{\Delta}\}$ and solve to obtain expressions in terms of the latter [Eq. (5.3a) to (5.5a)]. Interior nodes can be removed by constructing the full interpolation function, formulating the element strain energy, and "condensing out" the unwanted d.o.f. with use of procedures described in Section 2.8. Alternatively, one may construct the desired shape functions directly, via procedures we shall discuss in Section 8.7.

8.5 Triangular Elements

In the case of the triangle the interpolation procedures are intimately associated with the idea of *triangular coordinates*. These coordinates not only facilitate the construction of shape functions that refer directly to the joint d.o.f. rather than to generalized d.o.f., but they also comprise an organized method for designation of element node points and have other advantages whose significance becomes apparent only after their details have been described.

An organized method of designation of points in triangular coordinates can be constructed as illustrated in Fig. 8.12. In this development the sides of the element are identified by the opposite vertex; e.g., side 1 is opposite vertex point 1. We first seek a means of identifying a point within the triangle. If, as shown in Fig. 8.12, lines are drawn from a given point within the triangle to the three vertices, it is seen that the triangle is subdivided into three triangles of area A_1, A_2, and A_3, where the subscripts are identified with the adjacent sides. The *triangular coordinates* L_i ($i = 1, 2, 3$) are, by

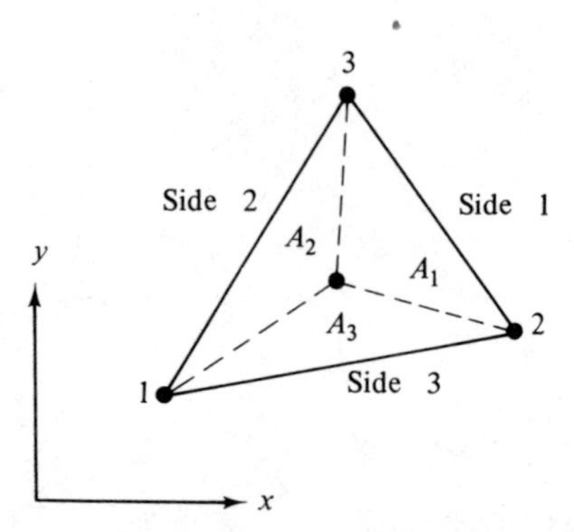

Fig. 8.12 Triangular coordinates.

definition, the ratios of the areas A_i to the total area A; i.e.,

$$L_1 = \frac{A_1}{A}, \qquad L_2 = \frac{A_2}{A}, \qquad L_3 = \frac{A_3}{A} \tag{8.19}$$

Also, the sum of these areas A_i must equal A:

$$A_1 + A_2 + A_3 = A$$

and, dividing both sides by A,

$$L_1 + L_2 + L_3 = 1 \tag{8.20}$$

Clearly, in correspondence with the developments of Section 8.3.1, these are *natural coordinates* of a triangular region.

The triangular coordinates can now be used to describe the rectangular coordinates, x and y, of the point identified in Fig. 8.12. We have

$$\begin{aligned} x &= L_1 x_1 + L_2 x_2 + L_3 x_3 \\ y &= L_1 y_1 + L_2 y_2 + L_3 y_3 \end{aligned} \tag{8.21}$$

We can establish the validity of this statement by a simple test. Suppose we move the point within the triangle until it is coincident with point 1. Then $A_1 = A$, $A_2 = A_3 = 0$ and $L_1 = 1$, $L_2 = L_3 = 0$. This confirms that $x = x_1$ when x is located at point 1. Obviously the triangular coordinates are identically equal to the shape functions for the simple three-node triangle.

The given element geometric data are the x and y coordinates of the vertices. To solve for L_1, L_2, and L_3 in terms of this data we collect Eq. (8.20) and (8.21) into the set of three equations:

$$\begin{Bmatrix} 1 \\ x \\ y \end{Bmatrix} = \begin{bmatrix} 1 & 1 & 1 \\ x_1 & x_2 & x_3 \\ y_1 & y_2 & y_3 \end{bmatrix} \begin{Bmatrix} L_1 \\ L_2 \\ L_3 \end{Bmatrix}$$

and, by inversion,

$$L_i = \frac{1}{2A}(b_{0_i} + b_{1_i}x + b_{2_i}y) \quad (i = 1, 2, 3) \tag{8.22}$$

where

$$b_{0_i} = x_{i+1}y_{i+2} - x_{i+2}y_{i+1}, \qquad b_{1_i} = y_{i+1} - y_{i+2}, \qquad b_{2_i} = x_{i+2} - x_{i+1} \tag{8.23}$$

with $i = 1, 2, 3$ taken cyclicly, and

$$A = \tfrac{1}{2}(x_2y_3 + x_3y_1 + x_1y_2 - x_2y_1 - x_3y_2 - x_1y_3) \tag{8.24}$$

(A, as the symbol denotes, is the area of the triangle.)

If the side joining points 1 and 2 is placed coincident with the x axis and point 1 is placed at the origin of coordinates ($x_1 = y_1 = y_2 = 0$), we obtain

$$L_1 = \frac{1}{x_2y_3}(x_2y_3 - xy_3 - x_2y + x_3y)$$

$$L_2 = \frac{1}{x_2y_3}(xy_3 - x_3y)$$

$$L_3 = \frac{y}{y_3}$$

These terms are identical to the shape functions N_1, N_2, N_3 given in Eq. (5.21a) for the constant strain triangle oriented in this way.

In order to form the shape functions for higher-order elements in terms of the triangular coordinates L_i, it is first necessary to establish a special method of construction and identification of the points of such elements. These considerations are illustrated in Fig. 8.13. The sides are identified by the opposite vertex, e.g., side 1 is opposite vertex 1. The normal to a side identifies the corresponding direction. The dashed lines of Fig. 8.13(a) subdivide the distance between side 1 and point 1 into m equal segments in direction 1. Each line is identified with a digit from 0 to m, the line 0 being coincident with side 1. [We do not use the scheme of numbering such points from 1 to $m + 1$, see Fig. 8.3(a), because the present range is associated with a simpler definition of the joints. This same shifting was invoked for the one-dimensional case in Fig. 8.3(b).] A typical line is denoted by p in this figure. This defines ($m + 1$) points on the sides 2 and 3 and clearly sets the stage for construction of a displacement field based on an mth-order polynomial. The same construction applies to directions 2 and 3, Fig. 8.13(b) and (c), where typical lines are designated as q and r, respectively.

The special mode of identification of a typical point is illustrated in Fig. 8.13(d). The point is identified by three digits p, q, r, consistent with the designation of typical lines in the three directions. Note that the sum of the

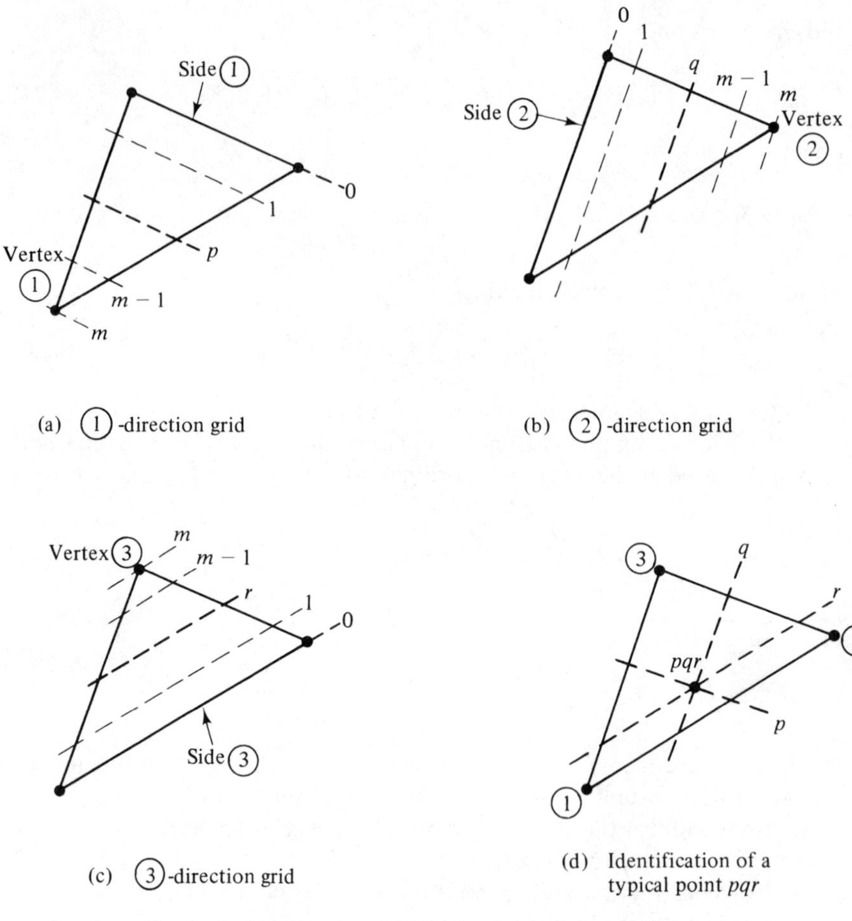

Fig. 8.13 Identification of grid points in triangular coordinates.

three digits $(p + q + r)$ must equal m. This system of point designation is illustrated for four different divisions of a triangular element in Fig. 8.14. It should be observed that the vertex points also carry the designations 1, 2, and 3 for purposes of identifying the triangular coordinates.

In defining displacement functions for the triangle we adopt the same designation for the node point displacements as for the node points themselves (i.e., pqr). In accordance with the custom already established, we seek element displacement functions of the form

$$\Delta = \lfloor \mathbf{N} \rfloor \{\boldsymbol{\Delta}\} = \sum^{\frac{1}{2}(m+1)(m+2)} N_{pqr}\Delta_{pqr} \tag{8.25}$$

Thus, for $m = 1$ the displacement function is

$$\Delta = N_{100}\Delta_{100} + N_{010}\Delta_{010} + N_{001}\Delta_{001} \tag{8.25a}$$

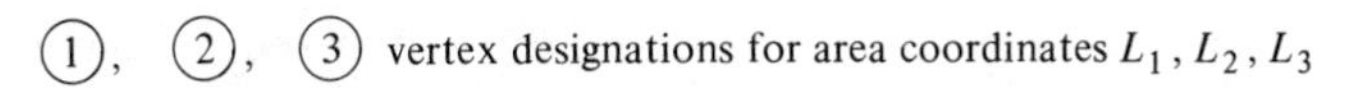

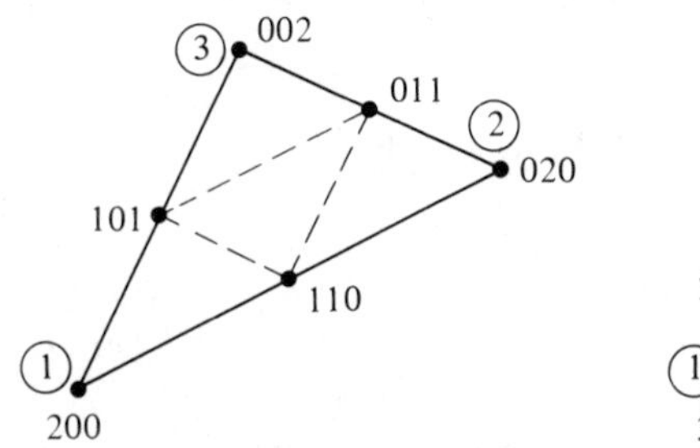

(a) Complete quadratic (m = 2)

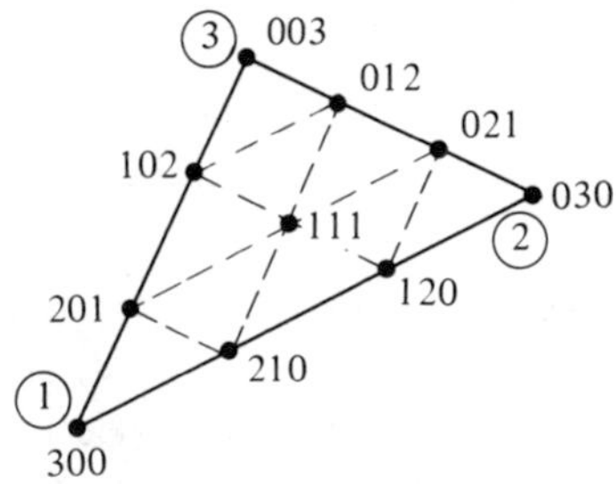

(b) Complete cubic (m = 3)

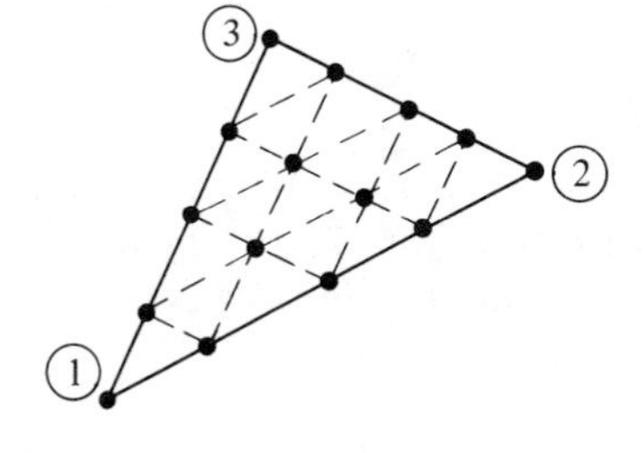

(c) Complete quartic (m = 4)

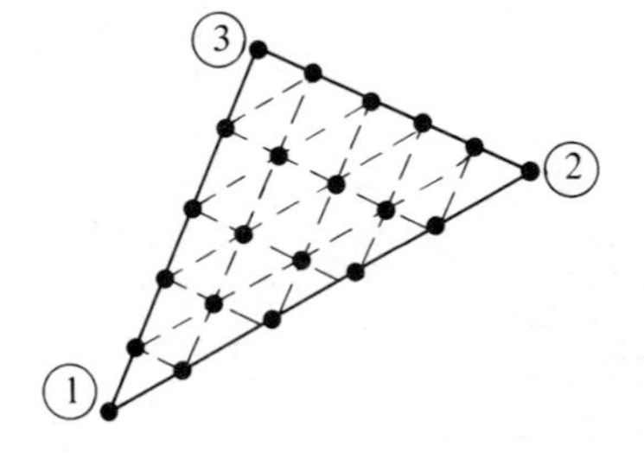

(d) Complete quintic (m = 5)

Fig. 8.14 Higher-order triangles.

and for $m = 2$, Fig. 8.14(a),

$$\Delta = N_{200}\Delta_{200} + N_{110}\Delta_{110} + \cdots + N_{101}\Delta_{101} \tag{8.25b}$$

and so on to any order of polynomial.

We must now establish a method of constructing shape functions N_{pqr} so as to meet the usual conditions on such functions (e.g., that $N_{pqr} = 1$ at point pqr; $N_{pqr} = 0$ at the other points). As we have seen in the one-dimensional case, we can do this by defining them as the product of functions of the respective coordinates and by performing a Lagrangian-type of interpolation in each direction. By analogy with the one-dimension case, the applicable formula is of the form

$$N_{pqr}(L_1, L_2, L_3) = N_p(L_1)N_q(L_2)N_r(L_3) \tag{8.12a}$$

where the terms on the right side are given by Eq. (8.11)

$$\begin{aligned} N_i(L_1) &= \prod_{j=1}^{i} \left(\frac{mL_1 - j + 1}{j} \right) \quad \text{for } i \geq 1 \\ &= 1 \quad \text{for } i = 0 \end{aligned} \tag{8.11}$$

with $i = p, q$, or r, as appropriate.

As an example, consider the development of the shape function N_{200}. We have $m = p = 2, q = r = 0$ so that $N_q = N_0(L_2) = 1$, $N_r = N_0(L_3) = 1$, and

$$N_2(L_1) = \frac{(2L_1 - 1 + 1)}{1} \times \frac{(2L_1 - 2 + 1)}{2} = L_1(2L_1 - 1)$$

hence $N_{200} = L_1(2L_1 - 1)$. It is readily established that $N_{020} = L_2(2L_2 - 1)$, $N_{002} = L_3(2L_3 - 1)$, $N_{110} = 4L_1L_2$, $N_{011} = 4L_2L_3$, and $N_{101} = 4L_3L_1$.

Occasions arise when it is desirable to work with generalized d.o.f. rather than with joint d.o.f. For such cases it is useful to observe that a complete polynomial of degree m can be written in triangular coordinates as

$$\Delta = \sum_{i=1}^{n} a_i L_1^p L_2^q L_3^r \quad (p + q + r = m) \tag{8.1a}$$

where the summation contains all homogeneous terms of total degree m; i.e., the exponents of the shape functions are exactly as defined by the three-digit identifiers of the joints, as in Fig. 8.14(a) and (b). For the cubic expansion, for example, we have

$$\begin{aligned}\Delta = (L_1)^3a_1 + (L_2)^3a_2 + (L_3)^3a_3 + (L_1)^2L_2a_4 + (L_2)^2L_1a_5 \\ + (L_3)^2L_1a_7 + (L_2)^2L_3a_7 + (L_1)^2L_3a_8 + (L_3)^2L_2a_9 + L_1L_2L_3a_{10}\end{aligned}$$

With this form, the convenience of explicit integration formulas for the triangle can be invoked in constructing a kernel stiffness matrix.

A second operation that must frequently be performed on triangular coordinate representations of fields is that of differentiation, particularly in the application of strain-displacement conditions. For example, consider the direct strain $\epsilon_x = \partial u/\partial x$. If we are dealing with the quadratic displacement field, where u is expressed as $N_{200}u_{200} + \ldots + N_{101}u_{101}$, the first term in the expression for ϵ_x is

$$\frac{\partial N_{200}}{\partial x} = 4L_1\frac{\partial L_1}{\partial x} - \frac{\partial L_1}{\partial x}$$

Noting, from Eq. (8.22), the definition of L_i, we have $\partial L_i/\partial x = b_{1_i}/2A$ $(i = 1, 2, 3)$ so that

$$\frac{\partial N_{200}}{\partial x} = \frac{b_{1_1}}{2A}(4L_1 - 1)$$

Finally, and perhaps most importantly, a simple explicit expression can be written for integration in area coordinates. This again represents an extension of one-dimensional formulations. The desired integral is of the form

$$\begin{aligned} \mathcal{I}(L_1, L_2, L_3) &= \int_A (L_1)^b (L_2)^c (L_3)^d \, dA \\ &= \frac{2A(b)!(c)!(d)!}{(b + c + d + 2)!} \end{aligned} \tag{8.26}$$

[Compare with Eq. (8.13).] In consequence of Eq. (8.20) only two coordinates are independent and the integral can always be transformed to

$$\mathcal{I}(L_1 L_2) = \int_A (L_1)^e (L_2)^f \, dA$$

and, since this is a special case of Eq. (8.26) with $d = 0$, $b = e$, $c = f$,

$$\mathcal{I}(L_1, L_2) = 2A \frac{e!f!}{(e + f + 2)!} \tag{8.26a}$$

With the detailed development above in hand, it is of interest to delineate some of the less apparent advantages of the triangular coordinate representation. For one, the definition of joints in a triangular coordinate representation of higher-order elements (Fig. 8.14) *automatically* positions the interior nodes. Note further that the arrays shown in these figures correspond identically to various complete levels of the Pascal triangle. Thus, *each order of interpolation in triangular coordinates produces a complete polynomial representation to the corresponding order*. We have already emphasized the significance of this completeness in a prior section and it is perhaps for this reason that triangular elements hold a special position in finite element analysis. Another reason, of course, is their flexibility in representation of geometric complexity.

8.6 Tetrahedronal Elements

The tetrahedron in Fig. 8.15 is the three-dimensional counterpart of the planar triangle. As in the case of the plane triangle, the definition of shape functions and the integration of the strain energy are facilitated by *tetrahedronal coordinates*, which are analogous to the triangular coordinates of Section 8.5.

The position of a point within a tetrahedron whose total volume is designated as (vol) can be defined by the following set of ratios:

$$L_1 = \frac{(\text{vol})_1}{(\text{vol})}, \qquad L_2 = \frac{(\text{vol})_2}{(\text{vol})}, \qquad L_3 = \frac{(\text{vol})_3}{(\text{vol})}, \qquad L_4 = \frac{(\text{vol})_4}{(\text{vol})} \tag{8.27}$$

where $(\text{vol})_i$ $(i = 1, \ldots, 4)$ designates the volume subtended by lines drawn

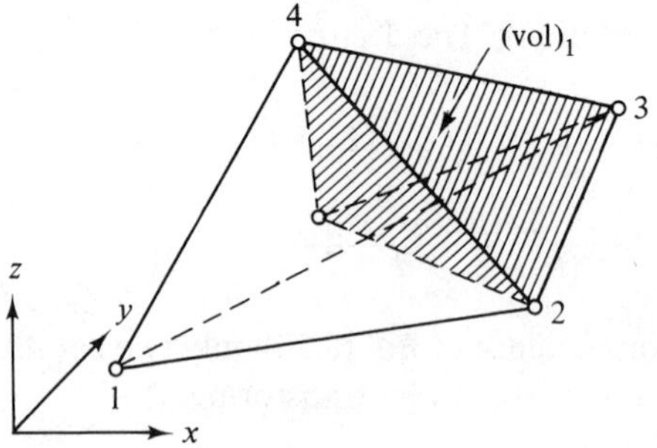

Fig. 8.15 Basic tetrahedron.

from the point to the vertices opposite the point i. $(\text{Vol})_1$ is illustrated in Fig. 8.15. $L_1, \ldots, L_4$ comprise the tetrahedronal coordinates. In view of Eq. (8.27),

$$L_1 + L_2 + L_3 + L_4 = 1 \tag{8.28}$$

and, supplementing this equation by equations that define the x, y, and z coordinates of a point in terms of tetrahedronal coordinates, we have

$$\begin{bmatrix} 1 & 1 & 1 & 1 \\ x_1 & x_2 & x_3 & x_4 \\ y_1 & y_2 & y_3 & y_4 \\ z_1 & z_2 & z_3 & z_4 \end{bmatrix} \begin{Bmatrix} L_1 \\ L_2 \\ L_3 \\ L_4 \end{Bmatrix} = \begin{Bmatrix} 1 \\ x \\ y \\ z \end{Bmatrix}$$

By inversion,

$$L_i = \frac{1}{6(\text{vol})}[(\text{VOL})_i + C_{1_{y_i}}x + C_{2_i}y + C_{3_i}z] \quad (i = 1, \ldots, 4) \tag{8.29}$$

where $(\text{VOL})_i$ is the volume subtended by side 1 and rays drawn from its vertices to the origin, (vol) is given by one-sixth times the determinant of the above 4×4 matrix and c_{1_i}, c_{2_i}, and c_{3_i} are determinants of 3×3 submatrices of this matrix appropriate to the definition of its inverse.

The analogy of tetrahedronal coordinates with triangular coordinates extends to the applicability of a *Pascal tetrahedron*, Fig. 8.16, for definition of element node point arrays for elements based on complete polynomials of various orders and to the definition of shape functions corresponding to these polynomials. A typical tetrahedronal shape function is designated with four subscripts, as N_{pqrs}, with the following form of dependence on $L_1, \ldots,$ L_4.

$$N_{pqrs}(L_1, L_2, L_3, L_4) = N_p(L_1)N_q(L_2)N_r(L_3)N_s(L_4) \tag{8.30}$$

i.e., the product of functions of the respective volume coordinates. The definition of the appropriate subscripts is analogous to the case of the triangle and is illustrated for the tetrahedron based on quadratic functions in Fig. 8.17. Note that the four subscripts must sum to the value $m = 2$ in this case and to the order (m) of the chosen function in the general case.

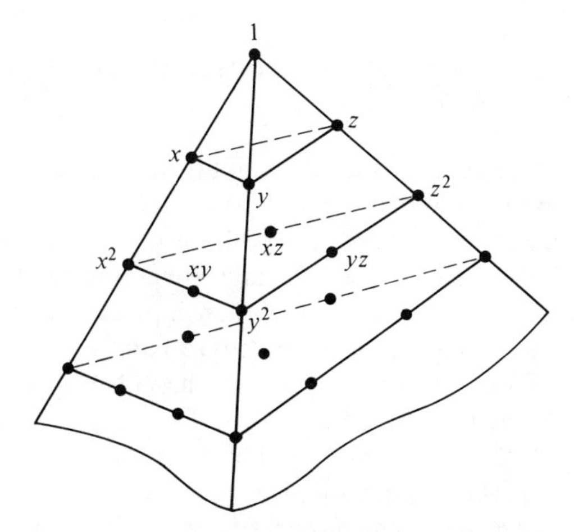

Fig. 8.16 Three-dimensional generalization of Pascal's triangle.

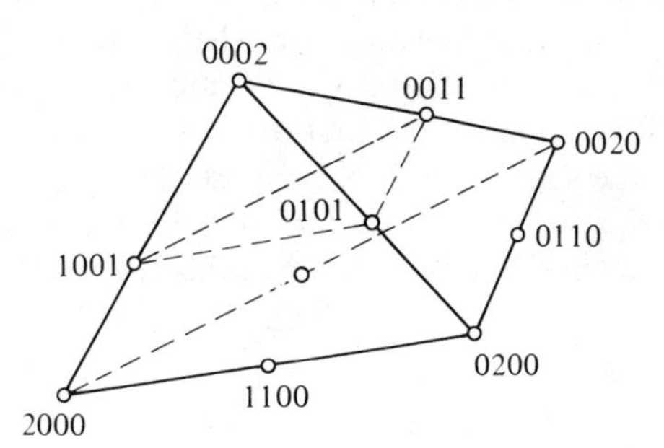

Fig. 8.17 Joint numbering system for quadratic tetrahedron.

The appropriate formula for N_i, $i = p, q$, or r, as appropriate, is again given by Eq. (8.11).

For integration in volume coordinates, we have the typical integral

$$\begin{aligned} \mathcal{I}(L_1, L_2, L_3, L_4) &= \int_{\text{vol}} (L_1)^a (L_2)^b (L_3)^c (L_4)^d \, d(\text{vol}) \\ &= \frac{6(\text{vol})\, a!b!c!d!}{(a+b+c+d+3)!} \end{aligned} \tag{8.31}$$

By virtue of Eq. (8.28) only three of the coordinates are independent and an even simpler formula is possible. Suppose that elimination of $(L_4)^d$ results in a transformation of the integral $\mathcal{I}(\quad)$ to the form

$$\mathcal{I}(L_1, L_2, L_3) = \int_{\text{vol}} (L_1)^e (L_2)^f (L_3)^g \, d(\text{vol})$$

we then have

$$\mathcal{I}(L_1, L_2, L_3) = \frac{6(\text{vol})\, e!f!g!}{(e+f+g+3)!} \tag{8.31a}$$

A detailed development of tetrahedronal coordinates can be found in Ref. 8.9.

8.7 Internal Modes and Reduction to Simpler Forms

We have noted the desirability of constructing element equations that pertain only to joints at the vertices and edges of the element. Interior d.o.f. are awkward to handle. We have also seen that interior d.o.f. arise naturally in the generation of shape functions for higher-order elements. A similar situation exists when relationships are constructed on the basis of generalized coordinates where such coordinates exceed in number the number of d.o.f. of the edges and vertices. These "surplus" generalized coordinates can be viewed as "internal" d.o.f. Two ways in which the internal modes can be eliminated are discussed in this section. The subsidiary problem of constructing shape functions for elements with different numbers of joints on the respective element edges is also examined.

Consider first the case where interior d.o.f. have arisen naturally in the development of a shape function for a higher-order element. The element stiffness matrix can first be constructed on the basis of all d.o.f. represented in the shape function. Suppose we designate the interior d.o.f. with the subscript b and the edge and vertex d.o.f. with the subscript c. The formulated element stiffness matrix can then be written in the form

$$\begin{bmatrix} \mathbf{k}_{bb} & \mathbf{k}_{bc} \\ \mathbf{k}_{cb} & \mathbf{k}_{cc} \end{bmatrix} \begin{Bmatrix} \mathbf{\Delta}_b \\ \mathbf{\Delta}_c \end{Bmatrix} = \begin{Bmatrix} \mathbf{F}_b \\ \mathbf{F}_c \end{Bmatrix} \tag{8.32}$$

Now, the forces at the interior points $\{\mathbf{F}_b\}$ of the element will be known values resulting from work-equivalent loads, applied concentrated loads, etc., or they will be of zero value, since these points do not connect to other elements of the structure. Consequently, the removal of internal modes is given exactly by the condensation scheme of Section 2.8.

It is pertinent to observe that the "internal modes" are clearly identified as "bubble modes", i.e., modes with an internal amplitude but zero value on the periphery of the element, since the amplitudes of a shape function are given by a unit value of the d.o.f. under study and zero values for all other d.o.f.

The second approach to the elimination of unwanted d.o.f. is to modify directly the shape function to include only the desired number of parameters. Perhaps the simplest scheme for elimination of d.o.f. from expressions constructed from interpolation procedures is to introduce relationships which tie the unwanted d.o.f. to those which are to be retained. Consider, for example, the triangle based on a quadratic displacement field, Fig. 8.14(a), for which the shape function was detailed in Section 8.5. Suppose the joint 110 is to be eliminated. We can require that the displacement along this edge be linear so that

$$\Delta_{110} = \frac{\Delta_{200} + \Delta_{020}}{2}$$

and, by substitution into the full expression for the displacement field, we obtain

$$\Delta = N^R_{200}\Delta_{200} + N^R_{020}\Delta_{020} + N_{002}\Delta_{002} + N_{011}\Delta_{011} + N_{101}\Delta_{101}$$

where

$$N^R_{200} = (N_{200} + \tfrac{1}{2}N_{110}), \qquad N^R_{020} = (N_{020} + \tfrac{1}{2}N_{110})$$

The approach above can be applied to rectangular regions as well. For example, the need to remove the interior point of a biquadratic displacement rectangle, Fig. 8.7(b), was noted above. This displacement can be expressed as the average of the displacements of the midpoints of the sides

$$\Delta_5 = \tfrac{1}{4}(\Delta_2 + \Delta_4 + \Delta_6 + \Delta_8)$$

Alternatively, one can introduce the corner point d.o.f. into the averaging process, using different proportions of the respective d.o.f. Thus, a variety of different expressions can be established in terms of a given reduced set of d.o.f.

A more elegant approach[8.10] to the construction of special displacement functions is through a procedure that involves the superposition of component shape functions. Before undertaking this approach, which we treat only for the case of rectangular elements, it is convenient to define the shape functions in terms of nondimensional coordinates (ξ, η) whose origin is located at the center of the rectangle. To this juncture, the shape functions for rectangles have been presented for a physical (x, y) coordinate system located at a corner of the rectangle.

To effect this transformation, see Fig. 8.18(a), we employ the relationships $\xi = (x - x_5)/(x_5 - x_1)$ and $\eta = (y - y_5)/(y_5 - y_1)$, where x_5 and y_5 are the coordinates of the center of the rectangle and x_1 and y_1 are the coordinates of the lower left-hand corner point. In this way the nondimensional coordinates of the four corner points are always $+1$ or -1.

Consider now the construction of shape functions for the biquadratic displacement rectangle that are intended to exclude an interior point, see Fig. 8.18(a). The shape function for the edge point 2 can be obtained as the simple product of the quadratic function $(1 - \xi^2)$ in the direction of the edge (the proper function for this direction for quadratic interpolation) and the linear function $\frac{1}{2}(1 - \eta)$ in the perpendicular direction. Thus, the total shape function for this point is $N_2 = \frac{1}{2}(1 - \xi^2)(1 - \eta)$.

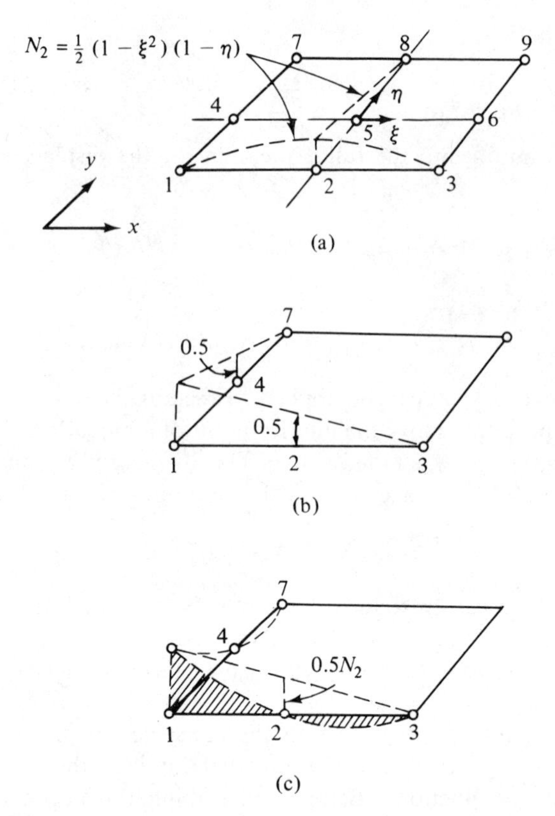

Fig. 8.18 Direct construction of eight-node rectangle displacement field.

The construction of the shape function for a corner point poses a more difficult problem. First, as shown in Fig. 8.18(b) for point 1, the bilinear function gives nonzero displacements at points 2 and 4. These can be reduced to zero by subtracting one-half the N_2 shape function, see Fig. 8.18(c), and one-half the N_4 shape function. [By the reasoning above, $N_4 = \frac{1}{2}(1 - \xi)(1 - \eta^2)$.] Thus,

$$N_1 = \tfrac{1}{4}(1 - \xi)(1 - \eta) - \tfrac{1}{4}(1 - \xi^2)(1 - \eta) - \tfrac{1}{4}(1 - \xi)(1 - \eta^2)$$

It can be shown that the polynomial coefficients embodied in the functions above are given by the regions defined in Fig. 8.8(c), i.e., the products of the quadratic expansions of the edges by the linear expansion in the perpendicular direction. The scheme is easily extended, using the procedure described above, to the formulation of expressions of any order along the edges of an element in two or three dimensions.

8.8 Isoparametric Representation[8.11]

Isoparametric elements are those for which the functional representation of deformational behavior is employed in representation of the element geometry. Isoparametric element construction represents the "mapping" of a nondimensionalized rectangular element with a specified number of nodes into the actual curved-boundary element with the same number of nodes. Thus, when the functional representation of the displacement field in a minimum potential energy formulation is a cubic polynomial, the faces of the element are described by the same cubic functions. If interelement compatible displacement fields are chosen for the geometric description, the distorted element will match any adjacent like-element without a discontinuity in the geometric description of the assembled analytical model.

In two-dimensional analysis, the simplest four-sided isoparametric element, Fig. 8.19(a), is one in which the linear field is used to generalize the rectangular form to an arbitrary quadrilateral. A better fit of curved boundaries is achieved with higher-order elements, Fig. 8.19(b) and (c), where the quadratic and cubic functional representations of displacement are applied to the representation of the boundaries. *Mixed-type elements*, Fig. 8.19(d) and (e), with different numbers of nodes on each side and with or without interior nodes, are also of practical value.

It is not essential that the same functional representations for displace-

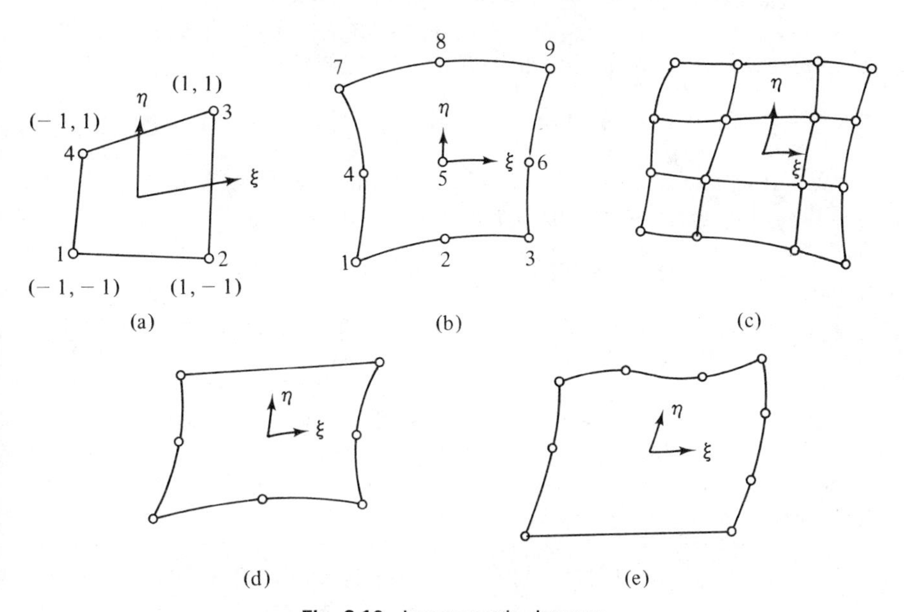

Fig. 8.19 Isoparametric elements.

ment be used in representation of geometry. When the geometric functional representation is of lower order than that for displacement, the element formulation is termed *subparametric*. If the geometric representation is of higher order, the element formulation is characterized as *superparametric*. Isoparametric, subparametric, and superparametric element representations are perhaps most important in the analysis of solids, a topic that we shall take up in Chapter 10. Three-dimensional analyses are usually extremely expensive in terms of problem size. If the structure features curved boundaries, many elements in a regular element representation are used exclusively to achieve geometric modeling without significant advantages in improved stress or displacement prediction. Thus, isoparametric element representations serve to reduce costs of idealization assignable strictly to geometric representation.

In describing the procedure for isoparametric formulations it is sufficient to consider the two-dimensional case. The first step in the formulation is to define the nondimensional coordinate system (ξ, η) with origin at the center of the element. This was done in Section 8.7 for the planar rectangular element. We note again that for any rectangle, $\xi = (x - x_c)/(x_c - x_1)$ and $\eta = (y - y_c)/(y_c - y_1)$, where x_c and y_c are the coordinates of the center of the rectangle and x_1 and y_1 are the coordinates of the lower left-hand corner point. We also again observe that in this way the nondimensionalized coordinates of the four corner points are always $+1$ or -1, see Fig. 8.19(a).

The next step is to express the shape functions $\lfloor \mathbf{N} \rfloor = \lfloor N_1, \ldots N_i \ldots N_n \rfloor$ in terms of the nondimensional coordinates. For bilinear interpolation, for example, we have [for points defined in Fig. 8.19(a)]

$$\lfloor \mathbf{N} \rfloor = \tfrac{1}{4}\lfloor (1-\xi)(1-\eta) \quad (1+\xi)(1-\eta) \quad (1+\xi)(1+\eta) \quad (1-\xi)(1+\eta) \rfloor$$

We designate shape functions that are thusly expressed as $\lfloor \mathbf{N}(\xi, \eta) \rfloor$, and describe the x and y coordinates of the element as follows:

$$x = \lfloor \mathbf{N}(\xi, \eta) \rfloor \{\mathbf{x}\}, \qquad y = \lfloor \mathbf{N}(\xi, \eta) \rfloor \{\mathbf{y}\} \tag{8.33}$$

where $\{\mathbf{x}\}$ and $\{\mathbf{y}\}$ list the x and y coordinates of the node points of the element; i.e.,

$$\{\mathbf{x}\} = \lfloor x_1 \ldots x_i \ldots x_n \rfloor^T, \qquad \{\mathbf{y}\} = \lfloor y_1 \ldots y_i \ldots y_n \rfloor^T \tag{8.34}$$

In this way $x = x_i$, $y = y_i$ at point i.

Correspondingly,

$$u = \lfloor \mathbf{N}(\xi, \eta) \rfloor \{\mathbf{u}\}, \qquad v = \lfloor \mathbf{N}(\xi, \eta) \rfloor \{\mathbf{v}\} \tag{8.35}$$

where $\{\mathbf{u}\} = \lfloor u_1 \;\; u_2 \;\; u_3 \;\; u_4 \rfloor^T$, $\{\mathbf{v}\} = \lfloor v_1 \;\; v_2 \;\; v_3 \;\; v_4 \rfloor^T$.

To construct an element stiffness matrix we must find the strains, which are derivatives of the displacements with respect to the x and y axes. The displacements, however, are now functions of the ξ and η coordinates. Hence, we must find a relationship between the derivatives with respect to x and y and the derivatives with respect to ξ and η. We can do this by the chain rule of differentiation,

$$\begin{aligned}\frac{\partial \mathbf{N}_i}{\partial \xi} &= \frac{\partial x}{\partial \xi}\frac{\partial N_i}{\partial x} + \frac{\partial y}{\partial \xi}\frac{\partial N_i}{\partial y}\\ \frac{\partial \mathbf{N}_i}{\partial \eta} &= \frac{\partial x}{\partial \eta}\frac{\partial N_i}{\partial x} + \frac{\partial y}{\partial \eta}\frac{\partial N_i}{\partial y}\end{aligned} \tag{8.36}$$

Using Eq. (8.33) we can determine $\partial x/\partial \xi = \lfloor \partial \mathbf{N}/\partial \xi \rfloor \{\mathbf{x}\}$ and similarly for $\partial x/\partial \eta$, etc. We can therefore write Eq. (8.36) as

$$\begin{Bmatrix} \dfrac{\partial \mathbf{N}_i}{\partial \xi} \\ \dfrac{\partial \mathbf{N}_i}{\partial \eta} \end{Bmatrix} = [\mathbf{J}] \begin{Bmatrix} \dfrac{\partial N_i}{\partial x} \\ \dfrac{\partial N_i}{\partial y} \end{Bmatrix} \tag{8.37}$$

where

$$[\mathbf{J}]_{2\times 2} = \begin{bmatrix} \dfrac{\partial \mathbf{N}}{\lfloor \partial \xi \rfloor} \\ \dfrac{\partial \mathbf{N}}{\lfloor \partial \eta \rfloor} \end{bmatrix}_{2\times n} [\{\mathbf{x}\}\{\mathbf{y}\}]_{n\times 2} \tag{8.38}$$

As is conventional in matrix algebra, the matrix of first derivatives, the 2×2 matrix $[\mathbf{J}]$, is termed the *Jacobian* matrix.

For the illustrative case of the bilinear element [Fig. 8.19(a)],

$$\begin{aligned}\frac{\partial \mathbf{N}}{\lfloor \partial \xi \rfloor} &= \frac{1}{4}\lfloor -(1-\eta) \quad (1-\eta) \quad (1+\eta) \quad -(1+\eta) \rfloor\\ \frac{\partial \mathbf{N}}{\lfloor \partial \eta \rfloor} &= \frac{1}{4}\lfloor -(1-\xi) \quad -(1-\xi) \quad (1+\xi) \quad (1+\xi) \rfloor\end{aligned}$$

Thus, the upper-left coefficient of $[\mathbf{J}]$ is

$$J_{11} = \tfrac{1}{4}[(1-\eta)(x_2 - x_1) + (1+\eta)(x_3 - x_4)]$$

and similarly for J_{12}, J_{21}, and J_{22}.

The left side of Eq. (8.37) is *given* and the vector on the right-hand side is *sought*. Thus, $[\mathbf{J}]$ must be inverted to yield the desired information. It is entirely possible that the analyst would specify a system of physical nodes (i.e., joint coordinates x_1, x_2, etc.) that would be inadmissible, a situation represented by a singular $[\mathbf{J}]$ matrix. The inverse of $[\mathbf{J}]$ is sensitive to certain types of distortion from the basic rectangular shape and also to the position of the node points along the sides.[8.13] In biquadratic elements, for example, the side nodes are best positioned midway between the adjacent corner points.

We are now in a position to examine the implications of isoparametric representation with respect to the formulation of the element stiffness matrix. The strain-displacement relationships are of the usual form $\boldsymbol{\epsilon} = [\mathbf{D}]\{\boldsymbol{\Delta}\}$ where the strains $\boldsymbol{\epsilon}$ refer to *cartesian* (x, y) coordinates. Thus, $[\mathbf{D}]$ contains derivatives of the shape functions with respect to the cartesian coordinates. For planar conditions [see Eq. (5.22)] this is given by

$$\begin{Bmatrix} \epsilon_x \\ \\ \epsilon_y \\ \\ \gamma_{xy} \end{Bmatrix} = \begin{bmatrix} \dfrac{\partial \mathbf{N}}{\lfloor \partial \mathbf{x} \rfloor} & \mathbf{0} \\ \mathbf{0} & \dfrac{\partial \mathbf{N}}{\lfloor \partial \mathbf{y} \rfloor} \\ \dfrac{\partial \mathbf{N}}{\lfloor \partial \mathbf{y} \rfloor} & \dfrac{\partial \mathbf{N}}{\lfloor \partial \mathbf{x} \rfloor} \end{bmatrix} = [\mathbf{D}]\begin{Bmatrix} \{\mathbf{u}\} \\ --- \\ \{\mathbf{v}\} \end{Bmatrix} \tag{8.39}$$

Thus, a transformation in accordance with Eq. (8.37) is necessary in order to evaluate $[\mathbf{D}]$.

The stress-strain relationships are $\boldsymbol{\sigma} = [\mathbf{E}]\boldsymbol{\epsilon}$, and the differential area $dx\,dy$ is replaced in all integrations by

$$dx\,dy = |\mathbf{J}|\,d\xi\,d\eta \tag{8.40}$$

where $|\mathbf{J}|$ symbolizes the determinant of $[\mathbf{J}]$. Also, limits of integration become -1 and $+1$. We can now write the usual formula for an element stiffness matrix, Eq. (6.12a), for unit dimension in the z-direction as

$$[\mathbf{k}] = \left[\int_A [\mathbf{D}]^T[\mathbf{E}][\mathbf{D}]\,dA\right] = \left[\int_{-1}^{+1}\int_{-1}^{+1} [\mathbf{D}]^T[\mathbf{E}][\mathbf{D}]\,\det|\mathbf{J}|\,d\xi\,d\eta\right] \tag{8.41}$$

It is apparent that $[\mathbf{J}]$ is rather complicated even for the simplest case, the bilinear element. Hence, explicit formulation of $[\mathbf{k}]$ is infeasible and the coefficients of this matrix must be evaluated through use of numerical integration.[8.12]

In connection with the choice of shape functions for isoparametric representation it is of interest to note that if rigid body motion and constant strain conditions are satisfied in the original (rectangular) behavior function, they will be preserved in the mapping. Under planar conditions the displacement corresponding to rigid body motion and constant strain can be written in the general form

$$\Delta = \alpha_1 + \alpha_2 x + \alpha_3 y$$

For the rectangle, we have

$$\Delta = \lfloor \mathbf{N}(\xi, \eta) \rfloor \{\boldsymbol{\Delta}\} = \alpha_1 + \alpha_2 x + \alpha_3 y \tag{8.42}$$

At each joint it is required that

$$\Delta_i = \alpha_1 + \alpha_2 x_i + \alpha_3 y_i$$

and, by substitution into Eq. (8.42), for n d.o.f.

$$\alpha_1 \sum_{i=1}^{n} N_i + \alpha_2 \sum_{i=1}^{n} N_i x_i + \alpha_3 \sum_{i=1}^{n} N_i y_i = \alpha_1 + \alpha_2 x + \alpha_3 y$$

or

$$\sum_{i=1}^{n} N_i = 1$$

$$\sum_{i=1}^{n} N_i x_i = x$$

$$\sum_{i=1}^{n} N_i y_i = y$$

The first of these conditions holds true in consequence of the basic properties of shape functions, while the second and third are true by virtue of Eq. (8.33).

The ideas of isoparametric representation extend naturally to three dimensions. No details are given here since the extension of the [**J**] and [**D**] matrices to the third dimension appears obvious. The extension to triangular and tetrahedronal representations is also straightforward, it being necessary to express one of the L_i coordinates in terms of the others before proceeding. Thus, for triangular coordinates, Eq. (8.20) must first be invoked and, for tetrahedronal coordinates, one must use Eq. (8.28).

REFERENCES

8.1 JOHNSON, M. and R. MCLAY, "Convergence of the Finite Element Method in the Theory of Elasticity," *J. Appl. Mech.*, **90**, pp. 274–289, June 1968.

8.2 TONG, P. and T. PIAN, "The Convergence of Finite Element Method in Solving Linear Elastic Problems," *Internat'l. J. Solids Struct.*, **3**, pp. 865–879, 1967.

8.3 HAISLER, W. and J. STRICKLIN, "Rigid-Body Displacements of Curved Elements in the Analysis of Shells by the Matrix-Displacement Method," *AIAA J.*, **5**, No. 8, pp. 1525–1527, Aug., 1967.

8.4 MURRAY, K. H., "Comments on the Convergence of Finite Element Solutions," *AIAA J.*, **8**, No. 4, pp. 815–816, 1970.

8.5 DUNNE, P., "Complete Polynomial Displacement Fields for Finite Element Method," *Aero. J.*, **72**, pp. 246–247, Mar., 1968.

8.6 EISENBERG, M. A. and L. E. MALVERN, "On Finite Element Integration in Natural Coordinates," *Internat'l. J. Num. Meth. Eng.*, **7**, pp. 574–575, 1973.

8.7 ABRAMOWITZ, M. and I. A. STEGUN, *Handbook of Mathematical Functions*, Nat'l. Bureau of Standards, Washington, D.C., 1964.

8.8 STRANG, G. and G. FIX, *An Analysis of the Finite Element Method*, Prentice-Hall, Inc., Englewood Cliffs, N.J., 1973.

8.9 SILVESTER, P., "Tetrahedronal Polynomial Finite Elements for the Helmholtz Equation," *Internat'l. J. Num. Meth. Eng.*, **4**, No. 4, pp. 405–413, 1972.

8.10 TAYLOR, R. L., "On the Completeness of Shape Functions for Finite Element Analysis," *Internat'l. J. Num. Meth. Eng.*, **4**, No. 1, pp. 17–22, 1972.

8.11 ZIENKIEWICZ, O. C., "Isoparametric and Allied Numerically Integrated Elements—A Review," in *Numerical and Computer Methods in Structural Mechanics*, S. J. Fenves, et al (eds.), Academic Press, New York N.Y., pp. 13–42, 1973.

8.12 IRONS, B. M., "Quadrature Rules for Brick Based Finite Elements," *Internat'l. J. Num. Meth. Eng.*, **3**, No. 2, pp. 293–294, 1971.

8.13 BOND, T. J., et al. "A Comparison of Some Curved Two-Dimensional Finite Elements," *J. Strain Anal.*, **8**, 3, 1973, pp. 182–190.

PROBLEMS

8.1 Examine the suitability of the displacement field given in Eq. (9.16) from the standpoint of the criteria defined in Section 8.1.

8.2 Discuss the suitability of the function shown below (Fig. P8.2) in representation of the displacement fields for the illustrated parallelogram finite element in plane stress, using the criteria of Section 8.1.

$$u = a_1x + a_2y + a_3\left(xy - \frac{x_4}{y_4}y^2\right) + a_4$$

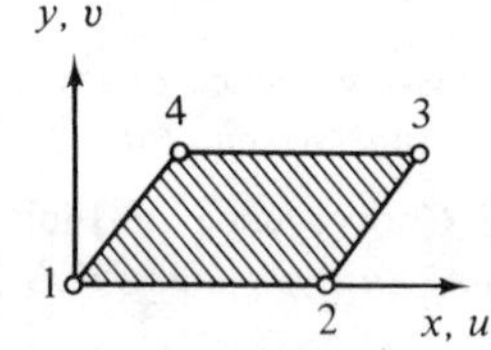

Fig. P8.2

8.3 Construct a displacement function (u) for the rectangular element of Fig. 8.7(b), based on the two-dimensional Lagrangian interpolation, but with point 5 removed via the imposition of a linear variation of displacement from points 4 to 6 and from points 2 to 8.

8.4 Redevelop the procedure to eliminate surplus d.o.f., given in Section 2.8, so as to allow for the inclusion of loads at the d.o.f. being eliminated.

8.5 Construct, in triangular coordinates, the shape function coefficients N_{300}, N_{210}, and N_{111} for a cubic displacement field.

8.6 Establish the N_{400} and N_{220} shape function coefficients, in triangular coordinates, for a quartic displacement field.

8.7 Confirm, by evaluation of the shape functions N_{200}, etc., as given in Section 8.5, that the expressions presented in Prob. 6.9 are correct.

8.8 Construct, for the case of $x_1 = y_1 = y_2 = z_1 = z_2 = z_3 = 0$, the shape functions for the tetrahedron of Fig. 8.15 in terms of the joint coordinates.

8.9 Evaluate the integral

$$\int_A \left(\frac{\partial N_{200}}{\partial x}\right)\left(\frac{\partial N_{011}}{\partial y}\right) dA$$

for the triangular element with vertex 200 at the origin of coordinates and with 110 and 020 on the x axis.

8.10 Calculate the shape functions for the four-jointed axial member [see Fig. 8.2] using Lagrange interpolation.

8.11 Construct a shape function for a rectangular plate in bending using Hermitian interpolation in the mode employed in forming Lagrangian elements. Does the resulting function meet all criteria described in Section 8.1?

8.12 Establish, via the Hermitian polynomial interpolation scheme, the one-dimensional shape function corresponding to the fifth-degree polynomial. The d.o.f. at each end of the segment will be the function itself and its first and second derivatives.

8.13 The biquadratic interpolation formula given by Eq. (8.17) is referenced to axes located at the lower left corner of the element. Transform this expression to one referenced to centroidal axes and $\xi - \eta$ coordinates.

8.14 Construct an appropriate u-displacement field for the element shown in Fig. P8.14.

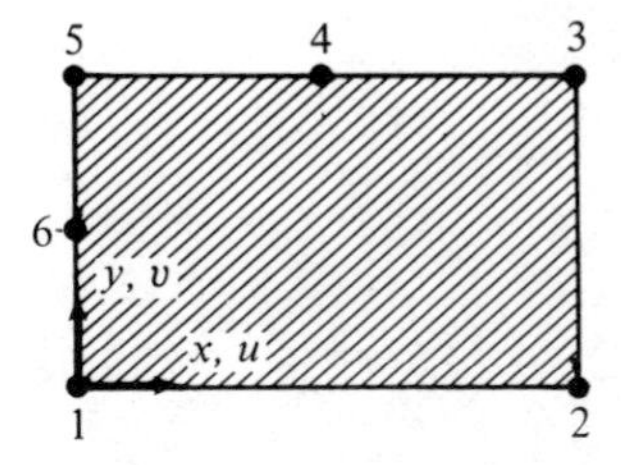

Fig. P8.14

8.15 Assume, as in the case of conduction heat transfer (see Section 5.4) that the behavior of a triangular element is described by the scalar Υ, given by

$$\Upsilon = L_1\Upsilon_1 + L_2\Upsilon_2 + L_3\Upsilon_3 + L_1L_2L_3\Upsilon_0 = \lfloor \mathbf{N} \rfloor \{\boldsymbol{\Upsilon}\}$$

where L_1, L_2, and L_3 are the triangular coordinates; Υ_1, Υ_2, and Υ_3 are vertex values; and Υ_0 is a generalized parameter. The behavior matrix is given by

$$[\mathbf{k}] = \left[\nu \int_A \left(\left\{ \frac{\partial N}{\partial \mathbf{x}} \right\} \left\lfloor \frac{\partial N}{\partial \mathbf{x}} \right\rfloor + \left\{ \frac{\partial N}{\partial \mathbf{y}} \right\} \left\lfloor \frac{\partial N}{\partial \mathbf{y}} \right\rfloor \right) dA \right]$$

where ν is the thermal conductivity. Formulate the behavior matrix for this case and eliminate the bubble mode $L_1 L_2 L_3$ by condensation.

9
Plane Stress

All ingredients for the formulation of a wide variety of finite element shapes and behavior representations are now at hand. This chapter therefore begins the description of specific types of elements for the analysis of continua. Four such chapters, covering plane stress elements, solid elements, special forms of solid elements, and plate flexure elements are presented. Three of these chapters, including the present one, open with a section that recapitulates the basic relationships for the type of behavior represented, i.e., the governing differential equations and specific forms of the relevant differential equations. Subsequent sections of these chapters, and the two other chapters in this group, take forms dictated by the type of element in question.

The present chapter deals with element representations for thin flat plates subjected to in-plane direct and shearing stresses, the plane stress condition. The plane stress condition is the simplest form of behavior for continuum structures and represents situations frequently encountered in practice. The element serves to represent component parts of thin-walled and stiffened sheet construction, box beams, and membrane action in shells.

The basic relationships for plane stress have served as the vehicle for the development of various fundamental theoretical considerations in the foregoing chapters. Thus, the opening section of this chapter is brief, consisting principally of referencing of where basic relationships are defined earlier in the text. Geometric form is the most significant demarcating aspect of plane stress elements. Although many forms are possible, the pertinent geometries are triangular and quadrilateral and each is studied in some detail in separate portions of the chapter.

Triangular element plane stress formulations are principally based upon assumed displacement fields and the potential energy integral. We review a number of alternative triangular element formulations of varying degrees of complexity in this chapter. Practical aspects of this arrangement of triangular elements are discussed at length, as are questions of the interpretation of stress computed by means of the simplest elements. Numerical results are presented as a function of grid refinement for two problems for which alternative analytical solutions are available. Some brief comments are given regarding the role of complementary and mixed variational principles in the construction of triangular finite elements.

The treatment of rectangular plane stress elements first concentrates upon formulations derived with use of interelement-compatible displacement fields. Numerical results are given for these elements, demonstrating that problems in which a state of flexure is to be described are best modeled by elements that contain special supplementary displacement functions. A special section is devoted to the treatment of the latter. The stress hybrid approach to element formulation possesses certain advantages in particular problems of plane stress analysis. This approach to the formulation is described in the concluding section of the chapter.

9.1 Basic Relationships

9.1.1 Differential and Constitutive Equations

In plane stress we consider a thin plate (Fig. 9.1) with x- and y-coordinate axes corresponding to the planar middle surface of the plate, subject to in-plane stress σ_x, σ_y, and τ_{xy} that are constant across the thickness t. The normal stress σ_z and the shear stresses τ_{xz} and τ_{yz} are assumed to be negligible. The equilibrium differential equations have been recorded as Eq. (4.2). The strain-displacement relationships are presented in Eq. (4.7). Equation (4.8) comprises the sole compatibility differential equation.

The constitutive relationships for plane stress are not often needed for situations more general than an orthotropic material. In such situations, with initial strains $\epsilon_x^{\text{init.}}$, $\epsilon_y^{\text{init.}}$, and $\gamma_{xy}^{\text{init.}}$, we have

$$\boldsymbol{\sigma} = [\mathbf{E}]\boldsymbol{\epsilon} - [\mathbf{E}]\boldsymbol{\epsilon}^{\text{init.}} \tag{4.15}$$

where

$$\boldsymbol{\sigma} = \lfloor \sigma_x \; \sigma_y \; \tau_{xy} \rfloor^{\mathrm{T}} \tag{9.1}$$

$$\boldsymbol{\epsilon} = \lfloor \epsilon_x \; \epsilon_y \; \gamma_{xy} \rfloor^{\mathrm{T}} \tag{9.2}$$

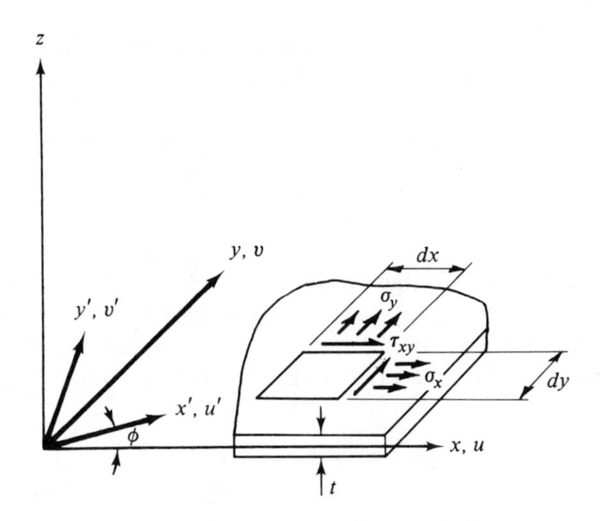

Fig. 9.1 Plane stress—basic representation.

and

$$[\mathbf{E}] = \frac{1}{(1 - \mu_{xy}\mu_{yx})}\begin{bmatrix} E_x & \mu_{yx}E_x & 0 \\ \mu_{xy}E_y & E_y & 0 \\ 0 & 0 & (1 - \mu_{xy}\mu_{yx})G \end{bmatrix} \tag{9.3}$$

with $\mu_{xy}E_y = \mu_{yx}E_x$. G is the shear modulus.

It is noteworthy that the orthotropic constitutive relationships are often given with reference to the *global* axes (x, y) and the element formulation must be accomplished in element axes (x', y'). If ϕ is the angle between the global and element axes as shown in Fig. 9.1, the transformation from global to element stress components is

$$\begin{Bmatrix} \sigma'_x \\ \sigma'_y \\ \tau'_{xy} \end{Bmatrix} = \begin{bmatrix} \cos^2\phi & \sin^2\phi & 2\sin\phi\cos\phi \\ \sin^2\phi & \cos^2\phi & -2\sin\phi\cos\phi \\ -\sin\phi\cos\phi & \sin\phi\cos\phi & \cos^2\phi - \sin^2\phi \end{bmatrix}\begin{Bmatrix} \sigma_x \\ \sigma_y \\ \tau_{xy} \end{Bmatrix} \tag{9.4a}$$

while the transformation for strain is

$$\begin{Bmatrix} \epsilon'_x \\ \epsilon'_y \\ \gamma'_{xy} \end{Bmatrix} = \begin{bmatrix} \cos^2\phi & \sin^2\phi & \sin\phi\cos\phi \\ \sin^2\phi & \cos^2\phi & -\sin\phi\cos\phi \\ -2\sin\phi\cos\phi & 2\sin\phi\cos\phi & \cos^2\phi - \sin^2\phi \end{bmatrix}\begin{Bmatrix} \epsilon_x \\ \epsilon_y \\ \gamma_{xy} \end{Bmatrix} \tag{9.4b}$$

The respective forms of boundary conditions have already been given;

the boundary conditions on stress appear as Eq. (4.5) and those on displacement are defined by Eq. (4.9). The governing equilibrium differential equations, in terms of displacements, are presented in Eq. (4.17). The governing compatibility differential equation is presented in terms of stress in Eq. (4.18) and in terms of the Airy stress function in Eq. (4.19).

9.1.2 Potential Energy

We examine here only the strain energy portion (U) of the potential energy. The potential of applied forces (V) is dependent on the form of such forces and can be discussed in a specific way only when the distribution of loads is defined. The strain energy in plane stress is given by

$$U = \frac{1}{2}\int_A \boldsymbol{\sigma}\,\boldsymbol{\epsilon}\,t\,dA \tag{9.5}$$

where $\boldsymbol{\sigma}$ and $\boldsymbol{\epsilon}$ are as defined in Eqs. (9.1) and (9.2). By introduction of the constitutive law for an orthotropic material (excluding initial strain) and the strain-displacement equations [Eq. (4.7)] we have

$$U = \frac{1}{2}\int_A \boldsymbol{\epsilon}[\mathbf{E}]\boldsymbol{\epsilon}\,t\,dA \tag{9.5a}$$

where $[\mathbf{E}]$ is defined by Eq. (9.3) and

$$\boldsymbol{\epsilon} = \begin{Bmatrix} \epsilon_x \\ \epsilon_y \\ \gamma_{xy} \end{Bmatrix} = \begin{Bmatrix} \dfrac{\partial u}{\partial x} \\ \dfrac{\partial v}{\partial y} \\ \dfrac{\partial u}{\partial y} + \dfrac{\partial v}{\partial x} \end{Bmatrix}$$

There is obtained, upon expansion,

$$\begin{aligned} U = \frac{1}{2(1-\mu_{xy}\mu_{yx})}\int_A \Big[& E_x\left(\frac{\partial u}{\partial x}\right)^2 + 2\mu_{xy}E_y\frac{\partial u}{\partial x}\frac{\partial v}{\partial y} + E_y\left(\frac{\partial v}{\partial y}\right)^2 \\ & + (1-\mu_{xy}\mu_{yx})G\left(\frac{\partial u}{\partial y} + \frac{\partial v}{\partial x}\right)^2\Big]t\,dA \end{aligned} \tag{9.5b}$$

The principle of minimum potential energy involves assumed displacement fields. In the present case the field of displacements generally consists of u and v components at each node point so we designate the vector of node point displacements as $\lfloor \boldsymbol{\Delta} \rfloor = \lfloor \lfloor \mathbf{u} \rfloor \lfloor \mathbf{v} \rfloor \rfloor$. Also, for stiffness formulations discussed explicitly in this chapter, the displacement fields u and v are written directly in terms of joint displacements, in shape function form, i.e.,

$$u = \lfloor \mathbf{N} \rfloor \{\mathbf{u}\}, \qquad v = \lfloor \mathbf{N} \rfloor \{\mathbf{v}\}$$

(generally, u is chosen as a function of $\{\mathbf{u}\}$ alone, and similarly for v). The d.o.f.-to-strain transformation for planar conditions has already been employed in Section 5.2, Eq. (5.22) as well as in Section 8.8, Eq. (8.39). Thus, consistent with prior symbolism, we write

$$\begin{Bmatrix} \epsilon_x \\ \epsilon_y \\ \gamma_{xy} \end{Bmatrix} = \begin{bmatrix} \left\lfloor \frac{\partial \mathbf{N}}{\partial \mathbf{x}} \right\rfloor & \mathbf{0} \\ \mathbf{0} & \left\lfloor \frac{\partial \mathbf{N}}{\partial \mathbf{y}} \right\rfloor \\ \left\lfloor \frac{\partial \mathbf{N}}{\partial \mathbf{y}} \right\rfloor & \left\lfloor \frac{\partial \mathbf{N}}{\partial \mathbf{x}} \right\rfloor \end{bmatrix} \begin{Bmatrix} \{\mathbf{u}\} \\ \{\mathbf{v}\} \end{Bmatrix} = [\mathbf{D}] \begin{Bmatrix} \{\mathbf{u}\} \\ \{\mathbf{v}\} \end{Bmatrix} \tag{9.6}$$

Equation (9.5a) therefore becomes

$$U = \frac{\lfloor \lfloor \mathbf{u} \rfloor \lfloor \mathbf{v} \rfloor \rfloor}{2} [\mathbf{k}] \begin{Bmatrix} \{\mathbf{u}\} \\ \{\mathbf{v}\} \end{Bmatrix} \tag{9.5c}$$

with

$$[\mathbf{k}] = \left[\int_A [\mathbf{D}]^{\mathrm{T}} [\mathbf{E}] [\mathbf{D}] \, t \, dA \right] \tag{9.7}$$

9.1.3 Complementary Energy

The complementary strain energy is, from Eq. (6.68a),

$$U^* = \frac{1}{2} \int_A \boldsymbol{\sigma} [\mathbf{E}]^{-1} \boldsymbol{\sigma} \, t \, dA \tag{9.8}$$

Utilizing stress-function concepts, the Airy stress function $\boldsymbol{\Phi}$ [Eq. (4.4)] is appropriate to the plane stress condition. The assumed field of the Airy stress function within the element can be described in shape function form

$$\boldsymbol{\Phi} = \lfloor \mathbf{N} \rfloor \{\boldsymbol{\Phi}\} \tag{6.77}$$

where $\{\boldsymbol{\Phi}\}$ is populated by joint values of the stress function and $\lfloor \mathbf{N} \rfloor$ lists the chosen shape functions. Then, by second differentiation of $\boldsymbol{\Phi}$ in accordance with the definition of the stress function [see Eq. (4.4)],

$$\boldsymbol{\sigma} = [\mathbf{N}''] \{\boldsymbol{\Phi}\} \tag{6.78}$$

where the coefficients of $[\mathbf{N}'']$ are, in general, functions of x and y. By substitution into the expression for complementary strain energy,

$$U^* = \frac{\lfloor \mathbf{\Phi} \rfloor}{2}[\mathbf{f}]\{\mathbf{\Phi}\} \tag{9.8a}$$

where, as in Eq. (6.72a),

$$[\mathbf{f}] = \left[\int_A [\mathbf{N}'']^T[\mathbf{E}]^{-1}[\mathbf{N}'']\, t\, dA\right] \tag{9.9}$$

Basic relationships for plane stress statements of mixed variational principles are not given here; the role of these principles in the subject class of element formulations is merely outlined in this chapter. Reference 9.1 contains details of the Reissner functional in terms of plane stress conditions.

9.2 Triangular Plane Stress Elements

9.2.1 Elements Based on Assumed Displacements

In this section we discuss plane stress triangles formulated under the assumption of complete linear, quadratic, and cubic displacement fields, respectively. Section 8.5 demonstrated that, in theory, there is no limitation of the degree of polynomial representation for triangular elements, it being a simple matter to array node points within and on the boundary of the element to accommodate the higher-order functions. In practice, however, the value of elements based on polynomials of order greater than three is questionable. On one hand there is a substantial increase in formulation cost for the element coefficients and on the other the requirements for element grid refinement to describe the geometry of the real structure often vitiates the advantage of more sophisticated representation of behavior within an element.

Elements discussed in this section are portrayed in Fig. 9.2. The basic element, with d.o.f. only at the vertex, Fig. 9.2(a), is formulated under the assumption of constant strains (which is synonymous with the assumption of constant stress, or linear displacement) and is often termed the *CST element*. The stiffness matrix for this element for an isotropic material was derived using alternative procedures in Sections 5.2 and 6.4 and was presented in Fig. 5.4. Since this derivation was given in considerable detail in Section 5.2, no further comments are offered here.

The next element in order of increasing sophistication is the six-jointed triangle, Fig. 9.2(b), whose formulation is based upon complete quadratic polynomials in description of the u and v displacements. As in Section 8.5 we have

$$u = N_{200}u_{200} + N_{020}u_{020} + N_{002}u_{002} + N_{110}u_{110} + N_{011}u_{011} + N_{101}u_{101} \tag{8.25b}$$

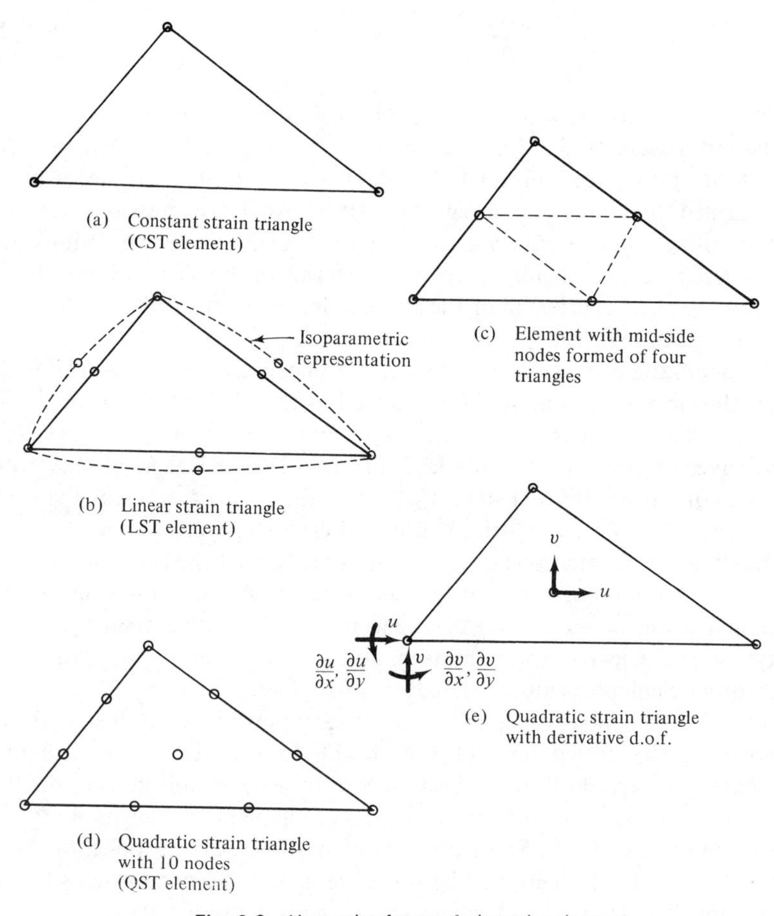

Fig. 9.2 Alternative forms of triangular element.

and similarly for v. In terms of triangular coordinates, from Eq. (8.11a, 8.12a),

$$N_{200} = L_1(2L_1 - 1), \qquad N_{020} = L_2(2L_2 - 1), \qquad N_{002} = L_3(2L_3 - 1)$$
$$N_{110} = 4L_1L_2, \qquad N_{011} = 4L_2L_3, \qquad N_{101} = 4L_3L_1 \tag{9.10}$$

In this case, upon application of the strain-displacement relationships [Eq. (4.7)], we obtain Eq. (9.6) wherein $\{\mathbf{u}\} = \lfloor u_{200} \ldots u_{101} \rfloor^T$, $\{\mathbf{v}\} = \lfloor v_{200} \ldots v_{101} \rfloor^T$,

$$\left\lfloor \frac{\partial N}{\partial \mathbf{x}} \right\rfloor = \left\lfloor \frac{\partial N_{200}}{\partial x} \quad \frac{\partial N_{020}}{\partial x} \quad \frac{\partial N_{002}}{\partial x} \quad \frac{\partial N_{110}}{\partial x} \quad \frac{\partial N_{011}}{\partial x} \quad \frac{\partial N_{101}}{\partial x} \right\rfloor \tag{9.11}$$

and similarly for the other vectors in Eq. (9.6). The first term in Eq. (9.11) is, as shown in Section 8.5

$$\frac{\partial N_{200}}{\partial x} = \frac{b_{1_1}}{2A}(4L_1 - 1) \tag{9.12}$$

All the other terms are also obtained easily by differentiation.

The stiffness matrix is formulated by use of Eq. (9.7). In practice the integration of the triple product $[\mathbf{D}]^T[\mathbf{E}][\mathbf{D}]$ over the area of the element is performed by numerical integration since explicit expressions for the element stiffness coefficients are rather complicated. Such expressions are nevertheless available, being expressible in terms of the CST stiffness coefficients.[9.2] Explicit expression of the stiffness matrix coefficients for elements of higher order than this proves too awkward.

The quadratic displacement field yields a linear strain (or stress) triangle so that the element is commonly designated as the *LST element*. It might be suggested that a grouping of four CST elements, as shown in Fig. 9.2(c), would have the same effect as the LST element. The LST element, however, defines a continuous (linear) state of stress in the element, whereas the CST array gives four different constant values of each stress component.

The differential equations of equilibrium are not satisfied within the LST element. This point was previously demonstrated in Section 4.5, using the polynomial form of the u- and v-displacement fields rather than the shape function form. Clearly, the condition of equilibrium at points within all higher-order elements is also violated.

The LST element is amenable to representation in isoparametric form, as shown by the dotted lines of Fig. 9.2(b). Procedures to establish this representation were outlined in Section 8.8. In general, all the specific elements to be discussed in this and subsequent chapters are amenable to representation in this form. Since the details in all cases correspond to those defined in Section 8.5, there will be no subsequent reference to the isoparametric geometric form except where special considerations arise.

The next level of refinement is the triangle based upon complete cubic (10-term) displacement functions for both the u and v components. In this case two alternative arrangements of the d.o.f. are encountered. In one, Fig. 9.2(d), an array of 10 node points is established in the conventional way and values of u and v are selected as d.o.f. at each such point. The other arrangement, Fig. 9.2(e), consists only of vertex nodes at which u and v and their derivatives ($du/dx = u_x$, etc.) are specified. This defines a total of 9 d.o.f. related to each component (u and v), for a total of 18 d.o.f. There is a total of 20 d.o.f. in the full expansion in both u and v. The additional 2 d.o.f. may be accommodated as values of the two displacement components at the element centroid. One may remove these 2 d.o.f. via the condensation process detailed in Section 2.8 or through more elegant procedures as described in detail in Refs. 9.4–9.6.

Still another approach is to subdivide the basic triangle into three triangular subregions, select 9- or 10-term polynomials for u and v within each subregion, and eliminate interior d.o.f., by imposition of conditions of displacement continuity. This approach is more popular for triangles in bending and is discussed in that connection in Section 12.3.2.

Higher-order elements with node point arrays that conform to the Pascal triangle, i.e., with nodes along the sides and in the interior of the element, produce more global stiffness equations with a wider bandwidth than those with d.o.f. concentrated at the vertices. We can see the reason for this by examining the addition of two triangular elements to an idealization whose boundary is initially defined by the points $E\ A\ B\ C\ F$ in Fig. 9.3. If, as in Fig. 9.3(a), we add elements of the quadratic strain type with 10 nodes [the type of element shown in Fig. 9.2(d)], then 18 d.o.f. are added with a corresponding half-bandwidth. The addition of two quadratic strain elements with derivative d.o.f. at the nodes and centroidal nodes [the type of element shown in Fig. 9.2(e)] adds just 10 d.o.f., however. The difference is due to the connectivity of d.o.f at point D in the elements with derivative d.o.f.

The effect of increased bandwidth is to increase the cost of the equation-solving portion of analysis. Another advantage of elements with derivatives as d.o.f. is that the derivatives employed as d.o.f. are directly proportional to strains and, therefore, to stresses so that certain stress boundary conditions may be specified directly. A disadvantage is that the joint force quantities corresponding to the displacement derivatives in the form of d.o.f. have little physical meaning in the analysis of plane stress.

9.2.2 Factors in the Definition of Triangular Gridworks

The triangular element owes its popularity to the simplicity of formulation of the constant strain form of this element and also to its usefulness in

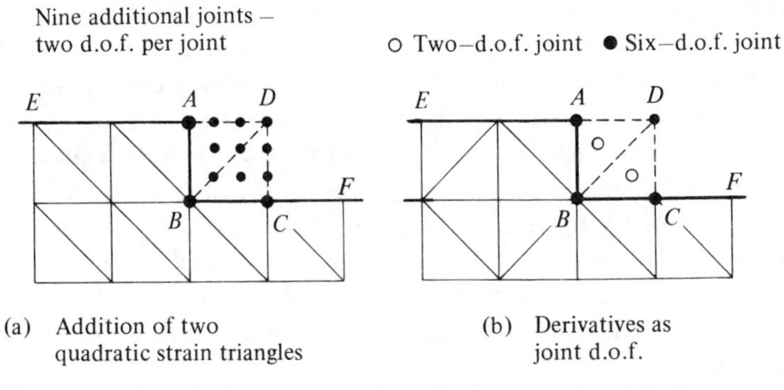

Fig. 9.3 Comparison of alternative arrangements of d.o.f.

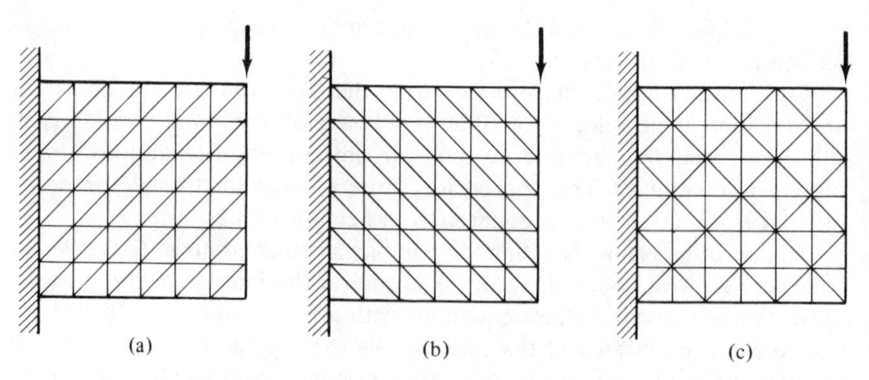

Fig. 9.4 Alternative arrays for a given problem.

the description of the geometry of complex structures. At the same time, in complex structures, certain difficulties arise in the choice of an appropriate gridwork from among numerous available alternatives.

The problem of a lack of "geometric isotropy" must first be recognized in connection with alternative triangular element arrays. To illustrate this point we examine the problem of stress and displacement analysis of the bracket shown in Fig. 9.4. The arrays shown in Fig. 9.4(a) and (b), respectively, although containing exactly the same number of elements of identical form, will produce different solutions for stress and displacement. These differences may be small for grids of the degree of refinement required in practical analysis but it is nevertheless desirable to take whatever steps are available to eliminate or reduce this inconsistency. In the present example the scheme shown in Fig. 9.4(c) is an obvious method of eliminating the problem. In many practical structures, however, the solution is not so easy and some deficiency in this regard must be accepted.

As we see in the example above, geometric isotropy can be preserved when the structure is of rectangular shape or contains many rectangular regions. The prevalence of rectangularity in practical structures motivates the use of "building block" elements, wherein the analyst specifies rectangular elements that are actually preformed of a number of triangular elements. Figure 9.5 shows two such elements, in each of which geometric isotropy is preserved.

The rectangular element of Fig. 9.5(a) is constructed of four triangles. A pair of triangles, whose thickness is half that of the actual plate, are arranged with the diagonal joining points 1 and 3, and upon this is superimposed a pair of triangles of the same thickness with the diagonal joining points 2 and 4. Alternatively, four triangles are arranged as in Fig. 9.5(b) with a central point (5), which is eliminated from the resulting rectangular element stiffness matrix through use of the condensation process described in Section 2.8.

One disadvantage of the schemes above is the difficulty of interpreting

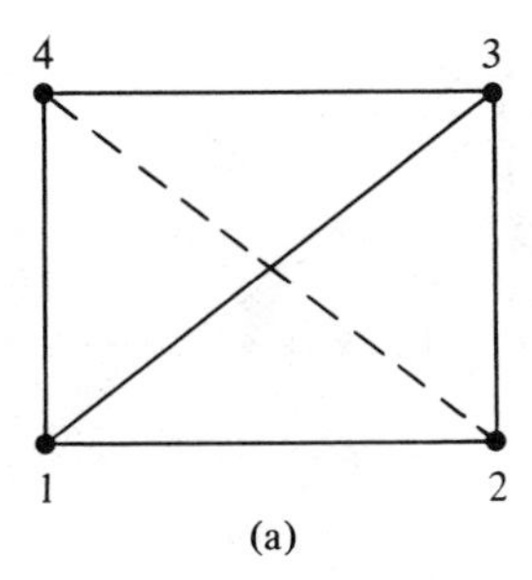

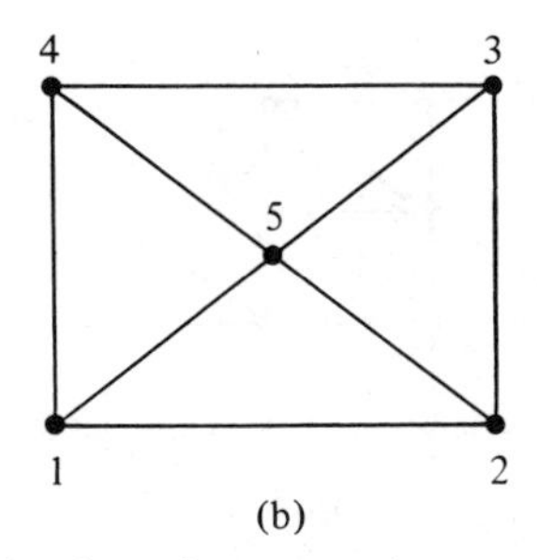

Fig. 9.5 Methods of combination of triangles to form rectangles.

the calculated stresses in a manner appropriate to the design process. Design procedures for rectangular panels require the definition of constant or linear stress fields throughout the element. In the arrangement of Fig. 9.5(b), however, the state of stress within the rectangle is described by four distinct values of each stress component. The common practice is to average these values to obtain a composite value for the complete rectangle. The problem in this approach is that the four discrete values may differ significantly, placing doubt upon the accuracy of the average value.

Theoretical studies have been performed[9.7] of the rate of convergence to the exact solution of the governing differential equation [Eq. (4.17)] of various types of triangular element arrays. The schemes studied are shown in Fig. 9.6. Pattern *A* was found to have the best rate of convergence. As noted previously, however, the problem of geometric isotropy arises with this arrangement of elements. The equilateral triangle gridwork (pattern *D*) demonstrates the same rate of convergence as pattern *A*. Poorer convergence characteristics are evidenced by arrays *B* and *C*. These patterns feature errors in different directions that compensate each other when the arrays are used in combination in a given analysis, as is usually the case.

Another practical question associated with the geometric arrangement of elements based on assumed displacement fields arises on account of the stress state at free edges. As Fig. 5.3 discloses, the stresses in a constant stress element "protrude" from the faces of the element. Thus, a stress-free state on the edge of a structure is merely approximated. The gridwork in the direction normal to such an edge must be of sufficient refinement to permit the very small stress state on the edge and a transition to what may be a significant intensity of this stress component at points in the interior of the structure.

To discuss further the implications of the above in practical applications it is necessary to note that care must be exercised to avoid elongated elements. The stiffness of the constant strain triangle does not reduce to an axial member in the limit of elongation and it can be shown[9.8] that solution accuracy decreases with increase of element aspect ratio (the ratio of the maximum

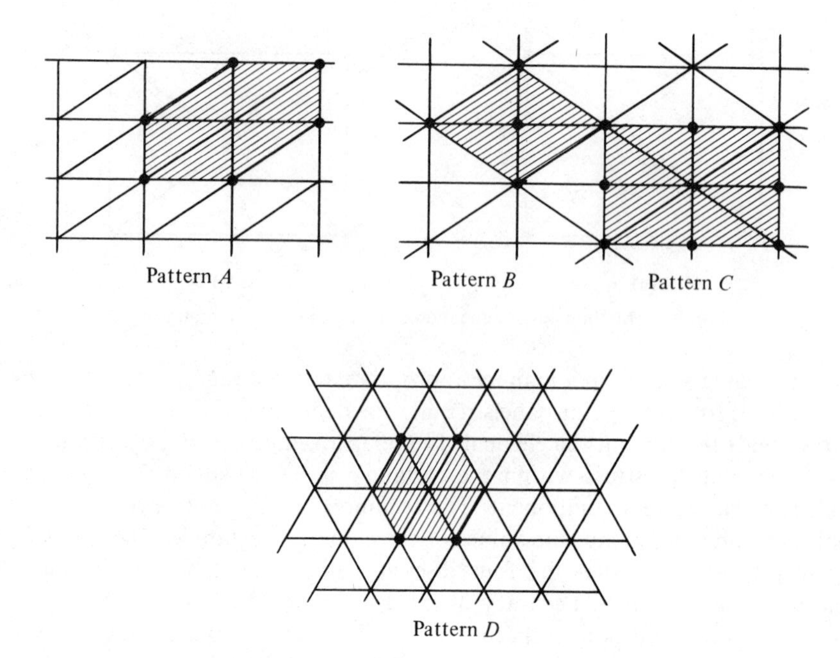

Fig. 9.6 Arrays examined in convergence study of Ref. 9.7.

element dimension in one coordinate direction to the maximum dimension in the other direction). One should strive for equilateral triangles.

It was shown in Chapters 6 and 7 that a minimum potential energy solution based on a finite number of d.o.f. gives a lower bound to the exact strain energy. Thus, for a given number of d.o.f., the objective should be to locate the grid points to achieve the maximum strain energy. It is theoretically possible to accomplish this positioning of the node points in connection with the total analysis process, in which the x and y coordinates of the points are treated as d.o.f. in the functional and participate in the definition of the extremum of the functional.[9.9] This process must, of course, be performed iteratively and is inordinately expensive for real problems.

Under the circumstances of practical analysis the engineer can estimate regions of sharp strain gradient and in these regions, if conventional elements are employed, it is important either to use a high degree of grid refinement of the simpler elements or to use higher-order elements. If the latter approach is used, one must form a transition from the higher-order elements in the regions of severe strain gradient to the simpler elements in regions where the strain distribution is essentially uniform or of less importance. To do so it is useful to employ higher-order elements with fewer node points on edges that join with the simpler element.[9.10] This situation is illustrated in Fig. 9.7 for classical problem of a circular disk subjected to a pair of radially directed concentrated loads. The mode of construction of element dis-

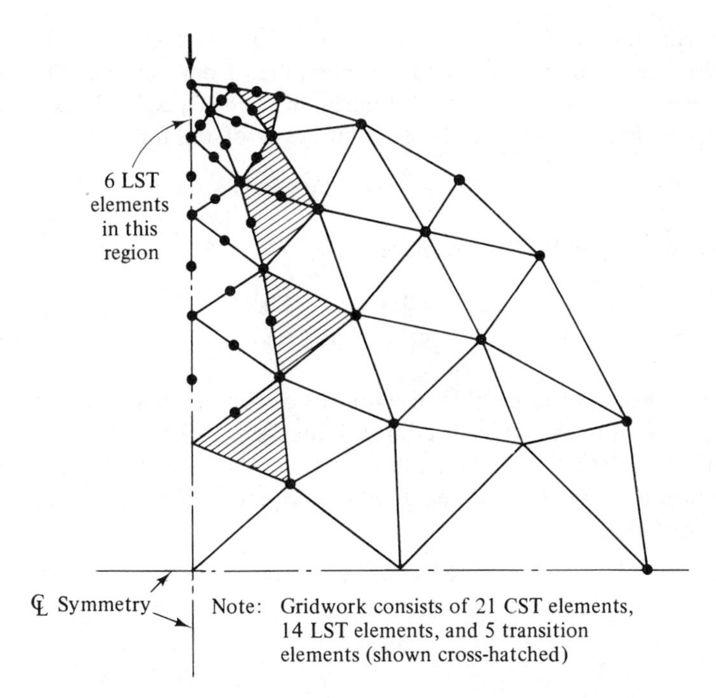

Fig. 9.7 Finite element representation of quadrant of radially loaded disk. Use of higher-order and lower-order elements in combination, with transition elements in between (from Ref. 9.10). Reproduced by permission of the Council of the Institution of Mechanical Engineers from the *Journal of Strain Analysis.*

placement fields with disparate numbers of node points on the respective edges of the element was described in Section 8.7. Under points of concentrated load, or at the tips of cracks in the material where the stress is theoretically infinite (a *stress singularity*), it is preferable to incorporate the singularity in the formulation of elements at and near the load points or crack tip. We shall allude to such formulations again at the close of Section 9.3.3.

9.2.3 Interpretation of Stress Fields

Since displacement-based finite element formulations stem from the principle of minimum potential energy, where in general the equations of equilibrium are satisfied only in the mean throughout the element and, in general, point-wise equilibrium is satisfied neither within the element nor along element juncture lines, it should be anticipated that difficulties will be confronted in the interpretation of calculated stresses.

Before reviewing these difficulties, it is important to examine the details of stress computation when such effects as initial strains, distributed loads, and inertial forces are present. It will be recalled from Section 5.1, Eq. (5.7a), that the element stress field $\boldsymbol{\sigma}$ can be computed from the element displacement vector $\{\boldsymbol{\Delta}\}$ by use of the relationship $\boldsymbol{\sigma} = [\mathbf{E}][\mathbf{D}]\{\boldsymbol{\Delta}\} = [\mathbf{S}]\{\boldsymbol{\Delta}\}$. If initial strains $\boldsymbol{\epsilon}^{\text{init.}}$ are present, the stress-strain law is of the form

$$\boldsymbol{\sigma} = [\mathbf{E}]\boldsymbol{\epsilon} - [\mathbf{E}]\boldsymbol{\epsilon}^{\text{init.}}$$

and, since $\boldsymbol{\epsilon} = [\mathbf{D}]\{\boldsymbol{\Delta}\}$, we now have

$$\begin{aligned}\boldsymbol{\sigma} &= [\mathbf{E}][\mathbf{D}]\{\boldsymbol{\Delta}\} - [\mathbf{E}]\boldsymbol{\epsilon}^{\text{init.}} \\ &= [\mathbf{S}]\{\boldsymbol{\Delta}\} - [\mathbf{E}]\boldsymbol{\epsilon}^{\text{init.}}\end{aligned} \tag{5.7b}$$

The inclusion of terms that account for distributed loads is not so easily perceived. It was shown in Section 6.1 that when distributed loads $\bar{\mathbf{T}}$ are present, the concept of "work-equivalent" or "consistent" loads results in the following form of the element stiffness equations (excluding other types of special forces):

$$\{\mathbf{F}\} = [\mathbf{k}]\{\boldsymbol{\Delta}\} - \{\mathbf{F}^d\} \tag{6.16a}$$

where

$$\{\mathbf{F}^d\} = \left\{\int_{S_\sigma} [\mathbf{\mathcal{Y}}]^{\mathrm{T}} \cdot \bar{\mathbf{T}}\, dS\right\} \tag{6.12f}$$

Now, the stress-strain law $\boldsymbol{\sigma} = [\mathbf{E}]\boldsymbol{\epsilon}$ would appear to be unchanged under the conditions above and $\boldsymbol{\epsilon} = [\mathbf{D}]\{\boldsymbol{\Delta}\}$ is unaffected so that $\boldsymbol{\sigma} = [\mathbf{S}]\{\boldsymbol{\Delta}\}$ would appear to apply in calculation of stress for distributed load.

Let us examine this circumstance relative to the problem shown in Fig. 9.8(a), an axial member subjected to a uniformly applied load of intensity q. The structure is divided into two elements, Fig. 9.8(b). Each element has work-equivalent loads $\{\mathbf{F}^d\} = qL/2 \lfloor 1 \quad 1 \rfloor^{\mathrm{T}}$. The stiffness equations of the whole structure are

$$\begin{Bmatrix} qL \\ q\dfrac{L}{2} \end{Bmatrix} = \frac{AE}{L}\begin{bmatrix} 2 & -1 \\ -1 & 1 \end{bmatrix}\begin{Bmatrix} u_2 \\ u_3 \end{Bmatrix}$$

and, solving

$$\begin{Bmatrix} u_2 \\ u_3 \end{Bmatrix} = \frac{L}{AE}\begin{bmatrix} 1 & 1 \\ 1 & 2 \end{bmatrix}\begin{Bmatrix} qL \\ q\dfrac{L}{2} \end{Bmatrix}$$

so that $u_2 = \frac{3}{2}(qL^2/AE)$, $u_3 = 2qL^2/AE$. The stress matrix for the axial member is the row matrix $E/L \lfloor -1 \quad 1 \rfloor$ and we presumably have the constant stress in each member $\sigma_{1-2} = \frac{3}{2}qL/A$, $\sigma_{2-3} = qL/A$. This stress state is plotted with dotted lines in Fig. 9.8(c). The exact distribution, which is determined by inspection, is shown by a solid line.

To take an entirely different view of stress calculation, consider the basic element force-displacement equations. For member 2-3,

$$\begin{Bmatrix} F_{2-3} \\ F_{3-2} \end{Bmatrix} = \frac{AE}{L}\begin{bmatrix} 1 & -1 \\ -1 & 1 \end{bmatrix}\begin{Bmatrix} \frac{3}{2}\frac{qL^2}{AE} \\ \frac{2qL^2}{AE} \end{Bmatrix} - \begin{Bmatrix} \frac{qL}{2} \\ \frac{qL}{2} \end{Bmatrix}$$

so that $F_{2-3} = -qL$ [minus is tension here; see Fig. 9.8(d)], $F_{3-2} = 0$. Similarly, we find for member 1-2 that $F_{2-1} = qL$ and $F_{1-2} = -2qL$ (again, minus denotes tension at left end). Dividing these forces by the cross-sectional area A we obtain the exact stress distribution† of Fig. 9.8(c). Thus, we obtain the proper stresses by first computing the joint forces and then transforming these into stresses. The expression for the joint forces might include not only the distributed loads but also inertia forces $[\mathbf{m}]\{\ddot{\mathbf{\Delta}}\}$, as in Eq. (6.16), or other special distributed effects.

We note that the joint force was *locally* (at the joint) transformed into a stress. This is a straightforward operation for axial elements where there is a one-to-one correspondence between the joint force component and the desired stress component. In plane stress, however, there are *three* stress components and only *two* forces at each joint, and this type of disparity exists in most multiaxial stress situations. Thus, the usual approach is to neglect the "correction" represented by the term $\{\mathbf{F}^d\}$ when calculating stresses and essentially revert to the formula $\boldsymbol{\sigma} = [\mathbf{S}]\{\mathbf{\Delta}\}$. The error in such an approach is represented in our numerical example by the crosshatched areas of Fig. 9.8(c), and it is clear that this error becomes small when many elements are employed, as is often the case in practical calculation.

The procedure based on initial calculation of joint forces also enables the appropriate interpretation of the procedure based on the stress matrix. In Eq. (6.16a), as we have implied above, the term $[\mathbf{k}]\{\mathbf{\Delta}\}$ corresponds to $[\mathbf{S}]\{\mathbf{\Delta}\}$ of Eq. (5.7b) and we can identify $\{\mathbf{F}^d\}$ as corresponding to $[\mathbf{E}]\boldsymbol{\epsilon}^{\text{init.}}$ in Eq. (5.7b). But $[\mathbf{E}]\boldsymbol{\epsilon}^{\text{init.}}$ can be viewed as an "initial stress" $\boldsymbol{\sigma}^{\text{init.}}$, which in this case is the stress due to the work-equivalent load. The difficulty remains, however, of establishing such a transformation for multiaxial stress states.

In the foregoing we have seen that under certain conditions (e.g., an axial member under distributed applied loads) a continuous, exact distribution of stresses can be calculated but that practical considerations may cause the definition of approximate "histogram" forms of stress distribution where

†It is of interest to note that the exact stresses will be obtained at the element *juncture points*, whatever the distribution of applied loads, when work-equivalent loads are calculated on the basis of shape functions that represent the exact solution of the homogeneous (zero load) form of the relevant governing differential equation. This is because all pertinent conditions (equilibrium, compatibility) are exactly satisfied at these points relative to the work-equivalent loads (see Ref. 9.11).

the stress is defined discontinuously across element boundaries. Elsewhere, as in the case of constant strain triangular elements in structures subjected to concentrated forces, the numerical solution basically gives a discontinuous definition of the overall stress distribution. There is consequently a need for a scheme that will produce, for design purposes, a continuous representation of the stress field. This can be done in a rational way by means of the *conjugate stress* idea.(9.12) This idea merely involves a smoothing technique that provides for continuous stress field representation in conforming finite element models.

If, in the case of a displacement field $\mathbf{\Delta} = [\mathbf{N}]\{\mathbf{\Delta}\}$, the shape functions $\lfloor \mathbf{N} \rfloor$ give a conforming representation, then the simplest and most natural *conforming stress* approximation will be of the form

$$\boldsymbol{\sigma}^c = [\mathbf{N}]\{\boldsymbol{\sigma}\}$$

where $\boldsymbol{\sigma}^c$ is the conforming stress field and $\{\boldsymbol{\sigma}\}$ lists the values of stress at specified points. We shall refer to this representation of stress as *displacement-consistent*. It is possible, of course, to select conforming stress representations that are not displacement-consistent, but this requires an independent consideration that does not make use of the data and calculations already performed on the displacement formulation. The "standard" conjugate stress calculation is based on the premise that the stress is conforming and displacement-consistent. Note that $\boldsymbol{\sigma}^c$ may possess three components in the plane stress case ($\boldsymbol{\sigma}_x$, $\boldsymbol{\sigma}_y$, τ_{xy}) so that $[\mathbf{N}]$ is a rectangular matrix.

In accordance with the theory of conjugate stress approximation, one forms two types of element matrices. The first is a square matrix

$$\left[\int_{\text{vol}} [\mathbf{N}]^T[\mathbf{N}]\, d(\text{vol})\right]$$

The columns correspond to the entries in $\{\boldsymbol{\sigma}\}$. This matrix can be loosely identified with the virtual work of forces corresponding to the conjugate stresses $\boldsymbol{\sigma}^c$ and the corresponding compatible virtual displacements. The second matrix is a row vector given by

$$\left\lfloor \int_{\text{vol}} \boldsymbol{\sigma}[\mathbf{N}]\, d(\text{vol}) \right\rfloor$$

where $\boldsymbol{\sigma}$, written here in row form, defines the nonconforming stress field. This too can be viewed from the virtual work perspective, being identified with the virtual work of forces corresponding to the nonconforming stress field $\boldsymbol{\sigma}$ and the compatible displacements.

The element matrices above are combined to form a representation of the complete structure by use of a summation procedure identical to that employed in direct stiffness analysis. Retaining the same nomenclature for the global representation as for the element representation we can write the solution for the vector of conjugate stress values as

$$\lfloor \boldsymbol{\sigma} \rfloor = \left\lfloor \int_{\text{vol}} \boldsymbol{\sigma} \lfloor \mathbf{N} \rfloor \, d(\text{vol}) \right\rfloor \left[\int_{\text{vol}} \lfloor \mathbf{N} \rfloor^{\text{T}} \lfloor \mathbf{N} \rfloor \, d(\text{vol}) \right]^{-1}$$

This expression can be viewed as the result of equating the alternative virtual work expressions.

To illustrate this idea we apply it to the problem shown in Fig. 9.8. In this case $d(\text{vol}) = A\,dx$ for each element and

$$\sigma_{1-2} = \left\lfloor \left(1 - \frac{x}{L}\right) \quad \frac{x}{L} \right\rfloor \begin{Bmatrix} \sigma_1 \\ \sigma_2 \end{Bmatrix}, \qquad \sigma_{2-3} = \left\lfloor \left(1 - \frac{x}{L}\right) \quad \frac{x}{L} \right\rfloor \begin{Bmatrix} \sigma_2 \\ \sigma_3 \end{Bmatrix}$$

Thus, for each element

$$\left[\int_{\text{vol}} \lfloor \mathbf{N} \rfloor^{\text{T}} \lfloor \mathbf{N} \rfloor \, d(\text{vol}) \right] = \frac{AL}{6} \begin{bmatrix} 2 & 1 \\ 1 & 2 \end{bmatrix}$$

Note that this matrix is identical to the element consistent mass matrix except for multiplying constants. We assemble these matrices to form the representation for the complete structure

$$\begin{matrix} \sigma_1 & \sigma_2 & \sigma_3 \end{matrix}$$
$$\begin{bmatrix} 2 & 1 & 0 \\ 1 & 4 & 1 \\ 0 & 1 & 2 \end{bmatrix}$$

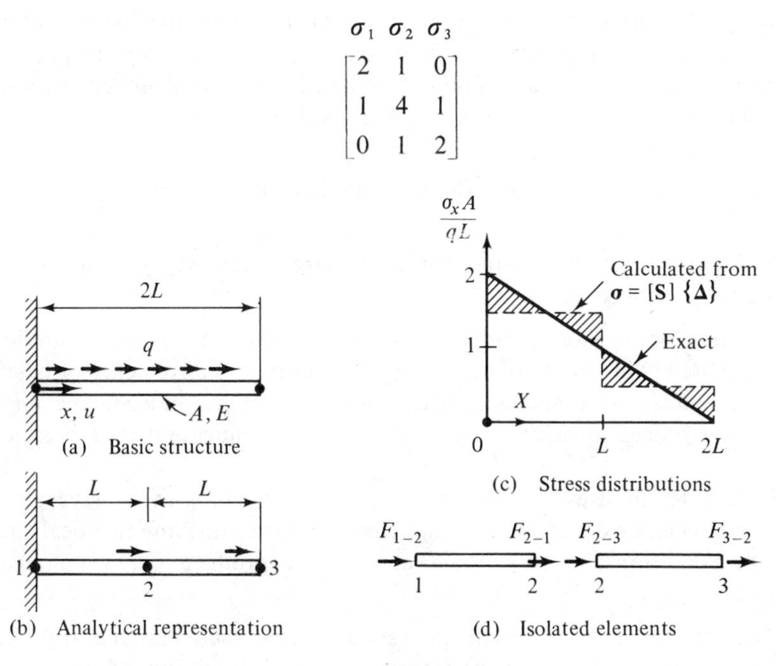

Fig. 9.8 Calculation of stress in uniformly loaded axial members.

Also, for each element, where the stress $\boldsymbol{\sigma} = \sigma_x$ is constant,

$$\int_{\text{vol}} \boldsymbol{\sigma} \lfloor \mathbf{N} \rfloor \, d(\text{vol}) = \frac{AL\sigma_x}{2} \lfloor 1 \quad 1 \rfloor$$

and, again by assembly for the whole structure, with $\sigma_x = \frac{3}{2}(qL/A)$ in member 1-2 and $\sigma_x = \frac{1}{2}(qL/A)$ in member 2-3

$$q\frac{L^2}{4} \lfloor 3 \quad 4 \quad 1 \rfloor$$

In accordance with the conjugate stress approach we therefore have

$$\lfloor \sigma_1 \quad \sigma_2 \quad \sigma_3 \rfloor = q\frac{L^2}{4} \lfloor 3 \quad 4 \quad 1 \rfloor \frac{6}{AL} \begin{bmatrix} 2 & 1 & 0 \\ 1 & 4 & 1 \\ 0 & 1 & 2 \end{bmatrix}^{-1}$$

$$= q\frac{L}{A} \lfloor 2 \quad 1 \quad \tfrac{1}{4} \rfloor$$

This distribution is identical to the exact stress distribution in the interval 1-2 and departs from the exact solution in the interval 2-3 because σ_3 is calculated above as $qL/4A$ rather than zero.

It can be shown[9.12] that the stress calculated in the manner above minimizes the mean-square error between the nonconforming stress $\boldsymbol{\sigma}$ and the conjugate stress field $\boldsymbol{\sigma}^c$. In other words, when $\{\boldsymbol{\sigma}\}$ is calculated as above, the following integral assumes a minimum value:

$$\int_{\text{vol}} \{[\mathbf{N}]\{\boldsymbol{\sigma}\} - \boldsymbol{\sigma}\}^2 \, d(\text{vol})$$

Several generalizations of the conjugate stress idea suggest themselves. Two are mentioned here:

1. Conforming but non-displacement-consistent stress representations could be used. This would of course result in a loss of convenience of having $\lfloor \mathbf{N} \rfloor$ on hand from previous displacement calculations. Indeed, one might refer to the procedure described above as the "iso-conjugate stress representation."
2. A different stress than $\boldsymbol{\sigma}$ might be used in forming the row vector. If some stress can be defined that comes closer toward satisfying the local equilibrium conditions, its use could conceivably produce better conjugate stresses.

The requirements of design analysis are often such that the approach above, involving as it does the formation and inversion of large-scale matrices, is economically unjustified and reliance is placed upon direct interpretation of the values of stress obtained with use of the element stress matrices. The format $\boldsymbol{\sigma} = [\mathbf{S}]\{\boldsymbol{\Delta}\}$, where $[\mathbf{S}]$ is in general a function of the spatial coordinates, defines the stress field in terms of these same coordinates. For the purposes of numerical calculation, however, it is necessary to establish the format $\{\boldsymbol{\sigma}\} = [\mathbb{S}]\{\boldsymbol{\Delta}\}$ where $\{\boldsymbol{\sigma}\}$ again defines stresses at specified points but now with no attention given to these values in the conjugate stress context. The definition of these points for the purpose of interpretation relative to design is a key challenge to the analyst.

The problem is particularly acute when constant stress triangular elements are employed. The best location for definition of the stress state would appear to be the centroid of each element but the results obtained are not easily interpretable without a large number of elements. One can also form the average of the stresses at the nodes, where a number of elements join. In either case the discreteness of the resulting data suggests the advisability of contour plots of the stress components.

Some approaches[9.13] to the interpretation of calculated stress fields for problems with triangular elements arrayed within a rectangular gridwork are portrayed in Fig. 9.9. The scheme shown in Fig. 9.9(a) circumvents entirely the use of element data and seeks a finite difference approximation to the strains by use of nodal displacements. Thus, at point 3

$$\epsilon_x = \frac{u_4 - u_2}{2a}, \qquad \epsilon_y = \frac{v_7 - v_6}{2b}, \qquad \text{etc.,}$$

and with the assistance of the constitutive equations the stresses are readily computed.

In another simple alternative scheme one can imagine the finite element model being separated along a grid line, Fig. 9.9(b). The interactive forces (F_{x_i} and F_{y_i}) acting at the nodes along this line are computed by multiplying the appropriate element joint displacements by the respective element stiffness matrices and summing the so-calculated forces at each node. These forces are then distributed as a stepped stress diagram, Fig. 9.9(c), dotted line, which can then be interpreted in polygonal form, solid lines. Proximity is used as the basis for distribution of the shear stresses. Thus, at point 2 for example

$$\sigma_{y_2} = \frac{F_{y_2}}{at}, \qquad \tau_{xy_2} = \frac{F_{x_2}}{at}$$

A refinement of this technique is as follows. For each point, one can write the equation of statics that relates F_y to the adjacent σ's; see Fig. 9.9(d). For example,

$$F_{y_3} = \tfrac{1}{3}(4\sigma_{yx_3} + \sigma_{yx_2} + \sigma_{yx_1})at$$

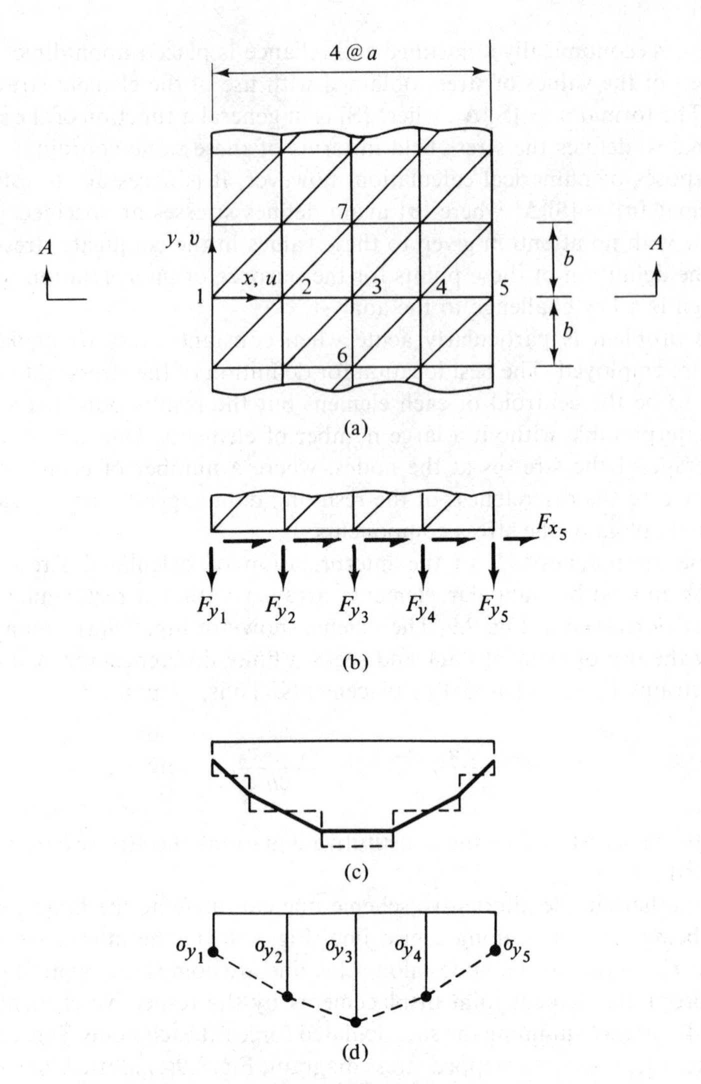

Fig. 9.9 Approaches to stress determination for rectangular grids.

There are as many such equations as there are unknown stresses. The solution of these equations then uniquely defines the stress distribution. This can be regarded as an elementary form of the conjugate stress approach.

9.2.4 Numerical Comparisons—Triangles

Two problems, which have long served as comparison bases for alternative plane stress element formulations, illustrate significantly different performance characteristics for the triangular elements as a group. The existence of these problems as comparison bases stems from the fact that they are among the relatively few plane stress problems that have been examined with thoroughness via more traditional solution approaches.

The first problem, Fig. 9.10, is a constant thickness rectangular plate subjected to parabolically distributed edge loadings. Details of the solution of this problem by assumed polynomial representation of stress and the least work principle are given in Ref. 9.14. The inset on Figure 9.10 shows a representative gridwork of triangles of the constant and linear strain types (CST and LST elements). (Due to symmetry about two axes, only a quarter of the plate need be studied.) Other gridworks, with differing numbers of d.o.f., derive from the arrangement shown.

The results shown in Fig. 9.10, which refer to the horizontal displacement of point A, indicate highly accurate solutions for relatively modest numbers of d.o.f. Similar trends and levels of accuracy are evidenced by the solution data for stress, although, as noted in the previous section, there are certain difficulties in the interpretation of the solution data for stress. The linear strain triangle solutions are a distinct improvement over the constant strain triangle solutions, although the latter are acceptable for even modest numbers of d.o.f.

Figure 9.11 illustrates the second problem, a cantilever beam of unit thick-

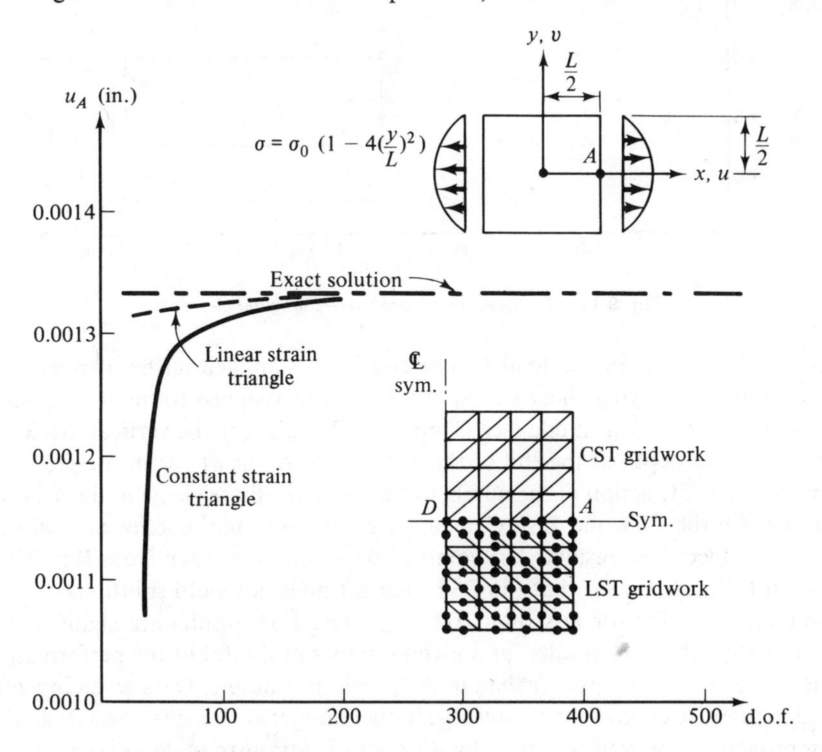

Fig. 9.10 Plates subjected to parabolically distributed edge loads. Comparison of results from triangular elements.

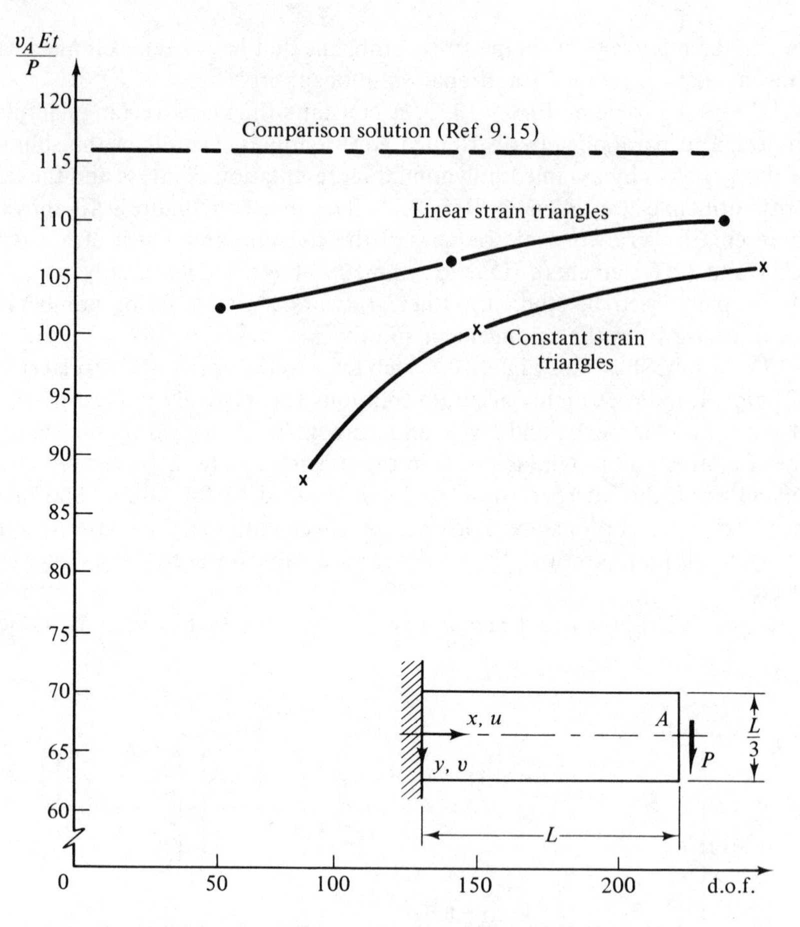

Fig. 9.11 Cantilever beam analysis—triangular elements.

ness subjected to an end load P. The end load is applied in the manner of a parabolically varying shear stress, and loads are assigned to the node points in a work-equivalent mode (see Chapter 6). Results for the vertical displacement at mid-depth at the end of the cantilever are shown, as in the previous problem, as a function of the number of d.o.f. in the finite element idealization and again the CST and LST elements are employed in the series of analyses that produce these results. The comparison solution is taken from Ref. 9.15.

In this case we see that the CST element does not yield solutions of reasonable accuracy for less than 200 d.o.f. The LST results are significantly better than the CST results for a given number of d.o.f. but the performance here is substantially poorer than in the previous example. Data given in Refs. 9.3 and 9.16 confirm these results. Other numerical results show that the improvement of results gained by a quadratic strain triangle in comparison with the linear strain triangle is not very great.

The results above indicate that it is preferable to employ a linear strain triangle in general plane stress applications but that triangles of higher order are not clearly more advantageous. Any such conclusion must be tempered by the consideration of cost of formation of element stiffnesses, equation bandwidth, and the details of the available global analysis capability and computer configuration. We note further that although the general plane stress problem has been dealt with adequately, the plane stress formulations described so far are not a feasible means for analyzing flexure. This matter will again be taken up in Section 9.3.

9.2.5 Alternative Variational Principles in Formulation of Triangular Elements

The simplicity and accuracy of minimum potential energy (displacement-based) finite element representations for plane stress situations inhibits the development of element representations based on alternative variational principles. The complementary energy formulation is valuable as a means of establishing upper bounds on certain solution parameters, as discussed in Chapter 7, but the relative awkwardness of the complementary formulation and the fact that practical considerations in the definition of loads and the approximation of the true geometry may result in violations on the conditions for upper bound solutions have limited its development and application.

The complementary energy formulation of plane stress problems involves a functional expressed in terms of second derivatives of the Airy stress function (Φ) is employed as the basic unknown value at the joints. Consequently, continuity of both Φ and its first derivatives across the element boundaries is required. This problem has been studied extensively in the context of plate flexure where the transverse displacement (w) must satisfy the same form of differential equations as Φ. The choice of representations for this type of field is therefore taken up in Chapter 12. A summary of applications in plane stress is given in Ref. 9.17.

The stresses or other force parameters can also be retained as the joint unknowns in a complementary energy formulation for plane stress. Some authors (e.g., Ref. 9.18) have adopted this type of scheme for the purpose of numerically verifying the upper bound properties of the solution. A constant stress state is assumed within the triangle and the element equations are developed in the stiffness matrix format so that the complete structure may be analyzed by the displacement method. This analytical model suffers from difficulties related to kinematic instability. (See Sect. 3.3.)

The complementary energy approach in plane stress also appears to be useful from the view of inelastic analysis. The "as measured" form of material properties gives the strain in terms of stress; i.e., $\boldsymbol{\epsilon} = [\mathbf{E}]^{-1} \boldsymbol{\sigma}$. The inverse is required for potential energy formulations, and this may prove difficult to calculate for time-dependent situations. Limit analysis studies (Ref. 9.19) have exploited the convenience of the complementary energy approach.

The hybrid stress approach to element formulation depends on a modified form of the complementary energy functional. The "bounding" properties no longer apply but at the same time it is possible to ensure that the solution will be within bounds defined by solutions obtained with the conventional energy principles. Furthermore, a convenient representation of stress singularities is feasible through this approach. We shall discuss these points further in Section 9.3.3, where the hybrid stress approach in plane stress is exemplified through the medium of the rectangular element.

Finally, we note that Reissner energy formulations for plane stress possess the same advantages and disadvantages as the complementary energy formulations. Reference 9.1 illustrates the Reissner energy approach in application to triangular elements.

9.3 Rectangular Elements

9.3.1 Assumed Displacement Representations

Even for the simplest form of the rectangular plane stress element, that with joints only at the four corner points, Fig. 9.12, a number of alternative stiffness matrices can be formulated. The number of independent parameters in representation of the fundamental deformational state is equal to the total number of generalized coordinates minus the number of independent rigid body motions. In this case we have eight generalized coordinates (u and v displacements at each of four corner points) and three rigid body modes. Thus, a total of *five* parameters are available for definition of the deformational state. With three of the five allotted to satisfying the condition that constant strains must be present, there is latitude in the selection of two additional parameters. We shall examine two of these possibilities in this section.

In developing in detail the first stiffness matrix for a rectangular element we make the choice of displacement fields (u, and v) that vary linearly along the element edges. The condition of interelement continuity of displacement will be met in the combination of such elements to form the representation

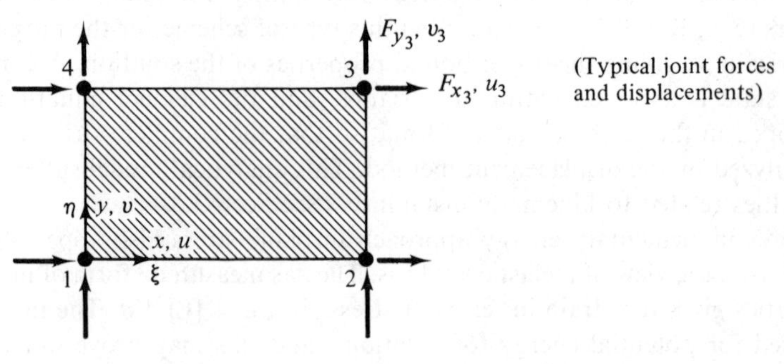

Fig. 9.12 Rectangular plane stress element.

of a complete planar structure or if the element is joined to CST triangular elements. It was shown in Section 8.4 that the chosen displacement fields $u = \lfloor \mathbf{N} \rfloor \{\mathbf{u}\}$ and $v = \lfloor \mathbf{N} \rfloor \{\mathbf{v}\}$ are given by the two-point Lagrangian interpolation function, where, with $\xi = x/x_2$, $\eta = y/y_3$,

$$\lfloor \mathbf{N} \rfloor = \lfloor (1-\xi)(1-\eta) \mid \xi(1-\eta) \mid \xi\eta \mid (1-\xi)\eta \rfloor \tag{9.13a}$$

$$\{\mathbf{u}\} = \lfloor u_1\ u_2\ u_3\ u_4 \rfloor^{\mathrm{T}} \tag{9.13b}$$

$$\{\mathbf{v}\} = \lfloor v_1\ v_2\ v_3\ v_4 \rfloor^{\mathrm{T}} \tag{9.13c}$$

Invoking the strain-displacement equations we obtain the matrix $[\mathbf{D}]$ of Eq. (9.6), Section 9.2.1 for which the coefficients are

$$\begin{aligned} \left\lfloor \frac{\partial \mathbf{N}}{\partial \mathbf{x}} \right\rfloor &= \frac{1}{x_2} \lfloor -(1-\eta) \mid (1-\eta) \mid \eta \mid -\eta \rfloor \\ \left\lfloor \frac{\partial \mathbf{N}}{\partial \mathbf{y}} \right\rfloor &= \frac{1}{y_3} \lfloor -(1-\xi) \mid -\xi \mid \xi \mid (1-\xi) \rfloor \end{aligned} \tag{9.14}$$

where $\{\mathbf{u}\}$ and $\{\mathbf{v}\}$ in the second expression on the right side of Eq. (9.6) are given by Eqs. (9.13b, c). With $[\mathbf{D}]$ in hand, and with the matrix $[\mathbf{E}]$ defined for a particular type of material (isotropic, orthotropic, or other), the stiffness matrix is calculated from Eq. (9.7) as

$$[\mathbf{k}] = \left[\int_A [\mathbf{D}]^{\mathrm{T}} [\mathbf{E}] [\mathbf{D}]\, t\, dA \right]$$

The final form of the stiffness matrix for the present case, for an isotropic material, is given in Fig. 9.13.

It is of interest to examine the basic character of this formulation. The chosen displacement field is everywhere (within the element and across element boundaries) continuous. What about the conditions of equilibrium? By substitution of the expressions for u and v into the governing equilibrium differential equations [Eq. (4.17)], we find the following respective remainders:

For x-direction equilibrium,

$$\frac{E}{2(1-\mu)x_2 y_3}[v_1 - v_2 + v_3 - v_4]$$

For y-direction equilibrium,

$$\frac{E}{2(1-\mu)x_2 y_3}[u_1 - u_2 + u_3 - u_4]$$

We see that if displacement takes place in uniform extension, ($u_1 = u_4$,

$$
\frac{Et}{3\gamma_1\gamma_2 x_2 y_3}
\begin{array}{cccccccc}
u_1 & v_1 & u_2 & v_2 & u_3 & v_3 & u_4 & v_4 \\
4(y_3^2+\gamma_1 x_2^2) & & & & & \text{Symmetric} & & \\
3\gamma_2 & 4(x_2^2+\gamma_1 y_3^2) & & & & & & \\
2(y_3^2-2\gamma_1 x_2^2) & -\gamma_3 & 4(y_3^2+\gamma_1 x_2^2) & & & & & \\
\gamma_3 & -2(2x_2^2-\gamma_1 y_3^2) & -3\gamma_2 & 4(x_2^2+\gamma_1 y_3^2) & & & & \\
-2(y_3^2+\gamma_1 x_2^2) & -3\gamma_2 & -2(2y_3^2-\gamma_1 x_2^2) & -\gamma_3 & 4(y_3^2+\gamma_1 x_2^2) & & & \\
-3\gamma_2 & -2(x_2^2+\gamma_1 y_3^2) & \gamma_3 & 2(x_2^2-2\gamma_1 y_3^2) & 3\gamma_2 & 4(x_2^2+\gamma_1 y_3^2) & & \\
-2(2y_3^2-\gamma_1 x_2^2) & \gamma_3 & -2(y_3^2+\gamma_1 x_2^2) & 3\gamma_2 & 2(y_3^2-2\gamma_1 x_2^2) & -\gamma_3 & 4(y_3^2+\gamma_1 x_3^2) & \\
-\gamma_3 & 2(x_2^2-2\gamma_1 y_3^2) & 3\gamma_2 & 2(x_2^2-\gamma_1 y_3^2) & \gamma_3 & -2(2x_2^2+\gamma_1 y_3^2) & -3\gamma_2 & 4(x_2^2+\gamma_1 y_3^2)
\end{array}
$$

$$\gamma_1 = \frac{1-\mu}{2}$$

$$\gamma_2 = \frac{1+\mu}{2}$$

$$\gamma_3 = \frac{3}{2}(1-3\mu)$$

Fig. 9.13 Stiffness matrix for rectangle-linear edge displacements.

$u_2 = u_3$, $v_1 = v_2$, $v_3 = v_4$), the expressions above are zero and equilibrium prevails. The departure from equilibrium is therefore proportional to the extent of shearing behavior. The shearing stresses vary linearly over the area of the element. The direct stresses are constant in their own directions but vary linearly at right angles to such directions.

The element of Fig. 9.12 is the basic component of the family of higher-order *Lagrangian* plane stress rectangles, so-named because the displacement fields are constructed by use of the Lagrangian interpolation formula. The biquadratic member of this family is illustrated in Fig. 8.7(b). Quadratic interpolation is employed in construction of the x and y multipliers, respectively, of the shape functions. Procedures for the removal of interior and edge d.o.f and for the transformation of the basic rectangle into isoparametric form are given in Sections 8.7, 8.8 and are not discussed here.

The second formulation that we give in detail is one of the earliest developments of a rectangular element stiffness matrix.[9.20] Fundamentally, this is based in the following assumed stress field:

$$\begin{aligned} \sigma_x &= \beta_1 + \beta_2 y \\ \sigma_y &= \beta_3 + \beta_4 x \\ \tau_{xy} &= \beta_5 \end{aligned} \tag{9.15}$$

This stress field is illustrated in Fig. 9.14(a). The plate is of thickness t. Substitution of Eq. (9.15) into the differential equations of equilibrium [Eq. (4.2)] confirms that it is an equilibrium stress field.

To form the stiffness matrix we seek displacement functions u and v that correspond to Eq. (9.15). These can be obtained by transforming the stresses into strains by use of the constitutive equations in the form $\boldsymbol{\epsilon} = [\mathbf{E}]^{-1}\boldsymbol{\sigma}$ and then by integration of the strain-displacement equations. We find, in this manner, that

$$\begin{aligned} u &= (1-\xi)(1-\eta)u_1 + \xi(1-\eta)u_2 + \xi\eta u_3 + (1-\xi)\eta u_4 \\ &\quad + \frac{1}{2}\left[\mu\frac{x_2}{y_3}(\xi - \xi^2) + \frac{y_3}{x_2}(\eta - \eta^2)\right](v_1 - v_2 + v_3 - v_4) \\ v &= (1-\xi)(1-\eta)v_1 + \xi(1-\eta)v_2 + \xi\eta v_3 + (1-\xi)\eta v_4 \\ &\quad + \frac{1}{2}\left[\frac{x_2}{y_3}(\xi - \xi^2) + \mu\frac{y_3}{x_2}(\eta - \eta^2)\right](u_1 - u_2 + u_3 - u_4) \end{aligned} \tag{9.16}$$

Figure 9.14(b) gives a description of the displaced state of the element for a unit value of u_3 with all other d.o.f. suppressed. These same characteristics are evidenced for unit values of the other d.o.f. Note that a linear u displacement occurs along the edge in the direction of the imposed displacement, while the v displacements vary quadratically along the perpendicular edges. Since only the two end point displacements are available on each edge

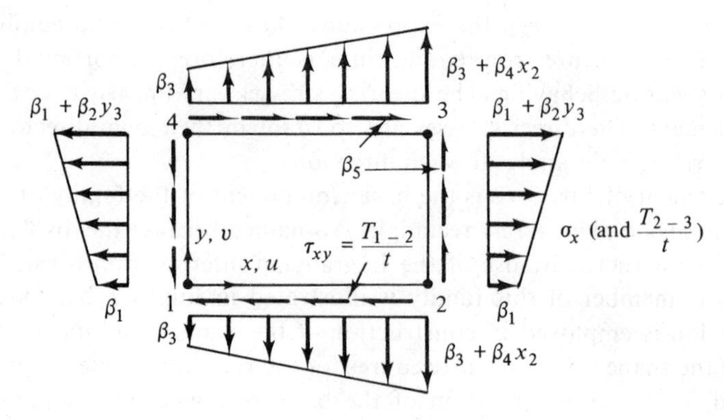

(a) Stress and edge traction patterns

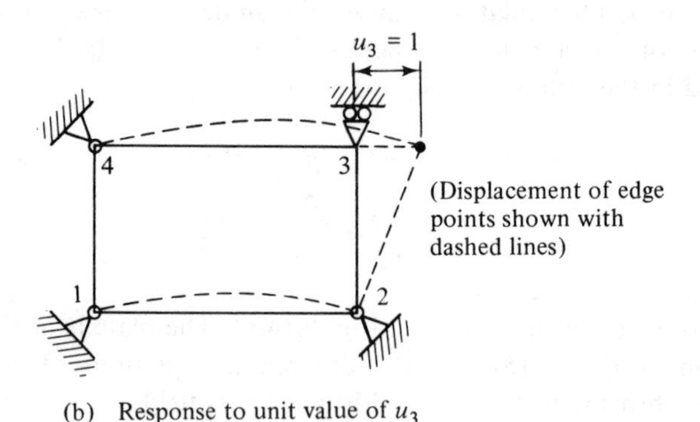

(b) Response to unit value of u_3

Fig. 9.14 Modes of behavior of rectangular element based on five-term stress field.

to define a particular displacement component, we find that interelement displacement continuity requirements are violated by the displacement field of this element.

Using the displacement functions of Eq. (9.16), the strain-displacement relationships [Eq. (4.7)] and the formula for the element stiffness relationships [Eq. (9.7)], we obtain the element stiffness matrix presented in Fig. 9.15.

9.3.2 *Incompatible Modes*[9.21]

It was seen in Section 9.2.4, Fig. 9.11, that triangular element stiffness matrices based on linear displacement fields were of poor accuracy in the analysis of beam flexure. We would expect, therefore, that rectangular element stiffness matrices are equally unsatisfactory in dealing with the same problem. This assumption will be borne out by numerical results to be presented in Section 9.3.4. The source of the inaccuracy can be discerned by study of a simple rectangle subjected to a pure bending condition [Fig. 9.16(a)].

$$[\mathrm{k}] = \frac{Et}{16\,\gamma_1\gamma_2 x_2 y_3}
\begin{bmatrix}
x_2^2\gamma_1 + y_3^2\gamma_4 & & & & & & & \\
\gamma_2 x_2 y_3 & y_3^2\gamma_1 + x_2^2\gamma_4 & & & & \text{Symmetric} & & \\
x_2^2\gamma_1 - y_3^2\gamma_4 & \gamma_3 x_2 y_3 & x_2^2\gamma_1 + y_3^2\gamma_4 & & & & & \\
-\gamma_3 x_2 y_3 & x_2^2\gamma_5 - y_3^2\gamma_1 & -\gamma_2 x_2 y_3 & y_3^2\gamma_1 + x_2^2\gamma_4 & & & & \\
-x_2^2\gamma_1 - y_3^2\gamma_5 & -\gamma_2 x_2 y_3 & y_3^2\gamma_5 - x_2^2\gamma_1 & \gamma_3 x_2 y_3 & x_2^2\gamma_1 + y_3^2\gamma_4 & & & \\
-\gamma_2 x_2 y_3 & -y_3^2\gamma_1 - x_2^2\gamma_5 & -\gamma_3 x_2 y_3 & y_3^2\gamma_1 - x_2^2\gamma_4 & \gamma_2 x_2 y_3 & y_3^2\gamma_1 + x_2^2\gamma_4 & & \\
y_3^2\gamma_5 - x_2^2\gamma_1 & -\gamma_3 x_2 y_3 & -x_2^2\gamma_1 - y_3^2\gamma_5 & \gamma_2 x_2 y_3 & x_2^2\gamma_1 - y_3^2\gamma_4 & \gamma_3 x_2 y_3 & x_2^2\gamma_1 + y_3^2\gamma_4 & \\
\gamma_3 x_2 y_3 & y_3^2\gamma_1 - x_2^2\gamma_4 & \gamma_2 x_2 y_3 & -y_3^2\gamma_1 - x_2^2\gamma_5 & -\gamma_3 x_2 y_3 & x_2^2\gamma_5 - y_3^2\gamma_1 & -\gamma_2 x_2 y_3 & y_3^2\gamma_1 + x_2^2\gamma_4
\end{bmatrix}$$

Columns: u_1, v_1, u_2, v_2, u_3, v_3, u_4, v_4

$$\gamma_1 = \frac{1-\mu}{2}, \qquad \gamma_4 = \frac{4-\mu^2}{3}$$

$$\gamma_2 = \frac{1+\mu}{2}, \qquad \gamma_5 = \frac{2+\mu^2}{3}$$

$$\gamma_3 = \frac{1-3\mu}{2},$$

Fig. 9.15 Stiffness matrix of rectangle—based on assumed stresses of Eq. (9.15).

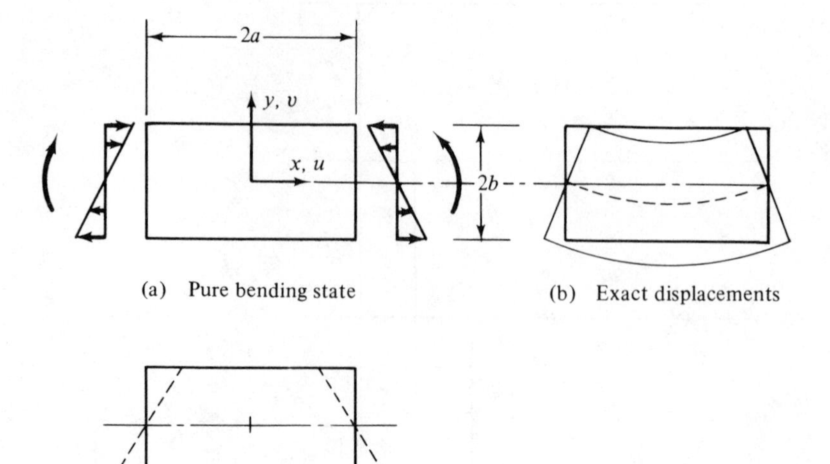

(a) Pure bending state

(b) Exact displacements

(c) Element displacements–linear fields

Fig. 9.16 Modes of representation of pure bending.

In this case the exact displacements [Fig. 9.16(b)] are described by

$$u = C_1 xy \tag{9.17a}$$

$$v = \tfrac{1}{2}C_1(a^2 - x^2) \tag{9.17b}$$

where C_1 and a are constants. By substitution of these equations into the shear strain-displacement equation, $\gamma_{xy} = \partial u/\partial y + \partial v/\partial x$, it can be verified that the condition of zero shear strain, which exists in pure bending, is satisfied. Since the displacement field of the linear-edge-displacement function [Eq. (9.13)] contains only the expression of Eq. (9.17a) and forms the description of displacement shown in Fig. 9.16(c), it is necessary to add a mode in the form of Eq. (9.17b) to the original linear displacement field. For generality it is desirable to preserve symmetry in x and y so we add such modes to both the u and v fields. Thus, we have

$$\begin{aligned} u &= \lfloor \lfloor \mathbf{N}_L \rfloor \lfloor \mathbf{N}_B \rfloor \rfloor \begin{Bmatrix} \mathbf{u}_L \\ \hdashline \mathbf{u}_B \end{Bmatrix} \\ v &= \lfloor \lfloor \mathbf{N}_L \rfloor \lfloor \mathbf{N}_B \rfloor \rfloor \begin{Bmatrix} \mathbf{v}_L \\ \hdashline \mathbf{v}_B \end{Bmatrix} \end{aligned} \tag{9.18}$$

where $\lfloor \mathbf{N}_L \rfloor$, $\{\mathbf{u}_L\}$, and $\{\mathbf{v}_L\}$ are the shape functions and joint displacement values for the linear field [Eq. (9.13)] and $\lfloor \mathbf{N}_B \rfloor$ and $\{\mathbf{u}_B\}$ and $\{\mathbf{v}_B\}$ are the shape functions and joint displacements of the incompatible "bubble" mode, added to describe flexural behavior in the manner of Eq. (9.17b) and Fig. 9.16(b).

The stiffness matrix for the element is first formed in terms of $\{\mathbf{u}_L\}$, $\{\mathbf{u}_B\}$,

$\{\mathbf{v}_L\}$, and $\{\mathbf{v}_B\}$. The stiffness matrix for this element, referenced only to $\{\mathbf{u}_L\}$ and $\{\mathbf{v}_L\}$ is obtained by elimination of the supplemental d.o.f. $\{\mathbf{u}_B\}$ and $\{\mathbf{v}_B\}$ by application of the condensation procedure of Section 2.8. Upon performance of these operations it is found that the resulting stiffness matrix is identical to that which is presented in Fig. 9.15. We can discern the reason for this by examination of Fig. 9.14(b), which shows the edge displacements associated with a unit displacement of the type treated in the present development. These same displacements are defined by Eq. (9.16), and it is clear that they are of the form of the bubble modes introduced in the present development.

9.3.3 Hybrid Stress Formulation[9.22]

The rectangle in plane stress affords an opportunity for exemplification of the hybrid stress approach that is more substantive than the example given in Section 6.7.

In the case of plane stress, a stress field which satisfies the differential equations of equilibrium is that which was defined by Eq. (9.15) and illustrated in Fig. 9.14(b). Adopting the symbolism of Section 6.7, we note that three basic matrices are required in the formation of a stiffness matrix, the [Z], [L], and $[\bar{\mathfrak{Y}}]$ matrices. The [Z] matrix [see Eq. (6.77)] merely lists the coefficients of the stress equations, so we have

$$[\mathbf{Z}] = \begin{array}{c} \begin{matrix} \beta_1 & \beta_2 & \beta_3 & \beta_4 & \beta_5 \end{matrix} \\ \begin{bmatrix} 1 & y & 0 & 0 & 0 \\ 0 & 0 & 1 & x & 0 \\ 0 & 0 & 0 & 0 & 1 \end{bmatrix} \end{array} \tag{9.19}$$

In this case

$$\{\boldsymbol{\beta}_f\} = \lfloor \beta_1\ \beta_2\ \beta_3\ \beta_4\ \beta_5 \rfloor^{\mathrm{T}}$$

The [L] matrix defines the distribution of the element edge tractions consistent with the element stress field. The illustration of the stress state, Fig. 9.14(a), also describes the state of edge tractions $T_{x_{1\text{-}4}}$, $T_{y_{2\text{-}3}}$, etc. For example,

$$T_{x_{2\text{-}3}} = t(\beta_1 + \beta_2 y)$$

In this manner one can construct the equations

$$\mathbf{T} = [\mathbf{L}]\{\boldsymbol{\beta}_f\} \tag{6.61}$$

where

$$\mathbf{T} = \lfloor T_{x_{1\text{-}2}}\ T_{y_{1\text{-}2}}\ T_{x_{2\text{-}3}}\ T_{y_{2\text{-}3}}\ T_{x_{3\text{-}4}}\ T_{y_{3\text{-}4}}\ T_{x_{4\text{-}1}}\ T_{y_{4\text{-}1}} \rfloor^{\mathrm{T}} \tag{9.20}$$

and the coefficients of [L] derive from Eq. (9.15).

For boundary displacements, linear representations are chosen. Thus, along edge 2-3, the displacement is described by

$$u_{2-3} = \left(1 - \frac{y}{y_3}\right)u_2 + \frac{y}{y_3}u_3$$

and when this type of expression is written for all edges, the result can be collected in the form

$$\bar{\mathbf{u}} = [\bar{\mathfrak{Y}}]\begin{Bmatrix}\{\mathbf{u}\}\\ \{\mathbf{v}\}\end{Bmatrix} \tag{6.17}$$

where

$$\bar{\mathbf{u}} = \lfloor u_{1-2}\, v_{1-2} \ldots v_{4-1}\rfloor^{\mathrm{T}} \tag{9.21}$$

and $\{\mathbf{u}\}$ and $\{\mathbf{v}\}$ are defined by Eq. (9.13b) and (9.13c). We now form, in accordance with the description of the hybrid method given in Section 6.7, the element stiffness matrix $[\mathbf{k}] = [\mathcal{G}]^{\mathrm{T}}[\mathcal{H}]^{-1}[\mathcal{G}]$ where

$$[\mathcal{G}] = \left[\int_{S_n} [\mathbf{L}]^{\mathrm{T}}[\bar{\mathfrak{Y}}]\, dS\right] \tag{6.62}$$

(S_n is the complete boundary of the element) and

$$[\mathcal{H}] = \left[\int_{\mathrm{vol}} [\mathbf{Z}]^{\mathrm{T}}[\mathbf{E}]^{-1}[\mathbf{Z}]\, d(\mathrm{vol})\right] \tag{6.78}$$

It turns out that the stiffness matrix obtained in the manner above is identical to the stiffness matrix given in Fig. 9.15, which was derived by use of the displacement field of Eq. (9.16). The displacement field of Eq. (9.16) of course corresponds to the stress field employed in the above. One distinction however, is that the displacement field of the pure stiffness formulation evidences *nonlinear* edge displacements, whereas the formulation above deals only with *linear* edge displacements. The explanation is that the nonlinear components of displacement of the pure stiffness formulation, see Fig. 9.14(a), are in a direction perpendicular to the forces that produce such displacements and therefore do no work.

The fact that in the example presented the stiffness matrix could have been obtained by a conventional stiffness formulative procedure rather than through hybrid analysis does not mean that this will always be the case in hybrid stress formulations for plane stress. An almost limitless variety of hybrid plane stress formulations can be established with use of differing stress and displacement fields. The hybrid stress approach is useful on at least two counts. Certain displacement parameters solved for in this manner will lie between the upper bound given by an "equilibrium" formulation and the lower bound of a "compatible" formulation when the interior stress field corresponds to the former and the boundary displacements to the latter.[9.23] Also, it is feasible to introduce expressions that represent singularities in the solution for stresses, as occur at the tips of cracks in the structure.[9.24]

9.3.4 Numerical Comparisons

Figure 9.17 shows the nondimensionalized tip deflection $v_A Et/P$ plotted as a function of the number of d.o.f. employed in analyses using rectangular elements for the problem of the end-loaded cantilever described in Section 9.2.4. Results are shown for both the rectangular element formulated on the basis of displacements that vary linearly along the element boundary (the stiffness matrix of Fig. 9.13) and the element based on the displacement field of Eq. (9.16) (the stiffness matrix of Fig. 9.15). The latter, as we have

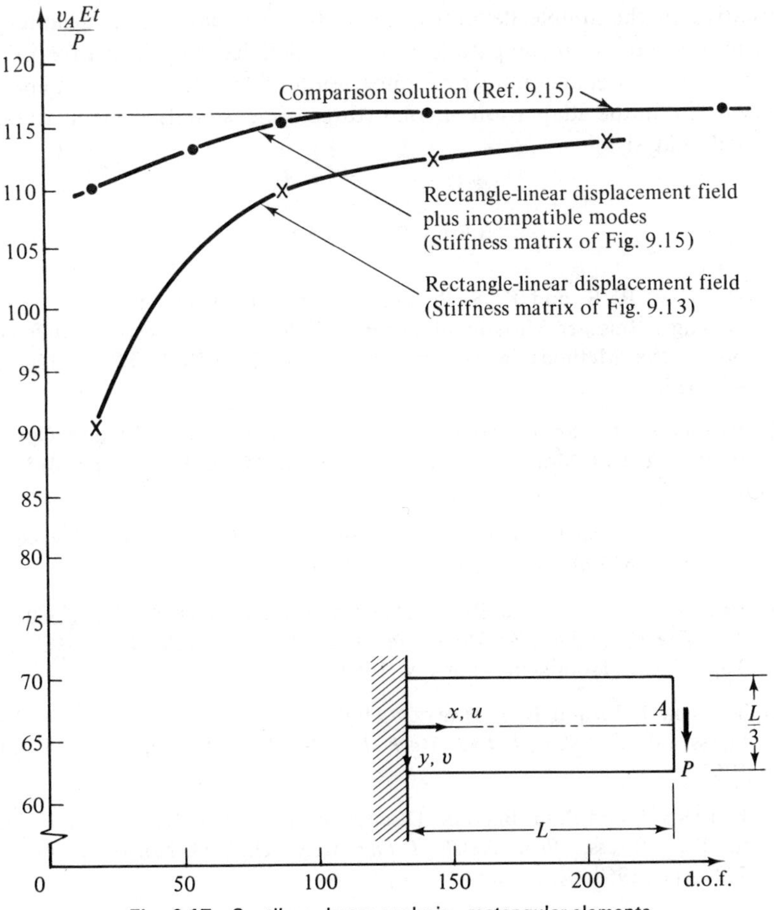

Fig. 9.17 Cantilever beam analysis—rectangular elements.

seen, is alternatively interpreted as an assumed stress hybrid formulation or as a formulation based on linear edge displacements supplemented by bubble modes.

The results demonstrate that the linear edge displacement (interelement compatible) formulation converges rather slowly to the comparison solution. This same characteristic was evidenced by the triangular elements (Fig. 9.11) In contrast, the formulation with incompatible modes proves highly accurate for this problem. Results for the lower end of the d.o.f. scale (< 60 d.o.f.) were obtained with grid refinement only in the spanwise (x) direction, i.e., with only one element across the depth of the beam. Thus, one is able to use the general plane stress formulation in representation of the special case of flexure, which usually requires the imposition of the assumption that cross sections that are plane before deformation will remain so in the deformed state. In beam bending there is not often a need to formulate an alternative to the simple flexural element but we shall see in Chapter 10 that the concept of incompatible modes, which has an alternative in the notion of using a coarse numerical integration of the element strain energy, is very useful in the adaptation of solid (three-dimensional) elements to plate and shell analysis.

REFERENCES

9.1 DUNHAM, R. S. and K. S. PISTER, "A Finite Element Application of the Hellinger-Reissner Variational Theorem," Proc. of the Second Conference on Matrix Methods in Structural Mechanics, AFFDL TR 68–150, pp. 471–487.

9.2 PEDERSON, P., "Some Properties of Linear Strain Triangles and Optimal Finite Element Models," *Internat'l. J. Num. Meth. Eng.*, **7**, pp. 415–430, 1973.

9.3 BREBBIA, C. and J. J. CONNOR, *Fundamentals of Finite Element Techniques*, Butterworths Publishers, London, 1973.

9.4 HOLAND, I., "The Finite Element Method in Plane Stress Analysis," Chapter 2 of *The Finite Element Method in Stress Analysis*, Holand and Bell (ed.), Tapir Press, Trondheim, Norway, 1969.

9.5 TOCHER, J. L. and B. J. HARTZ, "Higher-Order Finite Element for Plane Stress," Proc. ASCE, *J. Eng. Mech. Div.*, **93**, No. EM4, pp. 149–174, Aug., 1967.

9.6 HOLAND, I. and P. G. BERGAN, Discussion of "Higher-Order Finite Element for Plane Stress," Proc. ASCE, *J. Eng. Mech. Div.*, **94**, No. EM2, pp. 698–702, Apr., 1968.

9.7 WALZ, J. E., R. E. FULTON, N. J. CYRUS, and R. T. EPPINK, "Accuracy of Finite Element Approximations," NASA TN D-5728, Mar., 1970.

9.8 TAIG, I. C. and R. I. KERR, "Some Problems in Discrete Element Representation of Aircraft Structures," in *Matrix Methods of Structural Analysis*, B. Fraeijs de Veubeke (ed.), The MacMillan Co., New York, N.Y., 1964, pp. 282–284.

9.9 TURCKE, D. J. and G. M. MCNEICE, "Guidelines for Selecting Finite Element Grids Based on an Optimization Theory," *Internat'l. J. Comp. Struct.*, **4**, 1974.

9.10 MCNEICE, G. M. and S. F. HUNNISETT, "Mixed-Displacement Finite-Element Analysis with Particular Application Using Plane Stress Triangles," *J. Strain Analysis*, **7**, No. 4, pp. 243–252, 1972.

9.11 TONG, P., "Exact Solutions of Certain Problems by Finite Element Method," AIAA J., **7**, No. 1, pp. 178–180, 1969.

9.12 ODEN, J. T. and H. J. BRAUCHLI, "On the Calculation of Consistent Stress Distributions in Finite Element Approximations," *Internat'l. J. Num. Meth. Eng*. **3**, pp. 317–322, 1971.

9.13 HRENNIKOFF, A., "Precision of Finite Element Method in Plane Stress," *Pub. Internat'l. Assn. Bridge Struct. Eng.*, **29-II**, pp. 125–137, 1969.

9.14 TIMOSHENKO, S. and J. N. GOODIER, *Theory of Elasticity*, 2nd ed., McGraw-Hill Book, Co., New York, N.Y., pp. 167–171, 1951.

9.15 HOOLEY, R. F. and P. D. HIBBERT, "Bounding Plane Stress Solutions by Finite Elements," Proc. ASCE, *J. Struct. Div.*, **92**, No. ST. 1, pp. 39–48, Feb., 1966.

9.16 COWPER, G. R., "Variational Procedures and Convergence of Finite Element Methods," in *Numerical and Computer Methods in Structural Mechanics*, S. J. Fenves, et al (eds.), Academic Press, New York, N.Y., 1973. pp. 1–12.

9.17 GALLAGHER, R. H. and A. K. DHALLA, "Direct Flexibility-Finite Element Elastoplastic Analysis," Proc. of First Internat'l. Conf. on Structural Mechanics in Reactor Technology, Berlin, **6**, Part M, Sept., 1971.

9.18 FRAEIJS DE VEUBEKE, B., "Upper and Lower Bounds in Matrix Structural Analysis," in *Matrix Methods of Structural Analysis*, B. Fraeijs de Veubeke (ed.), The MacMillan Co., New York, N.Y., 1964, pp. 166–201.

9.19 BELYTSCHKO, T. and P. G. HODGE, "Plane Stress Limit Analysis by Finite Elements," Proc. ASCE, *J. Eng. Mech. Div.*, **96**, No. EM6, pp. 931–944, Dec., 1970.

9.20 TURNER, M. J., R. W. CLOUGH, H. C. MARTIN, and L. J. TOPP, "Stiffness and Deflection Analysis of Complex Structures," *J. Aero. Sci.*, **23**, No. 23, 9, pp. 805–824, Sept., 1956.

9.21 WILSON, E. L., et al., "Incompatible Displacement Models," in *Numerical and Computer Methods in Structural Mechanics,* S. J. Fenves, et al. (eds.), Academic Press, New York, N.Y., 1973, pp. 43–57.

9.22 PIAN, T. H. H., "Derivation of Element Stiffness Matrices by Assumed Stress Distributions," *AIAA J.*, **2**, pp. 1333–1335, 1964.

9.23 PIAN, T. H. H. and PIN TONG, "Basis of Finite Element Methods for Solid Continua," *Internat'l. J. Num. Meth. Eng.*, **1**, No. 1, pp. 3–28, 1969.

9.24 TONG, P., T. H. H. PIAN, and S. J. LASRY, "A Hybrid-Element Approach to Crack Problems in Plane Elasticity," *Internat'l J. Num. Meth. Eng.*, **7**, No. 3, pp. 297–308, 1973.

PROBLEMS

9.1 Formulate the stiffness matrix of the constant stress triangle (Fig. 5.4) with use of the assumed stress hybrid method.

9.2 Formulate a mixed force-displacement matrix for the constant strain triangle in plane stress using the Reissner variational principle. Transform the result to the stiffness matrix of the element in the manner done for the beam element in Section 6.8.

9.3 Validate the k_{11} (F_{x_1} versus u_1) term of the stiffness matrix of the rectangular element presented in Fig. 9.15.

9.4 Construct the stiffness matrix for an annular element, of the form in Fig. P9.4. All conditions are axisymmetric so that $\epsilon_r = du/dr$, $\epsilon_\theta = u/r$. Choose as the radial displacement function $u = \left[1 - \frac{(r-r_1)}{r_{2-1}}\right]u_1 + \frac{(r-r_1)}{r_{2-1}}u_2$, where $r_{2-1} = r_2 - r_1$.

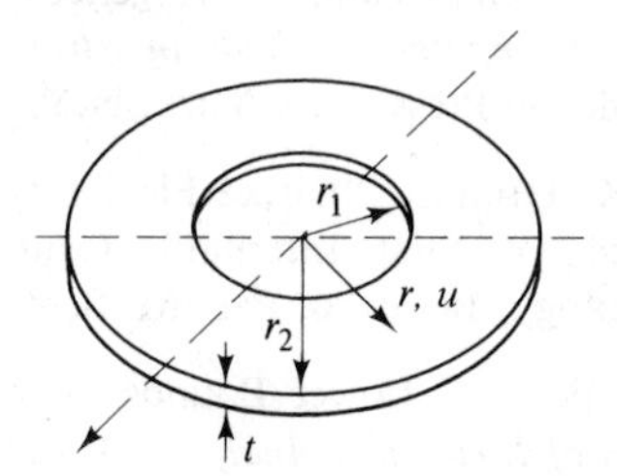

Fig. P9.4

9.5 Discuss the formulation of a potential energy hybrid rectangular element (flexibility matrix), using stress functions to describe the element edge tractions.

9.6 Calculate the work-equivalent loads for the parabolic stress distribution and indicated gridwork of elements in Fig. 9.10. Assume that the elements are constant strain rather than linear strain triangles.

9.7 Formulate the initial force vector for a linear strain triangle for a linearly distributed temperature change above the stress-free state; i.e., $\Upsilon = N_1\Upsilon_1$

$+ N_2\Upsilon_2 + N_3\Upsilon_3$ where N_1, etc., are the shape functions of a linear field and Υ_1, Υ_2, and Υ_3 are the vertex temperatures.

9.8 Devise an appropriate displacement field for formulation of an element stiffness matrix for the sector element shown in Fig. P9.8 and construct the joint displacement-to-strain matrix **[D]**.

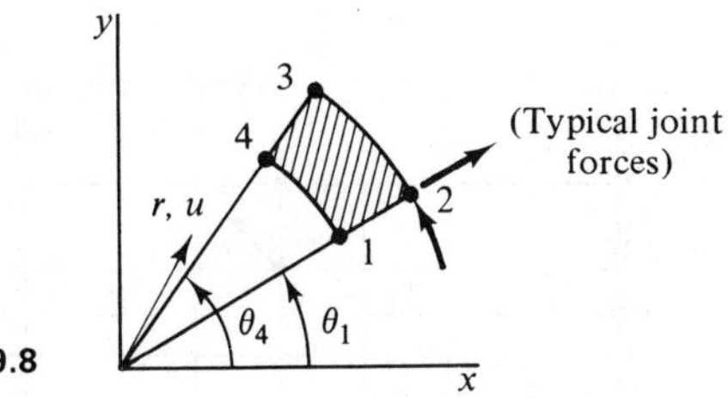

Fig. P9.8

9.9 (Problem for computational assignment.) Perform a finite element analysis of the square plate shown in Fig. P9.9 (same problem as Fig. 9.10) in accordance with individually assigned type of element and gridwork. For comparison solution for both stress and displacement, see "A Shallow Shell Finite Element of Triangular Shape" by G. R. Cowper, G. M. Lindberg, and M. D. Olson, *Internat'l. J. Solids Struct.*, **6**, pp.1133–1156, 1970.

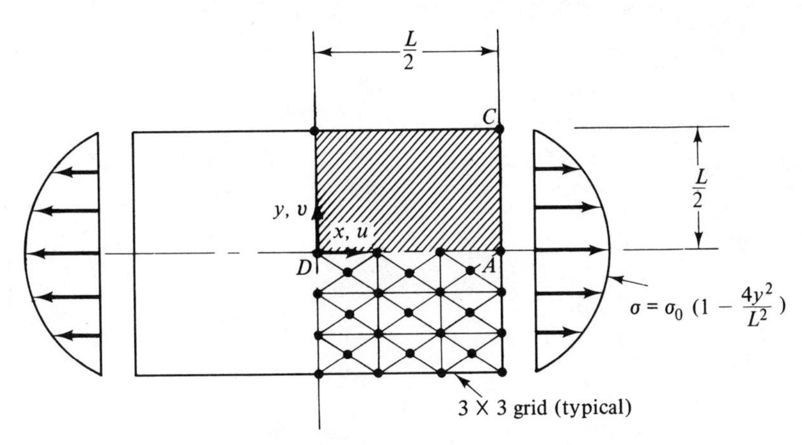

Fig. P9.9

Notes: Define the gridwork only within the crosshatched quadrant. Note that u and v displacements are zero at point D. v displacements along x axis are zero. u displacements along y axis are zero.

$$E = 10^7, \qquad \mu = 0.3, \qquad t = 1$$

9.10 (Problem for computational assignment.) Perform a finite element analysis of the rectangular section-unit thickness cantilever beam shown in Fig. P9.10 (same problem as Fig. 9.11) in accordance with individually assigned type of element and gridwork. Load P is distributed parabolically in the manner of shear stresses on a rectangular cross-section so that

$$\tau_{xy} = \frac{9P}{2L}\left(1 - 36\frac{y^2}{L^2}\right)$$

$$E = 10^7, \qquad \mu = 0.2, \qquad t = 1.0$$

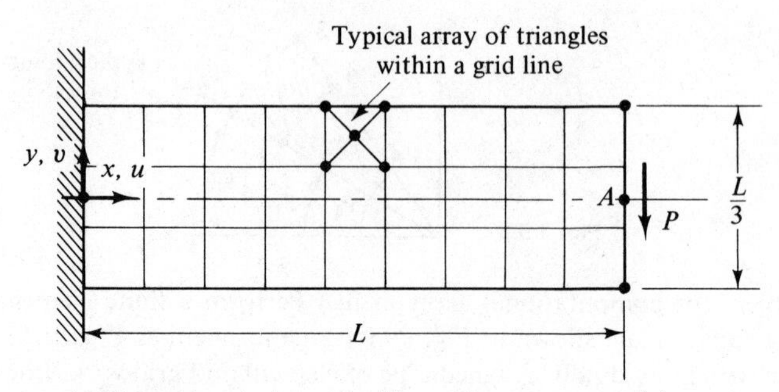

(3 X 9 gridwork shown for illustrative purposes only)

Fig. P9.10

10

Solid Elements: General Case

Solid, or three-dimensional, elements enable the solution of problems in the general three-dimensional theory of elasticity. Such problems earlier drew relatively little attention in design practice due to the intractability of conventional solution approaches. Thus, in this area finite element analysis achieves the role of a virtually unchallenged tool for the solution of all but the simplest of problems. Many problems indeed exist, extending from massive concrete dam structures, rock mechanics for underground borings, and stresses in soils, to the details of stresses in flanges and thick-walled pipe intersections.

The basic solid elements are merely three-dimensional generalizations of the planar elements. The tetrahedron, Fig. 10.1(a), is the extension of the triangle into a third dimension, while the hexahedron, Fig. 10.1(b), is the three-dimensional counterpart of the planar rectangle. Although special and alternative forms of solid elements have been formulated (e.g., pentagonal or wedge-shaped elements), practice has centered about the tetrahedron and hexahedron. This chapter restricts attention to these basic shapes.

Due to the "curse of dimensionality," illustrated in Fig. 10.2, finite element representations for solids require extremely large numbers of d.o.f. It is seen that if the requirements of solution accuracy demand 10 d.o.f. in a one-dimensional analytical model, accomplishment of corresponding accuracy in three dimensions requires 3,000 d.o.f. Economic feasibility is therefore a critical question in any practical application. Maximum efficiency must be gained in (1) data input and output procedures, (2) solution

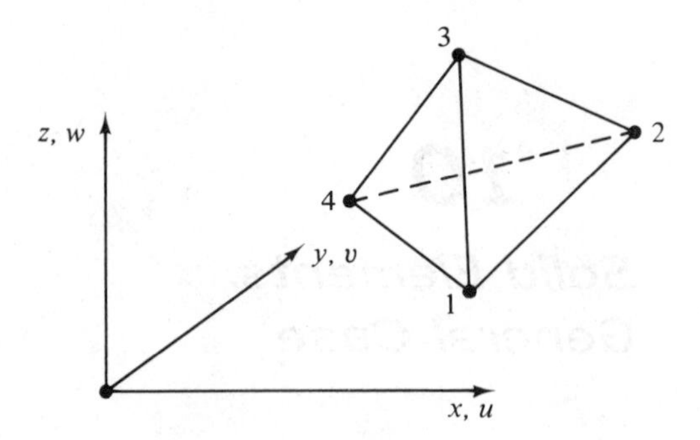

(a) Regular tetrahedronal geometry and coordinates

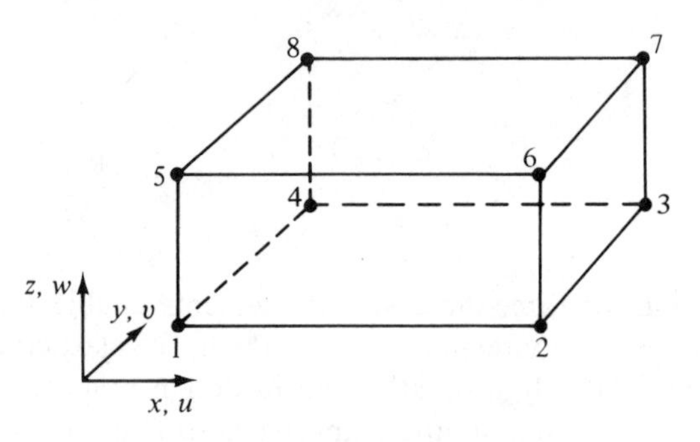

(b) Regular hexahedronal geometry and coordinates

Fig. 10.1 Solid elements.

of large-order systems of equations, and (3) representation of the actual structure by the finite elements.

Items 1 and 2 are beyond the scope of the present work; the reader is advised to consult references cited at the end of this chapter. Item 3 motivates the use of the most sophisticated procedures for geometric representation. Thus, the concepts of *isoparametric* representation of element geometry, i.e., the use of the shape functions for modeling of element boundaries, assumes special importance for solid elements. Such concepts were developed in Section 8.8 and are further discussed here.

Existing solid element formulations are almost exclusively based upon assumed displacement fields and the principle of minimum potential energy. Complementary and mixed formulations have yet to demonstrate an advantage for this class of problem. In a complementary approach expressed in terms of stress functions for three-dimensional elasticity theory, for example, the difficulty of working with functions that have continuity of both the func-

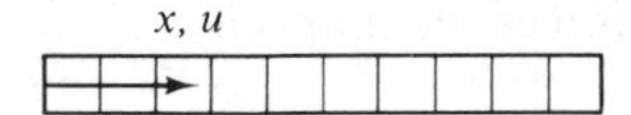

(a) One dimensional (10 d.o.f.)

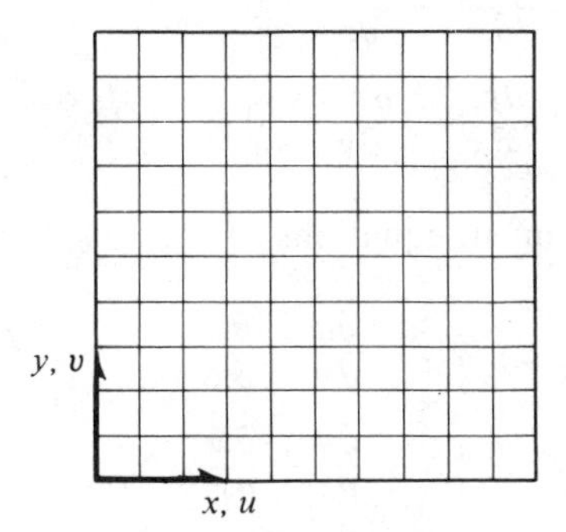

(b) Two dimensional (200 d.o.f.)

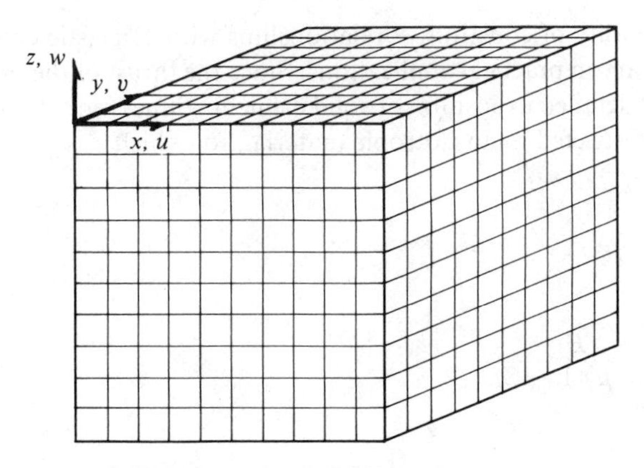

(c) Three dimensional (3,000 d.o.f.)

Fig. 10.2 Expansion of d.o.f. with dimensionality.

tion itself and certain of its first derivatives across the element interfaces must be confronted. The present chapter is therefore limited to assumed displacement formulations.

10.1 Basic Relationships

10.1.1 Equations of Elasticity

The basic relationships for solid elements are those of the linear theory of elasticity. From an examination of the equilibrium of a differential element, we have (excluding body forces for simplicity)

$$\begin{aligned}\frac{\partial \sigma_x}{\partial x} + \frac{\partial \tau_{xy}}{\partial y} + \frac{\partial \tau_{xz}}{\partial z} &= 0 \\ \frac{\partial \sigma_y}{\partial y} + \frac{\partial \tau_{yx}}{\partial x} + \frac{\partial \tau_{yz}}{\partial z} &= 0 \\ \frac{\partial \sigma_z}{\partial z} + \frac{\partial \tau_{zx}}{\partial x} + \frac{\partial \tau_{zy}}{\partial y} &= 0\end{aligned} \tag{10.1}$$

and for the linear strain-displacement equations

$$\begin{aligned}\epsilon_x &= \frac{\partial u}{\partial x}, & \gamma_{xy} &= \frac{\partial u}{\partial y} + \frac{\partial v}{\partial x} \\ \epsilon_y &= \frac{\partial v}{\partial y}, & \gamma_{yz} &= \frac{\partial v}{\partial z} + \frac{\partial w}{\partial y} \\ \epsilon_z &= \frac{\partial w}{\partial z}, & \gamma_{zx} &= \frac{\partial w}{\partial x} + \frac{\partial u}{\partial z}\end{aligned} \tag{10.2}$$

The generality of constitutive relationships with 21 elastic constants may prove necessary in practical applications. Since the thrust of the present chapter is toward detailed exposition of solid element stiffness equations, however, attention is restricted to an isotropic material, for which

$$\begin{Bmatrix} \sigma_x \\ \sigma_y \\ \sigma_z \\ \tau_{xy} \\ \tau_{yz} \\ \tau_{zx} \end{Bmatrix} = \frac{E}{(1+\mu)(1-2\mu)}$$

$$\times \begin{bmatrix} (1-\mu) & \mu & \mu & 0 & 0 & 0 \\ \mu & (1-\mu) & \mu & 0 & 0 & 0 \\ \mu & \mu & (1-\mu) & 0 & 0 & 0 \\ 0 & 0 & 0 & \frac{(1-2\mu)}{2} & 0 & 0 \\ 0 & 0 & 0 & 0 & \frac{(1-2\mu)}{2} & 0 \\ 0 & 0 & 0 & 0 & 0 & \frac{(1-2\mu)}{2} \end{bmatrix} \begin{Bmatrix} \epsilon_x \\ \epsilon_y \\ \epsilon_z \\ \gamma_{xy} \\ \gamma_{yz} \\ \gamma_{zx} \end{Bmatrix}$$

or

$$\boldsymbol{\sigma} = [\mathbf{E}]\boldsymbol{\epsilon} \tag{10.3}$$

10.1.2 Potential Energy

The strain energy component of this potential energy is simply

$$U = \frac{1}{2}\int_{\text{vol}} \varepsilon[\mathbf{E}]\varepsilon \, d(\text{vol}) - \int_{\text{vol}} \varepsilon[\mathbf{E}]\varepsilon^{\text{init.}} \, d(\text{vol}) + \mathbf{C}(\varepsilon^{\text{init.}}) \tag{10.4}$$

with $\boldsymbol{\epsilon}$ as defined by Eq. (10.3) and, for the case of thermal expansion,

$$\boldsymbol{\epsilon}^{\text{init.}} = \lfloor \alpha\Upsilon \; \alpha\Upsilon \; \alpha\Upsilon \; 0 \; 0 \; 0 \rfloor^{\text{T}} \tag{10.5}$$

where α is the coefficient of thermal expansion and Υ is the temperature change from the stress-free state.

As in Chapter 9, it will be convenient to write some of the element displacement fields in terms of the joint displacements $\{\boldsymbol{\Delta}\}$. In such cases the transformation from joint displacements-to-strains is again symbolized by [see Eq. (5.6c)].

$$\boldsymbol{\epsilon} = [\mathbf{D}]\{\boldsymbol{\Delta}\} \tag{10.6}$$

also, the strain energy is of the form (see Section 6.4)

$$U = \frac{\lfloor \boldsymbol{\Delta} \rfloor}{2}[\mathbf{k}][\boldsymbol{\Delta}] - \lfloor \boldsymbol{\Delta} \rfloor\{\mathbf{F}^{\text{init.}}\} + \mathbf{C}(\varepsilon^{\text{init.}}) \tag{10.7}$$

with

$$[\mathbf{k}] = \left[\int_{\text{vol}} [\mathbf{D}]^{\text{T}}[\mathbf{E}][\mathbf{D}] \, d(\text{vol})\right] \tag{10.8}$$

$$\{\mathbf{F}^{\text{init.}}\} = \left\{\int_{\text{vol}} [\mathbf{D}]^{\text{T}}[\mathbf{E}]\{\boldsymbol{\epsilon}^{\text{init.}}\} \, d(\text{vol})\right\} \tag{10.9}$$

In other cases to be discussed in this chapter it is inconvenient to express the displacement fields in terms of joint displacements and it is necessary to employ generalized displacement parameters $\{\mathbf{a}\}$. Then, from Eq. (5.6a) the generalized displacement-to-strain transformation is symbolized by $\boldsymbol{\epsilon} = [\mathbf{C}]\{\mathbf{a}\}$ and from Eq. (5.4a) we have the joint displacement-to-generalized displacement transformation $\{\mathbf{a}\} = [\mathbf{B}]^{-1}\{\boldsymbol{\Delta}\}$. The stiffness matrix is therefore given by

$$[\mathbf{k}] = ([\mathbf{B}]^{-1})^{\text{T}}\left[\int_{\text{vol}} [\mathbf{C}]^{\text{T}}[\mathbf{E}][\mathbf{C}] \, d(\text{vol})\right][\mathbf{B}]^{-1} \tag{10.8a}$$

10.2 Tetrahedronal Element Formulations

10.2.1 General Considerations

It was shown in Section 8.6 that the notions of tetrahedronal coordinates (L_1, L_2, L_3, L_4) and of the "Pascal tetrahedron" lead naturally to the definition of a family of tetrahedronal elements of first and higher orders. Such elements feature the translational displacement components (u, v, w) at each node point. As the treatment of planar triangular elements disclosed, however, it is feasible to employ displacement derivatives (e.g., $\partial u/\partial x$, $\partial u/\partial y$, etc.), as well as the translational components, as node point d.o.f. This alternative does not exist for first- and second-order elements, due to an insufficient total number of d.o.f., but it can be invoked for the tetrahedron based on cubic functions as illustrated in Fig. 10.3.

A complete cubic polynomial encompasses 20 terms so that an element that features only translational displacements, Fig. 10.3(a), will possess 20 node points. A total of 60 d.o.f. will be defined at these points to account for the u-, v-, and w-displacement components. If *completeness* is to be preserved in the underlying polynomial functions for the element that employs displacement derivatives as d.o.f., the element must also possess 60 d.o.f.. The way in which this can be done is illustrated in Fig. 10.3(b). Each of the four vertices features the three translational displacement components and all three derivatives of each of these components—a total of 12 terms per node. With reference to u component, for example, we have the following:

$$u_1, \ldots, u_4; \left.\frac{\partial u}{\partial x}\right|_1, \ldots, \left.\frac{\partial u}{\partial x}\right|_4; \left.\frac{\partial u}{\partial y}\right|_1, \ldots, \left.\frac{\partial u}{\partial y}\right|_4; \left.\frac{\partial u}{\partial z}\right|_1, \ldots, \left.\frac{\partial u}{\partial z}\right|_4 \tag{10.10}$$

and similarly for v and w. This yields 48 d.o.f. The additional 12 d.o.f. can be defined as the translational displacements at the centroids of each of the

Fig. 10.3 Alternative forms of tetrahedronal element based on complete cubic displacement field.

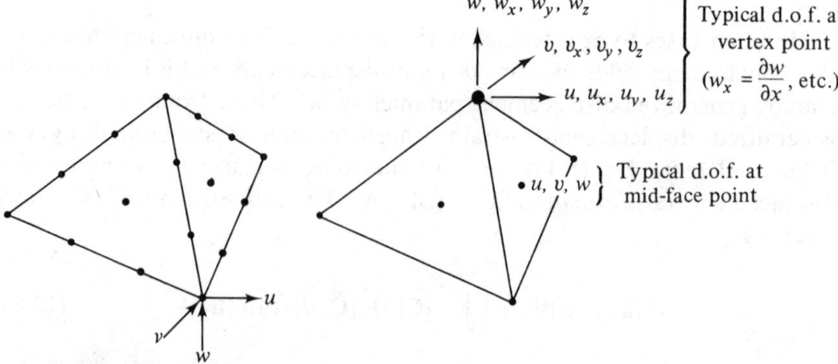

four faces [see Fig. 10.3(b)]. The existence of the node points at the centroids of the faces is awkward, however. Consequently, formulative efforts have been devoted to an element with d.o.f. only at the vertices, i.e., with 48 d.o.f. The following examination of the relative merit of alternative formulations of the tetrahedronal element discloses why such attention is given to the 48-d.o.f. element.

Extensive studies have been conducted of the relative computational efficiency of the different types of tetrahedronal elements.(10.1–10.3) Table 10.1, adapted from Ref. 10.3, lists the average number of d.o.f. per element for a system with an infinite number of elements; this number also is proportional to the average half-bandwidth of the associated global stiffness equations. When efficient methods of solution of the algebraic equations are employed, the cost of solution is proportional to approximately the square of the half-bandwith. Other factors must of course be considered in the definition of relative efficiency. Thus, the above implies a fixed number of elements for a given global geometric form; but, of course, account must be taken of the increased solution accuracy of higher-order elements. The T48 element is the most advantageous of the tetrahedronal elements shown in Table 10.1 on both counts. The average half-bandwidth associated with this element is small because of the presence of displacement-derivatives as d.o.f., a point which was discussed in Section 9.2 for planar elements with displacement-derivatives as d.o.f.

It should, at the same time, be noted that problems arise in conjunction with displacement-derivative d.o.f. for inhomogenous material situations, where single-valued strains are not, in theory, admissible. When this arises due to a modest variation in properties from point to point (e.g., due to temperature-dependent material properties in the presence of thermal gradients), the imposition of strain (displacement-derivative) continuity is not likely to be of significance. When there is marked separation of material constituents, however, the strain continuity condition is not admissible. One may then connect only translational displacement d.o.f.

One must also consider the factors relative to geometric representation, especially curved boundaries. It is possible of course to employ one mode of representation for the displacement fields and another for the description of the curved boundaries of the element. The orders of the functions employed may also differ. Nevertheless, it appears most convenient to employ the same form of approximation for both geometric and displacement fields.

With the factors above in mind, the following explicit developments for tetrahedronal elements are restricted to the linear-displacement representation and the T48 element. The former is fundamental to the entire tetrahedronal element family; the higher-order elements in this class (quadratic and cubic displacement fields) are easily formulated along the same lines. The tetrahedronal coordinates introduced in Section 8.4 permit the construction

Table 10.1 Tetrahedronal Finite Elements

Entry	*Representation*	*No. of Joints*	*d.o.f.* *Per Joint*	*d.o.f.* *Per Element*	*Remarks*	*Average Number of d.o.f. Per Element in Global Analysis*
a		4	3	12	Complete linear displacement fields (constant strain); for details, see Ref. 10.3–10.5 and 10.19	0.6
b		10	3	30	Complete quadratic displacement fields (linear strain); for details, see Ref. 10.4	4.2
c		20	3	60	Complete cubic displacement fields; for details, see Ref. 10.6	13.8
d		8		60	Complete cubic displacement fields; for details, see Ref. 10.6; 12 d.o.f. at vertices and along edges ($u, v, w, u_x, \cdots, w_z$) and 3 d.o.f. at centroid of faces (u, v, w)	8.4
e		16	3	48	Translational displacements as d.o.f.; incomplete cubic displacement fields; for details, see Ref. 10.2 and 10.6	—
f		4	12	48	Displacements and displacement derivatives as d.o.f.; incomplete cubic displacement fields; for details, see Ref. 10.1, 10.2, and 10.3	2.4

of the shape functions for any order of representation and give algebraic formulas for integrals of these functions over the volume of the element. (Explicit formulations for the quadratic element can be found in Ref. 10.4 and 10.5 and for the cubic displacement element in Ref. 10.6.) The T48 element is given attention because of its efficiency in application and because it illustrates the procedure to be followed when the displacement functions are described in terms of generalized coordinates rather than in shape function form.

10.2.2 Linear Displacement Element(10.19)

Linear displacement fields (**u**, **v**, **w**) can be written in terms of the tetrahedronal coordinates $\lfloor \mathbf{L} \rfloor = \lfloor L_1\ L_2\ L_3\ L_4 \rfloor$ (see Section 8.6) as follows:

$$\begin{aligned} u &= \lfloor \mathbf{L} \rfloor \{\mathbf{u}\} \\ v &= \lfloor \mathbf{L} \rfloor \{\mathbf{v}\} \\ w &= \lfloor \mathbf{L} \rfloor \{\mathbf{w}\} \end{aligned} \tag{10.11}$$

where

$$\{\mathbf{u}\} = \lfloor u_1\ u_2\ u_3\ u_4 \rfloor \tag{10.12}$$

and similarly for $\{\mathbf{v}\}$ and $\{\mathbf{w}\}$. Noting, from Eq. (8.29), that

$$L_i = \frac{1}{6(\text{vol})}[(\text{VOL})_i + c_{1_i}x + c_{2_i}y + c_{3_i}z]$$

where (vol), c_{1_i}, c_{2_i}, and c_{3_i} are defined in Section 8.6 in terms of the element joint coordinates $x_1 \ldots z_4$. Applying the strain-displacement relationships [Eq. (10.2)] to Eq. (10.11) we have the following evaluated form of Eq. (10.6)

$$\begin{Bmatrix} \epsilon_x \\ \epsilon_y \\ \epsilon_z \\ \gamma_{xy} \\ \gamma_{yz} \\ \gamma_{zx} \end{Bmatrix} = \frac{1}{6(\text{vol})} \begin{bmatrix} \lfloor \mathbf{c}_1 \rfloor & \lfloor \mathbf{0} \rfloor & \lfloor \mathbf{0} \rfloor \\ \lfloor \mathbf{0} \rfloor & \lfloor \mathbf{c}_2 \rfloor & \lfloor \mathbf{0} \rfloor \\ \lfloor \mathbf{0} \rfloor & \lfloor \mathbf{0} \rfloor & \lfloor \mathbf{c}_3 \rfloor \\ \lfloor \mathbf{c}_2 \rfloor & \lfloor \mathbf{c}_1 \rfloor & \lfloor \mathbf{0} \rfloor \\ \lfloor \mathbf{0} \rfloor & \lfloor \mathbf{c}_3 \rfloor & \lfloor \mathbf{c}_2 \rfloor \\ \lfloor \mathbf{c}_3 \rfloor & \lfloor \mathbf{0} \rfloor & \lfloor \mathbf{c}_1 \rfloor \end{bmatrix} \begin{Bmatrix} \{\mathbf{u}\} \\ \{\mathbf{v}\} \\ \{\mathbf{w}\} \end{Bmatrix} = [\mathbf{D}]\{\mathbf{\Delta}\} \tag{10.13}$$

where $\lfloor \mathbf{c}_1 \rfloor = \lfloor \mathbf{c}_{1_1}\ \mathbf{c}_{1_2}\ \mathbf{c}_{1_3}\ \mathbf{c}_{1_4} \rfloor$ and similarly for $\lfloor \mathbf{c}_2 \rfloor$ and $\lfloor \mathbf{c}_3 \rfloor$; $\lfloor \mathbf{0} \rfloor$ is a 1×4 null row. The linear displacement field produces a constant state of strain within the element. The element is therefore often referred to as the *constant strain tetrahedronal* (CSTh) element. With [**E**] given by Eq. (10.3) and with the constancy of all terms in the matrix [**D**], the stiffness matrix for the

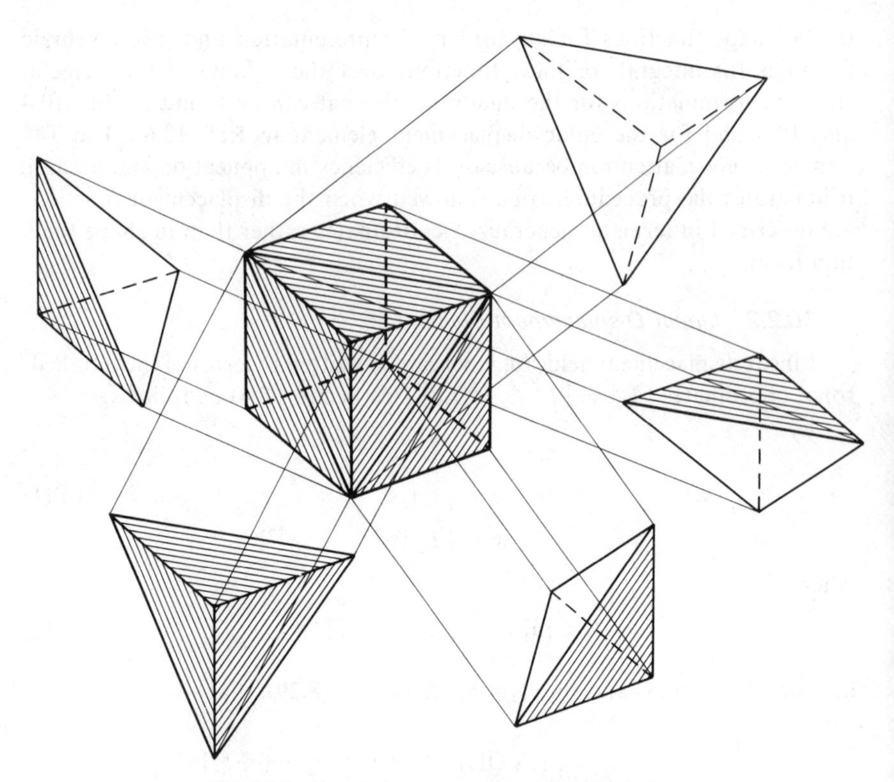

Fig. 10.4 Hexahedron composed of five tetrahedra.

tetrahedronal element follows directly from Eq. (10.8) as $[\mathbf{k}] = [\mathbf{D}]^T[\mathbf{E}][\mathbf{D}](\text{vol})$ where the element volume (vol) is calculated as discussed in Section 8.6.

It is extremely difficult in practical applications to define the geometry of the analytical model with this element alone. The problem is one of correctly assembling tetrahedra without leaving holes, etc. Hence, operating programs often permit direct specification of hexahedronal elements that are automatically constructed from a fixed number of tetrahedra. Figure 10.4, above, illustrates one such "superelement," comprised of five tetrahedra.

10.2.3 Higher-Order Element[10.6]

A simple approach to the formulation of the T48 element is on the basis of a series expressed in terms of tetrahedronal coordinates, as follows:

$$\begin{aligned} u = {} & L_1a_1 + L_2a_2 + L_3a_3 + L_4a_4 + L_1L_2a_5 + L_3L_4a_6 + L_4L_1a_7 \\ & + L_3L_2a_8 + L_2L_4a_9 + L_3L_1a_{10} + (L_1^2L_2 - L_1L_2^2)a_{11} \\ & + (L_3^2L_4 - L_3L_4^2)a_{12} + (L_4^2L_1 - L_4L_1^2)a_{13} \\ & + (L_3^2L_2 - L_3L_2^2)a_{14} + (L_2^2L_4 - L_2L_4^2)a_{15} + (L_3^2L_1 - L_3L_1^2)a_{16} \end{aligned} \tag{10.14}$$

The v and w displacement fields are defined in the same way so that the full set of generalized displacement parameters is $\{\mathbf{a}\} = \lfloor a_1 \ldots a_{48} \rfloor^T$.

The development of the stiffness matrix now follows the pattern we have established in preceding chapters. The displacement fields above are differentiated in accordance with the strain-displacement relationships [Eq. (10.2)], resulting in equations of the form $\boldsymbol{\epsilon} = [\mathbf{C}]\{\mathbf{a}\}$. Also, the displacement functions above and their first derivatives are evaluated at each of the four vertices resulting in 48 equations, which we symbolize as $\{\boldsymbol{\Delta}\} = [\mathbf{B}]\{\mathbf{a}\}$, where now $\{\boldsymbol{\Delta}\}$ is comprised of the d.o.f. given in Eq. (10.10) for the u field and by similar expressions for the v and w fields. It follows that the element stiffness matrix is given by Eq. (10.8a)

$$[\mathbf{k}] = ([\mathbf{B}]^{-1})^T \left[\int_{\text{vol}} [\mathbf{C}]^T[\mathbf{E}][\mathbf{C}]\, d(\text{vol}) \right] [\mathbf{B}]^{-1}$$

Clearly, the resulting expressions are too complicated for explicit representation.

The detailed procedures presented in Ref. 10.6 for the development of the stiffness of this element are considerably different than those given above. Explicit formulations of the stiffness matrix in generalized coördinates, and the transformation from generalized to nodal coordinates, are tabulated therein. The presentation of explicit relationships for the "kernel" stiffness matrix, as described in Section 8.2, where the stiffness matrix is formulated with reference to the generalized displacements $\{\mathbf{a}\}$, facilitates the construction of a variety of tetrahedronal element stiffness matrices for complete cubic displacement fields in terms of generalized parameters.

Other approaches have been taken to the construction of the stiffness matrix for the T48 element. Reference 10.1 first constructs a complete (20-term) cubic polynomial in volume coordinates and reduces this to 16 terms by the imposition of a constraint requiring quadratic variation of displacement on the element faces. Metrics are also developed for the definition of the element in curvilinear coordinates. Reference 10.2 supplements a complete quadratic in volume coordinates (10 terms) with 6 additional cubic terms.

10.3 Rectangular Hexahedronal Elements

10.3.1 General Considerations

The basic family of rectangular hexahedronal elements, with only translational d.o.f. at the node points, is portrayed in Fig. 10.5. These are termed elements of the *Lagrange* family because the displacement fields upon which they are based are constructed with use of the Lagrangian interpolation concepts described in Section 8.3.1. The simplest element in this family, Fig.

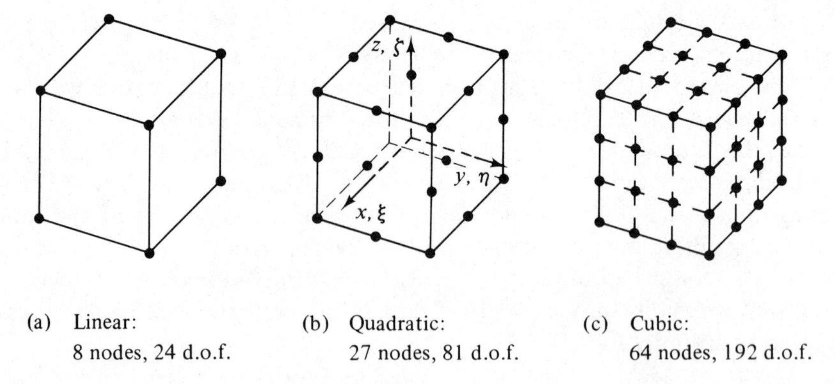

(a) Linear: 8 nodes, 24 d.o.f. (b) Quadratic: 27 nodes, 81 d.o.f. (c) Cubic: 64 nodes, 192 d.o.f.

Fig. 10.5 Lagrange family hexahedra (interior node points not shown).

10.5(a), is based upon linear displacement fields defined entirely by the eight corner point d.o.f. Extension to quadratic and cubic displacement fields introduces interior node points as shown in Figs. 10.5(b) and (c), respectively. Such interior nodes may be removed by use of the standard condensation procedure (Section 2.8). Alternatively, shape functions expressed entirely in terms of exterior node points may be constructed by operations on the Lagrangian interpolation functions in the manner described in Section 8.7.

The rectangular hexahedron may also be formulated with displacement-derivative d.o.f. at the corner points of the element, using Hermitian interpolation concepts (Section 8.3.2). The basic element of this family requires complete cubic displacement fields and features a total of 192 d.o.f.

Although the possibilities for formulation of rectangular hexahedronal elements to higher and higher order in any mode of representation of the d.o.f. is theoretically possible, practical applications have been limited to a few basic forms. These are summarized in Table 10.2. [10.3] Of these, we discuss in detail only the displacement fields for the entries *a*, *b*, and *c* since these have been the subjects of comparison studies to be discussed subsequently and are the most widely used in practice. Descriptions of the displacement fields for entries *d* and *e* can be found in Refs. 10.7 and 10.8, respectively.

10.3.2 Linear Displacement Rectangular Hexahedron

It is convenient in the following development to locate the origin of coordinates at the centroid of the element and to express all coordinates in non-dimensional form, with coordinates ξ, η and ζ formed in the manner described in Section 8.7 [see Fig. 10.5(b)]. We have for the displacement fields, via trivariate Lagrangian interpolation, in the manner described in Section

Table 10.2 Rectangular Hexahedronal Finite Elements

Entry	Representation	No. of Joints	d.o.f. Per Joint	d.o.f. Per Element	Remarks
a		8	3	24	Linear displacement fields, u, v, w, d.o.f. at each node; for details see Refs. 10.4, 10.9
b		20	3	60	Quadratic displacement fields, u, v, w, d.o.f. at each node; external nodes only; see Refs. 10.4, 10.10, and 10.17
c		32	3	96	Incomplete cubic displacement fields, u, v, w, d.o.f. at each node; see Refs. 10.4 and 10.14
d		64	3	192	Complete cubic displacement fields with internal nodes; u, v, w, d.o.f. at each node; termed *LUMINA* element when isoparametric; for details, see Ref. 10.7
e		8	12	96	Incomplete quintic displacement fields (Ref. 10.13) or Hermitian polynomial interpolation (Ref. 10.8); displacements and displacement derivatives as d.o.f.; isoparametric representations discussed in cited references

8.4 for the case of planar elements

$$u = \lfloor \mathbf{N} \rfloor \{\mathbf{u}\}, \qquad v = \lfloor \mathbf{N} \rfloor \{\mathbf{v}\}, \qquad w = \lfloor \mathbf{N} \rfloor \{\mathbf{w}\} \tag{10.15}$$

with

$$\lfloor \mathbf{N} \rfloor = \lfloor N_1 \ldots N_8 \rfloor,$$
$$\{\mathbf{u}\} = \lfloor u_1 \ldots u_8 \rfloor^{\mathrm{T}}, \qquad \{\mathbf{v}\} = \lfloor v_1 \ldots v_8 \rfloor,$$
$$\{\mathbf{w}\} = \lfloor w_1 \ldots w_8 \rfloor^{\mathrm{T}}$$

where

$$N_i = \tfrac{1}{8}(1 + \xi\xi_i)(1 + \eta\eta_i)(1 + \zeta\zeta_i) \tag{10.16}$$

The development of the stiffness matrix for this element follows the usual pattern. The displacement fields [Eq. (10.15)] are differentiated in accordance with the strain-displacement relationships [Eq. (10.2)] to yield the joint-displacement-to-strain transformation in the form

$$\{\boldsymbol{\epsilon}\} = [\mathbf{D}]\begin{Bmatrix} \{\mathbf{u}\} \\ \{\mathbf{v}\} \\ \{\mathbf{w}\} \end{Bmatrix} \tag{10.6a}$$

With [E] given by Eq. (10.3) the stiffness matrix for this element is represented by Eq. (10.8), i.e.

$$[\mathbf{k}] = \left[\int_{\mathrm{vol}} [\mathbf{D}]^{\mathrm{T}}[\mathbf{E}][\mathbf{D}]\, d(\mathrm{vol}) \right]$$

The explicit form of this matrix is presented in Ref. 10.9.

10.3.3 Higher-Order Rectangular Hexahedra

As we have indicated, the hexahedronal element is easily generalized to higher-order form by use of Lagrangian interpolation to the desired order. The difficulty with this approach is the presence of joints within the volume of the element and within the faces. Thus, representations that involve nodes only along the edges, such as the 20-node element shown as entry *b* in Table 10.2, are advantageous. For this case, the matrices of Eq. (10.15) have the form

$$\lfloor \mathbf{N} \rfloor = \lfloor N_1 \ldots N_{20} \rfloor, \qquad \{\mathbf{u}\} = \lfloor u_1 \ldots u_{20} \rfloor^{\mathrm{T}}$$

and similarly for {v} and {w}. As emphasized in Section 8.7, this form of shape function, i.e., with interior points removed, is not unique. A widely used alternative form is the following. At a vertex node (with origin of coordinates at centroid)

$$N_i = \tfrac{1}{8}(1 + \xi\xi_i)(1 + \eta\eta_i)(1 + \zeta\zeta_i)(\xi\xi_i + \eta\eta_i + \zeta\zeta_i - 2) \tag{10.17}$$

and for a typical mid-edge node

$$\xi_i = 0, \qquad \eta_i = \pm 1, \qquad \zeta_i = \pm 1$$
$$N_i = \tfrac{1}{4}(1 - \xi^2)(1 + \eta\eta_i)(1 + \zeta\zeta_i) \tag{10.18}$$

The relationship of this function to the coefficients of a polynomial expansion, as well as alternative shape functions for the 20-node element, are detailed in Ref. 10.10.

A more sophisticated representation, involving 32 nodes, is illustrated as entry *c* in Table 10.2. Here the vertex shape functions are of the form (with origin of coordinates at centroid)

$$N_i = \tfrac{1}{64}(1 + \xi\xi_i)(1 + \eta\eta_i)(1 + \zeta\zeta_i)[9(\xi^2 + \eta^2 + \zeta^2 - 19)] \tag{10.19}$$

and, for the typical edge node at $\xi_i = \pm\frac{1}{3}$, $\eta_i = \pm 1$, $\zeta_i = \pm 1$,

$$N_i = \tfrac{9}{64}(1 - \xi^2)(1 + 9\xi\xi_i)(1 + \eta\eta_i)(1 + \zeta\zeta_i) \tag{10.20}$$

The procedure for transformation of the displacement fields above into element stiffness matrices is identical to that described for the linear displacement hexahedron.

10.4 Numerical Comparisons

Studies[10.1, 10.3, 10.4, 10.9] have been performed on the relative accuracy and efficiency of certain of the tetrahedronal and hexahedronal elements described in Sections 10.3.1 and 10.3.2. Figure 10.6 shows a problem adopted in Ref. 10.4 for the purpose of developing comparison of solution accuracy. The structure is a cantilever beam subjected to an end moment, divided into a grid of 42 rectangular hexahedronal elements. Some of the solutions are obtained with hexahedronal elements that are combinations of tetrahedronal elements, e.g., the combination pictured in Fig. 10.4. The elements studied include the tetrahedra based on linear and quadratic displacement fields,

Fig. 10.6 Cantilever beam for solution accuracy study (from Ref. 10.4).

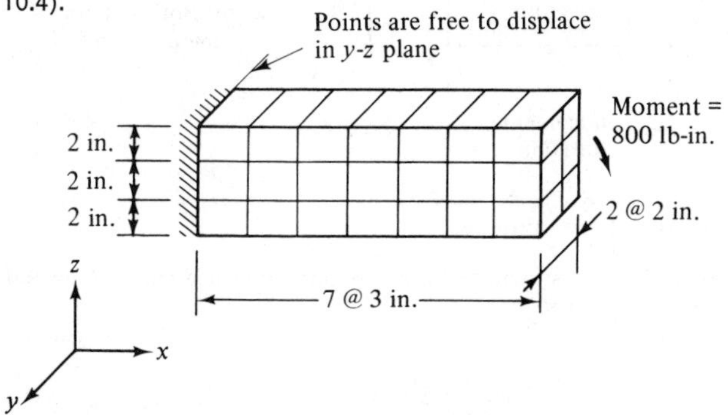

respectively (entries a and b of Table 10.1); the rectangular hexahedra based on the linear displacement field of Eq. (10.16) and on Eqs. (10.17), (10.18) (entry b, Table 10.2). The quadratic displacement tetrahedron and the 20-node hexahedron produce exactly the same result for the tip deflection as given by beam theory. The linear displacement tetrahedron (using 5 elements to form a hexahedron, as in Fig. 10.4) and linear displacement hexahedron give results that are, respectively, 39 and 10% less than the beam theory solution.

Figure 10.8 gives a comparison of the efficiency of solution, as measured by computer time expended, of the linear displacement hexahedron and the 20-node hexahedron (entry b, Table 10.2) for the two test problems shown in Fig. 10.7. The results show that for the slender beam, Figs. 10.7(a) and 10.8(a), where plate bending action appears, the 20-node element is clearly superior. This element is able to reproduce the linear strain distribution across the depth, which is characteristic of flexure. The precedence is reversed in the case of the deep cantilever, Figs. 10.7(b) and 10.8(b), where a more general stress state prevails. Many other results, beyond the scope of this text, are presented in Ref. 10.4.

Numerical comparisons given in Ref. 10.4 support a preference for directly formulated hexahedronal elements, rather than by combination of tetrahedronal elements in the manner of Fig. 10.4. Much more numerical

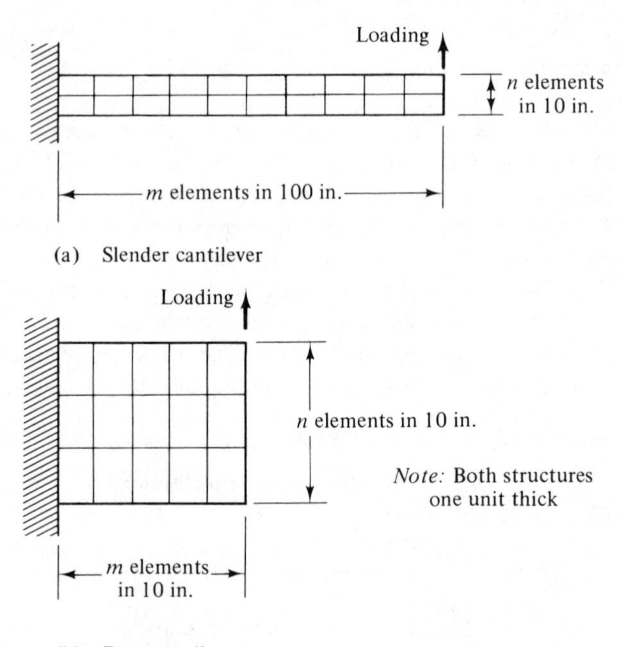

Fig. 10.7 Test structures for computing efficiency study (from Ref. 10.4).

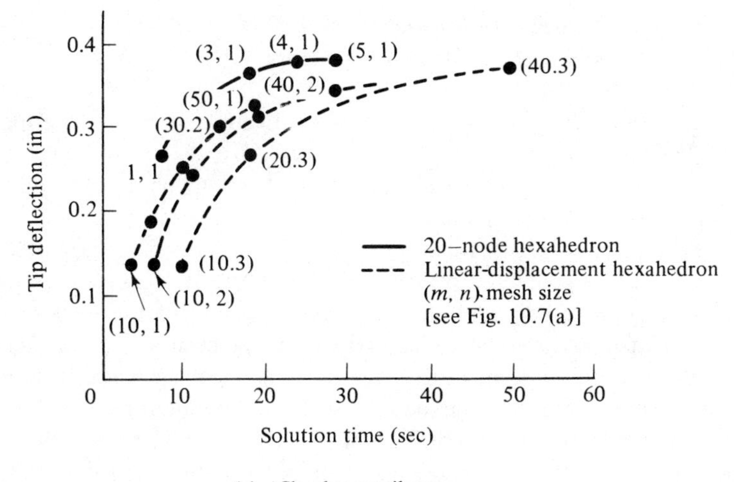

(a) Slender cantilever

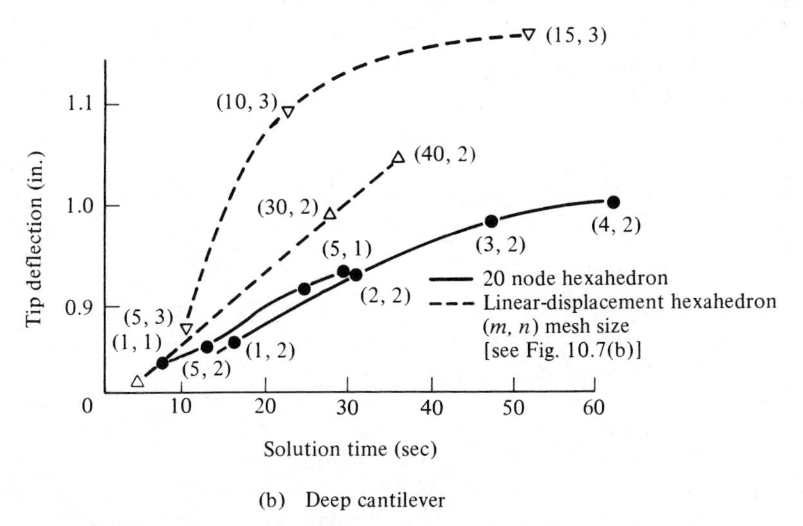

(b) Deep cantilever

Fig. 10.8 Results of cantilever beam efficiency studies (from Ref. 10.4).

experimentation is needed to confirm this conclusion. Differences in the details of the formulative approach, coding practices, and computer hardware influence the conclusions that different investigators reach, and many practitioners continue to favor tetrahedronal elements in the analysis of solid structures. An interesting summary of practice can be found in Ref. 10.11.

10.5 Isoparametric Representation and the Analysis of Shells with use of Solid Elements

Success in the application of solid elements is critically dependent on a high degree of efficiency in the total analysis approach. Use of the most efficient algorithms in construction of the algebraic equations to be solved and in the solution process itself is mandatory. [10.12]

Another factor in achieving maximum analysis efficiency is isoparametric representation of the geometry of the solid elements. If the elements are restricted to straight rather than curved faces, a portion of the number of d.o.f. in the total analysis must be viewed as having been allotted to geometric representation. This portion is reduced significantly when isoparametric concepts are employed to define curved element boundaries. Maximum use is then made of the d.o.f. in definition of structural response.

Figure 10.9 illustrates the arrangement of coordinate axes for isoparametric representation of the linear displacement hexahedron. The displacement field for the rectangular form of this element, presented as Eq. (10.16), is written directly in terms of the nondimensionalized coordinates (ξ, η, ζ) pictured in Fig. 10.9. Thus, by use of the procedures detailed in Section 8.8, the isoparametric form of this element can be constructed straightforwardly. The algebraic complexity of the formulation is such, however, that an explicit definition of the resulting stiffness coefficients is precluded for even this

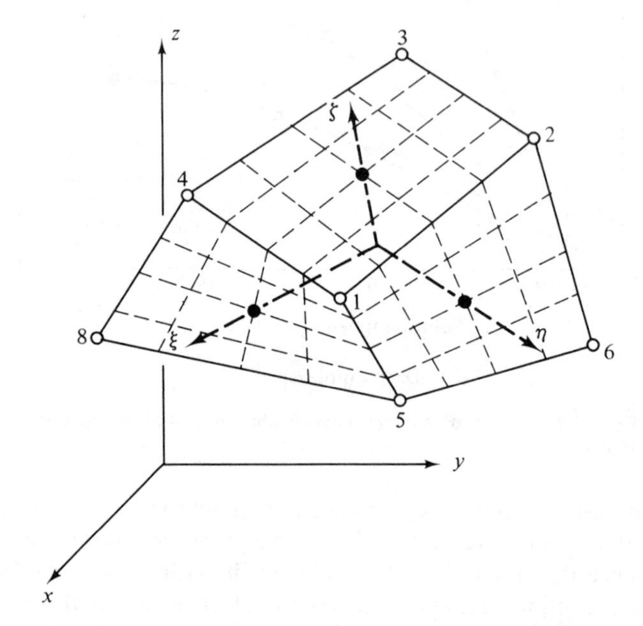

Fig. 10.9 Linear displacement hexahedron-isoparametric coordinates.

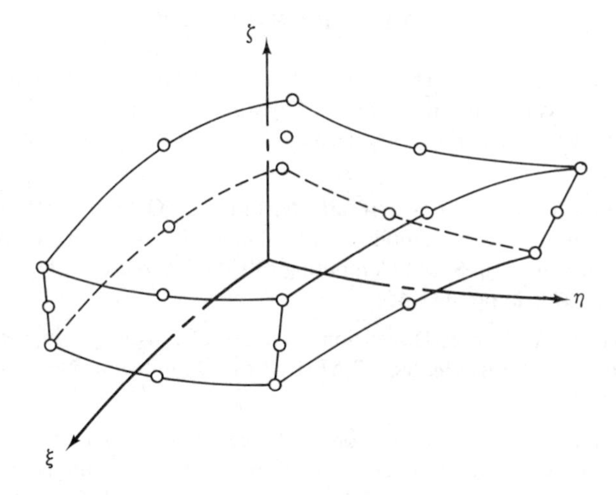

Fig. 10.10 Hexahedronal element that has been contracted to serve as curved shell element.

element, the simplest form of hexahedron. In general, numerical integration of the relevant energy integrals is essential. The reader is urged to consult Refs. 10.13–10.15 for descriptions of the many detailed aspects of the formulation of isoparametric forms of solid elements.

The isoparametric form of the solid element is also useful for representation of shell structures. Figure 10.10 shows the 20-node isoparametric element arranged in a form suitable for such analyses. Application of these elements to thick shell analysis problems yields excellent results, but the thin shell solutions are not approached as the thickness of the element is reduced. It was pointed out in Section 9.3.2 that this is due to an excessive stiffness in the mode of representation of the shear strain energy. References 10.16 and 10.17 demonstrate that good results can be obtained for thin shell applications if the shearing strain energy contribution to the element stiffness formulation is approximated while the accuracy in definition of the strain energy contribution for direct strains is retained. Since this type of element requires numerical integration in evaluation of the strain energy, it is a simple matter to do this by reducing the order of numerical integration of the shear strain energy relative to that of the remaining components of the strain energy. We discuss numerical results in Section 12.6.

An alternative approach to the analysis of shell structures with use of solid elements is by the addition of incompatible modes, which was also described in Section 9.3.2 and discussed in Ref. 10.18. This scheme enables the use of the simplest form of hexahedron, that which is fundamentally based upon linear displacement fields and which features only the eight-corner node points.

REFERENCES

10.1 HUGHES, J. R. and H. ALLIK, "Finite Elements for Compressible and Incompressible Continua," Proc. of Symp. on Application of Finite Element Methods in Civil Eng., Vanderbilt U., Nashville, Tenn., Nov., 1969, pp. 27–62.

10.2 RASHID, Y. R., P. D. SMITH, and N. PRINCE, "On Further Application of the Finite Element Method to Three-Dimensional Elastic Analysis," Proc. of Symp. on High Speed Computing of Elastic Structures, U. of Liege, Belgium, 1970, **2**, pp. 433–454.

10.3 FJELD, S. A., "Three-Dimensional Theory of Elasticity," in *Finite Element Methods in Stress Analysis*, TAPIR Press, Trondheim, Norway, 1969, pp. 333–364.

10.4 CLOUGH, R. W., "Comparison of Three Dimensional Finite Elements," Proc. of Symp. on Application of Finite Element Methods in Civil Eng., Vanderbilt U., Nashville, Tenn., Nov., 1969, pp. 1–26, printed by the American Society of Civil Engineers.

10.5 ARGYRIS, J. H., "Matrix Analysis of Three Dimensional Media—Small and Large Displacements," *AIAA J.*, **3**, No. 1, pp. 45–51, 1965.

10.6 ARGYRIS, J. H., I. FRIED, and D. W. SCHARPF, "The TET 20 and the TEA 8 Elements for the Matrix Displacement Method." *Aero. J.*, **72**, No. 691, pp. 618–623, July, 1968.

10.7 ARGYRIS, J. H., "The LUMINA Element for the Matrix Displacement Method," *Aero. J.*, **72**, No. 690, pp. 514–517, June, 1968.

10.8 ARGYRIS, J. H., I. FRIED, and D. W. SCHARPF, "The Hermes 8 Element for the Matrix Displacement Method," *Aero. J.*, **72**, No. 691, pp. 613–617, July, 1968.

10.9 MELOSH, R. J., "Structural Analysis of Solids," Proc. ASCE, *J. Struct. Div.*, **89**, No. ST-4, pp. 205–223, Aug., 1963.

10.10 RIGBY, G. L. and G. M. MCNEICE, "A Strain Energy Basis for Studies of Element Stiffness Matrices," *AIAA J.*, **10**, No. 11, pp. 1490–1493, 1972.

10.11 ANONYMOUS, "Three-Dimensional Continuum Computer Programs for Structural Analysis," ASME Special Publication, 1972.

10.12 RASHID, Y., "Three-Dimensional Analysis of Elastic Solids," *Internat'l. J. Solids Struct.*, Part I, **5**, pp. 1311–1332, 1969; Part II, **6**, pp. 195–207, 1970.

10.13 ZIENKIEWICZ, O., B. IRONS, F. C. SCOTT, and J. S. CAMPBELL, "Three Dimensional Stress Analysis." Proc. of Symp. on High Speed Computing of Elastic Structures, U. of Liege, Belgium, 1970, **1**, pp. 413–432.

10.14 ERGATOUDIS, J., B. M. IRONS, and O. C. ZIENKIEWICZ, "Three-Dimensional Analysis of Arch Dams and Their Foundations." Symp. on Arch Dams at the Institution of Civil Engs., London, Mar., 1968.

10.15 Irons, B. M., "Quadrature Rules for Brick Based Finite Elements," *Internat'l. J. Num. Meth. Eng.*, **3**, No. 2, pp. 293–294, 1971.

10.16 Pawsey, S. F., and R. W. Clough, "Improved Numerical Integration of Thick Shell Finite Elements," *Internat'l. J. Num. Meth. Eng.*, **3**, pp. 575–586, 1971.

10.17 Zienkiewicz, O. C., R. L. Taylor, and J. M. Too, "Reduced Integration Technique in General Analysis of Plates and Shells," *Internat'l. J. Num. Meth. Eng.*, **3**, pp. 275–290, 1971.

10.18 Wilson, E. et al., "Incompatible Displacement Models," in *Numerical and Computer Methods in Structural Mechanics,* S. J. Fenves et al. (eds.), Academic Press, New York, N. Y., 1973, pp. 43–57.

10.19 Gallagher, R. H., J. Padlog, and P. P. Bijlaard, "Stress Analysis of Heated Complex Shapes," *ARS J.*, **32**, No. 5, pp. 700–707, May, 1962.

11

Solid Elements: Special Cases

Two common situations in structural engineering practice where the state of stress is three dimensional but can be studied with use of a two-dimensional representation are those that feature the conditions of *plane strain* and *axisymmetry*, respectively. These situations and a special circumstance applicable to all classes of solid structures, that of *incompressibility* of the material under analysis, are studied in this chapter.

In plane strain, one deals with a situation in which the dimension of the structure in one direction, say the z-coordinate direction, is very large in comparison with the dimensions of the structure in the other two directions (x- and y-coordinate axes) and the applied forces act in the x-y plane and do not vary in the z direction. Perhaps the most important practical applications of this representation occur in the analysis of dams, tunnels, and other geotechnical works, although such small-scale problems as bars and rollers compressed by forces normal to their cross section are amenable to analysis in this way. The key aspects of finite element representations for plane strain analysis are described in Section 11.1.

Axisymmetric structures form another special class of three-dimensional analysis problem. Numerous civil, mechanical, nuclear, and aerospace engineering problems fall in the category of axisymmetric solids, including concrete and steel tanks, nuclear containment vessels, rotors, pistons, shells, and rocket nozzles. The distinction with the general three-dimensional problem is that a cylindrical rather than rectangular coordinate system is used in the definition of all pertinent relationships. The resulting simplifications are in some cases counterbalanced by difficulties in the integration of the strain energy to obtain the stiffness matrix.

Axisymmetric structures are often loaded axisymmetrically, enabling further simplifications in element formulation. This case is taken up in Section 11.2. Certain design analysis situations involve nonaxisymmetric loads, however. In such cases the analyst must decide between an expansion of the present formulation in terms of harmonics in the circumferential coordinate and the use of a completely general three-dimensional stress analysis. Economies of computation may favor the former, whose detailed features are described in Section 11.3.

Incompressible materials such as rubber feature a Poisson's ratio $\mu = 0.5$, causing difficulties in the conventional potential energy formulative procedure where the terms of the stress-strain matrix are divided by $(1 - 2\mu)$. Simple modifications can be made to the conventional potential energy approach, however, to circumvent this difficulty. Approaches based on complementary energy or on a functional of the Reissner type are also attractive in this case. Both classes of analysis procedures for compressible materials are studied in Section 11.4.

11.1 Plane Strain

A representative condition of plane strain is portrayed in Fig. 11.1. A rectangular bar, with z dimension greater than the x and y dimensions, is fixed against displacement in the z direction. The load $\bar{\mathbf{T}}$ is a function of x and y alone. Under this condition we note that the longitudinal strain (ϵ_z) is zero, as are the shear stresses τ_{xz} and τ_{yz}. Introducing the condition on ϵ_z into the relevant strain-stress equation [see Eq. (4.14)] we find, for an isotropic

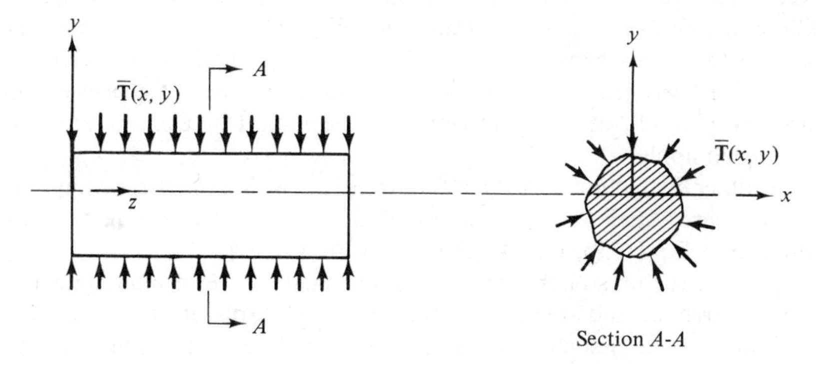

Fig. 11.1 Conditions for plane strain analysis.

material

$$\epsilon_z = \frac{\sigma_z}{E} - \frac{\mu\sigma_x}{E} - \frac{\mu\sigma_y}{E} = 0 \tag{11.1a}$$

$$\epsilon_x = \frac{\sigma_x}{E} - \frac{\mu\sigma_y}{E} - \frac{\mu\sigma_z}{E} \tag{11.1b}$$

$$\epsilon_y = \frac{\sigma_y}{E} - \frac{\mu\sigma_x}{E} - \frac{\mu\sigma_z}{E} \tag{11.1c}$$

Solving the first equation for σ_z and substituting into the final two equations yields (with the addition of the $\gamma_{xy} - \tau_{xy}$ equation)

$$\begin{Bmatrix} \epsilon_x \\ \epsilon_y \\ \gamma_{xy} \end{Bmatrix} = \frac{(1-\mu^2)}{E} \begin{bmatrix} 1 & \frac{-\mu}{(1-\mu)} & 0 \\ \frac{-\mu}{(1-\mu)} & 1 & 0 \\ 0 & 0 & \frac{2}{(1-\mu)} \end{bmatrix} \begin{Bmatrix} \sigma_x \\ \sigma_y \\ \tau_{xy} \end{Bmatrix} \tag{11.2}$$

and, by inversion,

$$\begin{Bmatrix} \sigma_x \\ \sigma_y \\ \tau_{xy} \end{Bmatrix} = \frac{E}{(1+\mu)(1-2\mu)} \begin{bmatrix} (1-\mu) & \mu & 0 \\ \mu & (1-\mu) & 0 \\ 0 & 0 & \frac{(1-2\mu)}{2} \end{bmatrix} \begin{Bmatrix} \epsilon_x \\ \epsilon_y \\ \gamma_{xy} \end{Bmatrix} \tag{11.3}$$

The linear strain-displacement equations refer only to the geometry of deformation and apply to both plane stress and plane strain. Thus, the pertinent relationships consist of Eq. (4.7), and the principal distinction between finite element plane strain and plane stress formulations resides in the differences between the stress-strain laws, Eq. (11.3) versus Eq. (9.3). The developments of Chapter 9, including concepts of higher-order elements, alternatives between additional element node points and displacement derivatives as d.o.f., and isoparametric representation of element geometry, once again apply.

Another distinction with the plane stress case is that the stress component σ_z is not zero. After computation of joint displacements, σ_z can be evaluated with use of Eq. (11.3), Eq. (4.7), and then Eq. (11.1a).

Often, structures of the type shown in Fig. 11.1 are of finite dimension in the z-direction and are not constrained against displacement in that direction, so that the assumption that $\epsilon_z = 0$ is not valid. In such cases it is customary to assume that $\epsilon_z =$ constant (the case of *generalized plane strain*).

To form a finite element idealization for this case we can employ the stress-strain law of three-dimensional elasticity (Eq. 10.3) with $\gamma_{xz} = \gamma_{yz} = 0$ and ϵ_z a constant to be determined. The strains ϵ_x, ϵ_y and γ_{xy} are related to the assumed u and v displacement fields in the usual way. The resulting global stiffness equations are then formulated in terms of nodal point values of u and v and the single constant ϵ_z.

11.2 Axisymmetric Solids

11.2.1 Basic Relationships

An axisymmetric finite element is in the form of a ring of constant cross section. The element is described in a cylindrical coordinate system whose axis of symmetry is the z axis and for which radial distances are defined by the coordinate r. A differential area of the cross section of such an element, including a portion of its external surface (ds), lies in the z-r plane as shown in Fig. 11.2. The circumferential coordinate, which does not appear in this sketch, is described by an angle θ. The nodal points of the element are in fact nodal circles. Thus, the analysis of axisymmetric solids for *axisymmetric loads* is also a two-dimensional problem since the displacement field can be described by just two components on the cross section, the radial (u) and axial (w) displacements.

The pertinent strain components in cylindrical coordinates are radial (ϵ_r), circumferential (ϵ_θ), axial (ϵ_z), and shearing (γ_{rz}) strains. The corresponding stress components are σ_r, σ_θ, σ_z, and γ_{rz}. Circumferential stresses and strains exist because the uniform radial displacement increases the circumferential (or hoop) length. The linear strain-displacement equations are[11.1]

$$\epsilon_r = \frac{\partial u}{\partial r}, \qquad \epsilon_\theta = \frac{u}{r}, \qquad \epsilon_z = \frac{\partial w}{\partial z}, \qquad \gamma_{rz} = \frac{\partial u}{\partial z} + \frac{\partial w}{\partial r} \tag{11.4}$$

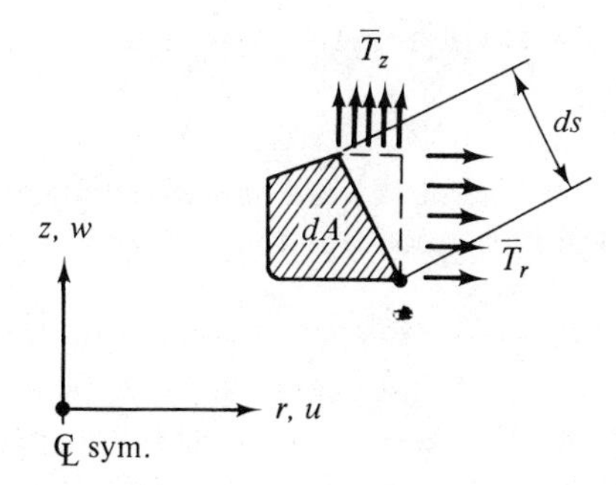

Fig. 11.2 Differential cross-sectional area of axisymmetric solid element.

The constitutive relationships are once again

$$\boldsymbol{\sigma} = [\mathbf{E}]\boldsymbol{\epsilon} - [\mathbf{E}]\boldsymbol{\epsilon}^{\text{init.}} \tag{4.15}$$

where now

$$\boldsymbol{\sigma} = \lfloor \sigma_r \ \sigma_\theta \ \sigma_z \ \tau_{rz} \rfloor^{\mathrm{T}} \tag{11.5}$$

$$\boldsymbol{\epsilon} = \lfloor \epsilon_r \ \epsilon_\theta \ \epsilon_z \ \gamma_{rz} \rfloor^{\mathrm{T}} \tag{11.6}$$

$$\boldsymbol{\epsilon}^{\text{init}} = \lfloor \epsilon_r^{\text{init.}} \ \epsilon_\theta^{\text{init.}} \ \epsilon_z^{\text{init.}} \ \gamma_{rz}^{\text{init.}} \rfloor^{\mathrm{T}} \tag{11.7}$$

In particular, for the important case of a temperature change (Υ) above the stress-free state in an isotropic material

$$\epsilon_r^{\text{init.}} = \epsilon_\theta^{\text{init.}} = \epsilon_z^{\text{init.}} = \alpha\Upsilon, \qquad \gamma_{rz}^{\text{init.}} = 0$$

The matrix of elastic constants is the same as in the plane strain case except that now a row and column must be added to accommodate a third direct stress component. For an isotropic material we have

$$[\mathbf{E}] = \frac{E}{(1+\mu)(1-2\mu)} \begin{bmatrix} (1-\mu) & & & \text{Symmetric} \\ \mu & (1-\mu) & & \\ \mu & \mu & (1-\mu) & \\ 0 & 0 & 0 & \dfrac{(1-2\mu)}{2} \end{bmatrix} \tag{11.8}$$

where the rows and columns are arranged in correspondence with the stress and strain vectors of Eq. (11.5) and (11.6).

Because of axial symmetry we can immediately transform the volume integral of the potential energy expression into an area integral. The differential element of volume represented by the differential area shown in Fig. 11.2 is $d(\text{vol}) = 2\pi r\, dA$ and the total surface area represented by the length ds is $dS = 2\pi r\, ds$. Thus, the potential energy expression is of the form

$$\Pi_p = \pi \int_A \boldsymbol{\epsilon}[\mathbf{E}]\boldsymbol{\epsilon} r\, dA - 2\pi \int_A \boldsymbol{\epsilon}[\mathbf{E}]\boldsymbol{\epsilon}^{\text{init.}} r\, dA - 2\pi \int_s (u \cdot \bar{T}_r + w \cdot \bar{T}_z) r\, ds \tag{11.9}$$

where $\boldsymbol{\epsilon}$, $\boldsymbol{\epsilon}^{\text{init.}}$, and $[\mathbf{E}]$ are as defined by Eq. (11.6) to (11.8), and $\bar{T}_r$ and $\bar{T}_z$ are the prescribed tractions per unit of surface area.

11.2.2 Axisymmetric Triangular Ring Element

Axisymmetric solid elements are generalizations of plane stress elements and, as in plane strain, many of the considerations of Chapter 9 are again

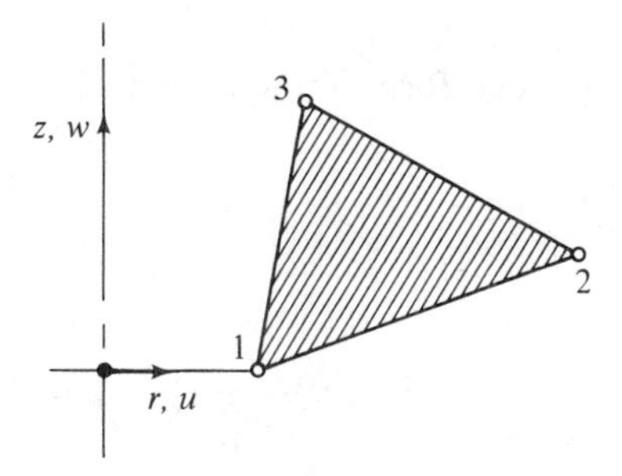

Fig. 11.3 Triangular ring element section.

applicable. We therefore detail here the relationships for only the simplest triangular element, Fig. 11.3. The element is positioned arbitrarily in the r-z plane rather than with one edge coincident with a coordinate axis. A linear displacement field is appropriate to this type of element. Because of the dependence of the ϵ_θ strain-displacement equations on the inverse of the radius, special difficulties exist in the construction of the element stiffness matrix, even for this simplest case of a linear displacement field. In order to understand these difficulties it is more convenient to work with the displacement field expressed in terms of generalized displacements than with shape function forms of these equations. Thus, we choose

$$\begin{aligned} u &= a_1 + a_2 r + a_3 z \\ w &= a_4 + a_5 r + a_6 z \end{aligned} \tag{11.10}$$

By differentiation of these displacement functions in accordance with the pertinent strain-displacement relationships [Eq. (11.4)], there is obtained

$$\boldsymbol{\epsilon} = \begin{Bmatrix} \epsilon_r \\ \epsilon_\theta \\ \epsilon_z \\ \gamma_{rz} \end{Bmatrix} = \begin{bmatrix} 0 & 1 & 0 & 0 & 0 & 0 \\ \frac{1}{r} & 1 & \frac{z}{r} & 0 & 0 & 0 \\ 0 & 0 & 0 & 0 & 0 & 1 \\ 0 & 0 & 1 & 0 & 1 & 0 \end{bmatrix} \begin{Bmatrix} a_1 \\ \cdot \\ \cdot \\ a_6 \end{Bmatrix} = [\mathbf{C}]\{\mathbf{a}\} \tag{11.11}$$

Inserting this in Eq. (11.9), we obtain (excluding, for simplicity, initial strains)

$$\Pi_p = \tfrac{1}{2}\lfloor \mathbf{a} \rfloor [\mathbf{k}^a]\{\mathbf{a}\} + V^a \tag{11.12}$$

where V^a symbolizes the potential of prescribed loads referenced to the parameters $\{\mathbf{a}\}$, and the kernel stiffness is

$$[\mathbf{k}^a] = \frac{2\pi E}{(1+\mu)(1-2\mu)}\left[\int_A [\mathbf{C}]^{\mathrm{T}}[\mathbf{E}]'[\mathbf{C}] r\, dA\right] \tag{11.13}$$

where the expanded form of the matrix $\left[\int_A [\mathbf{C}]^T[\mathbf{E}]'[\mathbf{C}] r\, dA\right]$ is

$$\begin{matrix} a_1 & a_2 & a_3 & a_4 & a_5 & a_6 \end{matrix}$$

$$\begin{bmatrix} (1-\mu)I_4 & & & & & \\ I_2 & 2I_1 & & & \text{Symmetric} & \\ (1-\mu)I_5 & I_3 & (1-\mu)I_6 + (\frac{1}{2}-\mu)I_1 & & & \\ 0 & 0 & 0 & 0 & & \\ 0 & 0 & (\frac{1}{2}-\mu)I_1 & 0 & (\frac{1}{2}-\mu)I_1 & \\ \mu I_2 & 2\mu I_1 & \mu I_3 & 0 & 0 & (1-\mu)I_1 \end{bmatrix} \quad (11.14)$$

in which

$$I_1 = \iint r\, dr\, dz \quad (11.15)$$

$$I_2 = \iint dr\, dz \quad (11.16)$$

$$I_3 = \iint z\, dr\, dz \quad (11.17)$$

$$I_4 = \iint \frac{dr\, dz}{r} \quad (11.18)$$

$$I_5 = \iint \frac{z}{r}\, dr\, dz \quad (11.19)$$

$$I_6 = \iint \frac{z^2}{r}\, dr\, dz \quad (11.20)$$

The terms I_1, I_2, and I_3 are easily evaluated to yield

$$I_1 = \frac{(r_1 + r_2 + r_3)[r_1(z_2 - z_3) + r_2(z_3 - z_1) + r_3(z_1 - z_2)]}{6} \quad (11.15a)$$

$$I_2 = \frac{[r_1(z_2 - z_3) + r_2(z_3 - z_1) + r_3(z_1 - z_2)]}{2} \quad (11.16a)$$

$$I_3 = \frac{(z_1 + z_2 + z_3)[r_1(z_2 - z_3) + r_2(z_3 - z_1) + r_3(z_1 - z_2)]}{6} \quad (11.17a)$$

The terms I_4, I_5, and I_6 involve the variable r in the denominator of the integrand and result in considerably more complicated expressions, as follows:

$$I_4 = \sum_{i=1}^{3} \frac{(r_i z_{i+1} - r_{i+1} z_i)}{(r_i - r_{i+1})} \ln \frac{r_i}{r_{i+1}} \quad (11.18a)$$

$$I_5 = H_{12} + H_{23} + H_{31} \quad (11.19a)$$

where, for $i, j = 1, 2, 3$,

$$H_{ij} = \frac{-(z_i - z_j)}{4(r_i - r_j)}[z_i(3r_j - r_i) - z_j(3r_i - r_j)] + \frac{1}{2}\left(\frac{r_i z_j - r_j z_i}{r_i - r_j}\right)^2 ln\,\frac{r_i}{r_j} \tag{11.21}$$

and

$$I_6 = G_{12} + G_{23} + G_{31} \tag{11.20a}$$

where, for $i, j = 1, 2, 3$,

$$\begin{aligned} G_{ij} = {} & \frac{(z_i - z_j)}{18(r_i - r_j)^2}[z_j^2(11r_i^2 - 7r_ir_j + 2r_j^2) + 2z_iz_j(2.5r_i^2 - 11r_ir_j + 2.5r_j^2) \\ & + z_i^2(11r_j^2 - 7r_ir_j + 2r_i^2)] + \frac{1}{3}\left(\frac{r_i z_j - r_j z_i}{r_i - r_j}\right)^3 ln\,\frac{r_i}{r_j} \end{aligned} \tag{11.22}$$

Expressions given above are referenced to the generalized parameters $\{\mathbf{a}\}$. The stiffness matrix referenced to physical coordinates is easily obtained by constructing the transformation from generalized to physical coordinates by evaluation of Eq. (11.10) at the d.o.f. and by application of this transformation to $[\mathbf{k}^a]$.

Special cases arise when joints are located on the axis of symmetry since the terms with $ln\,(r_i/r_j)$ and $(r_i - r_j)$ in the denominator become infinite. Evaluation can be accomplished using L'Hospital's rule. For the special case of $r_i = 0$, $r_j, r_k \neq 0$, we have $(i, j, k = 1, 2, 3)$

$$I_4 = \frac{[r_j(z_k - z_i) + r_k(z_i - z_j)]}{(r_j - r_k)} ln\,\frac{r_j}{r_k} \tag{11.18b}$$

$$I_5 = H_{jk} - \frac{1}{4}[(z_j - z_i)(3z_i + z_j) + (z_i - z_k)(3z_i + z_k)] - \frac{1}{2}z_i^2\,ln\,\frac{r_j}{r_k} \tag{11.19b}$$

$$\begin{aligned} I_6 = G_{jk} & - \frac{(z_j - z_i)}{18}(11z_i^2 + 5z_iz_j + 2z_j^2) \\ & - \frac{(z_i - z_k)}{18r_k^2}(11z_i^2 + 5z_iz_k + 2z_k^2) \\ & - \frac{1}{3}z_i^3\,ln\,\frac{r_j}{r_k} \end{aligned} \tag{11.20b}$$

Note also that $u_i = 0$, resulting in a contraction of the element stiffness matrix.

When $r_i = r_j = 0$, and $r_k \neq 0$, it can be shown[11.3] that terms containing I_4, I_5, and I_6 do not appear in the stiffness matrix that has been contracted by virtue of the fact that $u_i = u_j = 0$.

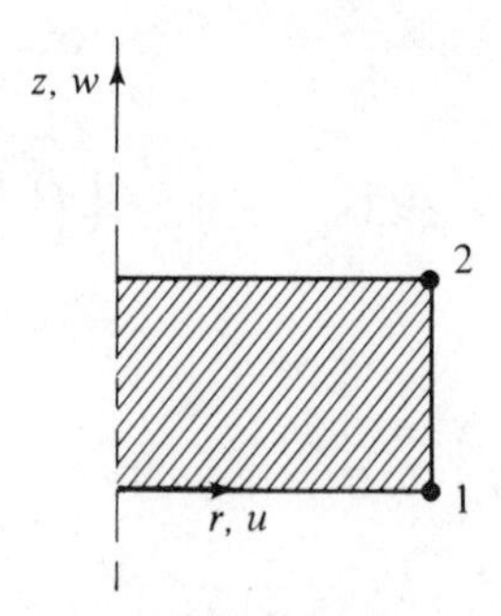

Fig. 11.4 Core element.

In certain applications it is useful to have available a "core" element for solid cylinders, as shown in Fig. 11.4. To formulate this, one may employ the displacement fields $u = a_1 r + a_2 z$ and $w = a_3 r + a_4 z$. The detailed derivation of stiffness equations for this element follows the pattern established above.

It is apparent that explicit "exact" formulas for the stiffness coefficients of linear displacement triangles are awkward. Nevertheless, relationships presented above have appeared in various forms, e.g., in Refs. 11.2-11.4. Formulas and tabulated coefficients for basic terms of higher-order elements are presented in Refs. 11.5 and 11.6. A simple approximate formulation for the axisymmetric triangle is based upon use of an "average" radius (e.g., the centroidal value) that is treated as a constant of integration. The accuracy of this approximation depends on the closeness of the element to the axis of rotation.

The triangular ring element proves to be highly accurate in comparison analyses of classical solutions and this circumstance presumably carries over into practical problems. Figure 11.5 shows the element formulation above in application to the problem of a thick-walled cylinder under internal pressure. A rather coarse grid of elements is employed. The differences between the closed form and finite element solutions are undiscernible. The second problem, Fig. 11.6, involves the analysis of a clamped spherical cap subjected to a concentrated load at the crown.[11.3] The solution of this problem on the basis of a thin shell finite element analysis is shown for comparison. Clearly, the axisymmetric solid finite element results converge to a solution that differs somewhat from the thin shell finite element solution. The difference is accounted for by the discrepancy between thick shell behavior and the simplified representation furnished by thin shell theory.

11.3 Nonaxisymmetric Loads

The loads applied to geometrically axisymmetric structures will not necessarily be distributed in an axisymmetric form. The wind acting on chimneys or other circular cylindrical structures is an example from practice. Also,

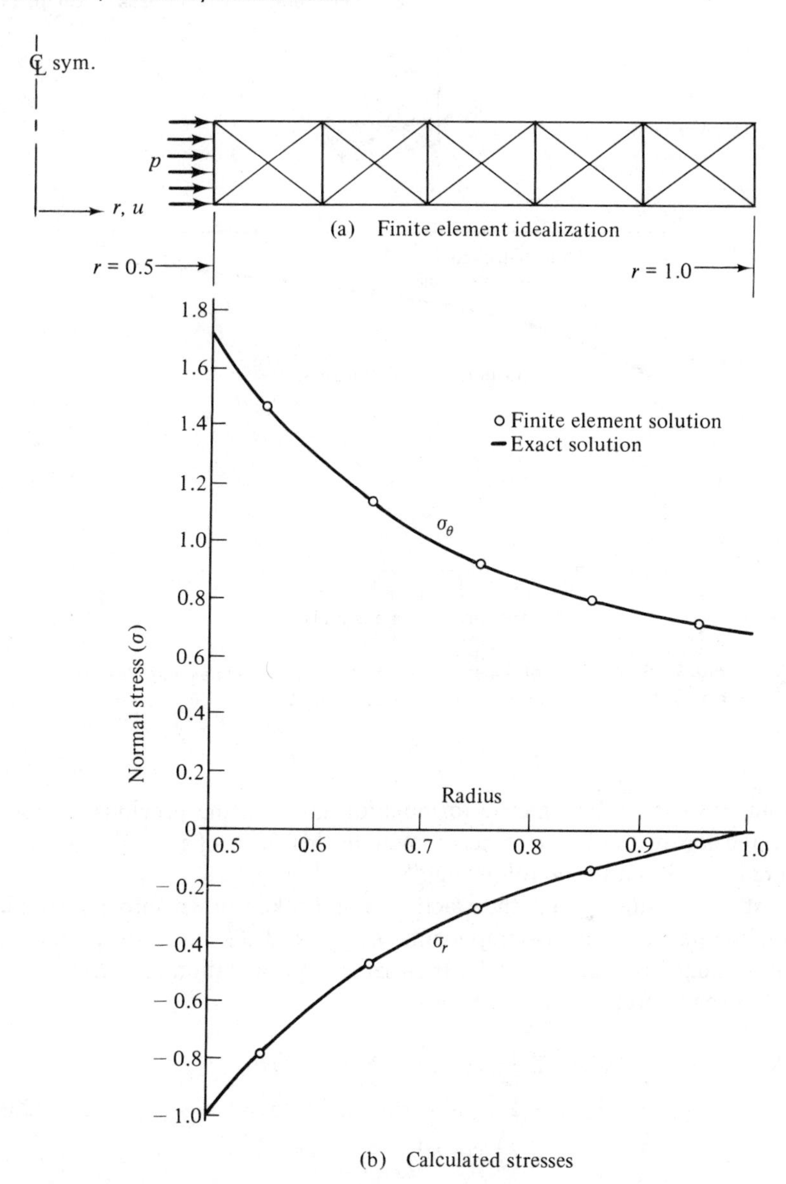

Fig. 11.5 Analysis of thick-walled cylinder under internal pressure.

in dealing with earthquakes, the inertia forces arising from ground accelerations will comprise nonaxisymmetric loadings on tanks and thick-walled cylindrical structures. In cases where the distributed load $\bar{\mathbf{T}}$ varies only with the circumferential coordinate (θ) and can be represented by a modest number of terms of a series expansion, it is possible to preserve many of the

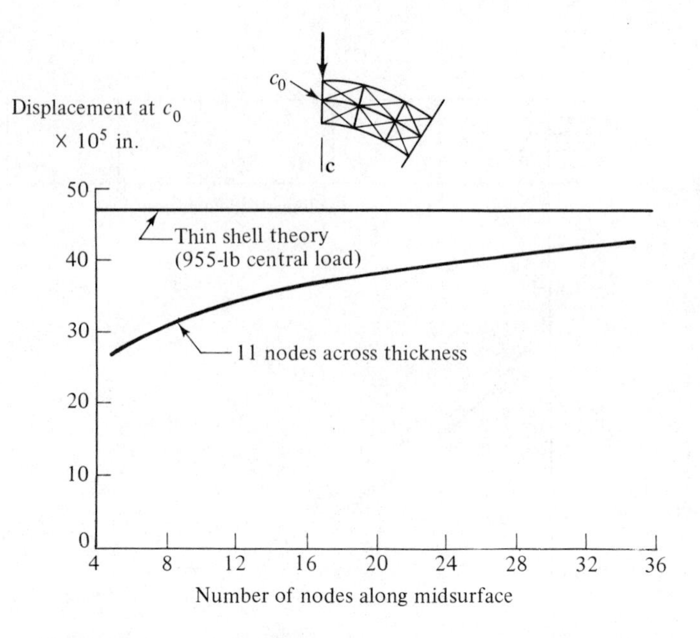

Fig. 11.6 Analysis of spherical cap with use of triangular ring elements (Ref. 11.3). Courtesy of the *AIAA Journal.*

advantages of the axisymmetric formulation given in the previous section. The extension of the latter to accommodate this form of nonaxisymmetric loading is described in the following.

First it is assumed that the traction $\bar{\mathbf{T}}$ is broken down into its radial, circumferential, and axial components T_r, T_θ, and T_z, respectively. Then, by the standard procedures of Fourier series representation (cf., Ref. 11.7) the following approximations are constructed:

$$\begin{aligned} \bar{T}_r &= \Sigma\, \bar{T}^s_{r_n} \cos n\theta + \Sigma\, \bar{T}^a_{r_n} \sin n\theta \\ \bar{T}_\theta &= -\Sigma\, \bar{T}^s_{\theta_n} \sin n\theta + \Sigma\, \bar{T}^a_{\theta_n} \cos n\theta \\ \bar{T}_z &= \Sigma\, \bar{T}^s_{z_n} \cos n\theta + \Sigma\, \bar{T}^a_{z_n} \sin n\theta \end{aligned} \tag{11.23}$$

where each term of each series is called a *harmonic* and n is the *order* of each harmonic. The summations are on n and extend over as many terms as are necessary for proper representation of the circumferential variation of loading. The superscripts s refer to symmetric components of loading, while a applies to antisymmetric components. Representative components of the radial load are sketched in Fig. 11.7.

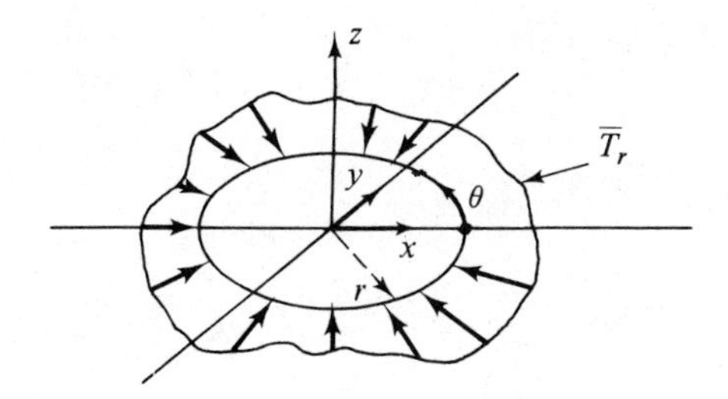

(a) Distribution of nonaxisymmetric radial load

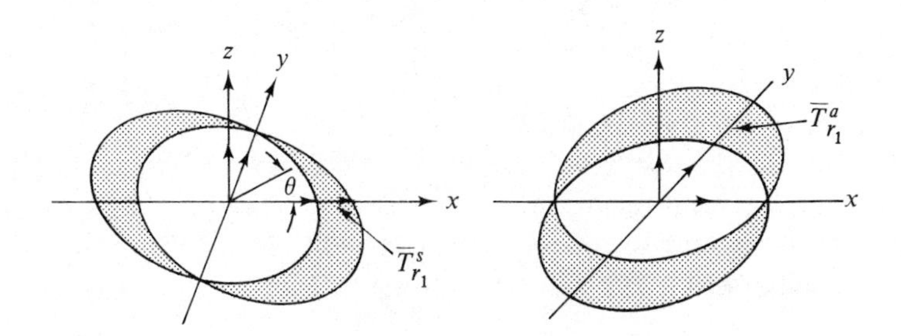

(b) First symmetric harmonic (c) First antisymmetric harmonic

Fig. 11.7 Nonaxisymmetric radial loading and first harmonics.

The resulting displacements have corresponding variation so we write

$$\begin{aligned} u &= \Sigma\, u^s_n \cos n\theta + \Sigma\, u^a_n \sin n\theta \\ v &= -\Sigma\, v^s_n \sin n\theta + \Sigma\, v^a_n \cos n\theta \\ w &= \Sigma\, w^s_n \cos n\theta + \Sigma\, w^a_n \sin n\theta \end{aligned} \tag{11.24}$$

where u^s, v^s, and w^s are the symmetric components of displacement and u^a, v^a, and w^a are the antisymmetric components. These components are defined in terms of d.o.f. by the conventional shape function representations

$$\begin{aligned} u^s_n &= \lfloor \mathbf{N}_u \rfloor \{\mathbf{u}^s_n\}, \qquad & u^a_n &= \lfloor \mathbf{N}_u \rfloor \{\mathbf{u}^a_n\} \\ v^s_n &= \lfloor \mathbf{N}_v \rfloor \{\mathbf{v}^s_n\}, \qquad & v^a_n &= \lfloor \mathbf{N}_v \rfloor \{\mathbf{v}^a_n\} \\ w^s_n &= \lfloor \mathbf{N}_w \rfloor \{\mathbf{w}^s_n\}, \qquad & w^a_n &= \lfloor \mathbf{N}_w \rfloor \{\mathbf{w}^a_n\} \end{aligned} \tag{11.25}$$

where $\{\mathbf{u}^s_n\}, \ldots, \{\mathbf{w}^a_n\}$ are vectors which list the element d.o.f. for symmetric and antisymmetric portions of the nth harmonic and $\lfloor \mathbf{N}_u \rfloor$, $\lfloor \mathbf{N}_v \rfloor$, and $\lfloor \mathbf{N}_w \rfloor$ are vectors which list the shape functions for u, v, and w displacement components, respectively. These are functions only of the r and z coordinates.

The strain-displacement equations, because of the requirement to account for circumferential behavior, are now of the form

$$\begin{aligned}
&\epsilon_r = \frac{\partial u}{\partial r}, \qquad \epsilon_z = \frac{\partial w}{\partial z}, \qquad \epsilon_\theta = \frac{u}{r} + \frac{1}{r}\frac{\partial v}{\partial \theta} \\
&\gamma_{rz} = \frac{\partial u}{\partial z} + \frac{\partial w}{\partial r}, \qquad \gamma_{z\theta} = \frac{\partial v}{\partial z} + \frac{1}{r}\frac{\partial w}{\partial \theta}, \qquad \gamma_{r\theta} = \frac{1}{r}\frac{\partial u}{\partial \theta} + \frac{\partial v}{\partial r} - \frac{v}{r}
\end{aligned} \tag{11.26}$$

We list these components of strain for the nth harmonic in a column vector as follows:

$$\boldsymbol{\epsilon}_n = \lfloor \epsilon_r^n \ \ \epsilon_\theta^n \ \ \epsilon_z^n \ \ \gamma_{r\theta}^n \ \ \gamma_{z\theta}^n \ \ \gamma_{zr}^n \rfloor^{\mathrm{T}}$$

The material stiffness matrix consistent with this strain vector is the 6×6 representation given by Eq. (10.3).

If strain-displacement relationships [Eq. (11.26)] are applied to the equations for displacement for the nth harmonic [Eq. (11.24)], a set of equations connecting the strains to the joint displacements is obtained. Utilizing our usual notation for this type of tranformation we write

$$\begin{aligned}
\boldsymbol{\epsilon}_n &= [\mathbf{D}_n^s]\begin{Bmatrix}\{\bar{\mathbf{u}}_n^s\}\\ \{\bar{\mathbf{v}}_n^s\}\\ \{\bar{\mathbf{w}}_n^s\}\end{Bmatrix} + [\mathbf{D}_n^a]\begin{Bmatrix}\{\bar{\mathbf{u}}_n^a\}\\ \{\bar{\mathbf{v}}_n^a\}\\ \{\bar{\mathbf{w}}_n^a\}\end{Bmatrix} \\
&= [\mathbf{D}_n^s]\{\boldsymbol{\Delta}_n^s\} + [\mathbf{D}_n^a]\{\boldsymbol{\Delta}_n^a\}
\end{aligned} \tag{11.27}$$

where $[\mathbf{D}_n^s]$ and $[\mathbf{D}_n^a]$ represent, respectively, the symmetric and antisymmetric displacement-to-strain transformations and

$$\begin{aligned}
\lfloor \boldsymbol{\Delta}_n^s \rfloor &= \lfloor \lfloor \mathbf{u}_n^s \rfloor \lfloor \mathbf{v}_n^s \rfloor \lfloor \mathbf{w}_n^s \rfloor \rfloor \\
\lfloor \boldsymbol{\Delta}_n^a \rfloor &= \lfloor \lfloor \mathbf{u}_n^a \rfloor \lfloor \mathbf{v}_n^a \rfloor \lfloor \mathbf{w}_n^a \rfloor \rfloor.
\end{aligned}$$

The element potential energy can now be constructed. $d(\text{vol}) = r\,d\theta\,dr\,dz$ and with use of Eqs. (10.3) and (11.27) the expression for potential energy in the nth harmonic, $\boldsymbol{\Pi}_{p_n}$, becomes (we exclude, for simplicity, initial strains)

$$\begin{aligned}
(\boldsymbol{\Pi}_{p_n}) = {} & \frac{\lfloor \boldsymbol{\Delta}_n^s \rfloor}{2}[\mathbf{K}_n^s]\{\boldsymbol{\Delta}_n^s\} + \frac{\lfloor \boldsymbol{\Delta}_n^a \rfloor}{2}[\mathbf{K}_n^a]\{\boldsymbol{\Delta}_n^a\} - \lfloor \boldsymbol{\Delta}_n^s \rfloor\{\mathbf{F}_n^{d^s}\} \\
& - \lfloor \boldsymbol{\Delta}_n^a \rfloor\{\mathbf{F}_n^{d^a}\} - \lfloor \boldsymbol{\Delta}_n^s \rfloor\{\mathbf{F}_n^s\} - \lfloor \boldsymbol{\Delta}_n^a \rfloor\{\mathbf{F}_n^a\}
\end{aligned} \tag{11.28}$$

where, for the symmetric terms, the stiffness matrix is

$$[\mathbf{K}_n^s] = \left[\int\int\int [\mathbf{D}_n^s]^{\mathrm{T}}[\mathbf{E}][\mathbf{D}_n^s] r\,d\theta\,dr\,dz\right] \tag{11.29}$$

$\{\mathbf{F}_n^{d^s}\}$ and $\{\mathbf{F}_n^s\}$ are the element distributed load and joint force vectors, respectively, for symmetric behavior. The terms with superscript a similarly define matrices for antisymmetric behavior. It should be noted that expansion of the expressions for these matrices yields terms of the type

$$\int_0^{2\pi} \cos^2 n\theta \, d\theta = \int_0^{2\pi} \sin^2 n\theta \, d\theta = \pi$$

which then eliminate the circumferential coordinate from the integral.

The analysis problem is thus divided into a group of separate analyses for each of the symmetric and antisymmetric modes for each harmonic. The response of the structure is given by the sum of the respective solutions. It is to be observed that in the solution for the harmonic conponents, those with $n = 1$ require three support conditions and for $n > 1$ only the axial rigid body freedom of the structure needs to be suppressed to make the global stiffness matrix nonsingular. For $n = 0$ the rigid body rotation about and displacement along the axis need to be suppressed.

Application of the approach described in this section is exemplified in Refs. 11.8–11.10.

11.4 Specified Volume Change—Incompressibility

A common problem in soil mechanics concerns the analysis of consolidation. This problem is characterized by the specification of volume changes within the soil. In cases where the volume change is required to be zero, the condition of *incompressibility* prevails. Both cases require modification of the previously described procedures for the analysis of solids.

The analysis of an undrained, fully saturated soil in accordance with linear theory assumes that the material under study is a porous solid with two phases. One phase consists of a porous material with linear elastic properties, while the other phase is a continuous incompressible liquid. We assume that the stresses in the first phase $\boldsymbol{\sigma}^0$, the porous material, are related to the strains $\boldsymbol{\epsilon}^0$ in the conventional linear mode

$$\boldsymbol{\sigma}^0 = [\mathbf{E}^0]\boldsymbol{\epsilon}^0 \tag{11.30}$$

The elastic coefficients $[\mathbf{E}^0]$ are known and Poisson's ratio is less than 0.5. Equation (11.30) represents the constitutive relationships of the *drained* soil. A system of stiffness equations can be constructed on the basis of Eq. (11.30) and whatever d.o.f.-to-strain transformation is appropriate to type of element being used. The presence of the pore pressures in the saturated state, however, require the condition of zero volumetric strain ϵ_v; i.e.,

$$\epsilon_v = \epsilon_x + \epsilon_y + \epsilon_z = 0 \tag{11.31}$$

Thus, with the strain-displacement equations and the usual shape function representations for the displacements ($u = \lfloor \mathbf{N}_u \rfloor \{\mathbf{u}\}$, $v = \lfloor \mathbf{N}_v \rfloor \{\mathbf{v}\}$, and $\mathbf{w} = \lfloor \mathbf{N}_w \rfloor \{\mathbf{w}\}$), we have

$$\epsilon_v = \left\lfloor \left\lfloor \frac{\partial \mathbf{N}_u}{\partial \mathbf{x}} \right\rfloor \left\lfloor \frac{\partial \mathbf{N}_v}{\partial \mathbf{y}} \right\rfloor \left\lfloor \frac{\partial \mathbf{N}_w}{\partial \mathbf{z}} \right\rfloor \right\rfloor \begin{Bmatrix} \{\mathbf{u}\} \\ \{\mathbf{v}\} \\ \{\mathbf{w}\} \end{Bmatrix} = 0 \tag{11.32}$$

and since it is required that the change of volume for each element be zero, we have for a single element

$$\int_{\text{vol}} \epsilon_v \, d(\text{vol}) = \lfloor \mathbf{G}_s^e \rfloor \begin{Bmatrix} \{\mathbf{u}\} \\ \{\mathbf{v}\} \\ \{\mathbf{w}\} \end{Bmatrix} = 0 \tag{11.33}$$

where

$$\lfloor \mathbf{G}_s^e \rfloor = \int_{\text{vol}} \left\lfloor \left\lfloor \frac{\partial \mathbf{N}_u}{\partial \mathbf{x}} \right\rfloor \left\lfloor \frac{\partial \mathbf{N}_v}{\partial \mathbf{y}} \right\rfloor \left\lfloor \frac{\partial \mathbf{N}_w}{\partial \mathbf{z}} \right\rfloor \right\rfloor d(\text{vol}) \tag{11.34}$$

Equation (11.33) is clearly a constraint equation that can be appended to the global system of equations by the Lagrange multiplier technique (see Chapter 7). Hence, the full set of global equations is of the form

$$\begin{Bmatrix} \mathbf{P} \\ \mathbf{0} \end{Bmatrix} = \begin{bmatrix} \mathbf{K}^0 & \mathbf{G}_s^T \\ \mathbf{G}_s & \mathbf{0} \end{bmatrix} \begin{Bmatrix} \mathbf{\Delta}^0 \\ \boldsymbol{\lambda} \end{Bmatrix} \tag{11.35}$$

in which $\lfloor \mathbf{\Delta}^0 \rfloor = \lfloor \lfloor \mathbf{u} \rfloor \lfloor \mathbf{v} \rfloor \lfloor \mathbf{w} \rfloor \rfloor$

$\{\boldsymbol{\lambda}\}$ = vector of Lagrange multipliers, one for each element

$[\mathbf{K}^0]$ = global stiffness matrix, constructed from elements whose stiffness matrix is derived using the constitutive law Eq. (11.30)

$[\mathbf{G}_s]$ = matrix of constraint equation coefficients, formed of the rows $\lfloor \mathbf{G}_s^e \rfloor$ given by Eq. (11.34)

$\{\mathbf{P}\}$ = vector of applied loads.

As might be expected from the discussion of Lagrange multipliers in Chapters 6 and 7, the λs are proportional to the pore pressures within the respective elements. Such pressures are of an amount sufficient to prevent a change of volume of the element.

In the case of a soil consolidation process the change in volume is not zero but rather varies in time. If a step-by-step solution process is adopted, nonzero volume changes are defined within each time step. The right side of Eq. (11.34) is therefore nonzero, as is the right side of Eq. (11.35). These and other aspects of soil consolidation analysis are taken up in Refs. 11.11–11.14.

In single-phase materials, when Poisson's ratio (μ) equals 0.5, the value associated with incompressible materials, the terms of the stress-strain form of the constitutive relationships become infinite because of the factor $(1 - 2\mu)$ in a denominator [see Eq. (10.3), (11.3), and (11.8)]. When μ is only slightly different from 0.5, the accuracy of the solution for the displacements may be compromised and this in turn will have serious effects on the stress calculations since the stresses are obtained by differentiation of the displacements.

In order to modify the potential energy approach to deal with this problem, one can note that in the presence of incompressible behavior only the deviatoric components of strain play a role in the stress-strain response. Thus, the deviatoric components of strain are separated from the dilatational components and are used as the exclusive basis for the element formulation.

Alternatively, a convenient direct approach to the analysis of single-phase incompressible materials is by use of a special form of Reissner's principle introduced by Herrmann.[11.15] Reissner's functional was discussed in Section 6.8. Dealing, for simplicity, with the plane strain condition for isotropic incompressible materials we note that the physical circumstances of the problem permit the stress terms of the functional to be supplanted by a single pressure parameter, $\bar{p}$, defined as

$$\bar{p} = \frac{(\sigma_x + \sigma_y + \sigma_z)}{E} \tag{11.36}$$

Also, the stress-strain law is of the form of the "dilatation" relationship

$$(\sigma_x + \sigma_y + \sigma_z) = \frac{E}{(1 - 2\mu)}(\epsilon_x + \epsilon_y + \epsilon_z) \tag{11.37}$$

In the plane strain case, it can be shown that the stationary value of the following functional governs the solution for the displacements u and v and the pressure parameter $\bar{p}$

$$\Pi_I = \int_A \left\{ \frac{E}{2(1 + \mu)}[(\epsilon_x^2 + \epsilon_y^2) + \gamma_{xy}{}^{2/2} + 2\mu\bar{p}(\epsilon_x + \epsilon_y)] - \mu(1 - 2\mu)\bar{p}^2 \right\} dA - \int_{s_\sigma} \bar{\mathbf{T}} \cdot \mathbf{u} \, dS \tag{11.38}$$

In discretizing this functional for finite element analysis it is desirable to describe $\bar{p}$ in the form of a shape function with joint d.o.f. $\{\bar{\mathbf{p}}\}$ that connect to the corresponding values of the adjacent elements. If, on the other hand, $\bar{p}$ is left "free" within the element for the $\mu = 0.5$ case, the same difficulties that arise in a conventional potential energy formulation may occur. In particular, for the choice of a linear displacement field and a constant value of $\bar{p}$ for the element, it can be shown[11.16] that the potential and Reissner energy formulations are identical. Numerical results[11.16] tend to show that the most efficient solutions are obtained when the order of the interpolation on the representations of the displacements and on $\bar{p}$ are the same. Discretizations of this functional in the finite element mode and the construction of similar mixed functionals that account for anisotropic properties are described in Refs. 11.15–11.19. Many of the materials that are incompressible, e.g., rubber, experience very large strains under load. These require special forms of large strain constitutive relationship and an associated modification of the element formulations. This question is taken up in Ref. 11.20.

REFERENCES

11.1 Den Hartog, J. P., *Advanced Strength of Materials*, McGraw-Hill Book Co., New York, N.Y., 1952.

11.2 Dunham, R. S., and R. E. Nickell, "Finite Element Analysis of Axisymmetric Solids with Arbitrary Loadings," Report 67–6, Dept. of Civil Eng., Structural Engineering Laboratory, U. of California, Berkeley, Calif., June, 1967.

11.3 Utku, S., "Explicit Expressions for Triangular Torus Element Stiffness Matrix," *AIAA J.* **6**, No. 6, pp. 1174–1175, June, 1968.

11.4 Belytschko, T., "Finite Elements for Axisymmetric Solids under Arbitrary Loadings with Nodes at Origin," *AIAA J.*, **10**, No. 11, pp. 1582–1584, 1972.

11.5 Chacour, S. "A High Precision Axisymmetric Triangular Element Used in the Analysis of Hydraulic Turbine Components," Trans. ASME, *J. Basic Eng.*, **92**, pp. 819–826, 1970.

11.6 Silvester, P., and A. Konrad, "Axisymmetric Triangular Elements for the Scalar Helmholtz Equation," *Internat'l. J. Num. Meth. Eng.*, **5**, No. 4, pp. 481–498, 1973.

11.7 Sokolnikoff, I. S. and R. M. Redheffer, *Mathematics of Physics and Modern Engineering*, McGraw-Hill Book Co., New York, N.Y., 1966, pp. 56–83.

11.8 Wilson, E., "Structural Analysis of Axisymmetric Solids," *AIAA J.*, **3**, No. 12, pp. 2267–2274, Dec., 1965.

11.9 Argyris, J. H., K. E. Buck, I. Grieger, and G. Mareczek, "Application of the Matrix Displacement Method to the Analysis of Pressure Vessels," *Trans. ASME*, **92**, Ser. B., pp. 317–329, 1970.

11.10 Zienkiewicz, O. C., *The Finite Element Method in Engineering Science*, McGraw-Hill Book Co., Ltd., London, 1971. Chapter 13.

11.11 Christian, J. T., "Undrained Stress Distribution by Numerical Methods," Proc. ASCE, *J. Soil Mech. Fdn. Div.*, **94**, No. SM6, pp. 1333–1345, Nov., 1968.

11.12 Christian, J. T. and J. W. Boehmer, "Plane Strain Consolidation by Finite

Elements," Proc. ASCE, *J. Soil Mech. Fdn. Div.*, **96**, No. SM4, pp. 1435–1457, July, 1970.

11.13 HWANG, C., N. MORGENSTERN, and D. MURRAY, "On Solutions of Plane Strain Consolidation Problems by Finite Element Methods," *Canadian Geotechnical J.*, **8**, pp. 109–118, 1971.

11.14 SANDHU, R. S., "Finite Element Analysis of Consolidation and Creep," Proc. of Conf. on Applications of the Finite Element Method in Geotechnical Eng., C. Desai. (ed.), U.S. Army Eng. Vicksburg Experiment Sta., Vicksburg, Miss., pp. 697–698, 1972.

11.15 HERRMANN, L. R., "Elasticity Equations for Incompressible and Nearly Incompressible Materials by a Variational Theorem," *AIAA J.*, **3**, No. 10, pp. 1896–1900, Oct., 1965.

11.16 HUGHES, T. and H. ALLIK, "Finite Elements for Compressible and Incompressible Continua," Proc. of Symp. on Application of Finite Element Methods in Civil Eng., W. Rowan and R. Hackett (eds.), Vanderbilt U., Nashville, Tenn., pp. 27–62, Nov., 1969.

11.17 HWANG, C., M. HO, and N. WILSON, "Finite Element Analysis of Soil Deformations," Proc. of Symp. on Application of Finite Element Methods in Civil Eng., W. Rowan and R. Hackett (eds.), Vanderbilt U., Nashville, Tenn., pp. 729–746, Nov., 1969.

11.18 KEY, S. W., "A Variational Principle for Incompressible and Nearly-Incompressible Anisotropic Elasticity," *Internat'l. J. Solids Struct.*, **5**, pp. 951–964, 1969.

11.19 TAYLOR, R. L., K. PISTER, and L. R. HERRMANN, "On a Variational Theorem for Incompressible and Nearly Incompressible Orthotropic Elasticity," *Internation'l. J. Solids Struct.*, **4**, pp. 875–883, 1968.

11.20 ODEN, J. T. and J. E. KEY, "Numerical Analysis of Finite Axisymmetric Deformations of Incompressible Elastic Solids of Revolution," *Internat'l. J. Solids Struct.*, **6**, pp. 497–518, 1970.

PROBLEMS

11.1 Formulate the stiffness matrix for the annular element shown in cross section in Fig. P11.1, using a linear radial displacement function.

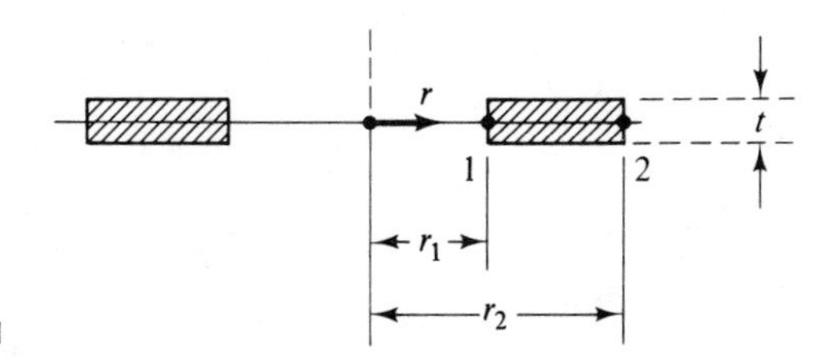

Fig. P11.1

11.2 Establish the stiffness matrix for the core element shown in cross section in Fig. 11.4, using bilinear displacement functions.

11.3 Verify the formula for I_4 given by Eq. (11.18b).

11.4 Using two-term harmonic expansions of the symmetric terms of the u displacement of Eq. (11.24) (i.e., u_1^s and u_2^s), formulate the corresponding stiffness matrices of the annular element defined in Prob. 11.1.

11.5 Starting with Reissner's functional for plane stress [adapted from Eq. (6.81)], establish the functional Π_I given by Eq. (11.38) for incompressible behavior.

11.6 Establish the discretized form of Π_I [Eq. (11.38)] and transform the resulting mixed matrix format to a stiffness matrix format by assuming that the parameters of the pressure field will not be connected to the parameters of this field of the adjacent elements.

11.7 Establish the matrix equations for generalized plane strain analysis, using the approach outlined at the close of Section 11.1.

11.8 Establish the explicit form of the matrix $[\mathbf{D}_n^s]$ of Eq. (11.27).

11.9 Formulate, for an incompressible elastic isotropic material, a stiffness matrix based on deviatoric strain components. Employ the simple (linear displacement) plane strain triangle.

11.10 Extend the plane strain material stiffness relationships [Eq. (11.3)] to account for initial strains.

12

Flat Plate Bending

Extensive efforts have been devoted to the formulation of plate flexure elements.[12.1] The requirements of an adequate formulation are often difficult to meet, and for this reason an exceptionally wide variety of alternative formulations have been proposed. Advances have been made, for the same reason, on the basis of a wider range of variational principles than for other types of elements. Potential energy formulations continue to predominate, however.

The significance of our study of flat plate bending extends well beyond a concern with applications that are directly treatable with the linear, static formulations discussed in this chapter. One effective method of conducting finite element curved thin shell analyses is by means of representations consisting of flat elements. Such elements derive from a superposition of plane stress and flexural behavior. The plane stress representations were described in Chapter 9; the present chapter completes development of the essential representation for this mode of shell analysis.

The importance of an understanding of plate flexure stems also from the practical significance of dynamic and instability effects in plate and shell structures. We deal with the latter topic in Chapter 13, and in the development of finite element equations to deal with such phenomena we shall draw upon the representations of element behavior established in this chapter.

Finally, it should be noted that difficulties alluded to above in selecting adequate displacement fields stem from the fact that thin plate flexure is governed by a *fourth-order* differential equation, rather than by a *second-*

order equation as in the case of the governing equilibrium differential equations of plane stress, plane strain, and three-dimensional analysis. Other formulations of the elasticity problem, as well as problems in a wide range of physical phenomena, are governed by fourth-order equations and it is important to understand the common basic difficulties in each.

We raise the argument above because difficulties in the establishment of admissible displacement fields are avoided by resorting to complementary or mixed variational principles, where the problems of stress field selection are minimal, or to the isoparametric solid elements, which are reduced to thin plate form by imposition of certain constraints and other operations. The latter were discussed in Chapters 9 and 10; the former approaches are described herein.

This chapter first deals with the simplest form of plate flexure situation, one that excludes transverse shear and initial strain. Also, the formulations and problems to be discussed are mainly concerned with isotropic materials. Following a brief review of the fundamental relationships of plate flexure, attention is given to many alternative formulations for quadrilateral and triangular elements. In contrast to Chapter 9, "Plane Stress," the triangle represents a more challenging formulative problem than the quadrilateral. The latter is therefore treated first.

In most respects the organization and style of this chapter are different from those of other chapters of the text. The details of formulation of just a few of the great variety of plate bending element formulations would require the length of a chapter. Our main objective is to give an overall view of finite element technology for plate bending and to do so a survey-style presentation is adopted. Some key aspects of the problem are detailed, however.

The basic relationships and energy functionals of plate bending are detailed and in this way it is possible to identify the important role of complementary and mixed functionals in this topic. Rectangular element formulations are explored in some depth. Close attention is also given to two widely used triangular elements. Finally, the problem of deflection due to transverse shear is examined. This aspect of plate bending has importance in its own right but also suggests approaches to flexure-without-transverse shear that are simpler, from a formulative standpoint, than the conventional assumed displacement approaches.

12.1 Plate Flexure Theory

12.1.1 Basic Relationships

The theory of thin plate flexure is developed in detail in many texts (e.g., Refs. 12.2–12.4) and is outlined here simply to furnish the necessary equations for subsequent illustrations of element formulations.

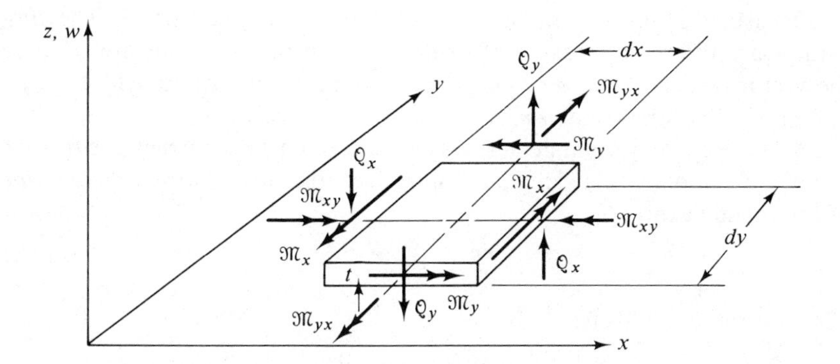

Fig. 12.1 Differential element of a thin flat plate.

A differential element of a thin flat plate, of thickness t, is portrayed in Fig. 12.1. A state of plane stress exists in the plate ($\sigma_z = \gamma_{xz} = \gamma_{yz} = 0$) and in accordance with the usual assumptions of thin plate flexure these stresses vary linearly across the thickness. Integration of the direct stresses across the thickness yields stress resultants in the form of *direct* ($\mathfrak{M}_x$, $\mathfrak{M}_y$) and *twisting* ($\mathfrak{M}_{xy}$) *moments* per inch. The positive sense of these moments is as shown in Fig. 12.1. For simplicity, we do not show the rates of change of these moments and the associated shears, which enter into the formulation of the differential equations of equilibrium. Thus,

$$\mathfrak{M}_x = \int_{-t/2}^{t/2} \sigma_x z\, dz$$

$$\mathfrak{M}_y = \int_{-t/2}^{t/2} \sigma_y z\, dz$$

and

$$\mathfrak{M}_{xy} = \mathfrak{M}_{yx} = \int_{-t/2}^{t/2} \tau_{xy} z\, dz$$

It is convenient to regard these *line loadings* $\boldsymbol{\mathfrak{M}} = \lfloor \mathfrak{M}_x\, \mathfrak{M}_y\, \mathfrak{M}_{xy} \rfloor^{\mathrm{T}}$ as supplanting the stress vector for plane stress, $\boldsymbol{\sigma} = \lfloor \sigma_x\, \sigma_y\, \tau_{xy} \rfloor^{\mathrm{T}}$.

The fundamental assumption of thin plate flexure is that lines that are initially normal to the plate middle surface remain normal during deformation of the plate. The rates of change of the angular displacements of these normals are the *direct* (κ_x, κ_y) and *twisting* (κ_{xy}) *curvatures*, and these are assumed to be adequately approximated by the following second derivatives of the transverse displacement, w:

$$\kappa_x = -\frac{\partial^2 w}{\partial x^2}, \qquad \kappa_y = -\frac{\partial^2 w}{\partial y^2}, \qquad \kappa_{xy} = -2\frac{\partial^2 w}{\partial x\, \partial y} \tag{12.1}$$

w is measured from the original state of the plate middle surface. The curvatures are the basic measure of deformation in thin plate flexure. Hence, the vector of curvatures $\boldsymbol{\kappa} = \lfloor \kappa_x \; \kappa_y \; \kappa_{xy} \rfloor$ supplants the strain field $\boldsymbol{\epsilon} = \lfloor \epsilon_x \; \epsilon_y \; \gamma_{xy} \rfloor$ of plane stress analysis.

With the analogies and definitions above in mind we construct one further analogy with plane elasticity and introduce the *constitutive relationships* of thin plate flexure:

$$\mathfrak{M} = [\mathbf{E}_f]\boldsymbol{\kappa} \tag{12.2}$$

where, for an orthotropic plate

$$[\mathbf{E}_f] = \begin{bmatrix} D_x & D_1 & 0 \\ D_1 & D_y & 0 \\ 0 & 0 & D_{xy} \end{bmatrix} \tag{12.3}$$

D_x, D_y, and D_1 are the orthotropic plate *flexural rigidities*. In the more familiar case of *isotropic* thin plate flexure

$$[\mathbf{E}_f] = D \begin{bmatrix} 1 & \mu & 0 \\ \mu & 1 & 0 \\ 0 & 0 & \dfrac{(1-\mu)}{2} \end{bmatrix} \tag{12.3a}$$

with

$$D = \frac{Et^3}{12(1-\mu^2)} \tag{12.4}$$

The governing equilibrium differential equation of plate flexure is important to an understanding of the choice of element displacement fields. The basis for this relationship is the equilibrium differential equations derived by examining, on a differential element, the equilibrium of forces with respect to the vertical direction and about the x and y axes, respectively. We thereby obtain

$$\frac{\partial \mathfrak{Q}_x}{\partial x} + \frac{\partial \mathfrak{Q}_y}{\partial y} + q = 0 \tag{12.5a}$$

$$\frac{\partial \mathfrak{M}_x}{\partial x} + \frac{\partial \mathfrak{M}_{xy}}{\partial y} - \mathfrak{Q}_x = 0 \tag{12.5b}$$

$$\frac{\partial \mathfrak{M}_y}{\partial y} + \frac{\partial \mathfrak{M}_{xy}}{\partial x} - \mathfrak{Q}_y = 0 \tag{12.5c}$$

where q is the transverse distributed load and $\mathfrak{Q}_x$ and $\mathfrak{Q}_y$ are the shear line loads.

Now, by substitution of the moment-curvature relationships [Eq. (12.2)] into Eqs. (12.5b) and (12.5c) and substitution of the resulting expressions into Eq. (12.5a), one obtains

$$D_x \frac{\partial^4 w}{\partial x^4} + 2(D_1 + 2D_{xy})\frac{\partial^4 w}{\partial x^2\, \partial y^2} + D_y \frac{\partial^4 w}{\partial y^4} = q \tag{12.6}$$

For an isotropic plate this simplifies to

$$D\left(\frac{\partial^4 w}{\partial x^4} + \frac{2\,\partial^4 w}{\partial x^2\, \partial y^2} + \frac{\partial^4 w}{\partial y^4}\right) = q \tag{12.6a}$$

We observe that the solution of the thin plate flexure problem, studied from the displacement point of view, depends entirely on the selection of the single component w, the transverse displacement.

12.1.2 Potential Energy

The majority of existing formulations of finite element plate flexure are obtained via application of this principle. We have, in further analogy with plane stress,

$$\Pi_p = \frac{1}{2}\int_A \boldsymbol{\kappa}^T[\mathbf{E}_f]\boldsymbol{\kappa}\, dA + V \tag{12.7}$$

where $\boldsymbol{\kappa}$ and $[\mathbf{E}_f]$ are as previously defined and V is the potential of the applied loads. In the case of specified distributed loads $\bar{q}$ that are normal to the plate surface, this is given by

$$-\int_A \bar{q}\cdot w\, dA \tag{12.8a}$$

while for specified edge line loads $\bar{\mathcal{Q}}$ and normal and twisting moments $\bar{\mathcal{M}}_n$ and $\bar{\mathcal{M}}_s$ (see Fig. 12.2), we have

$$-\int_{S_\sigma} (\bar{\mathcal{Q}}\cdot w + \bar{\mathcal{M}}_n\cdot\theta_n + \bar{\mathcal{M}}_s\cdot\theta_s)\, dS \tag{12.8b}$$

Fig. 12.2 Prescribed edge loads.

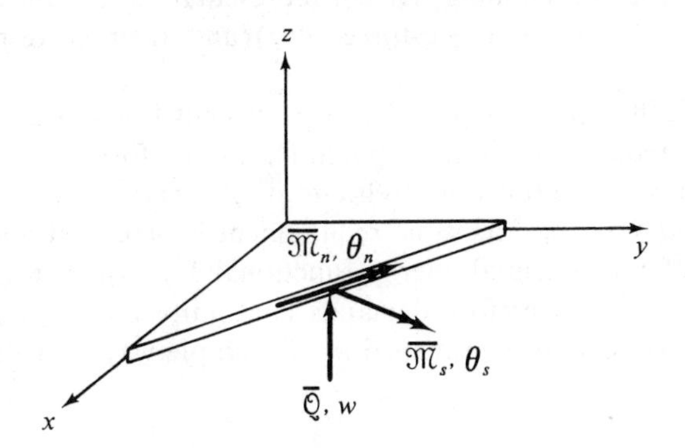

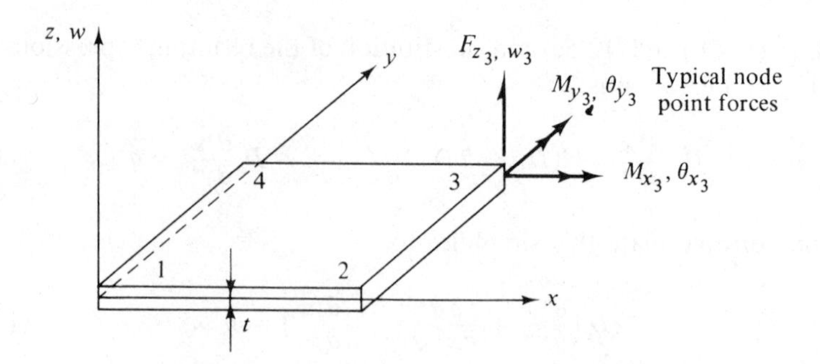

Fig. 12.3 Rectangular thin flat plate in flexure.

where S_σ is the portion of the edge on which these line loads are prescribed. Finally, for specified joint forces $\bar{F}_{z_i}$ and moments $\bar{M}_{x_i}$ and $\bar{M}_{y_i}$, we have

$$-\sum_{i=1}^{r} F_{z_i} w_i - \sum_{i=1}^{r} \bar{M}_{x_i}\theta_{x_i} - \sum_{i=1}^{r} \bar{M}_{y_i}\theta_{y_i} \tag{12.8c}$$

where the summation on i extends over the number of joints on the element.

Figure 12.3, which portrays a rectangular plate flexure element, clarifies the meaning of the concentrated forces and moments and their corresponding displacements. The basic assumptions of thin plate flexure theory (normals remain normal, neglect of transverse shear deformation) result in the equality of the slope of the middle surface and the angular displacement at a point. Thus

$$\theta_x = \left|\frac{\partial w}{\partial y}\right|, \qquad \theta_y = -\left|\frac{\partial w}{\partial x}\right|$$

In θ_y, the negative value of the derivative is due to the fact that a positive rotation about the y axis (as defined by the circular arrow, Fig. 12.3) causes a negative value of the w displacement. A further point to be noted is the distinction between the joint moments, which are *concentrated* values M_x, M_y in inch-pound (or meter-Newton) units and the *line moments* $\mathcal{M}_x$, $\mathcal{M}_y$, $\mathcal{M}_{xy}$ which are barred quantities in inch-pound per inch units. The joint parameters of the desired finite element representations are the concentrated moments (M_{x_i}, M_{y_i}) and direct forces (F_{z_i}) and their corresponding displacements.

It is physically apparent that the displacement field for a finite element in flexure, as required by a valid potential energy formulation, should be continuous in both the function itself and its slopes (first derivatives) across the element boundaries. The same requirement is perceived mathematically upon study of the potential energy functional Π_p, which involves *second derivatives* of w and therefore demands continuity of the *first derivatives.* This requirement cannot be satisfied in any simple way. For this reason alternative variational principles, which require continuity of only w itself, have proved very attractive in formulation of plate bending elements.

The form of the potential energy functional also discloses that the selected field must be at least quadratic; the second derivatives eliminate any contributions due to linear and constant terms in the displacement fields. As already seen for the case of the beam, it is customary to employ cubic functions in the description of flexure. The extent to which this approach is effective for plates will be disclosed in subsequent sections.

As a final comment regarding the potential energy we observe that for an isotropic material Eq. (12.6) is the *Euler equation* of this functional. The importance of this fact is that the same equation (with the Airy stress function, Φ, as independent variable) governs plate stretching in a complementary energy formulation. Consequently, our developments of acceptable displacement fields applies directly to the dual stretching formulation.

It is pertinent to examine the process of discretization of the potential energy to obtain the finite element force-displacement relationships. The approach taken closely follows the procedures applied in prior chapters. The expression for the chosen displacement field is first differentiated in accordance with Eq. (12.1) to form the $\boldsymbol{\kappa}$ field, yielding relationships of the form

$$\boldsymbol{\kappa} = [\mathbf{D}]\{\boldsymbol{\Delta}\} \tag{12.9}$$

In the rectangle shown in Fig. 12.3, for example,

$$\{\boldsymbol{\Delta}\} = \lfloor w_1\; w_2\; w_3\; w_4\; \theta_{x_1}\; \theta_{x_2}\; \theta_{x_3}\; \theta_{x_4}\; \theta_{y_1}\; \theta_{y_2}\; \theta_{y_3}\; \theta_{y_4} \rfloor^{\mathrm{T}} \tag{12.10}$$

Substitution of Eq. (12.9) into the strain energy portion of the potential energy expression yields

$$\Pi_p = \frac{\lfloor \boldsymbol{\Delta} \rfloor}{2}[\mathbf{k}]\{\boldsymbol{\Delta}\} + V \tag{12.7a}$$

where, as in prior chapters,

$$[\mathbf{k}] = \left[\int_A [\mathbf{D}]^{\mathrm{T}}[\mathbf{E}_f][\mathbf{D}]\, dA\right] \tag{12.11}$$

Note that the assumed displacement field plays a role in the determination of V when prescribed distributed loads $\bar{q}$, $\overline{\mathcal{Q}}$, $\overline{\mathfrak{M}}_n$, and $\overline{\mathfrak{M}}_{ns}$ are present.

It is important to take note of the definitions of θ_x and θ_y, given above, in cases where the assumed field is represented by a polynomial series. In such cases we have (see Section 8.2)

$$w = \lfloor \mathbf{p}(m) \rfloor \{\mathbf{a}\} \tag{12.12a}$$

so that

$$\theta_x = \left\lfloor \frac{\partial \mathbf{p(m)}}{\partial \mathbf{y}} \right\rfloor \{\mathbf{a}\} \tag{12.12b}$$

$$\theta_y = -\left\lfloor \frac{\partial \mathbf{p(m)}}{\partial \mathbf{x}} \right\rfloor \{\mathbf{a}\} \tag{12.12c}$$

Hence, after evaluation of Eq. (12.12) at the nodes, the full set of expressions is given, as usual, by

$$\{\mathbf{\Delta}\} = [\mathbf{B}]\{\mathbf{a}\} \tag{12.13}$$

After solving this equation for $\{\mathbf{a}\}$, the result is substituted back into Eq. (12.12). These expressions are then differentiated in accordance with Eq. (12.1), and Eq. (12.9) is formed. Alternatively, one may wish to differentiate Eq. (12.12) directly in accordance with the curvature-displacement relationships [Eq. (12.1)] and substitute the result into the strain energy expression. This gives a "kernel" stiffness matrix, which is referenced to the parameters $\{\mathbf{a}\}$. A stiffness matrix referenced to the joint displacements is produced by applying the inverse of the matrix $[\mathbf{B}]$ [Eq. (12.13)] to the kernel stiffness matrix in the manner of a coordinate transformation. We shall again have occasion to illustrate this procedure later in this chapter.

12.1.3 Complementary Energy

The complementary energy of an elastic structure is defined in Chapter 6 in terms of the direct stresses $\boldsymbol{\sigma} = \lfloor \sigma_x \, \sigma_y \, \tau_{xy} \rfloor$. Now, with use of the analogies defined in Sec. 12.1.1 we can construct the complementary energy expression for thin plate flexure. We have

$$\Pi_c = U^* + V^*$$

where, in this case,

$$U^* = \frac{1}{2} \int_A \mathfrak{M}^{\mathrm{T}} [\mathbf{E}_f]^{-1} \mathfrak{M} \, dA \tag{12.14}$$

and, for distributed and edge loads

$$V^* = -\int_A q \cdot \bar{w} \, dA - \int_{S_u} (\mathcal{Q} \cdot \bar{w} + \mathfrak{M}_n \cdot \bar{\theta}_n + \mathfrak{M}_s \cdot \bar{\theta}_s) \, dS \tag{12.15}$$

where now the transverse ($\bar{w}$) and angular ($\bar{\theta}_n$, $\bar{\theta}_s$) displacements are prescribed on the S_u portion of the boundary (the edge loadings are prescribed on the remaining portion of the boundary, S_σ).

A complementary energy function expressed in terms of stress functions is particularly useful in thin plate flexure analysis. In this case the relevant stress functions are Southwell's stress functions $\mathbf{\Phi}^u$ and $\mathbf{\Phi}^v$ (see Ref. 12.5), which are defined as follows:

$$\mathfrak{M}_x = -\left(\frac{\partial \mathbf{\Phi}^v}{\partial y} + \mathbf{\Omega}\right), \qquad \mathfrak{M}_y = -\left(\frac{\partial \mathbf{\Phi}^u}{\partial x} + \mathbf{\Omega}\right), \qquad \mathfrak{M}_{xy} = \frac{1}{2}\left(\frac{\partial \mathbf{\Phi}^v}{\partial x} + \frac{\partial \mathbf{\Phi}^u}{\partial y}\right)$$

$$\mathfrak{Q}_x = \frac{1}{2}\frac{\partial}{\partial y}\left(\frac{\partial \mathbf{\Phi}^u}{\partial y} - \frac{\partial \mathbf{\Phi}^v}{\partial x}\right) - \frac{\partial \mathbf{\Omega}}{\partial x},$$

$$\mathfrak{Q}_y = \frac{1}{2}\frac{\partial}{\partial x}\left(\frac{\partial \mathbf{\Phi}^v}{\partial x} - \frac{\partial \mathbf{\Phi}^u}{\partial y}\right) - \frac{\partial \mathbf{\Omega}}{\partial y} \tag{12.16}$$

in which $\mathbf{\Omega}$ is a parameter related to the distributed load in the following manner:

$$q = \left(\frac{\partial}{\partial x^2} + \frac{\partial}{\partial y^2}\right)\mathbf{\Omega} \tag{12.17}$$

If we substitute the expressions above for $\mathfrak{M}_x$, $\mathfrak{M}_y$, $\mathfrak{M}_{xy}$, $\mathfrak{Q}_x$, and $\mathfrak{Q}_y$ into Eq. (12.5), we confirm that the equations of equilibrium are satisfied, as must be the case with stress functions.

Most often the prescribed displacements are prescribed to be zero so that $V^* = 0$. In discussing discretization of $\mathbf{\Pi}_c$, therefore, we concentrate upon U^*. By substitution of Eq. (12.16) into Eq. (12.14), we obtain

$$U^* = \frac{1}{2}\left[\int_A \mathbf{\Phi}'^{\mathrm{T}}[\mathbf{E}_f]^{-1}\mathbf{\Phi}'\,dA - 2\int_A \mathbf{\Phi}'^{\mathrm{T}}[\mathbf{E}_f]^{-1}\begin{Bmatrix}1\\1\\0\end{Bmatrix}\mathbf{\Omega}\,dA + \int_A \lfloor 1\ 1\ 0 \rfloor[\mathbf{E}_f]^{-1}\begin{Bmatrix}1\\1\\0\end{Bmatrix}\mathbf{\Omega}^2\,dA\right] \tag{12.18}$$

where

$$\mathbf{\Phi}' = \left\lfloor -\frac{\partial \mathbf{\Phi}^v}{\partial y} \quad -\frac{\partial \mathbf{\Phi}^u}{\partial x} \quad \frac{1}{2}\left(\frac{\partial \mathbf{\Phi}^v}{\partial x} + \frac{\partial \mathbf{\Phi}^u}{\partial y}\right)\right\rfloor^{\mathrm{T}} \tag{12.19}$$

The third integral on the right will vanish upon differentiation of U^* and the second integral will produce a vector of constants. To deal with the basic properties of a finite element we consequently examine only the first integral. Clearly, except for differences in constants and the fact that $[\mathbf{E}_f]^{-1}$ supplants $[\mathbf{E}_f]$, this term is of a form identical to the strain energy for plane stress. Thus, we adopt the same type of approximation as in the plane stress case; i.e.,

$$\begin{aligned}\mathbf{\Phi}^u &= \lfloor \mathbf{N} \rfloor\{\mathbf{\Phi}^u\}\\ \mathbf{\Phi}^v &= \lfloor \mathbf{N} \rfloor\{\mathbf{\Phi}^v\}\end{aligned} \tag{12.20}$$

where $\{\boldsymbol{\Phi}^u\}$ and $\{\boldsymbol{\Phi}^v\}$ list values of Φ^u and Φ^v (and possibly their derivatives with respect to x and y) at the element joints and, after differentiation in accordance with Eq. (12.18), we obtain

$$\boldsymbol{\Phi}' = [\mathbf{D}]\begin{Bmatrix}\boldsymbol{\Phi}^u\\ \boldsymbol{\Phi}^v\end{Bmatrix} = [\mathbf{D}]\{\boldsymbol{\Phi}\} \tag{12.21}$$

Except for certain constants, the matrix $[\mathbf{D}]$ is the same as the strain-displacement transformation of plane stress [see Eq. (5.6b)]. Then, by substitution into the first integral of Eq. (12.18), we have

$$\frac{1}{2}\int_A \boldsymbol{\Phi}'^{\mathrm{T}}[\mathbf{E}_f]^{-1}\boldsymbol{\Phi}'\,dA = \frac{\lfloor\boldsymbol{\Phi}\rfloor}{2}[\mathbf{f}]\{\boldsymbol{\Phi}\} \tag{12.22}$$

where

$$[\mathbf{f}] = \left[\int_A [\mathbf{D}]^{\mathrm{T}}[\mathbf{E}_f]^{-1}[\mathbf{D}]\,dA\right] \tag{12.23}$$

By comparison with Eq. (9.7) we see that the flexibility matrix $[\mathbf{f}]$ for flexure, referenced to stress functions, is of the same general form of the plane stress stiffness matrix. (The thickness term, t, is included in $[\mathbf{E}_f]^{-1}$)

12.1.4 Reissner Energy

Reissner's functional for the general three-dimensional theory of elasticity was presented in Section 6.8. As in the case of the potential and complementary energy functionals we can establish the form of this energy for bending from the previous development by use of the analogies between direct stress and bending moment and direct strain and curvature. The form for plate bending, analogous to Eq. (6.81), is

$$\Pi_R = \int_A \mathfrak{M}\cdot\boldsymbol{\kappa}\,dA - U^* + V + V^* \tag{12.24}$$

where U^* and V are defined by Eqs. (12.14) and (12.18). Since we anticipate that the chosen displacement field (w) will not match the prescribed displacements ($\bar{w}$), we must express the displacement terms in V^* [Eq. (12.15)] as $(w - \bar{w})$, $(\partial w/\partial n - \partial\bar{w}/\partial n)$, and $(\partial w/\partial s - \partial\bar{w}/\partial s)$.

Equation (12.24) imposes no continuity requirements on $\mathfrak{M}$. Due to the presence of the second derivatives in $\boldsymbol{\kappa}$, however, it is necessary that the transverse displacement *and* its derivatives satisfy the interelement continuity conditions. This represents no advantage over the potential energy functional. To gain this advantage we integrate the expression above by parts, resulting in a functional Π_H of the type given by Herrmann[12.6, 12.7]

$$\Pi_H = \int_A \mathfrak{M}'\cdot\mathbf{w}'\,dA - U^* - \int_{S_n} \mathfrak{M}_s\frac{\partial w}{\partial s}\,dS + V + V^* \tag{12.24a}$$

where S_n is the total element boundary, and

$$\mathfrak{M}' \cdot \mathbf{w}' = \frac{\partial \mathfrak{M}_x}{\partial x} \cdot \frac{\partial w}{\partial x} + \frac{\partial \mathfrak{M}_y}{\partial y} \cdot \frac{\partial w}{\partial y} + \frac{\partial \mathfrak{M}_{xy}}{\partial x} \cdot \frac{\partial w}{\partial y} + \frac{\partial \mathfrak{M}_{xy}}{\partial y} \cdot \frac{\partial w}{\partial x}$$

Now, if the discretization of w is symbolized in the same manner as in previous chapters, i.e.,

$$\mathbf{w} = \lfloor \mathbf{N}_w \rfloor \{\mathbf{\Delta}\} \tag{12.25}$$

then, within the element, the slopes are with $\mathbf{w}' = \lfloor \partial w/\partial x \quad \partial w/\partial y \rfloor$

$$\mathbf{w}' = [\mathbf{N}'_w]\{\mathbf{\Delta}\}$$

and, on the boundary S_n,

$$\frac{\partial w}{\partial s} = [\mathbf{Y}]\{\mathbf{\Delta}\}$$

Also, for $\mathfrak{M}$ we write

$$\mathfrak{M} = [\mathbf{N}_M]\{\mathbf{M}\} \tag{12.26}$$

and, in the interior of the element,

$$\mathfrak{M}' = [\mathbf{N}'_M]\{\mathbf{M}\}$$

and, on the boundary,

$$\mathfrak{M}_s = [\mathbf{L}]\{\mathbf{M}\}$$

We obtain, upon substitution in Eq. (12.24),

$$\Pi_H = \lfloor \mathbf{M} \rfloor [\mathbf{\Omega}_{12}]\{\mathbf{\Delta}\} - \frac{\lfloor \mathbf{M} \rfloor}{2}[\mathbf{\Omega}_{11}]\{\mathbf{M}\} - \lfloor \mathbf{\Delta} \rfloor \{\bar{\mathbf{M}}\} + \lfloor \mathbf{M} \rfloor \{\bar{\mathbf{\Delta}}\} \tag{12.24b}$$

where

$$[\mathbf{\Omega}_{11}] = \left[\int_A [\mathbf{N}_M]^{\mathrm{T}}[\mathbf{E}_f]^{-1}[\mathbf{N}_M]\, dA\right]$$

$$[\mathbf{\Omega}_{12}] = [\mathbf{\Omega}_{21}]^{\mathrm{T}} = \left[\int_A [\mathbf{N}'_m]^{\mathrm{T}}[\mathbf{N}'_w]\, dA - \int_{S_n} [\mathbf{L}]^{\mathrm{T}}[\mathbf{Y}]\, dS\right]$$

and $\{\bar{\mathbf{M}}\}$ and $\{\bar{\mathbf{\Delta}}\}$ derive from the relevant prescribed loads and displacements, respectively, on the element surface and boundary.

Variation of Π_H with respect to $\lfloor \mathbf{M} \rfloor$ and $\lfloor \mathbf{\Delta} \rfloor$, in succession, yields

$$\begin{bmatrix} -\mathbf{\Omega}_{11} & \mathbf{\Omega}_{12} \\ \mathbf{\Omega}_{12}^{\mathrm{T}} & 0 \end{bmatrix} \begin{Bmatrix} \mathbf{M} \\ \mathbf{\Delta} \end{Bmatrix} = \begin{Bmatrix} \bar{\mathbf{\Delta}} \\ \bar{\mathbf{M}} \end{Bmatrix} \tag{12.27}$$

The interelement continuity conditions on w and $\mathfrak{M}$, from Π_H, are that w and $\mathfrak{M}_n$ be continuous, where $\mathfrak{M}_n$ is the line moment normal to the element boundary.

12.2 Rectangular Elements

12.2.1 Assumed Displacement Models—Single Field

Two general approaches to the formulation of plate bending stiffness matrices, based on assumed displacement fields, have been adopted. In one, termed here the *single-field* approach, the functional representations of displacement cover the total area of the element. Alternatively, in the subdomain approach, the element is divided into subregions and independent assumptions are made regarding the displacement fields within each subregion. The two approaches lead to the same stiffness matrix under certain conditions. We examine in this section the alternatives in single-field representation of rectangles and take up the same questions again for triangles in Sections 12.3.1 and 12.3.2.

When first examining the rectangular plate bending element (Fig. 12.3), it would appear that a simple generalization of the beam bending displacement function [Eq. (5.14a)] would suffice for the definition of the transverse displacement function w. It is recalled that this function gives continuity of both w and the angular displacement (θ) across the element joints. The conditions of continuity of displacement can be preserved across the boundaries of the rectangular element if we form the transverse displacement field from the beam functions, in the form

$$w = \lfloor \lfloor \mathbf{N}_w \rfloor \lfloor \mathbf{N}_{\theta_x} \rfloor \lfloor \mathbf{N}_{\theta_y} \rfloor \rfloor \begin{Bmatrix} \{\mathbf{w}\} \\ \{\boldsymbol{\theta}_x\} \\ \{\boldsymbol{\theta}_y\} \end{Bmatrix} = \lfloor \mathbf{N} \rfloor \{\boldsymbol{\Delta}\} \tag{12.28}$$

where $\{\boldsymbol{\Delta}\}$ is given by Eq. (12.10) and

$$\begin{aligned} \lfloor \mathbf{N}_w \rfloor &= \lfloor [N_1(x)N_1(y)]\,[N_2(x)N_1(y)]\,[N_2(x)N_2(y)]\,[N_1(x)N_2(y)] \rfloor \\ \lfloor \mathbf{N}_{\theta_y} \rfloor &= \lfloor [N_3(x)\cdot N_1(y)]\,[N_4(x)\cdot N_1(y)]\,[N_4(x)\cdot N_2(y)]\,[N_3(x)\cdot N_2(y)] \rfloor \\ \lfloor \mathbf{N}_{\theta_x} \rfloor &= \lfloor [N_1(x)\cdot N_3(y)]\,[N_2(x)\cdot N_3(y)]\,[N_2(x)\cdot N_4(y)]\,[N_1(x)\cdot N_4(y)] \rfloor \end{aligned} \tag{12.29a, b, c}$$

and

$$\begin{aligned} N_1(x) &= (1 + 2\xi^3 - 3\xi^2), & N_1(y) &= (1 + 2\eta^3 - 3\eta^2) \\ N_2(x) &= (3\xi^2 - 2\xi^3), & N_2(y) &= (3\eta^2 - 2\eta^3) \\ N_3(x) &= -x(\xi - 1)^2, & N_3(y) &= y(\eta - 1)^2 \\ N_4(x) &= -x(\xi^2 - \xi), & N_4(y) &= y(\eta^2 - \eta) \end{aligned} \tag{12.30}$$

where $\xi = x/x_2$ and $\eta = y/y_3$. We shall call this field the *crossed-beam* displacement function.

Testing of this function along the boundaries discloses that the continuity of linear (w) and angular (θ_x, θ_y) displacements is preserved when the element is joined to another based on the same function. Thus, interelement compatibility conditions are met. If the multipliers of the joint displacements in Eq. (12.29) are expanded and examined, it is found that the term representing constant shear strain, the simple twist function xy, is absent. As noted in Section 8.1, it is necessary to account for all constant strain states to ensure convergence to the correct result and for plate bending the simple twist represents constant twisting strain. Thus, we must reject this function.

The selection of an interelement-compatible displacement function that also includes all constant strain states can be accomplished by use of the interpolation concepts discussed in Chapter 8. Here, with the need to satisfy both the function and its derivatives along the boundaries, the Hermitian polynomial interpolation formula (Section 8.4) is applied. Based on this concept, an expression for the complete third-order polynomial can be written:(12.8)

$$w = [\text{Eq. (12.28)} + N_3(x)N_3(y)\mathbf{\Gamma}_1 + N_4(x)N_3(y)\mathbf{\Gamma}_2 + N_4(x)N_4(y)\mathbf{\Gamma}_3 + N_3(x)N_4(y)\mathbf{\Gamma}_4] \tag{12.31}$$

where the d.o.f. $\mathbf{\Gamma}_i$ are the cross-derivatives at the joints,

$$\mathbf{\Gamma}_1 = \left|\frac{\partial^2 w}{\partial x\,\partial y}\right|_1$$

etc. We note, therefore, that the crossed-beam function [Eq. (12.28)] is simply an incomplete Hermitian polynomial expansion and that the rectangle requires 16 terms in development of a representation of a single function. This same requirement has already been perceived in Section 8.4 where it was shown that the complete product of cubic functions encompasses 16 terms. The element stiffness matrix derived from Eq. (12.31) is presented in Ref. 12.8.

An alternative to these functions is the basic *12-term polynomial*, containing as many terms as there are "apparent" d.o.f. at the element joints. This polynomial is described in the Pascal triangle of Fig. 12.4. This can be transformed into shape function form. The shape function representation is of the same general form as in the crossed-beam case, Eq. (12.28). The row matrix $\lfloor \mathbf{N}_w \rfloor$ is again given by Eq. (12.29a). For the remaining terms of the shape function we have

$$\begin{aligned}\lfloor \mathbf{N}_{\theta_x} \rfloor &= \lfloor (1-\xi)N_3(y) \quad \xi N_2(y) \quad -\xi N_4(y) \quad -(1-\xi)N_4(y) \rfloor \\ \lfloor \mathbf{N}_{\theta_y} \rfloor &= \lfloor -(1-\eta)N_3(x) \quad (1-\eta)N_4(x) \quad \eta N_4(x) \quad -\eta N_3(x) \rfloor\end{aligned} \tag{12.32}$$

We note by observation of Fig. 12.4 that the 12-term function is not a complete polynomial in the sense defined in Section 8.1. The function is complete up to third order (10 terms) and a selection must be made of 2 terms from among the 5 terms of fourth order. The x^2y^2 term must be discounted since it cannot be logically paired with any other. The x^4 and y^4 terms give a quartic variation of displacement along the element edges and would give more serious discontinuities of displacement along the interelement boundaries than the x^3y and xy^3 terms. Thus, we choose the latter. It is interesting to note that this choice permits satisfaction of the governing differential equation [Eq. (12.6a)] over the unloaded region of the plate; it must again be emphasized, however, that satisfaction of this condition is not required in a stationary potential energy approximation.

Other properties of this function are of interest. The rigid-body-motion and constant-strain terms are present, and testing of it along the element boundaries discloses that the transverse displacements are interelement-compatible. The angular displacements, however, do not meet this condition.

To confirm this conclusion we need only evaluate the polynomial expansion (Fig. 12.4) along a typical edge. Choosing side 1-2 (the x axis) for this purpose, we have

$$w = a_1 + a_2x + a_4x^2 + a_7x^3$$

$$\frac{\partial w}{\partial x} = a_2 + 2a_4x + 3a_7x^2$$

$$\frac{\partial w}{\partial y} = a_3 + a_5x + a_8x^2 + a_{12}x^3$$

w and $\partial w/\partial x$ represent flexure in the x direction. Also, we note that the expansion for w is a cubic polynomial. From the experience with beam flexure it is clear that four of the d.o.f. at the end points (w_1, w_2, θ_{y_1}, θ_{y_2}) completely

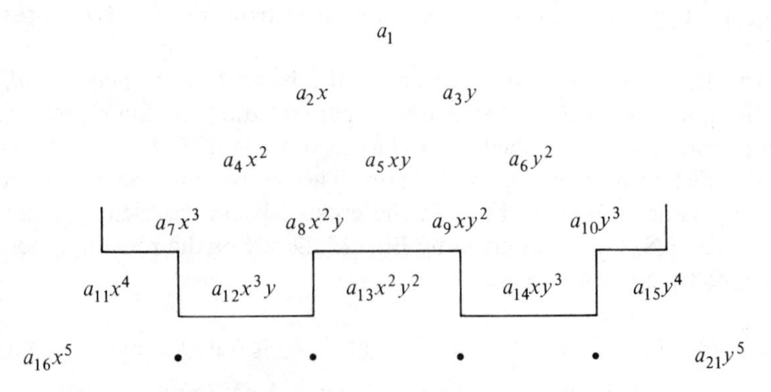

Fig. 12.4 Twelve-term polynomial in context of Pascal triangle.

define the variation of w and $\partial w/\partial x$ along this edge. The normal slope ($\partial w/\partial y$), however, is described by a cubic function and since only two d.o.f. (θ_{x_1} and θ_{x_2}) remain for definition of this variation, it is not uniquely defined. Thus, a solution obtained from a gridwork of elements of this type will not be a minimum potential energy solution. On the other hand, theoretical proofs of convergence for this element have been established (e.g., Ref. 12.13). The element stiffness matrix associated with this function is given in Table 12.1. (See pages 374–376.)

An examination of the Pascal triangle (Fig. 12.4) discloses that numerous alternative choices can be made of polynomial representations and there are corresponding alternatives when the displacement field is constructed in shape function form. Dawe(12.9) gives a number of shape functions for 12 d.o.f. representations. Gopalacharyulu(12.10) and Irons(12.11) discuss alternative 16-d.o.f. displacement fields. Bogner et al.(12.8) and Wegmuller and Kostem,(12.12) among others, formulate rectangular plates with greater than 16 d.o.f.

12.2.2 Assumed Displacement Models—Subdomain Formulations

Alternative developments based upon assumed displacement fields can be accomplished by subdividing the quadrilateral into four triangles and assuming independent displacement fields within each triangle. The triangles are then combined to form the quadrilateral by imposing conditions of displacement continuity along the "internal" boundaries as defined by the subregions.

Fraeijs de Veubeke(12.14) has adopted this approach in the formulation of a 16-d.o.f. interelement-compatible element. He assumes a complete cubic (10-term) polynomial displacement field within each triangle. The conventional 3-d.o.f. at the corners (translation and two rotations) are supplemented by a d.o.f. along each side in the form of an angular displacement.

A similar approach has been detailed by Clough and Felippa(12.15) who achieve a potential energy formulation of a quadrilateral element based upon definition of the element via appropriate combination of four triangles. Here the triangles themselves derive from a division of the element into three triangular subregions (formulation of these triangles is discussed in Section 12.3.2). It should be noted that prior to combination of the triangles to form the quadrilateral, constraints are placed on sides that are to be exterior boundaries so as to remove a d.o.f. at the midpoints of such sides. Thus, the final quadrilateral represents a total of 12 d.o.f. (three at each corner). The conditions of internal and interelement compatibility are met by this element.

12.2.3 Generalized Variational Approach

The generalized variational approach, which we described in Chapters 6 and 7, is especially attractive in plate flexure formulations. Since it is difficult to define and work with transverse displacement fields that are fully interelement-compatible, it is desirable to select a convenient field that does not meet this condition and subsequently enforce continuity by use of constraint equations. In the case of the 12-term function [Eq. (12.27)], for example, only the continuity of angular displacement along each boundary requires restoration. Rather extensive developments of this idea for quadrilateral bending elements may be found in the papers by Greene et al.[12.16] and Kikuchi and Ando.[12.17] We shall discuss this approach again for triangles in Section 12.3.

12.2.4 Mixed Stress-Displacement Formulations

Pian[12.18] and Severn and Taylor[12.19] have explored a number of hybrid stress formulations for the rectangle, and quadrilateral element formulations are given by Allwood and Cornes.[12.20] These are each single-field schemes. Cook[12.21] has presented two alternative subdomain formulations using hybrid stress concepts.

The Reissner energy approach, modified by Herrmann[12.7] as described in Section 12.2, has been utilized by Bron and Dhatt[12.22] for a variety of quadrilateral elements in both single-field and subdomain representations.

12.2.5 Assumed Stress Fields

It was pointed out that the complementary energy functional, in terms of Southwell stress functions, requires the same fields as used in description of displacements for plane stress analysis by the potential energy approach. Thus, the treatment given the latter topic in Section 9.3 applies here as well. Numerical results for this alternative are presented in Ref. 12.23.

12.2.6 Numerical Comparisons

Consider first the case of a square plate of dimension $2a \times 2a$ with fixed supports subjected to a concentrated load P_1 at the center (see Fig. 12.5). Due to symmetry about two axes we can analyze the problem with just one element in the quadrant of the plate, in which case there is only one d.o.f., w_1, the displacement under the load. In this case $P_1 = k_{11}w_1$, $P = P_1/4$, $x_2 = y_3 = a$. We have for the 12-term formulation, from Table 12.1 (see page 374).

$$\frac{P_1}{4} = \frac{Et^3}{360(1-\mu^2)a^2}[120(1+1) - 24\mu + 84]w_1$$

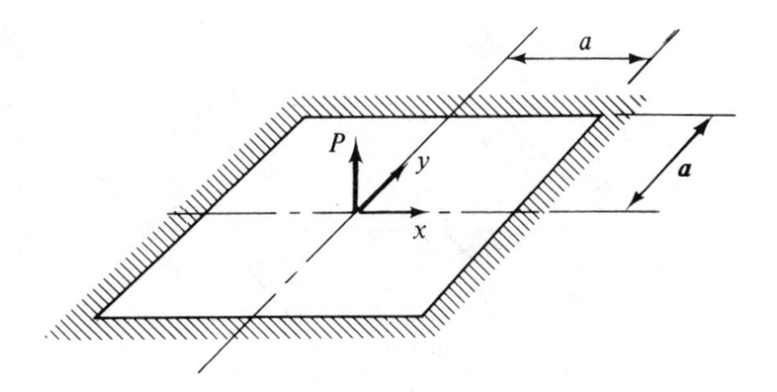

Fig. 12.5 Square plate with fixed edges.

or, with $D = Et^3/12(1 - \mu^2)$, $\mu = 0.3$

$$w_1 = 0.0237\frac{a^2}{D}P_1$$

For the 16-term formulation, using the stiffness coefficient given in Ref. 12.8, we find

$$w_1 = 0.0212\frac{a^2P_1}{D}$$

The exact solution(12.2) is $w_1 = 0.0224\,(a^2P_1/D)$ so that each solution is approximately 8% in error, but on different sides of the exact solution. As anticipated, the "conforming" (16-term) solution is a lower bound on the exact solution.

Figure 12.6 shows a problem we shall employ in comparing the various plate bending element formulations. It refers to the calculation of the displacement under a concentrated load applied to the center of a square, simply supported plate. Plots to be given will show the percentage error in the calculation of this displacement as a function of mesh size in a quadrant of the plate.

It should be emphasized that the results presented do not necessarily define the proper parameters for comparison of accuracy and efficiency nor is the abscissa (mesh size) necessarily the most valid measure of relative effort. Such quantities as stress or strain energy may very well be more significant as parameters of structural behavior. The most desirable measure of relative effort would account for such factors as programming effort, impact on the cost of equation-solving, and interpretability of results. For example, these same results have been plotted against a different measure of efficiency by Abel and Desai.(12.24) The principal objectives of plots

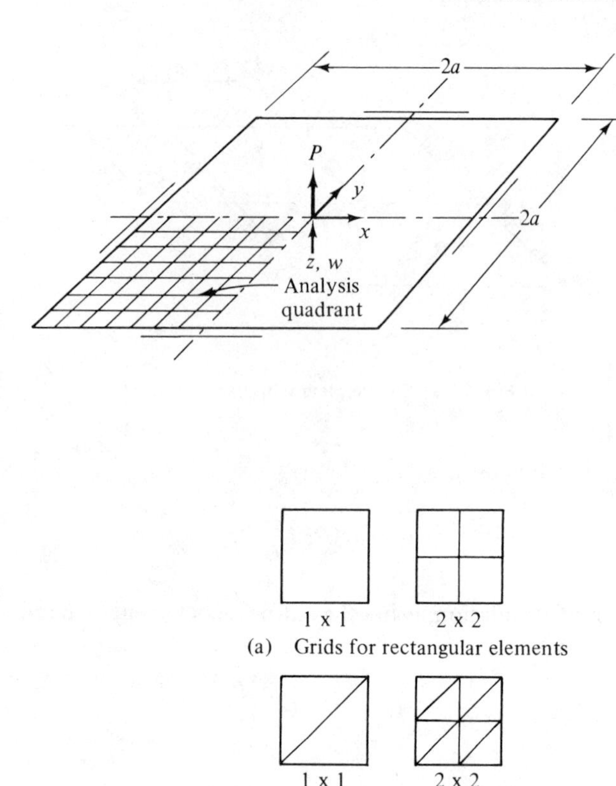

† Grids shown are representative. Orientations at 90° also employed in studies whose results are given in this chapter.

Fig. 12.6 Problem for numerical comparisons.

given in this chapter are to confirm upper or lower bound solution characteristics, demonstrate convergence, and assess alternatives within restricted classes of element shapes and formulative procedures.

Figure 12.7 gives results for the various rectangular element formulations. We note that the 12-term polynomial approaches the correct solution from above; since the conditions of interelement continuity of displacement are violated, the "lower bound" solution characteristic of a rigorous minimum potential energy formulation is not obtained. In contrast, the 16-term formulation and the subdomain formulation due to Fraeijs de Veubeke[12.14] are conforming and indeed give lower bound results. Two formulations based on a modified Reissner functional are also shown.[12.22] One of these em-

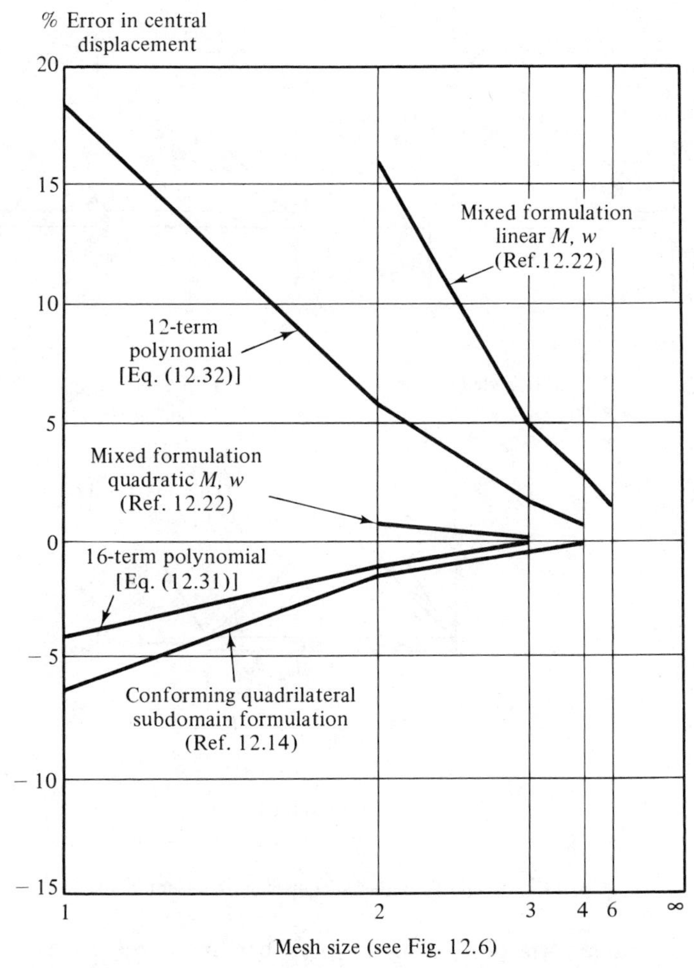

Fig. 12.7 Numerical comparisons: quadrilateral plate element formulations.

ploys linear bending moment and edge transverse displacement fields. The other employs quadratic functions. It is clear that a significant improvement in accuracy results from the increased order of these functions.

12.3 Triangular Elements

12.3.1 Assumed Displacement Formulations—Single Field

A series of different arrangements of d.o.f. for the triangular plate flexure element, each associated with a different choice of terms in a polynomial representation of the transverse displacement, w, is shown in Fig. 12.8

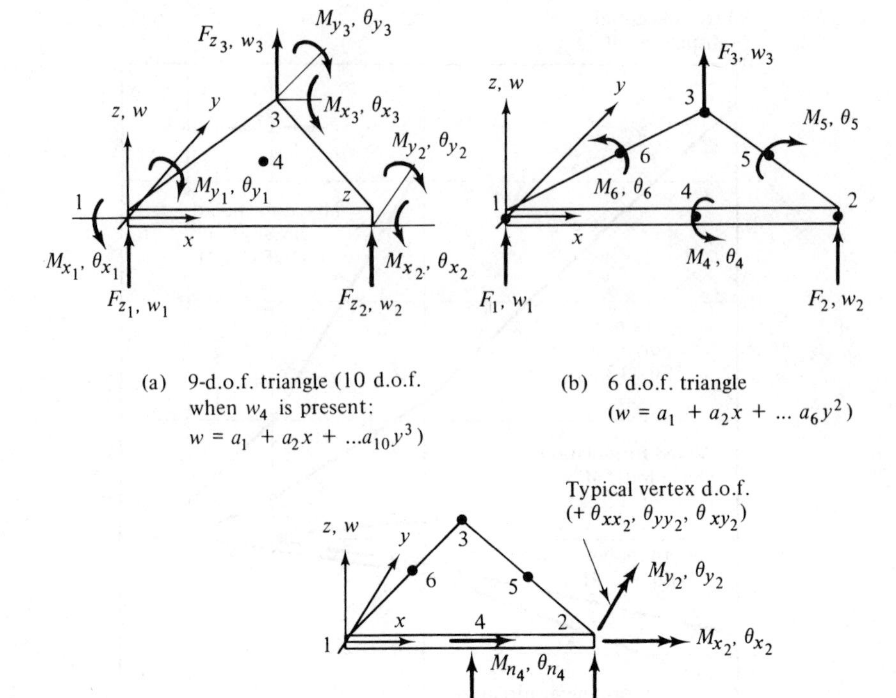

Fig. 12.8 Single-field triangular element representations.

The desirable form of a triangular plate bending element is sketched in Fig. 12.8(a). It has a z direction direct force and two bending moments at each vertex, with no interior points or points on the boundary between the vertices. It involves nine d.o.f. and therefore implies a 9-term expansion for w. We see from the Pascal triangle, however, that complete polynomials contain either 6 terms (quadratic) or 10 terms (cubic). A pair of terms could be combined to produce a 9-term representation [e.g., $a_9(x + y)xy$] but one finds that the transformation from generalized to joint d.o.f. is singular for certain element shapes. The choice of 9 terms therefore cannot be made with a polynomial series that is complete in a given order.

If 9-terms of the polynomial are chosen, without regard to completeness, then conditions of "geometric isotropy" will be violated. For example, the term xy^2 may be discarded. Alternatively, the x^2y term can be discarded. Elements formulated on the basis of these alternatives have been employed for some time and appear in various widely distributed programs but their solution accuracy is not good. More satisfactory alternative "single-field" approaches include the following:

1. An element may be defined to represent only 6 d.o.f. [Fig. 12.8(b)] and a complete quadratic function chosen to represent w. This formulation will violate the interelement displacement-continuity requirements.
2. A complete cubic function with 10 d.o.f. may be chosen, with the tenth d.o.f. positioned at an interior point. [Fig. 12.8(a)] This formulation will also violate interelement displacement-continuity requirements.
3. The number of d.o.f. may be expanded until a coincidence is obtained with a complete polynomial that also satisfies the interelement displacement-continuity condition. It will be found that this requires a complete *quintic* polynomial [Fig. 12.8(c)].
4. Construction of an interelement continuous displacement field, oriented toward a 9-d.o.f. representation [Fig. 12.8(b)], can be accomplished in terms of triangular coordinates by superposition of the proper systems of shape functions.[12.25] A significant improvement of this type of formulation is presented in Ref. 12.26.

We examine options 1, 2, and 3 in some detail in the following. Interested readers may consult Refs. 12.25 and 12.26 for details of option 4; some aspects of this approach will be taken up in the examination of numerical results, to be presented subsequently in this section.

The element representation for the 6-d.o.f. (complete quadratic) displacement field is given in Fig. 12.8(b). The transverse displacement field is described by

$$w = a_1 + a_2x + a_3y + a_4x^2 + a_5xy + a_6y^2 \tag{12.33}$$

Also, the joint displacements are defined as the transverse displacements at the vertices (points 1, 2, and 3) and the normal angular displacements at the midpoints of the sides (points 4, 5, and 6). Thus,

$$\lfloor \mathbf{\Delta} \rfloor = \lfloor w_1 \; w_2 \; w_3 \; \theta_4 \; \theta_5 \; \theta_6 \rfloor$$

By evaluation of each d.o.f. in $\lfloor \mathbf{\Delta} \rfloor$ in terms of the expansion represented by Eq. (12.33) we obtain the set of equations $\{\mathbf{\Delta}\} = [\mathbf{B}]\{\mathbf{a}\}$, [Eq. (12.13)], where $\{\mathbf{a}\} = \lfloor a_1 \ldots a_6 \rfloor^T$. The kernel stiffness matrix is formed on the basis of Eq. (12.33) [see Eq. (6.18)] and is transformed to the physical d.o.f. by use of the matrix $[\mathbf{B}]^{-1}$ as a transformation matrix.

Although this element formulation violates the conditions of interelement displacement continuity, it meets all conditions of equilibrium. The moment field meets the conditions of internal equilibrium and these conditions are preserved across element boundaries.[12.27]

The generalized potential energy approach can be used to advantage in the construction of an element stiffness matrix based upon a complete *cubic* polynomial.(12.28) The element is described in Fig. 12.8(a). In this case the value of w and of the angular displacements θ_x and θ_y is defined at each of the three vertices and the transverse displacement at the centroid is chosen as the tenth d.o.f. We have, for the vector of d.o.f.,

$$\lfloor \mathbf{\Delta} \rfloor = \lfloor w_1 \, \theta_{x_1} \, \theta_{y_1} \, w_2 \, \theta_{x_2} \, \theta_{y_2} \, w_3 \, \theta_{x_3} \, \theta_{y_3} \, w_4 \rfloor \tag{12.34}$$

The transverse displacement w is described by

$$w = \lfloor \mathbf{N} \rfloor \{\mathbf{\Delta}\}$$

where the entries in $\lfloor \mathbf{N} \rfloor$, in terms of triangular coordinates, are

$$\begin{aligned}
N_1 &= L_1^2(L_1 + 3L_2 + 3L_3) - 7L_1L_2L_3 \\
N_6 &= L_2^2(x_{23}L_3 - x_{12}L_1) + (x_{12} - x_{23})L_1L_2L_3 \\
N_2 &= L_1^2(y_{31}L_3 - y_{12}L_2) + (y_{12} - y_{31})L_1L_2L_3 \\
N_7 &= L_3^2(L_3 + 3L_1 + 3L_2) - 7L_1L_2L_3 \\
N_3 &= L_1^2(x_{12}L_2 - x_{31}L_3) + (x_{31} - x_{12})L_1L_2L_3 \\
N_8 &= L_3^2(y_{23}L_2 - y_{31}L_1) + (y_{31} - y_{23})L_1L_2L_3 \\
N_4 &= L_2^2(L_2 + 3L_3 + 3L_1) - 7L_1L_2L_3 \\
N_9 &= L_3^2(x_{31}L_1 - x_{23}L_2) + (x_{23} - x_{31})L_1L_2L_3 \\
N_5 &= L_2^2(y_{12}L_1 - y_{23}L_3) + (y_{23} - y_{12})L_1L_2L_3 \\
N_{10} &= 27L_1L_2L_3
\end{aligned}$$

$$x_{ij} = x_i - x_j \qquad y_{ij} = y_i - y_j \tag{12.35}$$

The element stiffness matrix is formed straightforwardly by differentiating Eq. (12.35) in accordance with Eq. 12.1 to produce $\boldsymbol{\kappa} = [\mathbf{D}]\{\mathbf{\Delta}\}$. Then, the element stiffness is obtained from Eq. (12.11).

Conditions of interelement compatibility are violated by the displacement field above. The w component is continuous across element boundaries, while the normal angular displacement, θ_n, is not. This displacement varies quadratically on each edge, requiring three displacement parameters for unique definition, but only two (θ_n at the end points of the edge) are available. The remaining four parameters at these points were allotted to the unique definition of w.

To resolve this situation we can write an equation that specifies continuity of normal slope at the midpoints of the sides. Suppose that two adjacent elements are designated as A and B and the normal slopes at the midpoint of their sides are denoted as θ_n^A and θ_n^B, respectively. Continuity of angular displacement is required so that

$$\theta_n^A - \theta_n^B = 0 \tag{12.36}$$

We can use this to form a constraint equation by first differentiating the displacement fields [Eq. (12.35)] of the adjacent elements with respect to n to obtain

$$\theta_n^A = \frac{\partial w^A}{\partial n} \quad \text{and} \quad \theta_n^B = \frac{\partial w^B}{\partial n}$$

and then by substitution of the resulting expressions into Eq. (12.36). The constraint equations produced in this manner can be accounted for in the global analysis through either the direct substitution or Lagrange multiplier methods, described in Sections 3.5 and 7.4, respectively.

An alternative way of constructing the interelement constraint condition in this case is by setting to zero the integral of the difference between the edge angular displacements of adjacent elements (see Section 7.4 and Ref. 12.29). Still another generalized variational approach is one[12.17] which forms a "corrective" element stiffness matrix which is added to the basic (interelement-incompatible) element stiffness matrix. The latter derives from a boundary integral on the element into which is inserted a simple interelement compatible function.

The next most refined representation is a complete quartic of 15 terms. Chu and Schnobrich[12.30] have formulated a triangular element on the basis of this function. The array of d.o.f. encompasses the conventional 3 d.o.f. at each vertex plus a transverse displacement and normal slope at the midpoint of each side. Interelement compatibility is violated because insufficient parameters are available for the unique definition of the normal slope on each edge. Irons[12.31] has also explored the formulation of a quartic element, but with 18 terms.

Proceeding to the complete quintic polynomial, the Pascal triangle discloses that this involves 21 terms. Complete satisfaction of interelement displacement continuity requirements can be achieved with the d.o.f. arrayed as shown in Fig. 12.8(c). This consists of six at each joint—the linear and angular displacements and the three curvatures, plus the angular displacement at the midpoint of each side. The formulation can be reduced to 18 d.o.f. by removing the angular displacement at the midpoints of the sides[12.33-12.35] by enforcing a cubic variation of the edge slope.

Single-field elements of even higher order, which retain adherence to the interelement compatibility conditions, have been formulated (e.g., Refs.

12.32, 12.36, 12.37). These extend to complete sextic (28-term) and septic (36-term) polynomials and in some cases involve special (higher-derivative) d.o.f. at the element node points.

12.3.2 Numerical Comparisons—Assumed Displacement Formulations—Single Field

Numerical results for the single-field formulations we have discussed, for the problem of the centrally loaded, square, simply supported plate, are shown in Fig. 12.9.

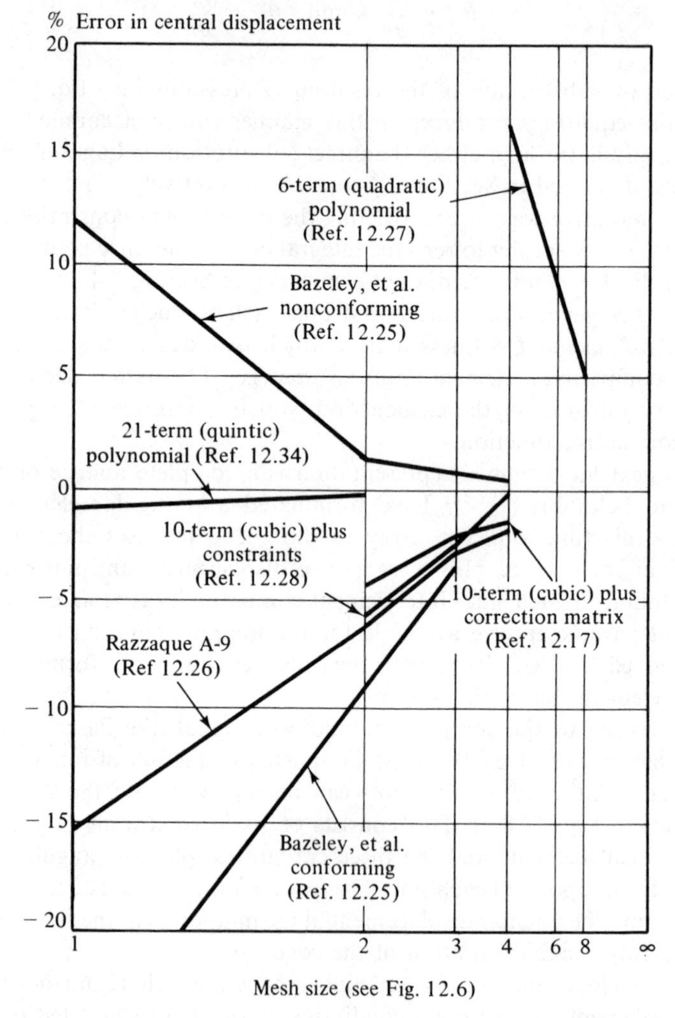

Fig. 12.9 Numerical comparisons: single-field triangular element formulations.

We note first that the 6-d.o.f. (complete quadratic) element(12.27) converges to the correct solution from above; i.e., it is an *upper bound* solution. This is because it is in complete conformity with the conditions of equilibrium. The results are rather poor for a given level of grid refinement but on the other hand the element stiffness matrix is rather simple of form. Indeed, it is possible to express this stiffness matrix in explicit form without excessive effort.

The only results shown for the element based on a complete cubic polynomial apply to the case where constraints are imposed to restore angular displacement continuity across the element boundaries. Results for the case where constraints are not imposed are so poor that they do not appear on the scales of this graph. As one might expect, the solutions we obtain either by enforcing a specific constraint on each interelement boundary(12.28) or by introducing a "corrective" stiffness matrix by use of a generalized variational principle(12.17) are very close.

We do, however, show the results obtained with use of the element based on cubic shape functions in terms of area coordinates [Eq. (12.35)] from which the term involving $L_1L_2L_3$ has been removed (i.e., the formulation described by Bazeley et al.).(12.25) When account is taken of the simplicity of this element, it can be concluded that the numerical results obtained are excellent. It must nevertheless be pointed out that the correctness of the results obtained depends on the geometry of the finite element gridwork.(12.25)

The results for the formulation based on a complete quintic (21-term) polynomial are highly accurate. Results for the case where the mid-side joints are eliminated before the analysis is performed are not shown because they are not significantly different. The formulative effort for this element is quite substantial and it must again be cautioned that the tradeoff between formulative effort and size of gridwork, which is not represented in Fig. 12.9, must be accounted for in any practical circumstance.

12.3.3 Assumed Displacement Formulations—Subdomain Approach

The formulation of triangular plate bending element stiffness matrices by means of the *subdomain* approach, through the device of subdividing the complete element into triangular subelements, was given impetus by Clough and Tocher.(12.38) They employed an incomplete (9-term) cubic polynomial in each of three subelements but with the coordinate system of each subelement so arranged that difficulties do not arise because of lack of geometric isotropy and in a manner that yields a quadratic variation of displacement along each edge that will be exterior in the complete element.

An improvement of the formulation above is to employ a complete cubic polynomial (10 terms) within each subregion. The basic d.o.f. are therefore

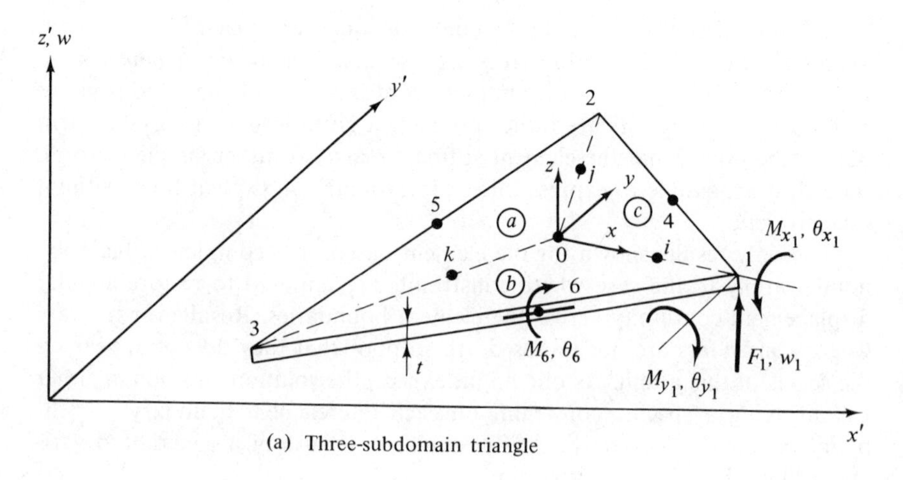

(a) Three-subdomain triangle

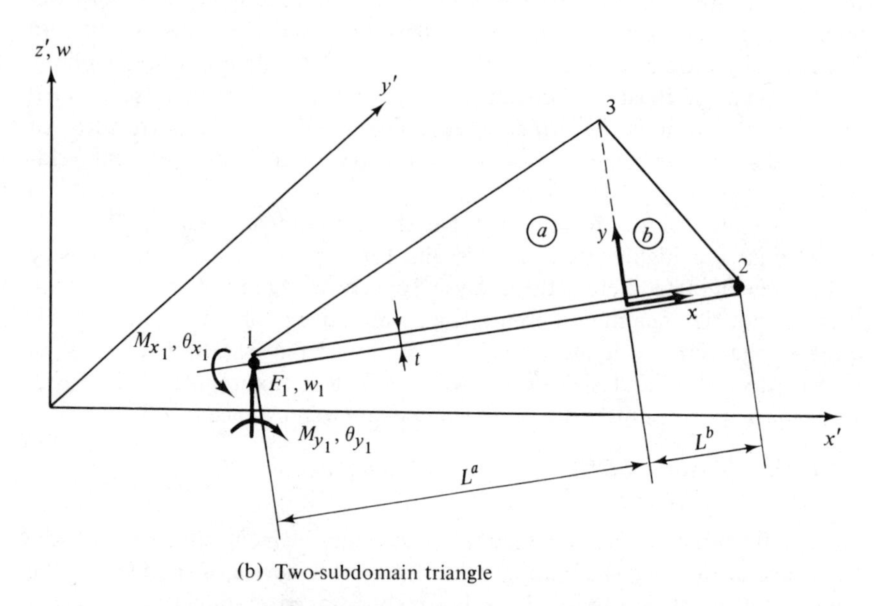

(b) Two-subdomain triangle

Fig. 12.10 Triangular elements—subdomain models.

30 in number and are subsequently reduced to 12—the 9 d.o.f. at the vertices and an angular displacement at the midpoint of each side—by imposition of conditions of continuity of displacement across the subdomain interfaces. We detail this formulation in the following.

The element is illustrated in Fig. 12.10a. The subregions are designated a, b, c. As we have noted, within each subregion the deflection is assumed to be described by a complete cubic polynomial:

$$\begin{aligned} w^a &= a_1 + a_2 x + a_3 y + \cdots + a_{10} y^3 = \lfloor \mathbf{p}(3) \rfloor \{\mathbf{a}\} \\ w^b &= b_1 + b_2 x + b_3 y + \cdots + b_{10} y^3 = \lfloor \mathbf{p}(3) \rfloor \{\mathbf{b}\} \\ w^c &= c_1 + c_2 x + c_3 y + \cdots + c_{10} y^3 = \lfloor \mathbf{p}(3) \rfloor \{\mathbf{c}\} \end{aligned} \tag{12.37}$$

The 30 undetermined parameters $\{\mathbf{a}\}, \{\mathbf{b}\}, \{\mathbf{c}\}$ are immediately reduced to 24 by noting the six conditions associated with the compatibility of displacement at the interior juncture point 0; i.e., $a_1 = b_1 = c_1$, $a_2 = b_2 = c_2$, $a_3 = b_3 = c_3$. The further reduction from 24 to 12 d.o.f. is accomplished by imposition of displacement continuity conditions and equality of "internal" and "external" d.o.f. at the vertices (1, 2, 3), the midpoints of the sides (4, 5, 6), and the interior interface midpoints (i, j, k).

To develop this approach in some detail, we define the following shorthand notation for the vector of displacements at a joint. At vertices the vector of displacements $\lfloor w \; \theta_x \; \theta_y \rfloor$ will be written as $\lfloor \boldsymbol{\Delta}_f^g \rfloor$, where the subscript f designates the joint ($f = 1, 2, 3$) and the superscript g designates the subregion ($g = a, b, c$) on which the vertex is located. For example, for joint 1, subelement a,

$$\begin{Bmatrix} w_1^a \\ \theta_{x_1}^a \\ \theta_{y_1}^a \end{Bmatrix} = \{\boldsymbol{\Delta}_1^a\}$$

When the vector refers to the displacements at a joint of the assembled element (combination of three subelements) or at the points at the midsides of the subelements (i, j, k, 4, 5, 6), there is no superscript. The 24 relevant compatibility equations can be written as

$$\begin{Bmatrix} \boldsymbol{\Delta}_1^c \\ \boldsymbol{\Delta}_2^a \\ \boldsymbol{\Delta}_3^b \\ --- \\ \boldsymbol{\Delta}_4^c \\ \boldsymbol{\Delta}_5^a \\ \boldsymbol{\Delta}_6^b \\ --- \\ \boldsymbol{\Delta}_1^c \\ \boldsymbol{\Delta}_2^a \\ \boldsymbol{\Delta}_3^b \\ --- \\ \boldsymbol{\Delta}_i^c \\ \boldsymbol{\Delta}_j^a \\ \boldsymbol{\Delta}_k^b \end{Bmatrix} = \begin{Bmatrix} \boldsymbol{\Delta}_1 \\ \boldsymbol{\Delta}_2 \\ \boldsymbol{\Delta}_3 \\ --- \\ \boldsymbol{\Delta}_4 \\ \boldsymbol{\Delta}_5 \\ \boldsymbol{\Delta}_6 \\ --- \\ \boldsymbol{\Delta}_1^b \\ \boldsymbol{\Delta}_2^c \\ \boldsymbol{\Delta}_3^a \\ --- \\ \boldsymbol{\Delta}_i^b \\ \boldsymbol{\Delta}_j^c \\ \boldsymbol{\Delta}_k^a \end{Bmatrix}$$

Nine equations—representing the equality of subelement and total element displacements at the vertices

Three equations—representing the identity of subelement and total element angular displacement at the midpoints of the sides points 4, 5, 6

Nine equations—representing the equality of displacements of the adjacent subelements at the vertices

Three equations—representing the equality of angular displacements at interior midside nodes i, j, k

By evaluation of these displacements at the node points with use of Eq. (12.37) we have

$$\begin{bmatrix} [\mathbf{B}_{aa}] & [\mathbf{B}_{a0}] \\ [\mathbf{B}_{0a}] & [\mathbf{B}_{00}] \end{bmatrix} \begin{Bmatrix} \{\mathbf{a}_a\} \\ \{\mathbf{a}_0\} \end{Bmatrix} = \begin{Bmatrix} \{\boldsymbol{\Delta}\} \\ \mathbf{0} \end{Bmatrix} \tag{12.38}$$

wherein the equations have been partitioned so that

$$\{\mathbf{a}_a\} = \{a_1, a_2, a_3, a_4, a_5, a_6, a_7, a_8, a_9, a_{10}, b_7, b_8\}^{\mathrm{T}}$$
$$\{\mathbf{a}_0\} = \{b_4, b_5, b_6, b_9, b_{10}, c_4, c_5, c_6, c_7, c_8, c_9, c_{10}\}^{\mathrm{T}}$$

and

$$\{\boldsymbol{\Delta}\} = \{w_1 \ \theta_{x_1} \ \theta_{y_1} \ w_2 \cdots \theta_{y_3} \ \theta_{n_4} \ \theta_{n_5} \ \theta_{n_6}\}^{\mathrm{T}} \tag{12.39}$$

where θ_{n_4}, θ_{n_5}, and θ_{n_6} are the angular displacements normal to the element edge at points 4, 5, 6. By the usual condensation procedure, (see Section 2.8)

$$[\hat{\mathbf{B}}_{aa}]\{\mathbf{a}_a\} = \{\boldsymbol{\Delta}\} \tag{12.40}$$

where

$$[\hat{\mathbf{B}}_{aa}] = [\boldsymbol{\Gamma}]^{\mathrm{T}} \begin{bmatrix} [\mathbf{B}_{aa}] & [\mathbf{B}_{a0}] \\ [\mathbf{B}_{0a}] & [\mathbf{B}_{00}] \end{bmatrix} [\boldsymbol{\Gamma}]$$

and

$$[\boldsymbol{\Gamma}] = \begin{bmatrix} \mathbf{I} \\ \hline -\mathbf{B}_{00}^{-1} \ \mathbf{B}_{0a} \end{bmatrix}$$

Finally, by solution of Eq. (12.40),

$$\begin{aligned} \{\mathbf{a}_a\} &= [\hat{\mathbf{B}}_{aa}]^{-1}\{\boldsymbol{\Delta}\} \\ \{\mathbf{a}_0\} &= -[[\mathbf{B}_{00}^{-1}][\mathbf{B}_{0a}]][\hat{\mathbf{B}}_{aa}]^{-1}\{\boldsymbol{\Delta}\} \end{aligned} \tag{12.41}$$

In forming the element stiffness the procedure is first to form a kernel stiffness in terms of the full set of parameters $\lfloor \lfloor \mathbf{a}_a \rfloor \lfloor \mathbf{a}_0 \rfloor \rfloor$ (see Section 8.2) and then to transform the result to physical coordinates by application of Eq. (12.41).

Perhaps the simplest subdomain formulation of the triangle is the *CPT* element of the STRUDL-II program.[12.39] This employs two subtriangles, Fig. 12.10b. It is basically assumed that cubic expansions govern the transverse displacements in regions a and b. In order to accomplish continuity of normal slope along side 1-2, where $\partial w/\partial n = \partial w/\partial y$, the x^2y terms of the expansions are discarded. (If these were retained, then the normal slope would be a quadratic function of x.) Also, to ensure continuity of w and normal slope $\partial w/\partial n = \partial w/\partial x$ across the boundary between region a and region b, we can set the coefficients of the constant and linear terms of all terms containing y to any power equal to one another in the respective expansions. Thus, we have the expansions for the respective subregions

$$w^a = a_1 + a_2 x + a_3 y + a_4 x^2 + a_5 xy + a_6 y^2 + a_7 x^3 + a_8 xy^2 + a_9 y^3$$
$$w^b = a_1 + a_2 x + a_3 y + b_4 x^2 + a_5 xy + a_6 y^2 + b_7 x^3 + a_8 xy^2 + a_9 y^3 \tag{12.42}$$

where the nine a_i's are the basic coefficients of the terms in the expansion for region a and also apply to the common terms in region b, and the two b_i's (b_4, b_7) have as yet no connection with the expansion in region a. To establish this connection, we form two constraint equations, as follows, which enforce a linear variation of the normal slope along sides 1-3 and 2-3:

$$\begin{aligned} a_7 &= \frac{1}{3}\left[2 - \left(\frac{y_3}{L^a}\right)^2\right]a_8 - \frac{y_3}{L^a}a_9 \\ b_7 &= \frac{1}{3}\left[2 - \left(\frac{y_3}{L^b}\right)^2\right]a_8 + \frac{y_3}{L^b}a_9 \end{aligned} \tag{12.43}$$

By substitution of these expressions into Eq. (12.42) the field is expressed in terms of nine parameters. Again, a kernel stiffness is formed in terms of these parameters, and the necessary transformation between them and the joint displacements is readily constructed. It is of interest to note that although the condition of normal slope continuity is enforced on the entire periphery of the element, the transverse displacement is not interelement-continuous along the side 1-2.

12.3.4 Numerical Comparisons—Subdomain Formulation of the Triangle Based on Assumed Displacements

Results for various subdomain formulations are given in Fig. 12.11. There is a significant improvement of the 10 d.o.f. per subdomain formulation over the 9 d.o.f. per subdomain results for coarse grids. It is of interest to note that the latter formulation produces a stiffness matrix (and, of course, numerical results) identical to those obtained by Bazeley, et al.[12.25] in a single-field development of a conforming triangular element (compare with solution shown in Fig. 12.9). The *CPT* element is of quite acceptable accuracy in this particular application.

12.3.5 Assumed Stress Fields

When the expression for complementary energy is written in terms of Southwell stress functions Φ^u and Φ^v, as done in Section 12.1.3, the problem of selecting representations for these stress functions is the same as that of selecting representations for the u and v displacement fields for plane stress. Thus, for a triangular element with d.o.f. at the vertices, one may approxi-

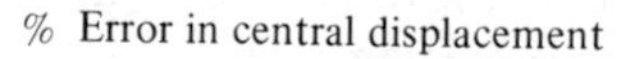

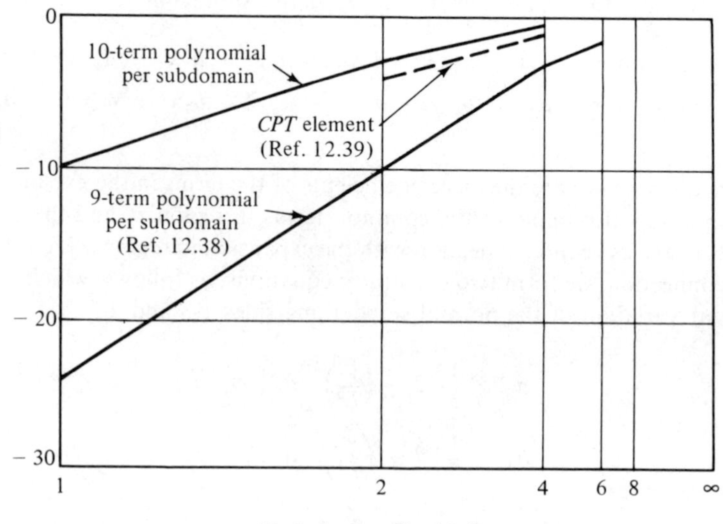

Fig. 12.11 Numerical comparison—subdomain triangular elements.

mate the stress function fields by[12.40]

$$\begin{aligned}\Phi^u &= N_1\Phi_1^u + N_2\Phi_2^u + N_3\Phi_3^u \\ \Phi^v &= N_1\Phi_1^v + N_2\Phi_2^v + N_3\Phi_3^v\end{aligned} \tag{12.44}$$

where $N_1 = L_1$, $N_2 = L_2$, and $N_3 = L_3$ and L_1, L_2, and L_3 are the triangular coordinates defined in Chapter 8. Alternatively if a triangular element is defined by d.o.f. at the vertices (designated as points 1, 2, 3) and by points at the midsides (points 4, 5, 6), then the appropriate expansions for Φ^u and Φ^v are of the forms[12.41]

$$\Phi^u = \lfloor \mathbf{N} \rfloor \{\mathbf{\Phi}^u\}, \qquad \Phi^v = \lfloor \mathbf{N} \rfloor \{\mathbf{\Phi}^v\} \tag{12.45}$$

with $\lfloor \mathbf{N} \rfloor = \lfloor N_1 \, N_2 \ldots N_6 \rfloor$, with terms as defined in Section 8.5, and

$$\begin{aligned}\{\mathbf{\Phi}^u\} &= \lfloor \Phi_1^u \; \Phi_2^u \ldots \Phi_6^u \rfloor^{\mathrm{T}} \\ \{\mathbf{\Phi}^v\} &= \lfloor \Phi_1^v \; \Phi_2^v \ldots \Phi_6^v \rfloor^{\mathrm{T}}\end{aligned}$$

The formulation of the element flexibility matrix now follows the procedure defined by Eqs. (12.19) to (12.23).

The apparent advantages of the approach to plate bending analysis above are compromised to a significant extent by difficulties in the definition of the

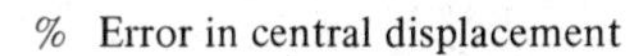

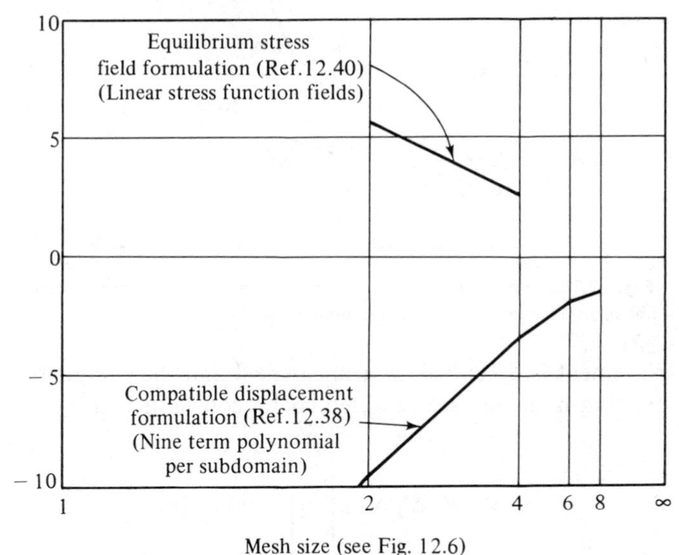

Fig. 12.12 Numerical comparisons: triangle based on equilibrium stress field versus triangle based on compatible displacements.

applied load state. It is to be recalled that the boundary integral in complementary energy pertains to the prescribed displacements, which are normally prescribed as zero. The boundary conditions on the loaded surface must be accounted for in special ways, usually through imposition of constraint equations. Calculation of displacements also presents certain difficulties. These and other aspects of a practical analysis by this method are detailed in Refs. 12.23 and 12.40–12.42.

Numerical results for the case of the simply supported square plate under central point load, based on the linear fields of Eq. (12.44) are shown in Fig. 12.12. For comparison we show numerical results for the interelement-compatible displacement formulation based on the subdomain approach using a nine-term polynomial in each subdomain.[12.38] The results illustrate the expected upper versus lower bound character of the alternative formulations and acceptable accuracy of the displacement in the complementary energy procedure.

12.3.6 Mixed Stress-Displacement Fields

The modified form of the Reissner variational principle, given by Eqs. (12.24) to (12.27) is representative of mixed stress displacement formulative procedures for plate bending elements and is detailed in the following for the case of the simplest triangular element (Fig. 12.13). Suppose the trans-

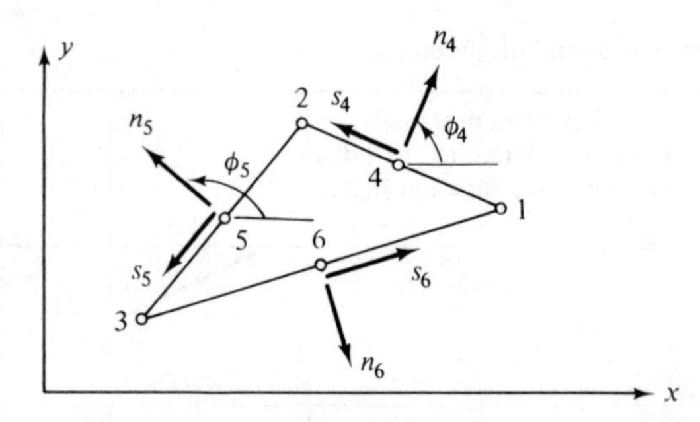

Fig. 12.13 Triangular bending element for mixed (Π_H) formulation—linear displacement and constant moment fields.

verse displacement is described by a linear field and the components of the internal bending moments are represented by constants. Thus

$$w = \lfloor N_1 \; N_2 \; N_3 \rfloor \begin{Bmatrix} w_1 \\ w_2 \\ w_3 \end{Bmatrix} = \lfloor \mathbf{N}_w \rfloor \{\mathbf{w}\}$$

where N_1, N_2, N_3 are the same shape functions defined above in Eq. (12.44). Also, $\mathfrak{M}_x = a_1$, $\mathfrak{M}_y = a_2$, $\mathfrak{M}_{xy} = a_3$, where a_1, a_2, and a_3 are constants. In order to define these constants in terms of physical parameters we define normal bending moments (M_4, M_5, M_6) at the midpoints of the sides. Upon evaluation of the bending moments M_x, M_y, and M_{xy} at these points and solution for a_1, a_2, and a_3, we obtain Eq. (12.26) (i.e., $\mathfrak{M} = [\mathbf{N}_m]\{\mathbf{M}\}$), in which

$$\mathfrak{M} = \lfloor \mathfrak{M}_x \; \mathfrak{M}_y \; \mathfrak{M}_{xy} \rfloor^{\mathrm{T}}$$

$$\{\mathbf{M}\} = \lfloor M_4 \; M_5 \; M_6 \rfloor^{\mathrm{T}}$$

$$[\mathbf{N}_M] = \begin{bmatrix} \cos^2 \phi_4 & \sin^2 \phi_4 & 2 \sin \phi_4 \cos \phi_4 \\ \cos^2 \phi_5 & \sin^2 \phi_5 & 2 \sin \phi_5 \cos \phi_5 \\ \cos^2 \phi_6 & \sin^2 \phi_6 & 2 \sin \phi_6 \cos \phi_6 \end{bmatrix}$$

The angles ϕ_4, ϕ_5, ϕ_6 are defined in Fig. 12.13. One must also obtain expressions for the tangential moments $\mathfrak{M}_s = \lfloor \mathfrak{M}_{s_1} \; \mathfrak{M}_{s_2} \; \mathfrak{M}_{s_3} \rfloor^{\mathrm{T}}$ as functions of $\{\mathbf{M}\}$. We can first write $\mathfrak{M}_s$ as a function of $\mathfrak{M}$ in the form $\mathfrak{M}_s = [\mathbf{\Gamma}_s]\mathfrak{M}$, where

$$[\mathbf{\Gamma}_s] = \begin{bmatrix} -\sin \phi_4 \cos \phi_4 & \sin \phi_4 \cos \phi_4 & \cos 2\phi_4 \\ -\sin \phi_5 \cos \phi_5 & \sin \phi_5 \cos \phi_5 & \cos 2\phi_5 \\ -\sin \phi_6 \cos \phi_6 & \sin \phi_6 \cos \phi_6 & \cos 2\phi_6 \end{bmatrix}$$

and it follows that

$$\begin{aligned}\mathfrak{M}_s &= [\boldsymbol{\Gamma}_s][\mathbf{N}_M]\{\mathbf{M}\} \\ &= [\mathbf{L}]\{\mathbf{M}\}\end{aligned}$$

As disclosed by Eq. (12.24b), the discretized form of the subject functional (Π_H) is comprised of the matrices $[\boldsymbol{\Omega}_{12}] = [\boldsymbol{\Omega}_{21}]^{\mathrm{T}}$ and $[\boldsymbol{\Omega}_{11}]$, which are in turn constructed from the matrices $[\mathbf{N}_M]$, $[\mathbf{N}'_m]$, $[\mathbf{N}_w]$, $[\mathbf{N}'_w]$, $[\mathbf{L}]$, and $[\mathbf{Y}]$, where the primes denote differentiation of the basic matrices $[\mathbf{N}_m]$ and $[\mathbf{N}_w]$ in accordance with definitions given in Section 12.1.4. Also, the matrix $[\mathbf{Y}]$ stems from a similarly defined differentiation of the displacement field. Since $[\mathbf{N}_m]$ is a matrix of constants, $[\mathbf{N}'_m] = [\mathbf{O}]$ and we have, from Eq. (12.24b),

$$[\boldsymbol{\Omega}_{11}] = \left[\int_A [\mathbf{N}_M]^{\mathrm{T}}[\mathbf{E}_f]^{-1}[\mathbf{N}_M]\,dA\right]$$

$$[\boldsymbol{\Omega}_{12}] = [\boldsymbol{\Omega}_{21}]^{\mathrm{T}} = \left[-\int_{S_n} [\mathbf{L}]^{\mathrm{T}}[\mathbf{Y}]\,dS\right]$$

It is of interest to note that the numerical results obtained with use of the formulation above are identical to those obtained with use of the stiffness matrix formulated on the basis of a complete quadratic displacement field.[12.27] See Section 12.3.1, Fig. 12.8(b), and Eq. (12.33). This might have been anticipated from the array of d.o.f., the scheme shown in Fig. 12.13 being identical to that represented in Fig. 12.8(b) because the normal bending moments (M_4, M_5, M_6) of the former correspond to the angular displacements ($\theta_4, \theta_5, \theta_6$) of the latter. The identity can be demonstrated algebraically, however. For a completely interior element (one surrounded by elements of the same type) $\{\bar{\boldsymbol{\Delta}}\} = 0$. Thus, as in Section 6.7, we can solve the upper partition of Eq. (12.27) for $\{\mathbf{M}\}$ in terms of $\{\boldsymbol{\Delta}\}$ and by substitution into the lower partition we find

$$[\mathbf{k}]\{\boldsymbol{\Delta}\} = \{\bar{\mathbf{M}}\}$$

where

$$[\mathbf{k}] = [\boldsymbol{\Omega}_{12}]^{\mathrm{T}}[\boldsymbol{\Omega}_{11}]^{-1}[\boldsymbol{\Omega}_{12}]$$

Detailed studies of the correspondence of the two formulations can be found in Refs. 12.43 and 12.44.

It is, of course, possible to employ higher-order representations of both the bending moment and displacement fields. A logical extension[12.45] is to choose a linear variation of moments and a quadratic variation of displacements. A wide range of alternative bending moment and displacement fields is explored in Ref. 12.22.

The hybrid schemes result in an entirely different class of mixed stress displacement formulations for the triangular bending elements. As in the case of plate stretching problems, the assumed stress hybrid approach, described in Section 6.6, dominates in application, and rather extensive studies of various combinations of internal bending moments and edge displacement fields are given in Refs. 12.43 and 12.46–12.48.

Figure 12.14 shows numerical results for various types of mixed formulations. The Π_H formulation based on constant moments and linear displacements, as we have already noted, gives results identical to those already presented in Fig. 12.9 for the six-term (quadratic) polynomial stiffness formu-

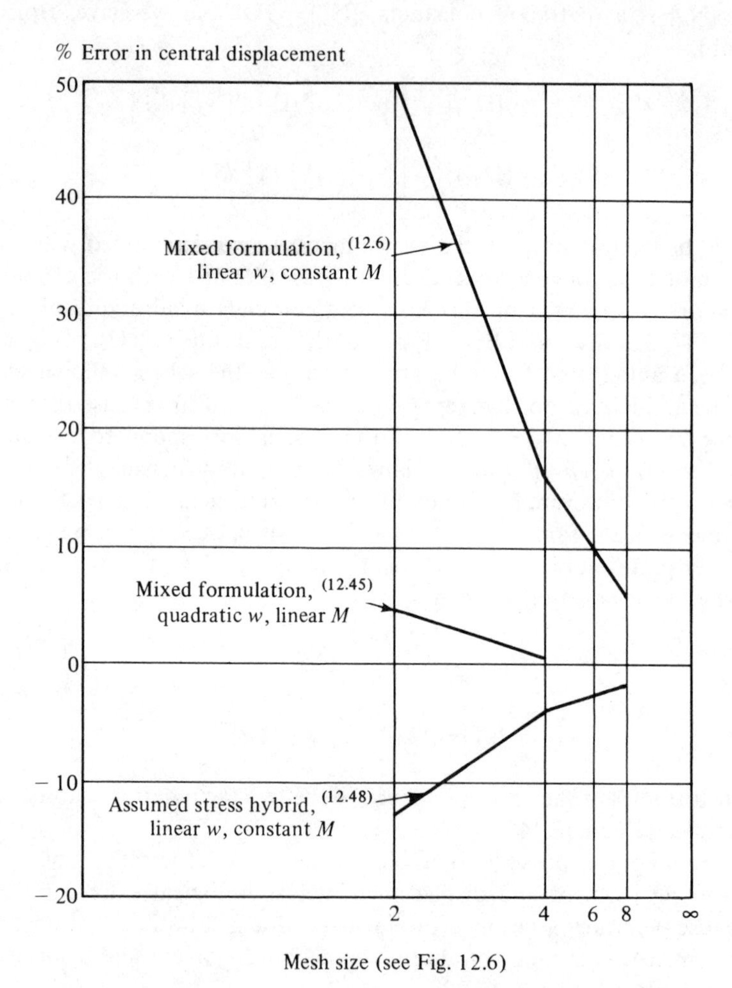

Fig. 12.14 Numerical comparisons: mixed and hybrid triangular elements.

lation. A mixed formulation based on a higher-order representation[12.45] (linear moments, quadratic displacements) substantially improves solution accuracy. Note, however, that twice as many node points (six versus three) are required of each element, a fact that is not represented by the horizontal scale of Fig. 12.14. Finally, we see that an assumed stress hybrid formulation[12.48] with fields comparable to the simplest Π_H formulation gives results which are to the opposite side of the correct solution and which are significantly more accurate for a given mesh size. We again caution, however, that many other factors must be taken into account when assessing relative merit.

12.4 Deflections Due to Transverse Shear

The formulation of a beam or plate element stiffness matrix to account for transverse shear deformation cannot be accomplished in exact form by substitution of the transverse displacement field Eq. (5.14a) into the sum of usual forms of the flexural and shear strain energies. It has been observed[12.49] that when shear deformations are not permitted in simple beam bending, the requirement that plane sections remain plane introduces internal restraint. When the restraint is removed to allow shear deformation, there is an addition to the internal energy and to preserve the equality of internal and external work the external work must be increased by the same amount. Thus, the nodal forces are associated with the increased, corresponding displacements and since a stiffness coefficient is defined as being associated with a unit displacement, the force required to produce a unit displacement is reduced when shear is permitted.

Consider, for example, the case of an element (Fig. 12.15) where it is assumed that the shear stress τ_{xz} is essentially constant on a portion of the area denoted as A_s and is negligible elsewhere (e.g., the case of a wide-flange

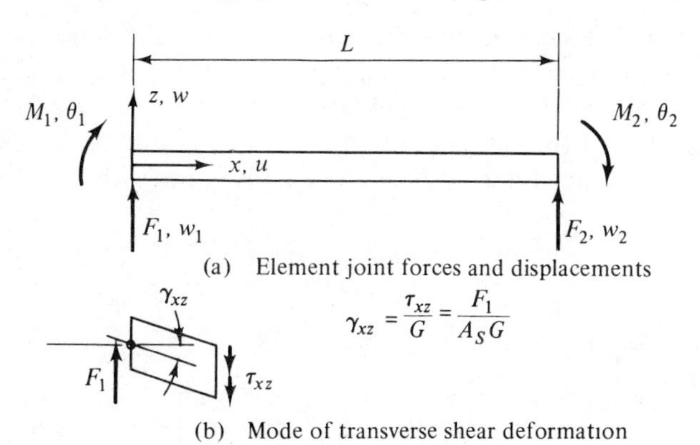

Fig. 12.15 Element sustaining both flexural and transverse shear deformation.

beam; the area can be adjusted to deal with other cross-sectional shapes by multiplication by a suitable constant). We exclude consideration of warping. Then, by Hooke's law, we have [see Fig. 12.15(b)]

$$\gamma_{xz} = \frac{F_1}{A_s G}$$

Also, from beam theory, $F_1 = EI(d^3w/dx^3)$, so that

$$\gamma_{xz} = \frac{EI}{A_s G}\left(\frac{d^3w}{dx^3}\right)$$

The contribution of shear deformation to the strain energy is

$$\frac{1}{2}\int_0^L (\gamma_{xz})^2 A_s G\, dx$$

so that with insertion of the expression for shear strain (γ_{xy}) above the total strain energy can be written as

$$U = \frac{1}{2}\int_0^L \left(\frac{d^2w}{dx^2}\right)^2 EI\, dx + \frac{1}{2}\int_0^L \left(\frac{d^3w}{dx^3}\right)^2 \frac{(EI)^2}{A_s G}\, dx$$

Now, if the transverse displacement w is described in the usual way, by a cubic polynomial [see Eq. (5.14a)], we can discretize this strain energy integral and by minimization of the strain energy one obtains the element stiffness matrix. A typical term of this matrix is k_{11}, which relates F_{z_1} to w_1

$$k_{11} = \frac{12EI}{L^3} + \frac{144(EI)^2}{A_s G L^5}$$

This term is in fact incorrect, for the reason stated at the outset.

A simple approach to the formulation of a correct element stiffness matrix[12.50] is to form directly a flexibility matrix that accounts for shear deformation. Thus, if the element is supported as a cantilever at point 2, then the deflection at point 1 due to shear alone is given by (with $G = E/2(1 + \mu)$)

$$\gamma_{xz} L = \frac{2(1 + \mu) F_1 L}{A_s E}$$

and the total flexibility equations are

$$\begin{Bmatrix} w_1 \\ \theta_1 \end{Bmatrix} = \begin{bmatrix} \dfrac{L^3}{3EI} + \dfrac{2(1 + \mu)L}{A_s E} & \dfrac{L^2}{2EI} \\ \dfrac{L^2}{2EI} & \dfrac{L}{EI} \end{bmatrix} \begin{Bmatrix} F_1 \\ M_1 \end{Bmatrix}$$

and, by use of the procedures detailed in Section 2.6, the stiffness matrix is constructed. It is then found that

$$k_{11} = \frac{12EI}{L^3} \frac{1}{[(1 + 24(1 + \mu)I/A_s L^2]}$$

which is the correct element stiffness matrix.

A general conclusion can be drawn from the above, to the effect that the inclusion of transverse shear deformation effects can be accomplished for beam, plate, and shell elements if they are formulated directly in terms of flexibility relationships (the complementary energy approach) or as mixed relationships which derive from functionals which contain the complementary strain energy [e.g., the Reissner functional, Eq. (12.24)].

We have emphasized that potential energy (assumed displacement) approaches predominate in practice. It is still possible to employ these approaches in the formulation of beam, plate, and shell stiffness equations that account for transverse shear deformation through a simple approximation in which the results of flexure-only and shear-only analyses are summed. To describe this procedure we examine the element 1-2 shown in Fig. 12.16, this being a segment of the total beam structure. We see that the transverse shear deformation is given by

$$\gamma_{xz} = \frac{w_2^s - w_1^s}{L}$$

where the superscript s has been applied to emphasize that these displacements occur due to transverse shear deformation alone. Also, since $\gamma_{xz} = 2(1 + \mu)F_1/A_s E$,

$$F_1 = \frac{A_s E}{2(1 + \mu)L}(w_2^s - w_1^s)$$

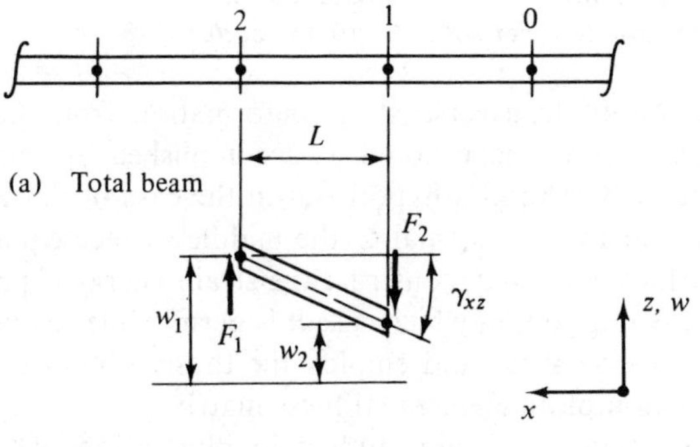

Fig. 12.16 Shear-only deflection of beam element.

This is an element stiffness equation and a stiffness equation for the remaining transverse shear force F_2 can be similarly constructed. By assembling these stiffness equations in the usual manner of a direct stiffness analysis, we can establish global shear stiffness equations. These we designate as

$$\{\mathbf{P}\} = [\mathbf{k}^s]\{\mathbf{\Delta}^s\} \tag{12.46}$$

and, by solution,

$$\{\mathbf{\Delta}^s\} = [\mathbf{k}^s]^{-1}\{\mathbf{P}\} \tag{12.47}$$

Denoting the global stiffness matrix for flexure-only as $[\mathbf{k}^f]$ and the corresponding displacements as $\{\mathbf{\Delta}^f\}$, we also have

$$\{\mathbf{\Delta}^f\} = [\mathbf{k}^f]^{-1}\{\mathbf{P}\} \tag{12.48}$$

The approximate total displacements are given by the sum of the shear-only and flexure-only displacements.

$$\{\mathbf{\Delta}\} = \{\mathbf{\Delta}^s\} + \{\mathbf{\Delta}^f\} \tag{12.49}$$

In applying the scheme above to flat plate or curved shell structures it is necessary to model the structure analytically as a system of fictitious shear-beam elements. When this is not an adequate approximation, it is possible to employ a plate theory that accounts for transverse shear deformation. There are a number of such theories(12.51–12.53) and most of them have been applied to finite element analysis.(12.54–12.57) It is likely however, that when the simplified approach is inadequate, the plate theories that account for transverse shear deformation will also be inadequate. The use of solid elements in a three-dimensional analysis would then be appropriate. We shall return to this point once again in Section 12.6.

12.5 *Removal of Constraint on Transverse Shear Deformation (Discrete-Kirchhoff Procedure)*(12.58)

The exclusion of transverse shear deformation from beam and plate bending analysis, as we have noted, is accomplished by imposition of the condition (one of Kirchhoff's hypotheses in the case of flat plates) that the angular displacement of a normal to the middle surface equals the slope of the middle surface. One may construct the strain energy expression without invoking this assumption, in which case it is permissible to append the strain energy for transverse shear and employ the thusly obtained energy expression in formulation of an element stiffness matrix.

The beam element is again studied in illustration of this procedure. Although we exclude the condition that the angular displacement (θ) equals the (negative) slope of the neutral axis, we retain the condition that plane sections taken on the undeformed state remain plane after deformation. Thus, the basic descriptor of flexural displacement is θ and the curvature is $\kappa = d\theta/dx$.

The total slope of the neutral axis, dw/dx, is now due to two effects: the angular displacement and the slope due to shear deformation. Retaining the simplification of a shear area A_s and the associated equivalent uniform state of transverse shear stress, γ_{xz}, we have

$$-\frac{dw}{dx} = \theta + \gamma_{xz}$$

or

$$-\gamma_{xz} = \theta + \frac{dw}{dx} \tag{12.50}$$

With these considerations in mind we can write the strain energy as

$$U = \frac{1}{2}\int_0^L \left(\frac{d\theta}{dx}\right)^2 EI\,dx + \frac{1}{2}\int_0^L \left(\theta + \frac{dw}{dx}\right)^2 A_s G\,dx \tag{12.51}$$

The strain energy is now defined by two independent variables θ and w. We therefore discretize the energy by assuming two independent fields

$$\begin{aligned} w &= \lfloor \mathbf{N}_w \rfloor \{\mathbf{w}\} \\ \theta &= \lfloor \mathbf{N}_\theta \rfloor \{\boldsymbol{\theta}\} \end{aligned} \tag{12.52}$$

where $\lfloor \mathbf{N}_w \rfloor$ and $\lfloor \mathbf{N}_\theta \rfloor$ are the respective shape function vectors and $\{\mathbf{w}\}$ and $\{\boldsymbol{\theta}\}$ are the corresponding d.o.f. By substitution into Eq. (12.51), we have

$$U = \frac{\lfloor \boldsymbol{\theta} \rfloor}{2}[\mathbf{k}^f]\{\boldsymbol{\theta}\} + \frac{\lfloor \boldsymbol{\theta} \rfloor}{2}[\mathbf{k}^{S_1}]\{\boldsymbol{\theta}\} + \lfloor \boldsymbol{\theta} \rfloor [\mathbf{k}^{S_2}]\{\mathbf{w}\} + \frac{\lfloor \mathbf{w} \rfloor}{2}[\mathbf{k}^{S_3}]\{\mathbf{w}\} \tag{12.53}$$

in which

$$[\mathbf{k}^f] = \int_0^L \{\mathbf{N}'_\theta\}\lfloor \mathbf{N}'_\theta \rfloor EI\,dx$$

$$[\mathbf{k}^{S_1}] = \int_0^L \{\mathbf{N}_\theta\}\lfloor \mathbf{N}_\theta \rfloor A_s G\,dx$$

$$[\mathbf{k}^{S_2}] = \int_0^L \{\mathbf{N}_\theta\}\lfloor \mathbf{N}'_w \rfloor A_s G\,dx$$

$$[\mathbf{k}^{S_3}] = \int_0^L \{\mathbf{N}'_w\}\lfloor \mathbf{N}'_w \rfloor A_s G\,dx$$

or

$$U = \frac{\lfloor \boldsymbol{\Delta} \rfloor}{2}[\mathbf{k}]\{\boldsymbol{\Delta}\} \tag{12.53a}$$

where

$$\lfloor \Delta \rfloor = \lfloor \lfloor \boldsymbol{\theta} \rfloor \lfloor \mathbf{w} \rfloor \rfloor$$

$$[\mathbf{k}] = \left[\begin{array}{c|c} \mathbf{k}^f + \mathbf{k}^{s_1} & \mathbf{k}^{s_2} \\ \hline \mathbf{k}^{s_2\mathrm{T}} & \mathbf{k}^{s_3} \end{array}\right]$$

It is important to observe that the energy above involves only the first derivatives of the independent variables, which means that the chosen functions need not meet the condition of slope continuity across the element boundaries. This suggests an approach to thin plate flexure analysis in which the angular displacement is employed as a basic variable and a small amount of shear strain energy, insufficient to produce a meaningful contribution to the transverse, is employed. Unfortunately, this is not an effective approach because the equations to be solved are ill-conditioned as the shear strain energy approaches zero and are singular when it is zero. Since this condition arises because the analytical model is unstable in the presence of independent displacement parameters $\lfloor \theta \rfloor$ and $\lfloor \mathbf{w} \rfloor$, we can restore stability by tying these d.o.f. together at *discrete points* in accordance with the Kirchoff assumption. That is, with $\gamma_{xz} = 0$, Eq. (12.50) requires

$$\theta = -\frac{dw}{dx}$$

A constraint equation among the joint parameters ($\lfloor \lfloor \boldsymbol{\theta} \rfloor \lfloor \mathbf{w} \rfloor \rfloor$) can be written on the basis of this condition since both θ and w are available in discretized form.

The concepts of a discrete Kirchhoff formulation are clarified by means of the example shown in Fig. 12.17.(12.54) A two-segment cantilever beam is acted upon by an end load P_3. We adopt the following displacement fields in representation of element A.

$$\theta = (1 - \xi)\theta_1 + \xi\theta_2$$
$$w = (1 - \xi)w_1 + \xi w_2$$

where ξ is the nondimensional axial coordinate within the element ($\xi = x/L$). Linear fields of this form are also chosen for element B. Note that the descrip-

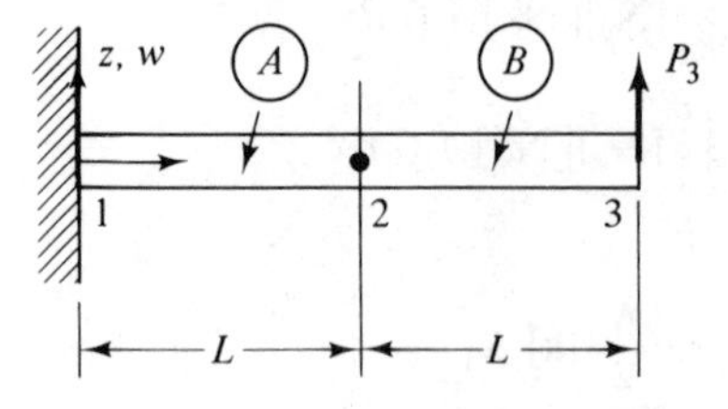

Fig. 12.17

tions of w and θ are approximate. If we neglect shear, we have the following strain energy [from Eq. (12.51)].

$$U = \frac{EI}{2L}\lfloor \theta_2\ \theta_3\ w_2\ w_3 \rfloor \begin{bmatrix} 2 & -1 & 0 & 0 \\ -1 & 1 & 0 & 0 \\ 0 & 0 & 0 & 0 \\ 0 & 0 & 0 & 0 \end{bmatrix} \begin{Bmatrix} \theta_2 \\ \theta_3 \\ w_2 \\ w_3 \end{Bmatrix}$$

Now, to form the constraint conditions that tie the θ's to the w's, we require that the shear strain γ_{xz} be zero at the center of each element, i.e., $dw/dx + \theta = 0$ at each element center point. Since θ varies linearly between the end points i and j, the value of θ at the center of the element is $\frac{1}{2}(\theta_i + \theta_j)$, so that

$$\frac{w_2}{L} + \frac{\theta_2}{2} = 0$$

$$\frac{w_3 - w_2}{L} + \frac{\theta_3 + \theta_2}{2} = 0$$

Solving these for θ_2 and θ_3, we find

$$\theta_2 = -\frac{2}{L}w_2$$

$$\theta_3 = -\frac{2w_3}{L} + \frac{4w_2}{L}$$

and, after substitution into U, we can identify the stiffness equations

$$\begin{Bmatrix} P_2 \\ P_3 \end{Bmatrix} = \frac{4EI}{L^3}\left[\begin{array}{c|c} 10 & -3 \\ \hline -3 & 1 \end{array}\right]\begin{Bmatrix} w_2 \\ w_3 \end{Bmatrix}$$

Solving these for $P_2 = 0$, $P_3 = P$, we find

$$w_3 = \frac{10PL^3}{4EI}$$

This is close to the exact solution of $8PL^3/3EI$. The error is due to the approximations of w and θ as linear functions.

In the general finite element formulation the constraints would be handled more elegantly, either by the matrix transformation method of Section 3.5 or by the Lagrange multiplier method of Section 7.3.

The discrete Kirchhoff approach has been applied effectively to problems of flat plates,(12.60, 12.61) axisymmetric shells,(12.59) and general thin shells.(12.58)

12.6 *Effectiveness of Solid Elements for Plate Bending Analysis*

It was shown in Section 9.3 that two-dimensional direct stress elements could be adapted to furnish efficient representations for beam flexure by addition of quadratic modes of displacement to the basic linear displacement field. Plate and shell analysis can be performed in like manner by adding quadratic modes to a hexahedronal element based on linear displacement fields; this point is taken up in Chapter 10. Alternatively, when using numerical integration in the construction of element stiffness matrices, one can employ the idea of reduced-order integration of the shear component of the strain energy, as also discussed in Chapters 9 and 10.

Figure 12.18 portrays the effectiveness of this approach in the analysis of the problem we have employed as a standard of comparison in this chapter. To illustrate the concept of supplementing a linear displacement (8-node) hexahedron with quadratic displacement modes we adopt results published by Cook.[12.48] The results obtained by applying the reduced integration to the shear strain energy of the 20-node hexahedronal element shown in Fig. 10.10 are those reported by Pawsey and Clough.[12.62] For comparison we show the numerical results for the triangular thin plate element whose stiffness formulation is based on one subdomain concept described in Section 12.3.3. (It should be noted that the basis for plotting results has been altered to account for differences in the manner of arrangement of elements. For the actual numerical results and the associated gridworks the reader should

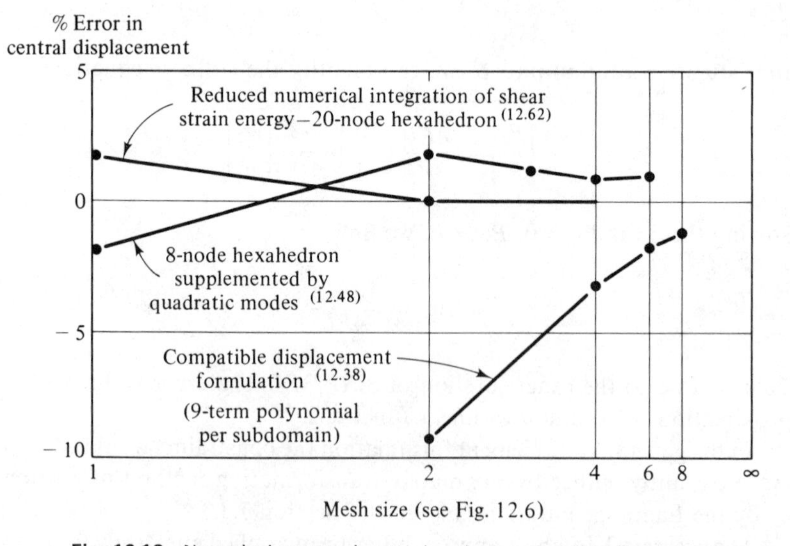

Fig. 12.18 Numerical comparisons: hexahedronal solid elements versus triangular plate bending element based on compatible displacements.

consult the cited references.) Figure 12.18 discloses that the modified solid elements are quite accurate in this application. The formulation of the 8-node hexahedron with supplemental quadratic modes gives results that apparently converge to a solution approximately $1\frac{1}{2}\%$ different from the correct solution. This would appear to be due to the representation of finite thickness effects in the element formulation.

12.7 Concluding Remarks

The status of finite element thin plate flexure is such that reliable and accurate formulations are available for assumed displacement (potential energy) models. The requirements upon these solutions are difficult to satisfy, however, and lead to extensive algebraic operations in the formation of the basic element stiffness coefficients. Thus, there is considerable interest in flat plate flexure formulations derived by use of other variational principles with less stringent requirements upon the assumed functions. Reliable and accurate formulations have also emerged from the latter but their scope is far from being fully explored.

Tradeoffs between element formulative effort, which is generally synonymous with sophistication in the definition of element behavior and geometry, and the global analysis solution effort, which decreases in size as the element formulative effort grows, are not yet clearly established for flat plate bending. Valid comparisons of these alternatives must include not only the operational costs to reach a desired level of solution accuracy but also must reflect an amortization of costs to develop the associated software.

With both flat plate stretching and flat plate bending formulations in hand it would appear that the analysis of thin shell structures, which evidence both stretching and flexural behavior, is feasible through a simple superposition of the two types of elements. This is indeed the case, although care must be exercised in formation of the global representation (see Section 3.5.3) and in the interpretation of the solution quantities. Many analysts prefer to formulate curved thin shell elements for the purpose of these calculations in order to avoid the shortcomings of flat elements, but this avenue gives rise to many new questions related to the appropriate choice of shell theory, geometric representation, choice of displacement functions, and many other factors. Treatment of the relevant considerations in the analysis of curved thin shells, by either flat or curved elements is beyond the scope of this text. The interested reader may find a review of this topic in Ref. 12.1.

REFERENCES

12.1 Gallagher, R. H., "Analysis of Plate and Shell Structures," Proc. of Conf. on Application of Finite Element Method in Civil Eng., Vanderbilt U., Nashville, Tenn., pp. 155–206, 1969.

12.2 TIMOSHENKO, S. and S. WOINOWSKY-KRIEGER, *Theory of Plates and Shells*, 2nd ed., McGraw-Hill Book Co., New York, N.Y., 1969.

12.3 MANSFIELD, E. H., *The Bending and Stretching of Plates*, Pergamon Press, Oxford, England, 1964.

12.4 MARGUERRE, K. and H. T. WOERNLE, *Elastic Plates*, Blaisdell Pub. Co., Waltham, Mass., 1969.

12.5 SOUTHWELL, R. V., "On the Analogues Relating Flexure and Extension of Flat Plates," *Quart. J. Mech. Appl. Math.*, **3**, pp. 257–270, 1950.

12.6 HERRMANN, L. R., "Finite Element Bending Analysis of Plates," *J. Eng. Mech. Div.*, ASCE, **93**, No. EM-5, pp. 13–25, 1967.

12.7 HERRMANN, L. R., "A Bending Analysis for Plates," Proc. (First) Conf. on Matrix Methods in Struct. Mech., AFFDL TR 66–80, pp. 577–604, Oct., 1965.

12.8 BOGNER, F. K., R. L. FOX, and L. A. SCHMIT, "The Generation of Interelement, Compatible Stiffness and Mass Matrices by the Use of Interpolation Formulas," Proc. (First) Conf. on Matrix Methods in Struct. Mech., AFFDL TR 66–80, Nov., 1965.

12.9 DAWE, D. J., "A Finite Element Approach to Plate Vibration Problems," *J. Mech. Eng. Sci.*, **7**, pp. 28–32, 1965.

12.10 GOPALACHARYULU, S., "A Higher Order Conforming Rectangular Element," *Internat'l. J. Num. Meth. Eng.*, **6**, No. 2, pp. 305–308, 1973.

12.11 IRONS, B. (Comment on Ref. 12.10), *Internat'l. J. Num. Meth. Eng.*, **6**, No. 2, pp. 308–309, 1973.

12.12 WEGMULLER, A. and C. KOSTEM, "Finite Element Analysis of Plates and Eccentrically Stiffened Plates," Fritz Eng. Lab. Report No. 378A.3, Lehigh U., Bethlehem, Pa., Feb., 1973.

12.13 WALZ, J. E., R. E. FULTON, and N. J. CYRUS, "Accuracy and Convergence of Finite Element Approximations," Proc. of 2nd Conf. on Matrix Methods in Struct. Mech., AFFDL TR 68–150, pp. 995–1027, Oct., 1968.

12.14 FRAEIJS DE VEUBEKE, B., "A Conforming Finite Element for Plate Bending," *Internat'l. J. Solids Struct.*, **4**, No. 1, pp. 95–108, 1968.

12.15 CLOUGH, R. and C. FELIPPA, "A Refined Quadrilateral Element for the Analysis of Plate Bending," Proc. of 2nd Conf. on Matrix Methods in Struct. Mech. AFFDL TR 68–150, pp. 399–440, Oct., 1968.

12.16 GREENE, B. E., R. E. JONES, R. W. MCLAY, and D. STROME, "Generalized Variational Principles in the Finite-Element Method," *AIAA J.*, **7**, pp. 1254–1260, July, 1969.

12.17 KIKUCHI, F. and Y. ANDO, "Some Finite Element Solutions for Plate Bending Problems by Simplified Hybrid Displacement Method," *Nuc. Eng. Design*, **23**, pp. 155–178, 1972.

12.18 PIAN, T. H. H., "Element Stiffness Matrices for Boundary Compatibility and for Prescribed Boundary Stresses," Proc. (First) Conf. on Matrix Methods in Struct. Mech., Wright-Patterson AFB, Ohio, AFFDL TR 65–80, pp. 457–478, Oct., 1965.

12.19 SEVERN, R., and P. TAYLOR, "The Finite Element Method for Flexure of Slabs when Stress Distributions are Assumed," *Proc. Inst. Civil Eng.*, **34**, pp. 153–163, 1966.

12.20 ALLWOOD, R. and G. CORNES, "A Polygonal Finite Element for Plate Bending Problems using the Assumed Stress Approach," *Internat'l. J. Num. Meth. Eng.*, **1**, No. 22, pp. 135–149, 1969.

12.21 COOK, R. D., "Two Hybrid Elements for the Analysis of Thick, Thin, and Sandwich Plates," *Internat'l. J. Num. Meth. Eng.*, **5**, No. 2, pp. 277–288, 1972.

12.22 BRON, J. and G. DHATT, "Mixed Quadrilateral Elements for Bending," *AIAA J.*, **10**, No. 10, pp. 1359–1361, Oct., 1972.

12.23 FRAEIJS DE VEUBEKE, B., G. SANDER, and P. BECKERS, "Dual Analysis by Finite Elements: Linear and Non Linear Applications," AFFDL TR 72–93, Dec., 1972.

12.24 ABEL, J. and C. DESAI, "Comparison of Finite Elements for Plate Bending," Proc. ASCE, *J. Struct. Div.*, **98**, No. ST9, pp. 2143–2148, Sept., 1972.

12.25 BAZELEY, G., Y. CHEUNG, B. IRONS, and O. ZIENKIEWICZ, "Triangular Elements in Plate Bending—Conforming and Non-Conforming Solutions," Proc. of (First) Conf. on Matrix Methods in Struct. Mech., AFFDL TR 66–80, pp. 547–576, Oct., 1965.

12.26 RAZZAQUE, A. Q., "Program for Triangular Elements with Derivative Smoothing." *Internat'l. J. Num. Meth. Eng.*, **6**, No. 3, pp. 333–344, 1973.

12.27 MORLEY, L. S. D., "The Constant-Moment Plate Bending Element," *J. Strain Analysis*, **6**, No. 1, pp. 20–24, 1971.

12.28 HARVEY, J. W. and S. KELSEY, "Triangular Plate Bending Elements with Enforced Compatibility," *AIAA J.*, **9**, pp. 1023–1026, 1971.

12.29 ANDERHEGGEN, E., "A Conforming Finite Element Plate Bending Solution," *Internat'l. J. Num. Meth. Eng.*, **2**, No. 2, pp. 259–264, 1970.

12.30 CHU, T. C. and W. C. SCHNOBRICH, "Finite Element Analysis of Translational Shells," *Comp. Struct.*, **2**, pp. 197–222, 1972.

12.31 IRONS, B., "A Conforming Quartic Triangular Element for Plate Bending," *Internat'l. J. Num. Meth. Eng.*, **1**, No. 1, pp. 29–46, 1969.

12.32 ARGYRIS, J. H., I. FRIED, and D. SCHARPF, "The TUBA Family of Plate Elements for the Matrix Displacement Method," *Aero. J.*, **72**, pp. 701–709, 1968.

12.33 BELL, K., "A Refined Triangular Plate Bending Finite Element," *Internat'l. J. Num. Meth. Eng.*, **1**, No. 1, pp. 101–122, 1969.

12.34 COWPER, G. R., E. KOSKO, G. LINDBERG, and M. OLSON, "Static and

Dynamic Applications of a High Precision Triangular Plate Bending Element," *AIAA J.*, **7**, No. 10, pp. 1957–1965, 1969.

12.35 BUTLIN, G. and R. FORD, "A Compatible Triangular Plate Bending Finite Element," *Internat'l. J. Solids Struct.*, **6**, pp. 323–332, 1970.

12.36 ZENISEK, A., "Interpolation Polynomials on the Triangle," *Num. Math.*, **15**, pp. 283–296, 1970.

12.37 SVEC, O. J. and G. Gladwell, "A Triangular Plate Bending Element for Contact Problems," *Internat'l. J. Solids Struct.*, **9**, pp. 435–446, 1973.

12.38 CLOUGH, R. W. and J. Tocher, "Finite Element Stiffness Matrices for the Analysis of Plate Bending," Proc. (First) Conf. on Matrix Methods in Struct. Mech., AFFDL TR 66–80, pp. 515–546, 1965.

12.39 CONNOR, J. and G. WILL, "A Triangular Flat Plate Bending Element" TR 68–3, Dept. of Civil Engineering, M.I.T., Cambridge, Mass., 1968.

12.40 ELIAS, Z. M., "Duality in Finite Element Methods," Proc. ASCE, *J. Eng. Mech. Div.*, **94**, No. EM 4, pp. 931–946, 1968.

12.41 MORLEY, L. S. D., "A Triangular Equilibrium Element with Linearly Varying Bending Moments for Plate Bending Problems," *J. Royal Aero. Soc.*, **71**, pp. 715–721, 1967.

12.42 MORLEY, L. S. D., "The Triangular Equilibrium Element in the Solution of Plate Bending Problems," *Aero. Quart.*, **XIX**, **4**, pp. 149–169, 1968.

12.43 ALLMAN, D., "Triangular Plate Element for Plate Bending with Constant and Linearly Varying Bending Moments," *High Speed Computing of Elastic Structures*, **1**, U. of Liege, Belgium, pp. 105–136, 1971.

12.44 HELLAN, K., "On the Unity of Constant Strain-Constant Moment Finite Elements," *Internat'l. J. Num. Meth. Eng.*, **6**, No. 2, pp. 191–209, 1973.

12.45 VISSER, W., "A Refined Mixed-Type Plate Bending Element," *AIAA J.*, **7**, No. 9, pp. 1801–1802, 1969.

12.46 DUNGER, R., R. T. SEVERN, and P. TAYLOR, "Vibration of Plate and Shell Structures Using Triangular Finite Elements," *J. Strain Analysis*, **2**, No. 1, pp. 73–83, 1967.

12.47 DUNGAR R. and R. T. SEVERN, "Triangular Finite Elements of Variable Thickness and their Application to Plate and Shell Problems," *J. Strain Analysis*, **4**, No. 1, pp. 10–21, 1969.

12.48 COOK, R. D., "Some Elements for Analysis of Plate Bending," Proc. ASCE, *J. Eng. Mech. Div.*, **98**, No. EM6, pp. 1452–1470, 1972.

12.49 SEVERN, R. T., "Inclusion of Shear Deflection in the Stiffness Matrix for a Beam Element," *J. Strain Analysis*, **5**, No. 4, pp. 239–241, 1970.

12.50 WILLIAMS, D., *An Introduction to the Theory of Aircraft Structures*, E. Arnold Pub., London, 1960.

12.51 LOVE, A. E. H., *A Treatise on the Mathematical Theory of Elasticity*, 4th ed., Dover Pub., Inc., New York, N.Y., 1927.

12.52 MINDLIN, R. D., "Influence of Rotatory Inertia and Shear on Flexural Motions of Isotropic Elastic Plates," *J. Appl. Mech.*, **18**, pp. 31–38, 1951.

12.53 REISSNER, E., "The Effect of Transverse Shear Deformation on the Bending of Elastic Plates," *J. Appl. Mech.*, **12**, pp. A.69–A.77, 1945.

12.54 SMITH, I., "A Finite Element Analysis for 'Moderately Thick' Rectangular Plates in Bending," *Internat'l. J. Mech. Sci.*, **10**, pp. 563–570, 1968.

12.55 GREIMANN, L. F. and P. P. LYNN, "Finite Element Analysis of Plate Bending with Transverse Shear Deformation," *Nuc. Eng. Design*, **14**, pp. 223–230, 1970.

12.56 PRYOR, C. W. and R. M. BARKER, "A Finite Element Analysis Including Transverse Shear Effects for Laminated Plates," *AIAA J.*, **9**, No. 5, pp. 912–917, 1971.

12.57 PRYOR, C. W., R. M. BARKER, and D. FREDERICK, "Finite Element Bending Analysis of Reissner Plates," Proc. ASCE, *J. Eng. Mech. Div.*, **96**, No. EM6, pp. 967–983, 1970.

12.58 WEMPNER, G., J. T. ODEN, and D. KROSS, "Finite-Element Analysis of Thin Shells," Proc. ASCE, *J. Eng. Mech. Div.*, **94**, No. EM6, pp. 1273–1294, 1968.

12.59 WEEKS, G. A., "A Finite Element Model for Shells Based on the Discrete Kirchhoff Hypothesis", *Internat'l. J. Num. Meth. Eng.*, **5**, No. 1, pp. 3–16, 1972.

12.60 STRICKLIN, J. A., W. E. HAISLER, P. R. TISDALE, and R. GUNDERSON, "A Rapidly Converging Triangular Plate Element," *AIAA J.*, **7**, No. 1, pp. 180–181, 1969.

12.61 FRIED, I., "Shear in C^0 and C^1 Plate Bending Elements," *Internat'l. J. Solids Struct.*, **9**, No. 4, pp. 449–460, 1973.

12.62 PAWSEY, S. F. and R. W. CLOUGH, "Improved Numerical Integration of Thick Shell Finite Elements," *Internat'l. J. Num. Meth. Eng.*, **3**, No. 4, pp. 575–586, 1971.

PROBLEMS

12.1 Prove, through integration by parts, that the functional Π_H (Eq. 12.24a) derives from the Reissner functional Π_R (Eq. 12.24).

12.2 Using the stiffness coefficients of Table 12.1 (the 12-term rectangle), calculate the central deflection of a simply supported square plate subjected to a central concentrated load. Use one element in a quadrant of the plate (invoke symmetry) and verify the result through comparison with the solution point shown in Fig. 12.7.

12.3 Formulate the work-equivalent corner force F_{z_1} for uniformly distributed load q on the 16-d.o.f. rectangular element. Calculate the central deflection of a square plate with fixed edges subjected to this loading using one element (by invoking symmetry) and compare the result with the exact solution.

Table 12.1 Stiffness matrix for rectangle in flexure—based on 12-term polynomial. (See Fig. 12.3 for description of element.)

$$[\mathbf{k}_f] = \frac{Et^3}{360(1-\mu^2)x_2y_3}$$

w_1	θ_{x_1}	θ_{y_1}	w_2
$120(\beta^2+\gamma^2)-24\mu+84$			Symmetric
$[10\beta^2+(1+4\mu)]6y_3$	$40x_2^2+8(1-\mu)y_3^2$		
$-[10\gamma^2+(1+4\mu)]6x_2$	$-30\mu x_2y_3$	$40y_3^2+8(1-\mu)x_2^2$	
$60(\gamma^2-2\beta^2)+24\mu-84$	$-[10\beta^2+(1-\mu)]6y_3$	$[-5\gamma^2+(1+4\mu)]6x_2$	$120(\beta^2+\gamma^2)-24\mu+84$
$[10\beta^2+(1-\mu)]6y_3$	$20x_2^2-2(1-\mu)y_3^2$	○	$-[10\beta^2+(1+4\mu)]6y_3$
$[-5\gamma^2+(1+4\mu)]6x_2$	○	$20y_3^2-8(1-\mu)x_2^2$	$-[10\gamma^2+(1+4\mu)]6x_2$
$-60(\gamma^2+\beta^2)-24\mu+84$	$[-5\beta^2+(1-\mu)]6y_3$	$[5\gamma^2-(1-\mu)]6x_2$	$-60(2\gamma^2-\beta^2)+24\mu-84$
$[5\beta^2-(1-\mu)]6y_3$	$10x_2^2+2(1-\mu)y_3^2$	○	$[-5\beta^2+(1+4\mu)]6y_3$
$[-5\gamma^2+(1-\mu)]6x_2$	○	$10y_3^2+2(1-\mu)x_2^2$	$-[10\gamma^2+(1-\mu)]6x_2$
$-60(2\gamma^2-\beta^2)+24\mu-84$	$[-5\beta^2+(1+4\mu)]6y_3$	$[10\gamma^2+(1-\mu)]6x_2$	$-60(\beta^2+\gamma^2)-24\mu+84$
$[5\beta^2-(1+4\mu)]6y_3$	$20x_2^2-8(1-\mu)y_3^2$	○	$[-5\beta^2+(1-\mu)]6y_3$
$-[10\gamma^2+(1-\mu)]6x_2$	○	$20y_3^2-2(1-\mu)x_2^2$	$[-5\gamma^2+(1-\mu)]6x_2$

Table 12.1 *(Cont.)*

$$\beta = \frac{x_2}{y_3} \qquad \gamma = \frac{y_3}{x_2}$$

Symmetric

θ_{x_2}	θ_{y_2}	w_3	θ_{x_3}
$40x_2^2 + 8(1-\mu)y_3^2$			
$30\mu x_2 y_3$	$40y_3^2 + 8(1-\mu)x_2^2$		
$[-5\beta^2 + (1+4\mu)]6y_3$	$[10\gamma^2 + (1-\mu)]6x_2$	$120(\beta^2+\gamma^2) - 24\mu + 84$	
$20x_2^2 - 8(1-\mu)y_3^2$	$\bigcirc$	$-[10\beta^2 + (1+4\mu)]6y_3$	$40x_2^2 + 8(1-\mu)y_3^2$
$\bigcirc$	$20y_3^2 - 2(1-\mu)x_2^2$	$[10\gamma^2 + (1+4\mu)]6x_2$	$-30\mu x_2 y_3$
$[5\beta^2 - (1-\mu)]6y_3$	$[5\gamma^2 - (1-\mu)]6x_2$	$60(\gamma^2 - 2\beta^2) + 24\mu - 84$	$[10\beta^2 + (1-\mu)]6y_3$
$10x_2^2 + 2(1-\mu)y_3^2$	$\bigcirc$	$-[10\beta^2 + (1-\mu)]6y_3$	$20x_2^2 - 2(1-\mu)y_3^2$
$\bigcirc$	$10y_3^2 + 2(1-\mu)x_2^2$	$[5\gamma^2 - (1+4\mu)]6x_2$	$\bigcirc$

Table 12.1 (*Cont.*)

θ_{y_3}	w_4	θ_{x_4}	θ_{y_4}
	(Symmetric)		
$40y_3^2 + 8(1-\mu)x_2^2$			
$[5\gamma^2 - (1+4\mu)]6x_2$	$120(\beta^2+\gamma^2) - 24\mu + 84$		
$\bigcirc$	$[10\beta^2 + (1+4\mu)]6y_3$	$40x_2^2 + 8(1-\mu)y_3^2$	
$20y_3^2 - 8(1-\mu)x_2^2$	$[10\gamma^2 + (1+4\mu)]6x_2$	$30\mu x_2 y_3$	$40y_3^2 + 8(1-\mu)x_2^2$

12.4 Formulate the shape function representation $w = \lfloor \mathbf{N} \rfloor \{\boldsymbol{\Delta}\}$ where $\{\boldsymbol{\Delta}\} = \lfloor w_1 \; w_2 \; w_3 \; \theta_4 \; \theta_5 \; \theta_6 \rfloor^T$ for the triangle shown in Fig. 12.8(b).

12.5 Discuss the formulation of a flexural stiffness matrix for a rectangular plate element by use of the assumed stress hybrid method. Select appropriate functions for the interior stress field and boundary displacements.

12.6 The stiffness matrix for a triangular plate element in bending is to be formulated by means of the subdomain approach, using three triangles of the form shown in Fig. 12.10a but with quadratic expansions in each subdomain. Reduce the 18 independent parameters of the two expansions (6 in each) to 12 by invoking conditions of displacement consistency. Then, reduce these to 9 by imposing consistency of angular displacement at the vertices. Discuss the adherence of the resulting function to displacement compatibility conditions along the element boundaries and along the subdomain interface.

12.7 Formulate the stiffness matrix for an annular flexural element, shown in Fig. P12.7 in cross section. Use the beam element shape function, but written in terms of the radial coordinate r.

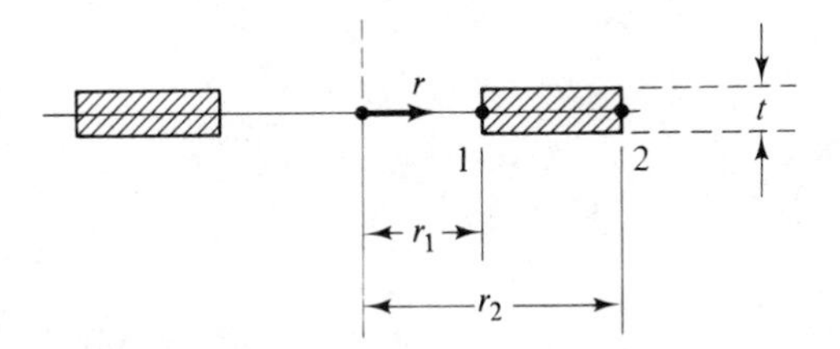

Fig. P12.7

12.8 A widely employed transverse displacement function for plate bending is obtained from Eq. (12.35) by constraining the centroidal displacement as follows:

$$\begin{aligned} w_4 = \tfrac{1}{3}(w_1 + w_2 + w_3) + \tfrac{1}{18}[&(x_2 + x_3 - 2x_1)\theta_{y_1} + (x_1 + x_3 - 2x_2)\theta_{y_2} \\ &+ (x_1 + x_2 - 2x_3)\theta_{y_3} + (y_2 + y_3 - 2y_1)\theta_{x_1} + (y_1 + y_3 - 2y_2)\theta_{x_2} \\ &+ (y_1 + y_2 - 2y_3)\theta_{x_3}] \end{aligned}$$

Define the resulting form of displacement function and examine its properties with respect to interelement displacement continuity.

12.9 A beam element is subjected to a temperature change above the stress-free state that varies linearly across the depth and along the length but is symmetric about the vertical axis of the cross section. The variation on the cross section is transformed into a thermal moment M^α whose lengthwise variation can be described as

$$M^\alpha = \left(1 - \frac{x}{L}\right) M_1^\alpha + \frac{x}{L} M_2^\alpha,$$

where M_1^α and M_2^α are the thermal moments at the respective end points. Formulate the thermal force vector for the beam element for this condition.

12.10 Using a two-segment cantilever beam of constant rectangular cross section, calculate the tip deflection under concentrated end load on the basis of the superposition method described in Section 12.4. Compare with the exact solution.

13
Elastic Instability Analysis

This chapter extends the finite element method to deal with problems of *linear* elastic instability. Linear elastic instability analysis is defined herein as an approach to the calculation of the *intensity* of applied loads for buckling of an elastic structure, where the internal distribution of forces due to the specified *distribution* of the applied loads is calculated in an independent linear analysis. Although the physical conditions associated with the collapse of structures involve the nonlinear aspects of instability as well as inelastic deformation, a linear instability analysis accurately describes the circumstances of failure that are of design importance for many structural forms, particularly beams and flat plates. Linear elastic instability theory therefore furnishes the basis for a large share of practical design formulations; even where nonlinear phenomena must be taken into account to define accurately the magnitude of load to cause failure, the form of the solution is frequently given adequately by linear analysis.

The finite element method has an important role to play in the solution of linear instability problems because it is able to cope with irregularities in loading and geometry that defy adequate treatment by classical means. Structures composed of materials with nonisotropic properties confront the same difficulties in classical analysis. Additionally, the concepts and element relationships pertinent to linear stability theory constitute the foundation for nonlinear stability theory.

As in other aspects of the finite element method, the treatment of elastic instability consists of two component parts: (1) formulation of element relationships and (2) solution of the complete system. Chapters 5 and 6 demonstrated that there are many ways of achieving the formulation of finite element relationships and a corresponding number of forms in which these can be stated. We consider only the principle of stationary potential energy, assumed displacement fields, and the associated statement of the element relationships in the form of stiffness equations. Correspondingly, treatment of the complete structural system is accomplished with use of stiffness equations, via the matrix displacement method. The highly automated and versatile capabilities of finite element stiffness analysis programs developed for stable analysis are applicable with only slight modification to linear elastic instability analysis.

The general theory of finite element linear elastic stability analysis is established in Section 13.1, followed by sections that deal specifically with prismatic and flat plate elements.

13.1 General Theory—Linear Stability Analysis

Consider first the uniform-section prismatic element (Fig. 13.1), which is frequently encountered in space frame structures and as a stiffener in stiffened flat plates and shells. This element furnishes the simplest illustration of the formulation of finite element stiffness properties for linear stability analysis while giving insight into certain key aspects of analysis common to all structural forms.

The element is assumed to sustain only axial and flexural deformation; shear deformation is disregarded. Thus only axial strains are present and these are described by the following strain-displacement equation:

$$\epsilon_x = \frac{du}{dx} - z\left(\frac{d^2w}{dx^2}\right) + \frac{1}{2}\left(\frac{dw}{dx}\right)^2 \tag{13.1}$$

The first and second terms are the familiar components of axial and flexural strain, respectively, while the third term, which is nonlinear in the transverse

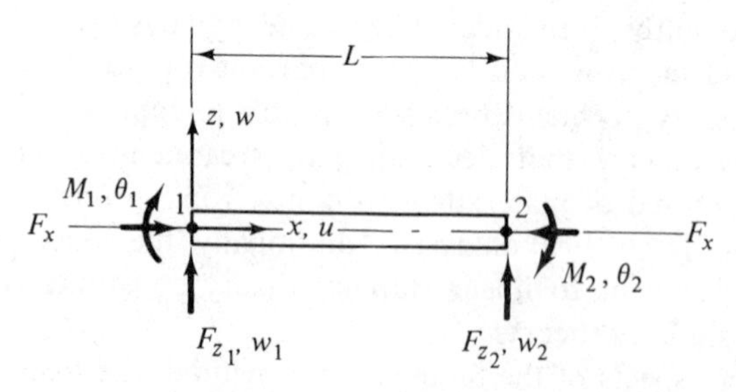

Fig. 13.1 Prismatic member.

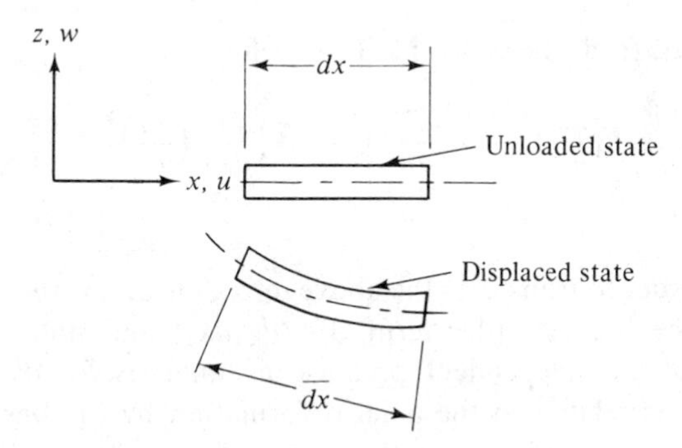

Fig. 13.2 Displacement geometry.

displacement w, represents a strain component associated with the coupling of flexural and axial action. The origin of this term can be seen from Fig. 13.2, where a differential length of fiber is sketched in the displaced position. The stretched length, $\overline{dx}$, is approximated by

$$\overline{dx} = \left[1 + \left(\frac{dw}{dx}\right)^2\right]^{1/2} dx$$

By expansion of this expression in series by use of the binomial theorem, we have

$$\overline{dx} = \left[1 + \frac{1}{2}\left(\frac{dw}{dx}\right)^2 + \cdots\right] dx$$

The series is truncated after the second term, which becomes the contribution of this effect to the total axial strain, the third term of Eq. (13.1).

The strain energy for the element is given by

$$U^e = \frac{1}{2}\int_{\text{vol}} E\epsilon_x^2 \, d(\text{vol}) \tag{13.2}$$

and, by substitution of Eq. (13.1) into (13.2) [with $d(\text{vol}) = dA\,dx$],

$$\begin{aligned} U^e = \frac{1}{2}\int_L\int_A\Bigg[\left(\frac{du}{dx}\right)^2 + z^2\left(\frac{d^2w}{dx^2}\right)^2 + \frac{1}{4}\left(\frac{dw}{dx}\right)^4 - 2z\left(\frac{du}{dx}\right)\left(\frac{d^2w}{dx^2}\right) \\ - z\left(\frac{d^2w}{dx^2}\right)\left(\frac{dw}{dx}\right)^2 + \left(\frac{du}{dx}\right)\left(\frac{dw}{dx}\right)^2\Bigg] E\,dA\,dx \end{aligned} \tag{13.3}$$

Next, we integrate across the depth of the member, noting that

$$\int_A dA = A, \qquad \int_A z\,dA = 0, \qquad \int_A z^2\,dA = I \tag{13.4}$$

for z measured from the centroid. This yields

$$U^e = \frac{1}{2}\int_L \left[A\left(\frac{du}{dx}\right)^2 + I\left(\frac{d^2w}{dx^2}\right)^2 + A\left(\frac{du}{dx}\right)\left(\frac{dw}{dx}\right)^2 + \frac{A}{4}\left(\frac{dw}{dx}\right)^4\right] E\,dx \tag{13.5}$$

Now, in order to transform the above into a linear instability formulation, we discard the highest-order term $A/4\,(dw/dx)^4$ and note that under the assumption of an independent prebuckling analysis for axial loading the axial load F_x is related to the axial deformation by the linear relationship

$$F_x = EA\frac{du}{dx} \tag{13.6}$$

where a tension value for F_x is taken as positive. Thus, Eq. (13.5) becomes

$$U^e = \frac{1}{2}\int_L \left[EA\left(\frac{du}{dx}\right)^2 + EI\left(\frac{d^2w}{dx^2}\right)^2 + F_x\left(\frac{dw}{dx}\right)^2\right] dx \tag{13.7}$$

The strain energy is thusly reduced to a form in which the axial and flexural strain energies are uncoupled:

$$U^e = U^e_a + U^e_f \tag{13.8}$$

with

$$U^e_a = \frac{1}{2}\int_L EA\left(\frac{du}{dx}\right)^2 dx \tag{13.9}$$

$$U^e_f = \frac{1}{2}\int_L \left[EI\left(\frac{d^2w}{dx^2}\right)^2 + F_x\left(\frac{dw}{dx}\right)^2\right] dx \tag{13.10}$$

The strain energy U_a refers to the independent prebuckling analysis for axial behavior.

Attention can now be restricted to the flexural behavior, it being assumed that the solution for axial load is accomplished through independent application of the principle of minimum potential energy.

The Euler equation for Eq. (13.10) can be established with use of the concepts of Chapter 6 and we find for this case that it is

$$EI\frac{d^4w}{dx^4} - F_x\frac{d^2w}{dx^2} = 0 \tag{13.11}$$

which is the well-known equation which governs beam buckling.

We assume a functional representation for the transverse displacement in terms of the four joint displacements; i.e.,

$$w = \lfloor N_1\ N_2\ N_3\ N_4 \rfloor \begin{Bmatrix} w_1 \\ w_2 \\ \theta_1 \\ \theta_2 \end{Bmatrix} = \lfloor \mathbf{N} \rfloor \{\boldsymbol{\Delta}_f\} \tag{13.12}$$

At this point we imply no particular form for the shape function $\lfloor \mathbf{N} \rfloor$, recognizing that the shape function that applies exactly to beam flexure does not satisfy exactly the differential equation pertinent to this case [Eq. (13.11)].

By substitution of Eq. (13.12) into Eq. (13.10),

$$U_f^e = \frac{\lfloor \boldsymbol{\Delta}_f \rfloor}{2}[\mathbf{k}_f]\{\boldsymbol{\Delta}_f\} + \frac{\lfloor \boldsymbol{\Delta}_f \rfloor}{2}[\mathbf{k}_g]\{\boldsymbol{\Delta}_f\} \tag{13.13}$$

where

$$[\mathbf{k}_f] = \left[\int_L EI\{\mathbf{N}''\}\lfloor \mathbf{N}'' \rfloor\, dx\right] \tag{13.14}$$

$$[\mathbf{k}_g] = \left[F_x \int_L \{\mathbf{N}'\}\lfloor \mathbf{N}' \rfloor\, dx\right] \tag{13.15}$$

The matrix $[\mathbf{k}_f]$ is the conventional element flexural stiffness matrix. The matrix $[\mathbf{k}_g]$ introduces the considerations related to elastic instability effects and may be viewed as an "increment" on the basic flexural stiffness. It is therefore often termed the *incremental* stiffness matrix. It can be discerned from Eq. [13.15] and will be verified subsequently in the derivation of the explicit form of $[\mathbf{k}_g]$ that the individual terms of this matrix are solely dependent on the geometric parameters (e.g., length). This matrix is therefore also commonly designated as the *geometric* stiffness matrix.

For flexure, the potential of the applied loads (in this case the joint forces —see Fig. 13.1) is given by

$$V^e = -\lfloor w_1\ w_2\ \theta_1\ \theta_2 \rfloor \begin{Bmatrix} F_{z_1} \\ F_{z_2} \\ M_1 \\ M_2 \end{Bmatrix} = -\lfloor \boldsymbol{\Delta}_f \rfloor \{\mathbf{F}\} \tag{13.16}$$

so that the flexural portion of the element potential energy is now represented by

$$\begin{aligned} \Pi_p^e &= U^e + V^e \\ &= \frac{\lfloor \boldsymbol{\Delta}_f \rfloor}{2}[\mathbf{k}_f]\{\boldsymbol{\Delta}_f\} + \frac{\lfloor \boldsymbol{\Delta}_f \rfloor}{2}[\mathbf{k}_g]\{\boldsymbol{\Delta}_f\} - \lfloor \boldsymbol{\Delta}_f \rfloor \{\mathbf{F}\} \end{aligned} \tag{13.17}$$

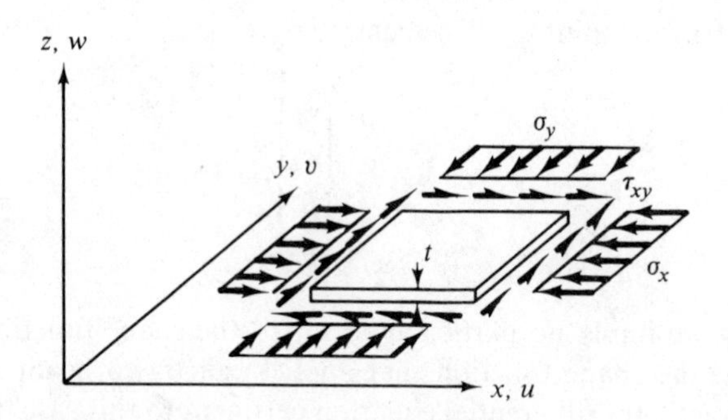

Fig. 13.3 Middle surface stresses acting on plate in flexure.

In thin plate behavior the strain-displacement equations corresponding to Eq. (13.1) are (see Fig. 13.3)

$$\begin{aligned} \epsilon_x &= \frac{\partial u}{\partial x} - z\frac{\partial^2 w}{\partial x^2} + \frac{1}{2}\left(\frac{\partial w}{\partial x}\right)^2 \\ \epsilon_y &= \frac{\partial v}{\partial y} - z\frac{\partial^2 w}{\partial y^2} + \frac{1}{2}\left(\frac{\partial w}{\partial y}\right)^2 \\ \gamma_{xy} &= \frac{\partial u}{\partial y} + \frac{\partial v}{\partial x} - 2z\frac{\partial^2 w}{\partial x\,\partial y} + \frac{\partial w}{\partial x}\frac{\partial w}{\partial y} \end{aligned} \tag{13.18}$$

By the same process described above one obtains the extended flexural strain energy for an isotropic plate

$$\begin{aligned} U^e = \frac{D}{2}\int_A &\left[\left(\frac{\partial^2 w}{\partial x^2} + \frac{\partial^2 w}{\partial y^2}\right)^2 + 2\mu\frac{\partial^2 w}{\partial x^2}\frac{\partial^2 w}{\partial y^2} + 2(1-\mu)\left(\frac{\partial^2 w}{\partial x\,\partial y}\right)^2\right] dA \\ &+ \frac{1}{2}\int_A \sigma_x t\left(\frac{\partial w}{\partial x}\right)^2 dA + \frac{1}{2}\int_A \sigma_y t\left(\frac{\partial w}{\partial y}\right)^2 dA \\ &+ \int_A \tau_{xy} t\left(\frac{\partial w}{\partial x}\right)\left(\frac{\partial w}{\partial y}\right) dA \end{aligned} \tag{13.19}$$

where t is the thickness of the plate and $\sigma_x t$, $\sigma_y t$, and $\tau_{xy} t$ are the middle surface (membrane) forces expressed as line loads, i.e., forces per unit width.

The choice of a functional representation for the transverse displacement can again be symbolized by $w = \lfloor \mathbf{N} \rfloor \{\boldsymbol{\Delta}_f\}$. Upon substitution of the assumed displacement function into Eq. (13.19), there is obtained

$$U^e = \frac{\lfloor \boldsymbol{\Delta}_f \rfloor}{2}[\mathbf{k}_f]\{\boldsymbol{\Delta}_f\} + \frac{\lfloor \boldsymbol{\Delta}_f \rfloor}{2}[\mathbf{k}_{g_x}]\{\boldsymbol{\Delta}_f\} + \frac{\lfloor \boldsymbol{\Delta}_f \rfloor}{2}[\mathbf{k}_{g_y}]\{\boldsymbol{\Delta}_f\} + \frac{\lfloor \boldsymbol{\Delta}_f \rfloor}{2}[\mathbf{k}_{g_{xy}}]\{\boldsymbol{\Delta}_f\} \tag{13.20}$$

where $[\mathbf{k}_f^e]$ is the conventional plate flexure stiffness matrix and

$$[\mathbf{k}_{gx}] = \left[\int_A (\sigma_x t)\{\mathbf{N}'_x\}\lfloor \mathbf{N}'_x \rfloor \, dA\right] \tag{13.21}$$

with

$$\frac{\partial w}{\partial x} = \left\lfloor \frac{\partial \mathbf{N}}{\partial x} \right\rfloor \{\mathbf{\Delta}_f\} = \lfloor \mathbf{N}'_x \rfloor \{\mathbf{\Delta}_f\} \tag{13.22}$$

Expressions similar to Eq. (13.21) hold for $[\mathbf{k}_{gy}]$ and $[\mathbf{k}_{gxy}]$. If we define

$$[\mathbf{k}_g] = [[\mathbf{k}_{gx}] + [\mathbf{k}_{gy}] + [\mathbf{k}_{gxy}]] \tag{13.23}$$

then Eq. (13.20) assumes the same form as Eq. (13.13).

It is interesting to note that since the terms representative of instability effects are geometric in nature, and independent of material properties, the extension to include such effects is identical for both isotropic and orthotropic plates; i.e., $[\mathbf{k}_g]$ is independent of the degree of anisotropy of the plate.

13.2 Global Formulation

The potential energy for the system is simply the sum of the potential energies of the component elements. The global potential energy is therefore of the same form as the element energies, as follows:

$$\begin{aligned}\Pi_{p_{\text{system}}} &= \Sigma \Pi_p^e \\ &= \frac{\lfloor \mathbf{\Delta}_f \rfloor}{2}[\mathbf{K}_f]\{\mathbf{\Delta}_f\} + \frac{\lfloor \mathbf{\Delta}_f \rfloor}{2}[\mathbf{K}_g]\{\mathbf{\Delta}_f\} + V\end{aligned} \tag{13.24}$$

where now

$$[\mathbf{K}_f] = [\Sigma \, \mathbf{k}_f] \tag{13.24a}$$

$$[\mathbf{K}_g] = [\Sigma \, \mathbf{k}_g] \tag{13.24b}$$

where the summations are taken over all elements in the system. The vectors $\{\mathbf{\Delta}_f\}$ and $\{\mathbf{P}\}$ are the displacements and applied loads, respectively. Note that $\{\mathbf{P}\}$ refers to the loads associated with flexure. We do not use the subscript f because this has already been used to denote loads associated with points free to displace. In the present work we emphasize that other loads $\{\mathbf{P}_a\}$ are present in the problem, associated with the axial behavior, but these do not appear in $\{\mathbf{P}\}$. We assume in this discussion that *conservative loading* takes place, defined by the condition that its potential (V) in any kinematically admissible displacement of the structure depends solely on the initial and final configurations of the system. We therefore exclude such cases as where the direction of load "follows" the deflected direction of the member upon which it acts.

For conditions of stable equilibrium, i.e., where the axial load is of a value less than the critical value, application of the *principle of stationary potential energy* in the form of the first variation of Π_p (i.e., $\delta\Pi_p = 0$) to Eq. (13.24) results in the stiffness equation

$$\{\mathbf{P}\} = [\mathbf{K}_f]\{\mathbf{\Delta}_f\} + [\mathbf{K}_g]\{\mathbf{\Delta}_f\} \tag{13.25}$$

One may proceed to solve Eq. (13.25) for the displacements $\{\mathbf{\Delta}_f\}$ by the usual means, yielding results that account for "beam-column" action, i.e., the stiffness-modifying effects of the axial loading. It should be noted that the action of tensile loading in increasing the flexural stiffness may be accounted for in this manner.

To deal with elastic instability, where the *intensity* of the axial load system to cause buckling is as yet unknown, the incremental stiffness matrix must first be numerically evaluated using an arbitrarily chosen load intensity (we assume that the *distribution* of axial forces is fixed). For buckling, the intensity of the axial load system is taken to be λ times the arbitrarily chosen intensity of $\{\mathbf{P}_a\}$ used in the construction of $[\mathbf{k}_g]$ and the equilibrium equation becomes

$$\delta\Pi_p = [\mathbf{K}_f]\{\mathbf{\Delta}_f\} + \lambda[\mathbf{k}_g]\{\mathbf{\Delta}_f\} - \{\mathbf{P}\} = 0 \tag{13.26}$$

At this point we must consider the conditions for neutral stability; they determine the eigenvalue λ and its associated buckling mode, $\{\mathbf{\Delta}_{CR}\}$. The necessary information cannot be drawn from the first variation of potential energy and instead we must progress to a determination of the second variation, $\delta^2\Pi_p = \delta\,(\delta\Pi_p)$.

It is clear from Fig. 6.2, which shows schematically the potential energy as a function of some representative displacement parameter $\mathbf{\Delta}$, that for stable equilibrium $\delta^2\Pi_p > 0$ and for neutral equilibrium $\delta^2\Pi_p = 0$. The latter condition defines the bifurcation of equilibrium, which we are seeking. Now, after application of this condition to Eq. (13.24), the second variation of the potential energy reads

$$\delta^2\Pi_p = \lfloor \delta\mathbf{\Delta}_f \rfloor [\mathbf{K}]\{\delta\mathbf{\Delta}_f\} = 0$$

and with $[\mathbf{K}] = [\mathbf{K}_f] + [\mathbf{K}_g]$ in the present case, this requires

$$|[\mathbf{K}_f] + [\mathbf{K}_g]| = 0 \tag{13.27}$$

where $|\quad|$ symbolizes the determinant.

Thus we have the condition that the determinant of the matrix Eq. (13.27) is equal to zero. An alternative argument for this condition is that there is no unique solution (the bifurcation condition) to Eq. (13.26) if there is a vector $\{\mathbf{\Delta}_f\}$ and scalar λ such that $[\mathbf{K}_f + \lambda \mathbf{K}_g]\{\mathbf{\Delta}_f\} = 0$.

Evaluation of the determinant in order to obtain the value of ω is inefficient for large-order systems. The usual approach is based upon the forms of Eq. (13.27) resulting from multiplication of it by $\{\mathbf{\Delta}\}$. Rearranging, one obtains

$$\frac{1}{\lambda}\{\mathbf{\Delta}_f\} = [\mathbf{K}_f]^{-1}[\mathbf{K}_g]\{\mathbf{\Delta}_f\} \tag{13.27a}$$

By iteration, or other means, the lowest value of λ and the associated vector $\{\mathbf{\Delta}_{f_{cr}}\}$ are obtained.

The vector of d.o.f. ($\{\mathbf{\Delta}_f\}$) contains both translational (w_i) and angular (θ_i) displacements in the beam and plate problems discussed so far. By intuition one would expect that the translational displacements are sufficient to define adequately the buckling mode and that this mode would in turn enable the calculation of a sufficiently accurate intensity of critical load. An exact condensation of the eigenvalue problem represented by Eq. (13.27) is not convenient since the result would yield matrices from which λ cannot be factored out. An iterative process would be necessary. On the other hand, the condensation procedure described in Section 2.8, which strictly applies only to the conventional stiffness matrix, can be applied to the geometric stiffness matrix to produce an approximate condensed matrix.

We adapt the procedure of Section 2.8 to the present problem as follows. First we assume that the vector of displacements has been partitioned into translational and angular displacements; i.e., $\lfloor \mathbf{\Delta} \rfloor = \lfloor \mathbf{w} : \boldsymbol{\theta} \rfloor$. For the matrix $[\mathbf{k}_f]$, for example, we have

$$[\mathbf{k}_f] = \left[\begin{array}{c|c} \mathbf{K}_{f_{ww}} & \mathbf{K}_{f_{w\theta}} \\ \hline \mathbf{K}_{f_{\theta w}} & \mathbf{K}_{f_{\theta\theta}} \end{array}\right]$$

(In reality *any* division of the total set of d.o.f. is possible but for simplicity of concept and notation we retain the division into $\{\mathbf{w}\}$ and $\{\boldsymbol{\theta}\}$ d.o.f.) In accordance with Section 2.8, the transformation from the $\lfloor \mathbf{w} \rfloor$ to the $\lfloor \mathbf{w} \,|\, \boldsymbol{\theta} \rfloor$ d.o.f., based on the conventional stiffness matrix $[\mathbf{k}_f]$, is

$$\left\{\begin{array}{c} \mathbf{w} \\ \hline \boldsymbol{\theta} \end{array}\right\} = \left[\begin{array}{c} \mathbf{I} \\ \hline -\mathbf{k}_{f_{\theta\theta}}^{-1}\mathbf{k}_{f_{\theta w}} \end{array}\right]\{\mathbf{w}\} = [\mathbf{\Gamma}_0]\{\mathbf{w}\} \tag{13.28}$$

The transformation matrix $[\mathbf{\Gamma}_0]$ is now applied to the conventional stiffness matrix $[\mathbf{k}_f]$ and to the geometric stiffness matrix $[\mathbf{k}_g]$ to yield

$$[\hat{\mathbf{k}}_f] + \lambda[\hat{\mathbf{k}}_g]\{\mathbf{w}\} = 0 \tag{13.29}$$

where

$$[\hat{\mathbf{k}}_f] = [\boldsymbol{\Gamma}_0]^T[\mathbf{k}_f][\boldsymbol{\Gamma}_0] \quad (13.29a)$$

$$[\hat{\mathbf{k}}_g] = [\boldsymbol{\Gamma}_0]^T[\mathbf{k}_g][\boldsymbol{\Gamma}_0] \quad (13.29b)$$

The eigenvalue ω and the associated buckling mode shape are now extracted from the reduced system Eq. (13.29). The effectiveness of this scheme will be described subsequently.

13.3 Prismatic Members

13.3.1 Flexural Buckling

Our objective now is the transformation of the strain energy [Eq. (13.13)] into an explicit element stiffness matrix formulation via the selection of a functional representation for w.

An exact representation can be formulated for this case, using the displacement function that satisfies the corresponding differential equation.[13.1] Pursuing the usual motivation in finite element analysis, however, that of choosing a simple, approximate displacement field, we alternatively select (with $\xi = x/L$) the displacement field for a flexural element without axial load [See Eq. (5.14a)].

$$w = \lfloor (1 - 3\xi^2 + 2\xi^3) \quad (3\xi^2 - 2\xi^3) - (1 - 2\xi + \xi^2)x \quad (\xi - \xi^2)x \rfloor \begin{Bmatrix} w_1 \\ w_2 \\ \theta_1 \\ \theta_2 \end{Bmatrix}$$

Use of this in Eq. (13.13) results in the well-known representation for the basic flexural matrix $[\mathbf{k}_f]$ given by Eq. (5.17), plus the following explicit form for the matrix $[\mathbf{k}_g]$:

$$[\mathbf{k}_g] = \frac{F_x}{30L} \begin{bmatrix} 36 & & & \text{(Symmetric)} \\ -36 & 36 & & \\ -3L & 3L & 4L^2 & \\ -3L & 3L & -L^2 & 4L^2 \end{bmatrix} \quad (13.30)$$

(columns labeled w_1, w_2, θ_1, θ_2)

As an example of the use of the consistent incremental stiffness matrix in the solution of beam instability problems, we solve for the buckling load of a simple beam, using one element, see Fig. 13.4.

$$w_1 = \theta_2 = 0$$

$$F_x = -P_{x_{cr}}$$

$$L = \frac{l}{2}$$

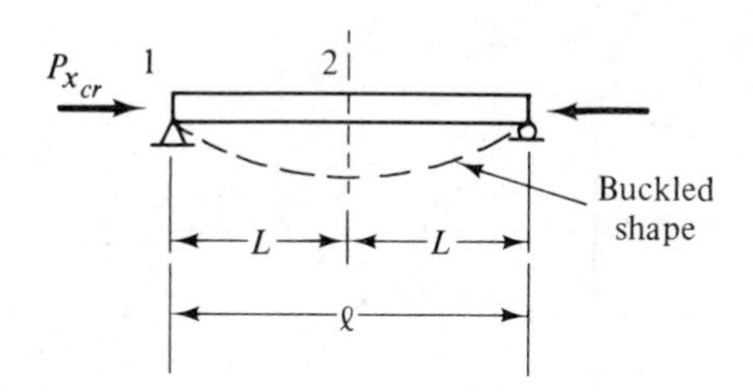

Fig. 13.4

Thus, the characteristic equation becomes

$$\left(\frac{8EI}{l} - \frac{P_{x_{cr}}l}{15}\right)\left(\frac{96EI}{l^3} - \frac{12P_{x_{cr}}}{5l}\right) - \left(\frac{24EI}{l^2} - \frac{P_{x_{cr}}}{10}\right)^2 = 0$$

Solving, one obtains

$$P_{x_{cr}} = 9.94\left(\frac{EI}{l^2}\right)$$

This solution is only 0.752% different from the exact result, $\pi^2(EI/l^2)$.

Figure 13.5 shows the percentage error in the solution of this problem as a function of the number of elements in the span. The corresponding results for a finite difference solution(13.2) are shown for comparison. The lessened accuracy of the finite difference model is due to the assumption of a linear variation of w between the joints, in contrast with the present assumption of a cubic variation. Note, however, that the finite difference equations employ only a single d.o.f. per joint, the transverse displacement w.

Figure 13.6 shows the percentage error in the prediction of the buckling load of a tapered cantilever beam as a function of the number of elements in the representation. The results given by a solid line(13.3) are for a stepped representation in which the geometric properties in the center of the element are used to describe the element. Two aspects of the results are to be noted. First the accuracy at any level of refinement is strikingly less than for the uniform beam. Second, rather than converging upon the solution from the high side, the results converge from the low side. Thus we find that the increased flexibility due to geometric approximation far outbalances the approximation in the element displacement function representation. Furthermore, it will be found that there is no "representative" cross section for the element.

It is difficult and often impossible to formulate the exact stiffness matrix for a nonuniform element. On the other hand, one can employ the principle of minimum potential energy in derivation of an approximate tapered element formulation, using a geometric representation that is exact (as done in Sections 6.4 and 7.2) or that closely approximates an actual taper, and assumed displacement representation, such as that employed for the uniform-section element.

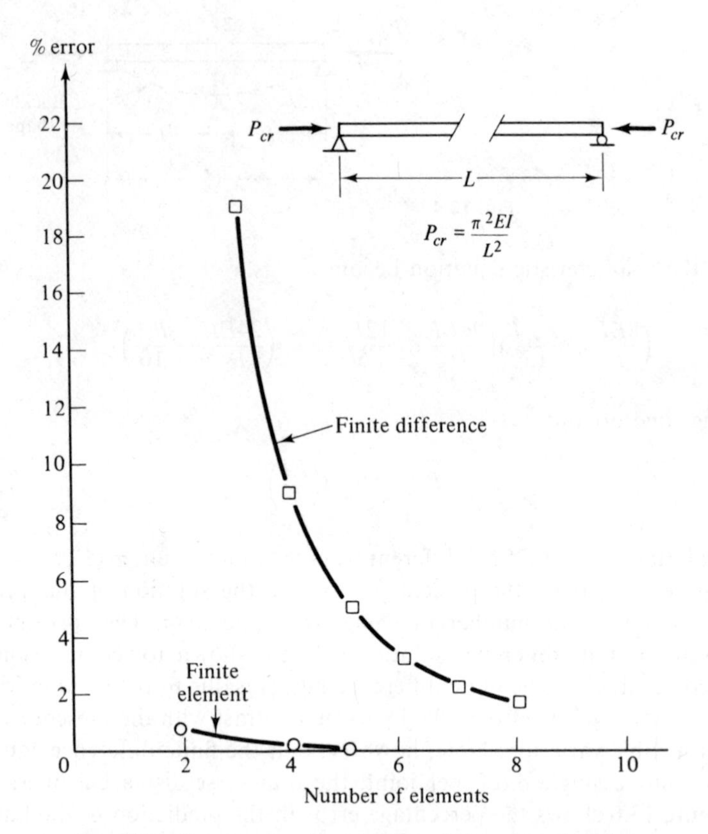

Fig. 13.5 Convergence characteristics: Euler column buckling.

By using this approach, which preserves the continuity of geometry across the joints, the highly accurate results shown by the dotted curve in Fig. 13.6 are obtained. Perhaps the most significant conclusion to be drawn from this data, since it applies to all classes of finite element representation, is that as much care must be exercised in geometric representation as in the selection of assumed displacement functions.

13.3.2 *Torsional-Flexural Buckling*

When the prismatic member forms a part of a space framework, it is generally subjected to flexural action in two planes, twisting about the member axis, and axial loading. The interaction of these components gives rise to elastic instability phenomena that are more complex than the simple modes of buckling described in the previous sections. Extension to this more general case follows the same lines as the previous development and details are to be found in Ref. 13.4. To illustrate the considerations involved, we examine

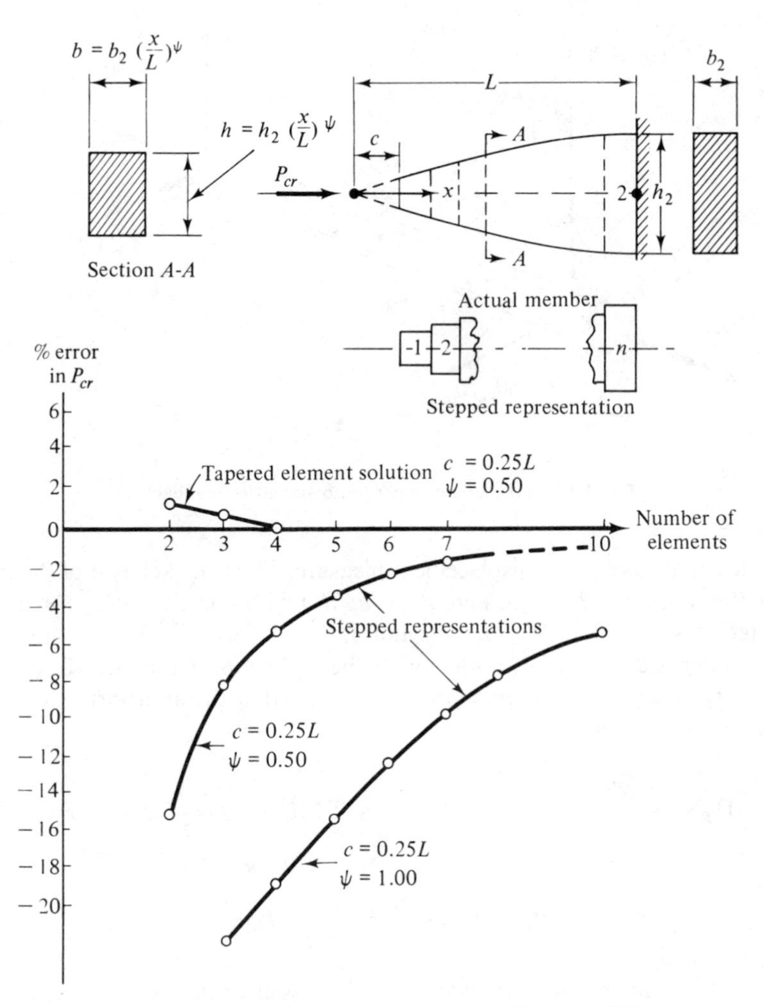

Fig. 13.6 Convergence characteristics—tapered beam stability.

here a special aspect of the more general case, the condition of torsional-flexural buckling.

The prismatic member that forms the basis of this development is shown in Fig. 13.7. It is assumed that an independent prebuckling analysis has been performed to determine the flexural and axial behavior in the y-z plane. Thus the relative amplitudes of the end moments M_{x_1} and M_{x_2}, the end shears F_{z_1} and F_{z_2}, and the axial load F_y are known. The buckling phenomena in question involve torsional deformation (about the y axis) and flexure in the x-y plane. In order to account for warping effects it is necessary to include the derivative of the angle of twist (θ_y) with respect to the y coordinate, which

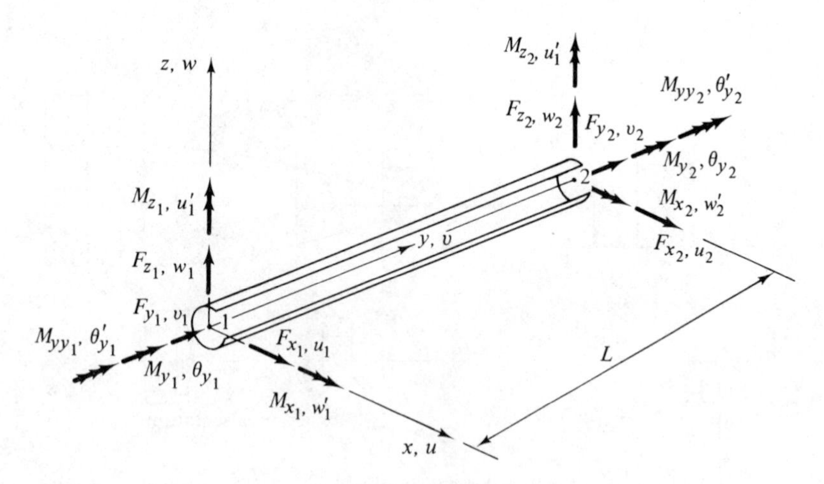

Fig. 13.7 Prismatic member for torsional-flexural buckling.

we designate as θ_y', as a displacement measure. Thus, at each end of the member the d.o.f. are θ_y, θ_y', u, and u' ($= du/dy$) with corresponding force parameters designated as M_y, M_{yy}, F_x, and M_z.

Under the conditions above, with the inclusion of torsional-flexural instability effects, it has been shown (Ref. 13.4) that the appropriate potential energy expression is

$$\Pi_p = \frac{1}{2}\int_L [EI_z(u'')^2 + GJ(\theta_y')^2 + E\Gamma(\theta_y'')^2 - \frac{F_y I_p}{A}(\theta_y')^2 - F_y(u')^2$$
$$+ (F_{z_1}y + F_{z_2}(L-y) + M_{x_1} - M_{x_2})\theta_x u''] \, dy \qquad (13.31)$$
$$- \sum_{i=1}^{2} (M_{y_i}\theta_y + M_{yy_i}\theta_y' + F_{x_i}u_i + M_{z_i}u_i')$$

where the primes denote derivatives with respect to the axial coordinate (y), G is the shear modulus, and

$$J = \text{St. Venant torsion constant,}$$
$$I_p = \text{polar moment of inertia,}$$
$$\Gamma = \text{warping constant,}$$

and A, I_z are as defined previously. $M_{x_{1-2}} = M_{x_1} - M_{x_2}$.

Each of the displacement variables, u and θ_y, is required to meet two boundary conditions at each joint, one on the variable itself (u_i, θ_{y_i}) and the other on the first derivative of the variable (u_i', θ_{y_i}'). This type of condition was met in simple flexure. Thus, the form of the function employed in representation of simple flexure can also be employed for this case:

$$u = \lfloor \mathbf{N} \rfloor \{\mathbf{\Delta}_u\} \tag{13.32}$$

$$\theta_y = \lfloor \mathbf{N} \rfloor \{\mathbf{\Delta}_{\theta_y}\} \tag{13.33}$$

where $\lfloor \mathbf{N} \rfloor$ is the shape function vector given in Eq. (5.14a) and

$$\{\mathbf{\Delta}_u\} = \lfloor u_1 \; u_2 \; u_1' \; u_2' \rfloor^{\mathrm{T}} \tag{13.34}$$

$$\{\mathbf{\Delta}_{\theta_y}\} = \lfloor \theta_{y_1} \; \theta_{y_2} \; \theta_{y_1}' \; \theta_{y_2}' \rfloor^{\mathrm{T}} \tag{13.35}$$

Substitution of these expressions into Eq. (13.31) and performance of the indicated integrations and application of the principle of stationary potential energy, i.e., differentiation with respect to each of the joint d.o.f., yields the same form of stiffness equation as in the prior sections of this chapter:

$$\{\mathbf{F}\} = [[\mathbf{k}_f] + [\mathbf{k}_g]]\{\mathbf{\Delta}\}$$

where in this case

$$\{\mathbf{F}\} = \lfloor F_{x_1} \; M_{z_1} \; M_{y_1} \; M_{yy_1} \; F_{x_2} \; M_{z_2} \; M_{y_2} \; M_{yy_2} \rfloor^{\mathrm{T}}$$

$$\{\mathbf{\Delta}\} = \lfloor u_1 \; u_1' \; \theta_{y_1} \; \theta_{y_1}' \; u_2 \; u_2' \; \theta_{y_2} \; \theta_{y_2}' \rfloor^{\mathrm{T}}$$

The basic stiffness ($[\mathbf{k}_f]$) and geometric stiffness matrices for this case are shown in Fig. 13.8.

To illustrate the accuracy of this formulation we examine the problem of lateral instability of a simply supported beam with ends also fixed against torsional rotation (see sketch, Fig. 13.9) under equal end moments $M_{x_1} = -M_{x_2} = M_{cr}$. A one-element solution is attempted, for which, due to symmetry, $u_2' = -u_1'$ and $\theta_{y_1}' = -\theta_{y_2}'$. All other joint displacement parameters ($u_1, u_2, \theta_{y_1}, \theta_{y_2}$) are zero. Applying these conditions to the stiffness equations of Fig. 13.8 we have

$$\begin{vmatrix} \dfrac{2EI_z}{L} & \dfrac{L}{6}M_{cr} \\ \dfrac{L}{6}M_{cr} & \left(\dfrac{5}{30}GJL + \dfrac{2E\Gamma}{L}\right) \end{vmatrix} = 0$$

from which we obtain

$$M_{cr} = \frac{3}{L}\sqrt{(EI_z)(GJ)\left(\frac{4}{3} + \frac{E\Gamma}{GJ}\frac{16}{L^2}\right)}$$

For this case the exact solution is

$$M_{cr} = \frac{\pi}{L}\sqrt{(EI_z)(GJ)\left(1 + \frac{E\Gamma}{GJ}\frac{\pi^2}{L^2}\right)}$$

$$\{\mathbf{F}\} = [[\mathbf{k}_f] + [\mathbf{k}_g]]\{\Delta\}$$

$$\{\mathbf{F}\} = \lfloor F_{x_1}\ M_{z_1}\ M_{y_1}\ M_{yy_1}\ F_{x_2}\ M_{z_2}\ M_{y_2}\ M_{yy_2} \rfloor^T$$

$$\{\Delta\} = \lfloor u_1\ u_1'\ \theta_{y_1}\ \theta_{y_1}'\ u_2\ u_2'\ \theta_{y_2}\ \theta_{y_2}' \rfloor^T$$

$[\mathbf{k}_f] = \frac{1}{L^3}$

u_1	u_1'	θ_{y_1}	θ_{y_1}'	u_2	u_2'	θ_{y_2}	θ_{y_2}'
$12EI_z$							
$-6EI_zL$	$4EI_zL^2$						
0	0	$\frac{12GJL^2}{10} + 12E\Gamma$			Symmetric		
0	0	$-\frac{GJL^3}{10} - 6E\Gamma L$	$\frac{4GJL^4}{30} + 4E\Gamma L^2$				
$-12EJ_z$	$6EI_zL$	0	0	$12EI_z$			
$-6EI_zL$	$2EI_zL^2$	0	0	$6EI_zL$	$4EI_zL^2$		
0	0	$-\frac{12GJL^2}{10} - 12E\Gamma$	$\frac{GJL^3}{10} + 6E\Gamma L$	0	0	$\frac{12GJL^2}{10} + 12E\Gamma$	
0	0	$-\frac{GJL^3}{10} - 6E\Gamma L$	$-\frac{GJL^4}{30} + 2E\Gamma L^2$	0	0	$\frac{GJL^3}{10} + 6E\Gamma L$	$\frac{4GJL^4}{30} + 4E\Gamma L^2$

$[\mathbf{k}_g] = \frac{1}{60}$

u_1	u_1'	θ_{y_1}	θ_{y_1}'	u_2	u_2'	θ_{y_2}	θ_{y_2}'
$\frac{72F_y}{L}$							
$-6F_y$	$8F_yL$						
$\frac{36M_{x_{1-2}}}{L} + 3F_{z_1} + 33F_{z_2}$	$-33M_{x_{1-2}} - 6F_{z_1}L - 27F_{z_2}L$	$\frac{72F_yI_p}{LA}$			Symmetric		
$-3M_{x_{1-2}} - 3F_{z_2}L$	$4M_{x_{1-2}}L + F_{z_1}L^2 + 3F_{z_2}L^2$	$-\frac{6F_yI_p}{A}$	$\frac{8F_yI_pL}{A}$				
$-\frac{72F_y}{L}$	$6F_y$	$-\frac{36M_{x_{1-2}}}{L} - 3F_{z_1} - 33F_{z_2}$	$3M_{x_{1-2}} + 3F_{z_2}L$	$\frac{72F_y}{L}$			
$-6F_y$	$-2F_yL$	$-3M_{x_{1-2}} + 3F_{z_1}L - 6F_{z_2}L$	$-M_{x_{1-2}}L - F_{z_1}L^2$	$6F_y$	$8F_yL$		
$-\frac{36M_{x_{1-2}}}{L} - 33F_{z_1} - 3F_{z_2}$	$3M_{x_{1-2}} + 6F_{z_1}L - 3F_{z_2}L$	$-\frac{72F_yI_p}{AL}$	$\frac{6F_yI_p}{A}$	$\frac{36M_{x_{1-2}}}{L} + 33F_{z_1} + 3F_{z_2}$	$33M_{x_{1-2}} + 27F_{z_1}L + 6F_{z_2}L$	$\frac{72F_yI_p}{LA}$	
$-3M_{x_{1-2}} - 3F_{z_1}L$	$-M_{x_{1-2}}L - F_{z_2}L^2$	$-\frac{6F_yI_p}{A}$	$-\frac{2F_yI_pL}{A}$	$2M_{x_{1-2}} + 3F_{z_1}L$	$4M_{x_{1-2}}L + 3F_{z_1}L^2 + F_{z_2}L^2$	$+\frac{6F_yI_p}{A}$	$\frac{8F_yI_pL}{A}$

Fig. 13.8 Stiffness matrices for torsional-flexural buckling of prismatic member.

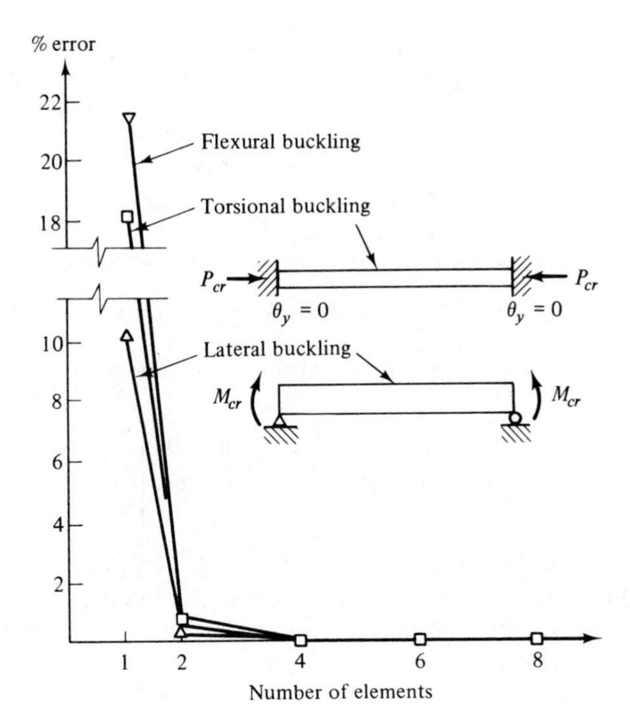

Fig. 13.9 Convergence characteristics: Flexural, torsional, and lateral buckling.

If the St. Venant torsional stiffness is small in comparison with the warping stiffness (i.e., $GJ \ll E\Gamma$), the error is approximately 20%; for the converse, for a rectangular section, the error is approximately 10%.

Figure 13.9 shows the convergence characteristics of torsional and lateral buckling solutions obtained with use of the derived formulations. The previous results from the flexural stability case are also shown for comparison. All three cases demonstrated an error of less than 1% for the two-element idealizations.

13.3.3 Framework Stability

The framework represents a more complex situation than one-directional (e.g., continuous beam) stability analysis because the distribution of axial loading generally depends on a coupled axial-flexural behavior. Thus, the flexural stability problem statement cannot be formulated independently of the axial deformation analysis and the problem becomes nonlinear. A simple example serves to illustrate this consideration. Figure 13.10 portrays the joint (j) with forces (and displacements) on a member referenced to the local (element) coordinates. These quantities are designated with primes. For

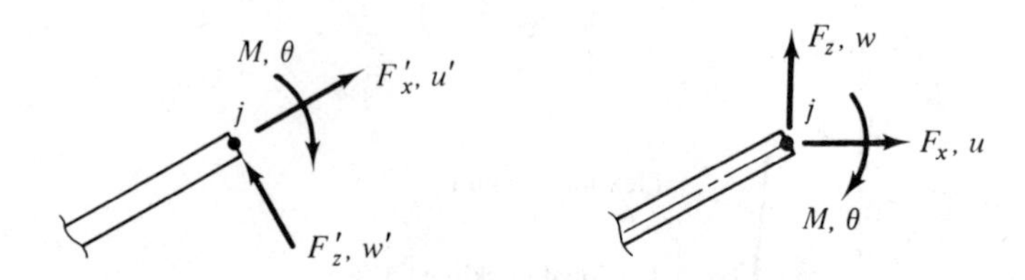

(a) Element coordinates (b) Global coordinates

Fig. 13.10 Framework element—joint forces.

each element, therefore, the matrices for linear stability analysis are

$$\begin{Bmatrix} \mathbf{F}'_x \\ \mathbf{F}'_z \\ \mathbf{M} \end{Bmatrix} = \begin{bmatrix} \mathbf{k}'_a & \mathbf{0} \\ \mathbf{0} & \mathbf{k}'_f \end{bmatrix} \begin{Bmatrix} \mathbf{u}' \\ \mathbf{w}' \\ \boldsymbol{\theta}' \end{Bmatrix} + \begin{bmatrix} \mathbf{0} & \mathbf{0} \\ \mathbf{0} & \mathbf{k}'_g \end{bmatrix} \begin{Bmatrix} \mathbf{u}' \\ \mathbf{w}' \\ \boldsymbol{\theta}' \end{Bmatrix} \tag{13.36}$$

where we now distinguish between axial $[\mathbf{k}'_a]$ and flexural $[\mathbf{k}'_f]$ behavior and the primes indicate that these matrices are referred to element axes. Upon transformation into global (unprimed) coordinates, u and w, the matrices are populated as follows:

$$\begin{Bmatrix} \mathbf{F}_x \\ \mathbf{F}_z \\ \mathbf{M} \end{Bmatrix} = \begin{bmatrix} \mathbf{k}_{11} & \mathbf{k}_{12} \\ \mathbf{k}_{21} & \mathbf{k}_{22} \end{bmatrix} \begin{Bmatrix} \mathbf{u} \\ \mathbf{w} \\ \boldsymbol{\theta} \end{Bmatrix} + \begin{bmatrix} \mathbf{k}_{g_{11}} & \mathbf{k}_{g_{12}} \\ \mathbf{k}_{g_{21}} & \mathbf{k}_{g_{22}} \end{bmatrix} \begin{Bmatrix} \mathbf{u} \\ \mathbf{w} \\ \boldsymbol{\theta} \end{Bmatrix} \tag{13.37}$$

It is seen that the element axial and flexural behavior representations are coupled, and we no longer subscript the linear stable terms. This condition is preserved after all elements are assembled in formation of the equations for the complete structure. Thus, the coefficients of the geometric stiffness are a function of the flexural behavior and cannot be determined independently.

To illustrate this situation, consider the framework shown in Fig. 13.11(a). Figure 13.11(b) shows the element idealization. Only point 2 is free to move so that stiffness equations need be written for only that point. With reference to the basic stiffness equations for flexure and axial loading, the desired expressions are

$$F_x^A = 12\frac{EI}{L^3}u_2 - 6\frac{EI}{L^2}\theta + \frac{5}{6}\frac{F_z^A}{L}u_2 - \frac{F_z^A}{10}\theta$$

$$F_z^B = 12\frac{EI}{L^3}w_2 - 6\frac{EI}{L^2}\theta + \frac{5}{6}\frac{F_x^B}{L}w_2 - \frac{F_x^B}{10}\theta$$

$$F_z^A = \frac{AE}{L}w_2, \qquad F_x^B = \frac{AE}{L}u_2$$

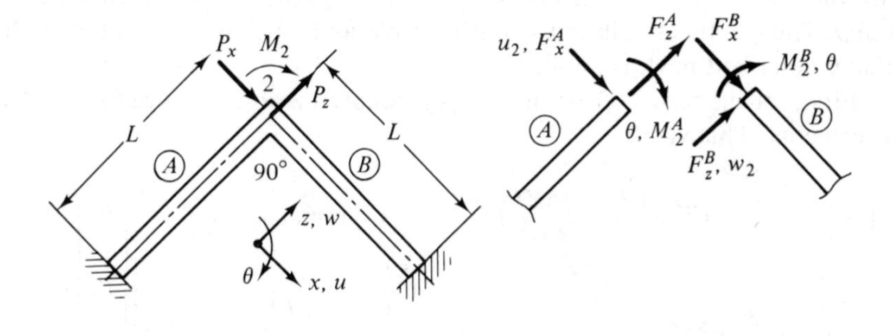

(a) Frame (b) Idealization

Fig. 13.11 Simple frame for illustration of instability analysis procedure.

$$M_z^A = -6\frac{EI}{L^2}u_2 + 4\frac{EI}{L}\theta - \frac{F_z^A}{10}u_2 + \frac{2}{15}LF_z^A\theta$$

$$M_z^B = -6\frac{EI}{L^2}w_2 + 4\frac{EI}{L}\theta - \frac{F_x^B}{10}w_2 + \frac{2}{15}LF_x^B\theta$$

Note that the element axes correspond to the global axes so that primes are not needed to distinguish between the two coordinate systems. Upon assembly of the elements, the resulting force-displacement relationships become

$$\begin{Bmatrix} P_x \\ P_z \\ M_z \end{Bmatrix} = \begin{bmatrix} \frac{AE}{L} + 12\frac{EI}{L^3} & & (\text{Symmetric}) \\ 0 & \frac{AE}{L} + 12\frac{EI}{L^3} & \\ -6\frac{EI}{L^2} & -6\frac{EI}{L^2} & 8\frac{EI}{L} \end{bmatrix} \begin{Bmatrix} u_2 \\ w_2 \\ \theta \end{Bmatrix}$$

$$+ \begin{bmatrix} \frac{5}{6}\frac{F_z^A}{L} & & (\text{Symmetric}) \\ 0 & \frac{5}{6}\frac{F_x^B}{L} & \\ -\frac{F_z^A}{10} & -\frac{F_x^B}{10} & \frac{2L}{15}(F_z^A + F_x^B) \end{bmatrix} \begin{Bmatrix} u_2 \\ w_2 \\ \theta \end{Bmatrix}$$

or, in summary form,

$$\begin{Bmatrix} P_x \\ P_z \\ M_z \end{Bmatrix} = [\mathbf{K}_f]\begin{Bmatrix} u_2 \\ w_2 \\ \theta \end{Bmatrix} + [\mathbf{K}_g]\begin{Bmatrix} u_2 \\ w_2 \\ \theta \end{Bmatrix}$$

Study of these equations discloses that the forces F_z^A and F_x^B, which populate the matrix $[\mathbf{K}_g]$, are themselves a function of the displacements u_2, w_2, and θ. Thus, a direct solution is not possible and an iterative solution technique, described next, is applied.

First, set the terms of the matrix $[\mathbf{K}_g]$ equal to zero and solve for the displacements. This yields

$$\begin{Bmatrix} u_2 \\ w_2 \\ \theta \end{Bmatrix} = \frac{1}{\text{Det.}} \left[\begin{array}{c|c|c} (\Psi)\dfrac{8EI}{L} - \left(\dfrac{6EI}{L^2}\right)^2 & \text{(Symmetric)} & \\ \hline \dfrac{6EI}{L^2} & (\Psi)\dfrac{8EI}{L} - \left(\dfrac{6EI}{L^2}\right)^2 & \\ \hline (\Psi)\dfrac{6EI}{L^2} & (\Psi)\dfrac{6EI}{L^2} & (\Psi)^2 \end{array} \right] \begin{Bmatrix} P_x \\ P_z \\ M_z \end{Bmatrix}$$

where $\Psi = (AE/L + 12EI/L^3)$,

$$\text{Det.} = \left(\frac{AE}{L} + \frac{12EI}{L^3}\right)\left[\left(\frac{AE}{L} + \frac{12EI}{L^3}\right)\frac{8EI}{L} - 2\left(\frac{6EI}{L^2}\right)^2\right]$$

This solution for u_1, w_1, and θ can be substituted into the element equations to yield first-pass solutions for F_z^A and F_x^B. The latter can then be employed in the $[\mathbf{K}_g]$ matrix to yield improved solutions for u_2, w_2, and θ. The process is continued until convergent solutions are obtained.

It is instructive to compare the approach above, even in the first-pass solution, with solutions obtained with axial loads based upon an independent membrane solution. In the latter approach, with $P_x = 1{,}000$ lb, $P_z = M_z = 0$,

$$F_z^A = 0$$
$$F_x^B = 1{,}000 \text{ lb}$$

Alternatively, with use of the equation given in the first-pass solution above, for the case of $L = 12$ in. and with a square cross section of 1×1 in., the result is

$$F_x^B = 995.7$$

The difference between the solution above and one that disregards the interaction of axial-flexural behavior is not significant (0.5%) for this case. To examine this more closely, however, a series of analyses were conducted for the structure shown in Fig. 13.12, with the same cross-sectional properties as above and for different values of the "shallowness" or knee-angle parameter ϕ. Table 13.1 gives the percentage improvement in accuracy of determination of $F_x^B = F_z^A$ due to recognition of axial flexural behavior, as opposed to a simplified approach.

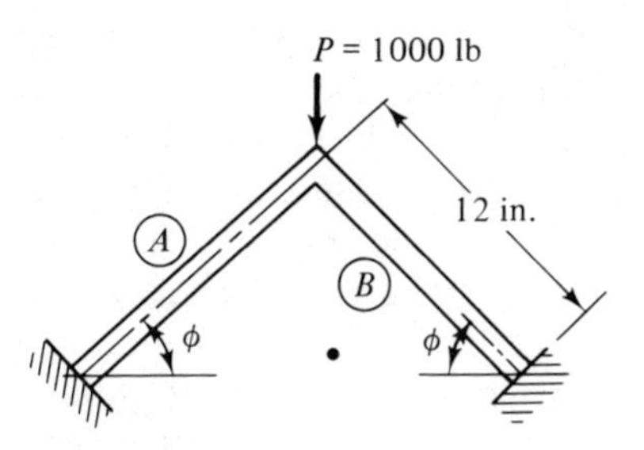

Fig. 13.12 Frame for numerical analysis—variable knee angle.

Table 13.1

ϕ (degrees)	*Uncoupled Membrane-Flexure Behavior* $F_x^B = P/2 \sin \phi$	F_x^B *Coupled Membrane-Flexure Behavior*	*Difference (%)*
45	707.0	702.2	0.8
30	1,000.0	980.0	2.0
15	1,932.0	1,761.0	8.9
5	5,737.0	3,008.0	47.5

It is apparent that axial-flexural coupling may, in general, have significant influence on elastic instability.

The influence of flexural behavior on the axial load distribution is usually disregarded in the so-called *classical* approach to linear stability analysis. An analysis is first performed for the axial load, disregarding flexural behavior. The geometric stiffness matrix is then formed directly. It is difficult if not impossible, however, to employ the highly automated procedures of an available finite element analysis computer program in a special way.

Examples of the matrix formulation of the finite element framework instability problem are to be found in Refs. 13.5–13.6.

13.4 Flat Plate Elements

The menu of elements for plate flexure is quite large, encompassing many geometric forms and displacement fields, and for the most part the stiffness relationships for elastic instability analysis are too complex for development here. The construction of such relationships in explicit form for triangular elements is especially complicated. We therefore limit our attention to two rectangular elements. The reader is advised to consult Refs. 13.7–13.12 for details of formulation of other plate bending elements and for comparison studies.

The two principal rectangular element displacement functions are the 12-term polynomial [Eq. (12.32)] and the 16-term function resulting from Hermitian polynomial interpolation [Eq. (12.31)]. Basic theoretical relationships for the plate element geometric stiffness matrices are given by Eq.

(13.21) and the comments following that equation. By substitution of the 12-term polynomial into these equations and performance of the indicated integration, one obtains the $[\mathbf{k}_{g_x}]$, $[\mathbf{k}_{g_y}]$, and $[\mathbf{k}_{g_{xy}}]$ matrices presented in Table 13.2. For details of the formulation of these matrices, see Ref. 13.13. Corresponding matrices for the 16-term formulation can be found in Ref. 13.14.

It is of interest to compare results obtained with these alternative formulations for the buckling load of a clamped square plate under uniaxial compression, using a single element within a quadrant as shown in Fig. 12.5. All joint displacements are zero except w_1; the behavior is therefore described by a single equation. For the 12-term formulation, using the basic flexural stiffness term from Table 12.1 and the geometric stiffness from Table 13.2, (with $x_2 = y_3 = a$)

$$\frac{D}{30a^2}[120(1 + 1) - 24(0.3) + 84] = \frac{552}{1260}t\sigma_{x_{cr}} \quad \text{or} \quad \sigma_{x_{cr}} = \frac{24.0D}{a^2t}$$

Using geometric stiffness coefficients presented in Ref. 13.14 for the 16-term formulation we obtain $\sigma_{x_{cr}} = 26.5D/a^2t$, while the analytical solution(13.15) is $24.8D/a^2t$. Thus, both solutions are relatively accurate for this extremely coarse grid.

Figure 13.13 portrays the percentage error as a function of idealization refinement in computation of the critical load of a uniaxially loaded, simply supported square plate on the basis of these alternatives. Both the 12-term and 16-term representations yield accurate solutions that converge to the correct result. The convergence trend of a valid stationary potential energy solution, i.e., convergence from the high side, is evidenced by the 16-term formulation. The higher accuracy of the interelement-compatible 16-term formulation is only partly counterbalanced by the additional computational effort occasioned by the extra d.o.f. per joint.

The latter view is substantiated by results, shown in Fig. 13.13, that were obtained by the static condensation procedure described in Section 13.2. The condensation procedure removed all but the joint transverse displacements, w_i, and the accuracy is of the same order as the 12-term polynomial model with three d.o.f. per joint.

The example above masks one of the most significant advantages of the finite element method in flat plate instability analysis. Because the applied inplane forces are constant, no analysis was needed to obtain their distribution in the interior of the plate. When the inplane forces are nonuniform or feature concentrated loads, however, or when the plate geometry is irregular (e.g., with stiffened cutouts or of special shape in planform), the problem is virtually intractable via classical analytical approaches. The finite element approach, on the other hand, suffers no disability because the distribution of middle surface forces is readily obtained from a plane stress finite element analysis of the type described in Chapter 9.

Table 13.2 Geometric stiffness matrices for rectangular plate element based on 12-term polynomial. Constant inplane stresses $\sigma_x, \sigma_y, \tau_{xy}$. (See Table 12.1 for details of element and $[\mathbf{k}_f]$ matrix).

$$\{\mathbf{F}\} = [[\mathbf{k}_f] + [\mathbf{k}_{g_x}] + [\mathbf{k}_{g_y}] + [\mathbf{k}_{g_{xy}}]]\{\mathbf{\Delta}\}$$
$$\{\mathbf{F}\} = \lfloor F_{z_1}\ M_{x_1}\ M_{y_1}\ F_{z_2}\ M_{x_2}\ M_{y_2}\ F_{z_3}\ M_{x_3}\ M_{y_3}\ F_{z_4}\ M_{x_4}\ M_{y_4} \rfloor$$
$$\{\mathbf{\Delta}\} = \lfloor w_1\ \theta_{x_1}\ \theta_{y_1}\ w_2\ \theta_{x_2}\ \theta_{y_2}\ w_3\ \theta_{x_3}\ \theta_{y_3}\ w_4\ \theta_{x_4}\ \theta_{y_4} \rfloor$$

$[\mathbf{k}_{g_x}] = \dfrac{\sigma_x t y_3}{1{,}260 x_2}$

w_1	θ_{x_1}	θ_{y_1}	w_2	θ_{x_2}	θ_{y_2}	w_3	θ_{x_3}	θ_{y_3}	w_4	θ_{x_4}	θ_{y_4}
552											
$66y_3$	$12y_3^2$										
$-42x_2$	○	$56x_2^2$				Symmetric					
204	$39y_3$	$-21x_2$	552								
$-39y_3$	$-9y_3^2$	○	$-66y_3$	$12y_3^2$							
$-21x_2$	○	$28x_2^2$	$-42x_2$	○	$56x_2^2$						
-204	$-39y_3$	$21x_2$	-552	$66y_3$	$42x_2$	552					
$39y_3$	$9y_2^2$	○	$66y_3$	$-12y_3^2$	○	$-66y_3$	$12y_3^2$				
$-21x_2$	○	$-7x_2^2$	$-42x_2$	○	$-14x_2^2$	$42x_2$	○	$56x_2^2$			
-552	$-66y_3$	$42x_2$	-204	$39y_3$	$21x_2$	204	$-39y_3$	$21x_2$	552		
$-66y_3$	$-12y_3^2$	○	$-39y_3$	$9y_3^2$	○	$39y_3$	$-9y_3^2$	○	$66y_3$	$12y_3^2$	
$-42x_2$	○	$-14x_2^2$	$-21x_2$	○	$-7x_2^2$	$21x_2$	○	$28x_2^2$	$42x_2$	○	$56x_2^2$

Table 13.2 (*Cont.*)

$$[\mathbf{k}_{gv}] = \frac{\sigma_y t y_3}{1{,}260 x_2}$$

w_1	θ_{x_1}	θ_{y_1}	w_2	θ_{x_2}	θ_{y_2}	w_3	θ_{x_3}	θ_{y_3}	w_4	θ_{x_4}	θ_{y_4}
552											
$42y_3$	$56y_3^2$										
$-66x_2$	○	$12x_2^2$				Symmetric					
-552	$-42y_3$	$66x_2$	552								
$42y_3$	$-14y_3^2$	○	$-42y_3$	$56y_3^2$							
$66x_2$	○	$-12x_2^2$	$-66x_2$	○	$12x_2^2$						
-204	$-21y_3$	$39x_2$	204	$-21y_3$	$-39x_2$	552					
$21y_3$	$-7y_3^2$	○	$-21y_3$	$28y_3^2$	○	$-42y_3$	$56y_3^2$				
$-39x_2$	○	$9x_2^2$	$39x_2$	○	$-9x_2^2$	$66x_2$	○	$12x_2^2$			
204	$21y_3$	$-39x_2$	-204	$21y_3$	$39x_2$	-552	$42y_3$	$-66x_2$	552		
$21y_3$	$28y_3^2$	○	$-21y_3$	$-7y_3^2$	○	$-42y_3$	$-14y_3^2$	○	$42y_3$	$56y_3^2$	
$39x_2$	○	$-9x_2^2$	$-39x_2$	○	$9x_2^2$	$-66x_2$	○	$-12x_2^2$	$66x_2$	○	$12x_2^2$

Table 13.2 *(Cont.)*

$[\mathbf{k}_{g_{xy}}] = \frac{\tau_{xy} t y_3}{1{,}260 x_2}$

w_1	θ_{x_1}	θ_{y_1}	w_2	θ_{x_2}	θ_{y_2}	w_3	θ_{x_3}	θ_{y_3}	w_4	θ_{x_4}	θ_{y_4}
180											
○	○										
○	$-5x_2y_3$	○					Symmetric				
○	○	$-36x_2$	-180								
○	○	$5x_2y_3$	○	○							
$36x_2$	$5x_2y_3$	○	○	$-5x_2y_3$	○						
-180	$-36y_3$	$36x_2$	○	$36y_3$	○	180					
$36y_3$	$6y_3^2$	$-5x_2y_3$	$-36y_3$	○	$5x_2y_3$	○	○				
$-36x_2$	$-5x_2y_3$	$6x_2^2$	○	$5x_2y_3$	○	○	$-5x_2y_3$	○			
○	$36y_3$	○	180	$-36y_3$	$-36x_2$	○	○	$36x_2$	-180		
$-36y_3$	○	$5x_2y_3$	$36y_3$	$-6y_3^2$	$-5x_2y_3$	○	○	$5x_2y_3$	○	○	
○	$5x_2y_3$	○	$36x_2$	$-5x_2y_3$	$-6x_2^2$	$-36x_2$	$5x_2y_3$	○	○	$-5x_2y_3$	○

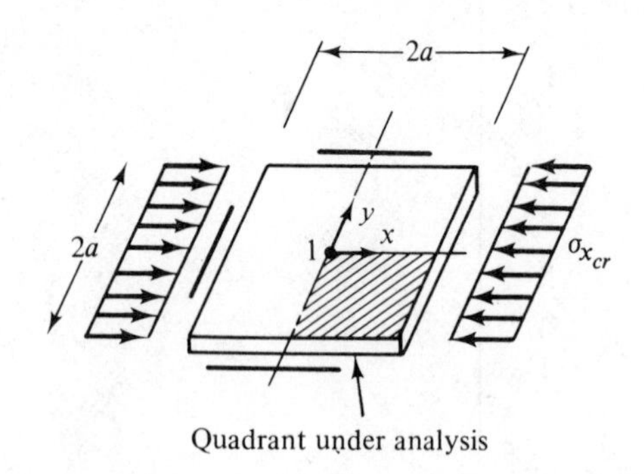

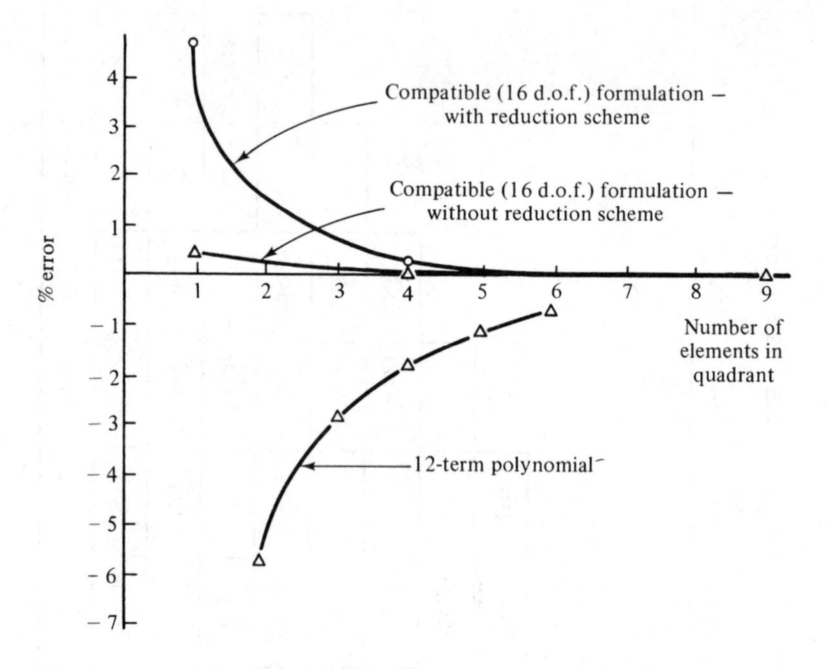

Fig. 13.13 Numerical comparisons in plate buckling—rectangular elements.

REFERENCES

13.1 LIVESLEY, R. K., *Matrix Methods of Structural Analysis*, Pergamon Press, Oxford, England, 1965 (Chapter 10).

13.2 WANG, C. T., *Applied Elasticity*, McGraw-Hill Book Co., New York, N.Y., 1954.

13.3 GALLAGHER, R. and B. LEE, "Matrix Dynamic and Instability Analysis with Nonuniform Elements," *Internat'l. J. Num. Meth. Eng.*, **2**, No. 2, pp. 265–276, 1970.

13.4 R. BARSOUM and GALLAGHER, R., "Finite Element Analysis of Torsional and Lateral Stability Problems," *Internat'l. J. Num. Meth. Eng.*, **2**, No. 3, pp. 335–352, 1970.

13.5 HALLDORSSON, O. and C. K. WANG, "Stability Analysis of Frameworks by Matrix Methods," Proc. ASCE, *J. Struct. Div.*, **94**, No. ST7, pp. 1745–1760, July, 1968.

13.6 HARTZ, B. J., "Matrix Formulation of Structural Stability Problems," Proc. ASCE, *J. Struct. Div.*, **91**, No. ST6, pp. 141–158, Dec., 1965.

13.7 GALLAGHER, R., R. GELLATLY, R. MALLETT, and J. PADLOG, "A Discrete Element Procedure for Thin Shell Instability Analysis," *AIAA J.*, **5**, No. 1, pp. 138–144, Jan., 1967.

13.8 ANDERSON, R. G., B. M. IRONS, and O. C. ZIENKIEWICZ, "Vibrations and Stability of Plates Using Finite Elements," *Internat'l. J. Solids Struct.*, **4**, pp. 1031–1035, Oct., 1968.

13.9 ARGYRIS, J. H., et al., "Some New Elements for the Matrix Displacement Method," Proc. of the 2nd Air Force Conference on Matrix Methods in Structural Mechanics, Dayton, Ohio, Oct., 1968.

13.10 KABAILA, A. P. and B. FRAEIJS DE VEUBEKE, "Stability Analysis by Finite Elements," AFFDL TR 70–35, Mar., 1970.

13.11 VOS, R. G. and W. P. VANN, "A Finite Element Tensor Approach to Plate Buckling and Postbuckling," *Internat'l. J. Num. Meth. Eng.*, **5**, No. 3, pp. 351–366, 1973.

13.12 CLOUGH, R. W. and C. A. FELIPPA, "A Refined Quadrilateral Element for Analysis of Plate Bending," Proc. of 2nd Conf. on Matrix Methods in Struct. Mech. AFFDL TR 68–160, Oct., 1968.

13.13 PRZEMIENIECKI, J. S., "Discrete-Element Methods for Stability Analysis of Complex Structures," *Aero. J.*, **72**, pp. 1077–1086, Dec., 1968.

13.14 PIFKO, A. and ISAKSON, G., "A Finite-Element Method for the Plastic Buckling Analysis of Plates," *AIAA J.*, **7**, No. 10, pp. 1950–1957, Oct., 1969.

13.15 TIMOSHENKO, S. and J. GERE, *Theory of Elastic Stability*, 2nd ed., McGraw-Hill Book Co., New York, N.Y., 1961.

PROBLEMS

13.1 Confirm, by application of the variational procedures described in Chapter 6, that Eq. (13.11) is the Euler equation of the functional represented by Eq. (13.10). ($V = 0$).

13.2 Develop the Euler equations corresponding to the functional for torsional-flexural behavior given by Eq. (13.31).

13.3 The solution to the governing differential equation for flexure in the presence of an axial load F_x [Eq. (13.11)] is of the form (see Fig. 13.1):

$$w = \left\{\frac{\cos\omega(L-x) + (\cos\omega L - \cos\omega x) + \omega L\sin\omega L[1-(x/L)] - 1}{\omega L\sin\omega L - 2(1-\cos\omega L)^2}\right\}w_1 + [\quad]w_2 + [\quad]\theta_1 + [\quad]\theta_2$$

where $\omega^2 = F_x/EI$. Utilize this displacement function in the determination of the stiffness coefficient relating w_1 to F_{z_1}.

13.4 When the stiffness matrix for a beam-column element is formulated on the basis of the "exact" displacement function, the stiffness coefficient that relates F_{z_1} to θ_1 (see Fig. 13.1) is given by

$$k_{12} = -\frac{EI}{L^2}\frac{\omega^2 L^2(1 - \cos \omega L)}{2(1 - \cos \omega L) - \omega L \sin \omega L}$$

where ω is as defined in Prob. 13.3. Establish a polynomial form of this coefficient by expansion in series of the trigonometric terms, preserving the first two terms of the expansion, and compare these with the formulation given in this chapter.

13.5 Perform analyses of a simply supported beam column, using the $[\mathbf{k}_g]$ matrix of this chapter, for a central concentrated load P and axial load intensities $F_x/F_{x_{cr}} = 0.5, 0.75$, and 0.9 $[F_{x_{cr}} = (\pi^2 EI/L^2)]$, and compare with the exact solutions.

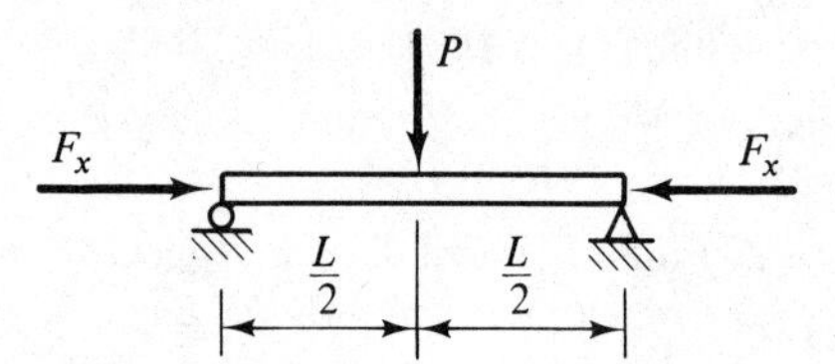

Fig. P13.5

13.6 Calculate the critical load of the tapered beam shown in Fig. P13.6 using one element.

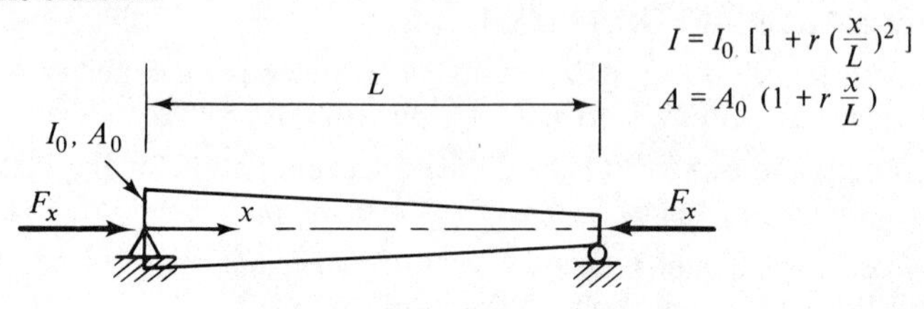

Fig. P13.6

13.7 Partition the instability analysis equations with reference to the sets of displacements $\{\mathbf{w}\}$ and $\{\boldsymbol{\theta}\}$, as in Eq. (13.28). Form an "exact" reduction of these equations by solving the lower partition for $\{\boldsymbol{\theta}\}$. Then, by a series expansion of the term $[\mathbf{k}_{f\theta\theta} + \omega\mathbf{k}_{g\theta\theta}]^{-1}$, which appears in the result, and by discard of terms of order higher than ω, show that the result obtained is the same as Eq. 13.29.

13.8 Calculate P_{cr} for the frame shown in Fig. P13.8. Disregard axial stiffness in constructing the force-displacement equations. $[P_{cr} = 0.135 I_0 E$ (from Ref. 13.15)]

13.9 Calculate P_{cr} for the stepped beam shown in Fig. P13.9.

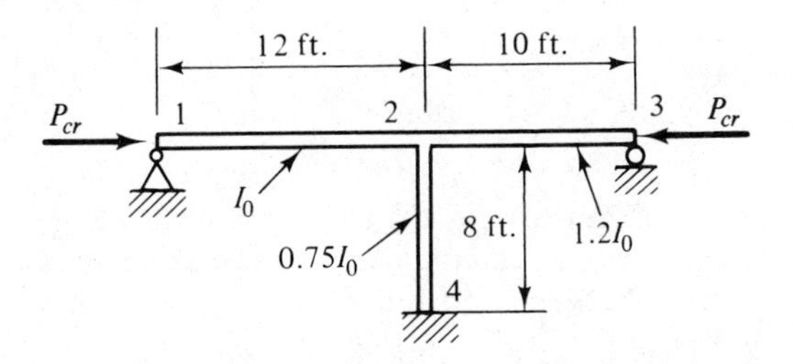

Fig. P13.8

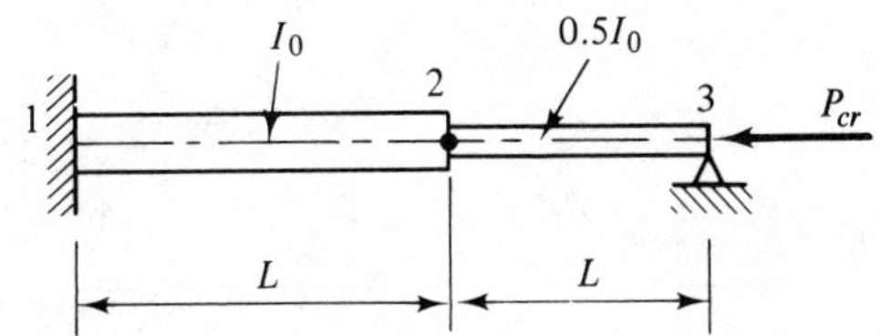

Fig. P13.9

13.10 Calculate P_{cr} for the beam shown in Fig. P13.10.

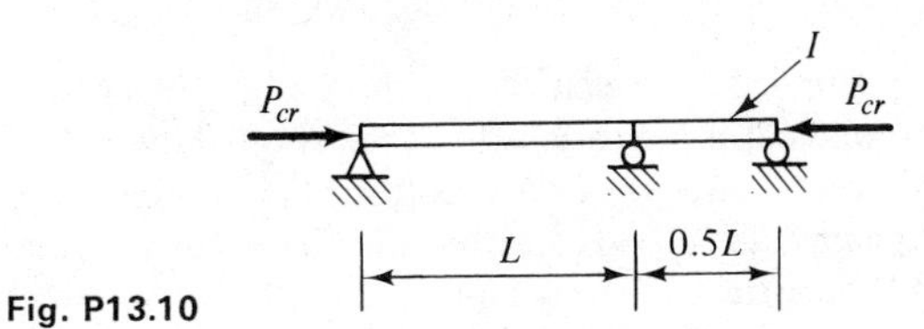

Fig. P13.10

13.11 (Computer-oriented solution) Calculate P_{cr} for the continuous beam shown in Fig. P13.11. ($E = 30 \times 10^6$ p.s.i., $I_0 = 10.0$ in.^{4}, $A_0 = 2.0$ in.^{2}).

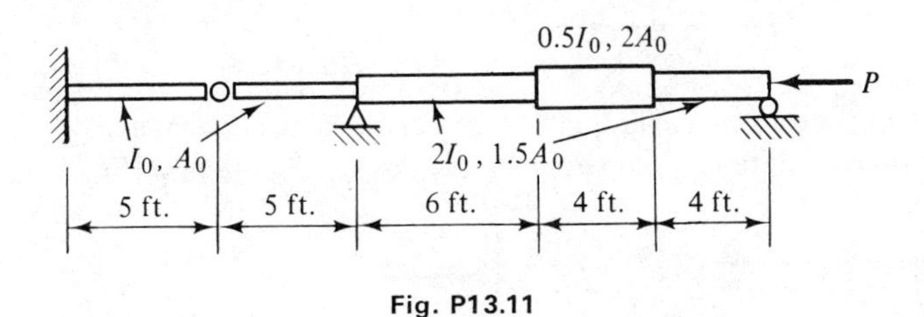

Fig. P13.11

13.12 Compute the critical load for a pin-ended column, using a single element, with an element stiffness representation formulated on the basis of the fifth-order Hermitian polynomial representation of transverse displacement. This representation requires the use of w, $dw/dx = -w_x$, and $d^2w/dx^2 = w_{xx}$ as the d.o.f. at each joint.

$$L^5 w = (L^5 - 10L^2x^3 + 15Lx^4 - 6x^5)w_1$$
$$- L(L^4x - 6L^2x^3 + 8Lx^4 - 3x^5)w_{x_1}$$

$$+ \tfrac{1}{2}L^2(L^3x^2 - 3L^2x^3 + 3Lx^4 - x^5)w_{xx_1}$$
$$+ (10L^2x^3 - 15Lx^4 + 6x^5)w_2$$
$$- L(7Lx^4 - 4L^2x^3 + 3x^5)w_{x_2}$$
$$+ \tfrac{1}{2}L^2(L^2x^3 - 2Lx^4 + x^5)w_{xx_2}$$

13.13 Formulate the $[\mathbf{k}_g]$ matrix for the triangular plate element shown in Fig. P13.13 for imposition of a constant middle-surface stress of σ_x. The element stiffness matrix is based on a quadratic (six-term) function for w (see Chapter 12, Fig. 12.8b)

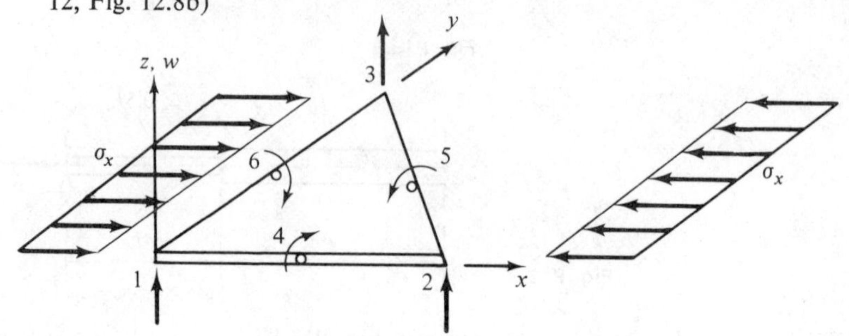

Fig. P13.13

13.14 The annular plate element shown in Fig. P13.14, whose stiffness matrix was formulated in the assigned problems of Chapter 12, sustains a temperature change (Υ) above the stress-free state. Using one element, calculate the critical value of Υ to cause buckling. (Use linear radial displacement function and linearize all integrations.)

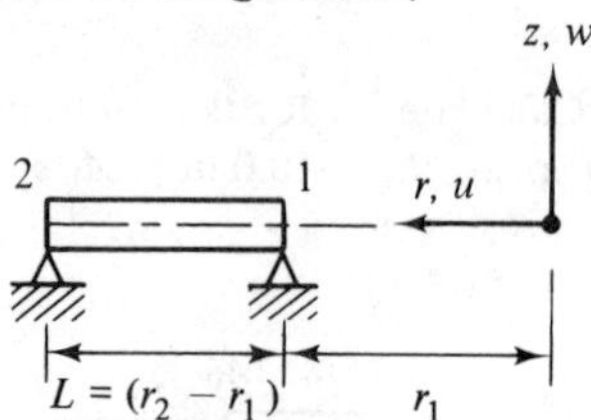

Fig. P13.14

13.15 For the beam shown in Fig. P13.15, determine the critical load P_x for a two-element solution. Compare with the "exact" solution, $\pi^2EI/4L^2$.

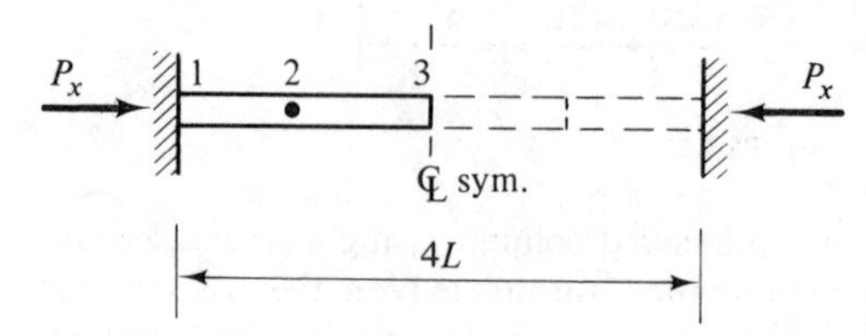

Fig. P13.15

13.16 For the beam of Prob. 13.15, determine the influence on the fundamental frequency of the following ratios of applied axial load (P_x) to "effective critical load," $P_{cr} = EI/L^2$; $P_x/P_{cr} = 10$, $P_x/P_{cr} = 20$, where L is the length of the member.

Index

Author Index

(Boldface designates chapter-end references)

Subject Index

R

S

T

U

V

W